WORLD HEALTH ORGANIZATION

INTERNATIONAL AGENCY FOR RESEARCH ON CANCER

IARC MONOGRAPHS
ON THE
EVALUATION OF THE CARCINOGENIC RISK OF CHEMICALS TO HUMANS

Some Halogenated Hydrocarbons and Pesticide Exposures

VOLUME 41

This publication represents the views and expert opinions
of an IARC Working Group on the
Evaluation of the Carcinogenic Risk of Chemicals to Humans
which met in Lyon,

4-11 February 1986

1986

IARC MONOGRAPHS

In 1969, the International Agency for Research on Cancer (IARC) initiated a programme on the evaluation of the carcinogenic risk of chemicals to humans involving the production of critically evaluated monographs on individual chemicals. In 1980, the programme was expanded to include the evaluation of the carcinogenic risk associated with exposures to complex mixtures.

The objective of the programme is to elaborate and publish in the form of monographs critical reviews of data on carcinogenicity for chemicals and complex mixtures to which humans are known to be exposed, and on specific occupational exposures, to evaluate these data in terms of human risk with the help of international working groups of experts in chemical carcinogenesis and related fields, and to indicate where additional research efforts are needed.

This project was supported by PHS Grant No. 2 UO1 CA33193-04 awarded by the US National Cancer Institute, Department of Health and Human Services.

©International Agency for Research on Cancer 1986

ISBN 92 832 1241 X

ISSN 0250-9555

All rights reserved. Application for rights of reproduction or translation, in part or *in toto*, should be made to the International Agency for Research on Cancer.

Distributed for the International Agency for Research on Cancer by the Secretariat of the World Health Organization

PRINTED IN THE UK

CONTENTS

NOTE TO THE READER 5

LIST OF PARTICIPANTS 7

PREAMBLE
    Background 13
    Objective and Scope 13
    Selection of Chemicals and Complex Mixtures for Monographs 14
    Working Procedures 14
    Data for Evaluations 15
    The Working Group 15
    General Principles 15
    Explanatory Notes on the Monographs Contents 23

GENERAL REMARKS ON THE SUBSTANCES AND EXPOSURES
    CONSIDERED 33

THE MONOGRAPHS
    Dichloromethane 43
    1,1,1,2-Tetrachloroethane 87
    Pentachloroethane 99
    1,3-Dichloropropene 113
    1,2-Dichloropropane 131
    Bis(2-chloro-1-methylethyl)ether 149
    Methyl chloride 161
    Methyl bromide 187
    Methyl iodide 213
    Chlorofluoromethane 229
    Chlorodifluoromethane 237
    2-Chloro-1,1,1-trifluoroethane 253
    Polybrominated biphenyls 261
    Amitrole 293
    Chlorophenols (occupational exposures to) 319
    Chlorophenoxy herbicides (occupational exposures to) 357

APPENDIX: SUMMARY OF FINAL EVALUATIONS 407

CUMULATIVE INDEX TO THE MONOGRAPHS SERIES 409

NOTE TO THE READER

The term 'carcinogenic risk' in the *IARC Monographs* series is taken to mean the probability that exposure to the chemical will lead to cancer in humans.

Inclusion of a chemical in the *Monographs* does not imply that it is a carcinogen, only that the published data have been examined. Equally, the fact that a chemical has not yet been evaluated in a monograph does not mean that it is not carcinogenic.

Anyone who is aware of published data that may alter the evaluation of the carcinogenic risk of a chemical to humans is encouraged to make this information available to the Unit of Carcinogen Identification and Evaluation, Division of Environmental Carcinogenesis, International Agency for Research on Cancer, 150 cours Albert Thomas, 69372 Lyon Cedex 08, France, in order that the chemical may be considered for re-evaluation by a future Working Group.

Although every effort is made to prepare the monographs as accurately as possible, mistakes may occur. Readers are requested to communicate any errors to the Unit of Carcinogen Identification and Evaluation, so that corrections can be reported in future volumes.

IARC WORKING GROUP ON THE EVALUATION OF THE CARCINOGENIC RISK OF CHEMICALS TO HUMANS: SOME HALOGENATED HYDROCARBONS AND PESTICIDE EXPOSURES

Lyon, 4-11 February 1986

Members[1]

A. Abbondandolo, Laboratory of Mutagenesis, National Institute for Cancer Research, viale Benedetto XV, 10, 16132 Genoa, Italy

A. Aitio, Institute of Occupational Health, Topeliuksenkatu 41 a A, SF-00250 Helsinki, Finland

M.W. Anders, Department of Pharmacology, University of Rochester Medical Center, 601 Elmwood Avenue, Rochester, NY 14642, USA

D. Anderson, Department of Genetic Toxicology, British Industrial Biological Research Association, Woodmansterne Road, Carshalton, Surrey SM5 4DS, UK

A. Brøgger, Department of Genetics, Institute for Cancer Research, The Norwegian Radium Hospital, Montebello, 0310 Oslo 3, Norway

P. Chambon, Institut Pasteur, 77 rue Pasteur, 69007 Lyon, France

P. Comba, Istituto Superiore di Sanita, viale Regina Elena 299, 00161 Rome, Italy

G. Della Porta, Division of Experimental Oncology A, Istituto Nazionale per lo Studio e la Cura dei Tumori, via Venezian 1, 20133 Milan, Italy

L. Fishbein, National Center for Toxicological Research, Jefferson, AR 72079, USA (*Vice-chairman*)

J.K. Haseman, Research Mathematical Statistician, National Institute of Environmental Health Sciences, PO Box 12233, Research Triangle Park, NC 27709, USA

S. Hernberg, Institute of Occupational Health, Topeliuksenkatu 41 a A, SF-00250 Helsinki, Finland (*Chairman*)

C. Hogstedt, National Board of Occupational Safety and Health, 17184 Solna, Sweden

---

[1]Unable to attend: G.L. Plaa, Département de Pharmacologie, Université de Montréal, C.P. 6128, Montréal, Québec, Canada; J.S. Wassom, EMCT Information Program, Oak Ridge National Laboratory, Building 9224, Room 102, Oak Ridge, TN 37831, USA

H. Kappus, Department of Dermatology, Free University of Berlin, Augustenburger Platz 1, 1000 Berlin (West) 65, Federal Republic of Germany

D.G. Kaufman, Department of Pathology, The University of North Carolina at Chapel Hill, Brintehous-Bullitt Building 228H, Chapel Hill, NC 27514, USA

R.J. Kavlock, Perinatal Toxicology Branch, Developmental Biology Division, Health Effects Research Laboratory, Environmental Protection Agency, Research Triangle Park, NC 27711, USA

F. Kuper, TNO-CIVO, Toxicology and Nutrition Institute, PO Box 360, 3700 AJ Zeist, The Netherlands

O. Møller-Jensen, Danish Cancer Registry, Institute of Cancer Epidemiology, Landskronagade 66, 2100 Copenhagen Ø, Denmark

A. Pinter, Department of Morphology, National Institute of Hygiene, Gyali ut 2-6, 1966 Budapest, Hungary

C. Rappe, Department of Organic Chemistry, University of Umeå, 90187 Umeå, Sweden

A.H. Smith, Occupational Health Epidemiology, 315 Warren Hall, School of Public Health, University of California, Berkeley, CA 94720, USA

M.D. Waters, Genetic Toxicology Division (MD-68), US Environmental Protection Agency, Health Effects Research Laboratory, Research Triangle Park, NC 27711, USA

E. Zeiger, National Toxicology Program, National Institute of Environmental Health Sciences, PO Box 12233, Research Triangle Park, NC 27709, USA

Representative of the National Cancer Institute

E.K. Weisburger, Division of Cancer Etiology, Building 31, Room 11A05, National Cancer Institute, Bethesda, MD 20892, USA

Representative of Tracor Jitco, Inc.

S. Olin, Tracor Jitco, Inc., 1601 Research Boulevard, Rockville, MD 20850, USA

Observers

Representative of the Chemical Manufacturers' Association

J. Norris, The Dow Chemical Co., 1803 Building, Midland, MI 48674, USA

Representative of the Commission of the European Communities

M.-T. van der Venne, Commission of the European Communities, Health and Safety Directorate, Bâtiment Jean Monnet (C4/83), 2920 Luxembourg, Grand Duchy of Luxembourg

PARTICIPANTS

Representative of CONCAWE

E. Longstaff, Imperial Chemical Industries PLC, Central Toxicology Laboratory, Alderley Park, Macclesfield, Cheshire SK10 4TT, UK

Representative of the European Chemical Industry, Ecology and Toxicology Centre

R. Jung, Hoechst, Pharma Forschung Toxikologie, Postfach 80 03 20, 6230 Frankfurt-am-Main 80, Federal Republic of Germany

Representative of the National Board of Occupational Safety and Health

P. Lundberg, National Board of Occupational Safety and Health, Research Department, 17184 Solna, Sweden

Secretariat

H. Bartsch, Division of Environmental Carcinogenesis
J.R.P. Cabral, Division of Environmental Carcinogenesis
B. Dodet, Division of Environmental Carcinogenesis
M. Friesen, Division of Environmental Carcinogenesis
L. Haroun, Division of Environmental Carcinogenesis (*Secretary*)
E. Heseltine, Editorial and Publishing Services
E. Johnson, Division of Epidemiology and Biostatistics
B. McKnight[1], Division of Epidemiology and Biostatistics
D. Mietton, Division of Environmental Carcinogenesis
R. Montesano, Division of Environmental Carcinogenesis
I. O'Neill, Division of Environmental Carcinogenesis
C. Partensky, Division of Environmental Carcinogenesis
I. Peterschmitt, Division of Environmental Carcinogenesis, Geneva, Switzerland
S. Poole, Birmingham, UK
R. Saracci, Division of Epidemiology and Biostatistics
L. Shuker, Division of Environmental Carcinogenesis
L. Simonato, Division of Epidemiology and Biostatistics
A. Tossavainen, Division of Environmental Carcinogenesis
H. Vainio, Division of Environmental Carcinogenesis (*Head of the Programme*)
J.D. Wilbourn, Division of Environmental Carcinogenesis
H. Yamasaki, Division of Environmental Carcinogenesis

Secretarial assistance

J. Cazeaux
M. Lézère
J. Nyairo
S. Reynaud

---

[1] Present address: University of Washington, Department of Biostatistics SC-32, Seattle, WA 98195, USA

PREAMBLE

IARC MONOGRAPHS PROGRAMME ON THE EVALUATION OF THE CARCINOGENIC RISK OF CHEMICALS TO HUMANS[1]

PREAMBLE

1. BACKGROUND

In 1969, the International Agency for Research on Cancer (IARC) initiated a programme to evaluate the carcinogenic risk of chemicals to humans and to produce monographs on individual chemicals. Following the recommendations of an ad-hoc Working Group, which met in Lyon in 1979 to prepare criteria to select chemicals for *IARC Monographs*(1), the *Monographs* programme was expanded to include consideration of exposures to complex mixtures which may occur, for example, in many occupations or as a result of human habits.

The criteria established in 1971 to evaluate carcinogenic risk to humans were adopted by all the working groups whose deliberations resulted in the first 16 volumes of the *IARC Monographs* series. This preamble reflects subsequent re-evaluation of those criteria by working groups which met in 1977(2), 1978(3), 1982(4) and 1983(5).

2. OBJECTIVE AND SCOPE

The objective of the programme is to elaborate and publish in the form of monographs critical reviews of data on carcinogenicity for chemicals, groups of chemicals, industrial processes and other complex mixtures to which humans are known to be exposed, to evaluate the data in terms of human risk with the help of international working groups of experts, and to indicate where additional research efforts are needed. These evaluations are intended to assist national and international authorities in formulating decisions concerning preventive measures. No recommendation is given concerning legislation, since this depends on risk-benefit evaluations, which seem best made by individual governments and/or other international agencies.

---

[1] This project is supported by PHS Grant No. 2 U01 CA33193-04 awarded by the US National Cancer Institute, Department of Health and Human Services.

The *IARC Monographs* are recognized as an authoritative source of information on the carcinogenicity of environmental and other chemicals. A users' survey, made in 1984, indicated that the monographs are consulted by various agencies in 45 countries. As of July 1986, 41 volumes of the *Monographs* had been published or were in press. Five supplements have been published: two summaries of evaluations of chemicals associated with human cancer, an evaluation of screening assays for carcinogens, and two cross indexes of synonyms and trade names of chemicals evaluated in the series(6).

3. SELECTION OF CHEMICALS AND COMPLEX EXPOSURES FOR MONOGRAPHS

The chemicals (natural and synthetic including those which occur as mixtures and in manufacturing processes) and complex exposures are selected for evaluation on the basis of two main criteria: (a) there is evidence of human exposure, and (b) there is some experimental evidence of carcinogenicity and/or there is some evidence or suspicion of a risk to humans. In certain instances, chemical analogues are also considered. The scientific literature is surveyed for published data relevant to the *Monographs* programme; and the IARC *Survey of Chemicals Being Tested for Carcinogenicity*(7) often indicates those chemicals that may be scheduled for future meetings.

As new data on chemicals for which monographs have already been prepared become available, re-evaluations are made at subsequent meetings, and revised monographs are published.

4. WORKING PROCEDURES

Approximately one year in advance of a meeting of a working group, a list of the substances or complex exposures to be considered is prepared by IARC staff in consultation with other experts. Subsequently, all relevant biological data are collected by IARC; recognized sources of information on chemical carcinogenesis and on-line systems such as CANCERLINE, MEDLINE and TOXLINE are used in conjunction with US Public Health Service Publication No. 149(8). Bibliographical sources for data on mutagenicity and teratogenicity are the Environmental Mutagen Information Center and the Environmental Teratology Information Center, both located at the Oak Ridge National Laboratory, TN, USA.

The major collection of data and the preparation of first drafts for the sections on chemical and physical properties, on production and use, on occurrence, and on analysis are carried out by Tracor Jitco, Inc., and its subcontractor, Technical Resources, Inc., both in Rockville, MD, USA, under a separate contract with the US National Cancer Institute. Most of the data so obtained refer to the USA and Japan; IARC attempts to supplement this information with that from other sources in Europe. Representatives from industrial associations may assist in the preparation of sections describing industrial processes.

Six months before the meeting, articles containing relevant biological data are sent to an expert(s), or are used by IARC staff, to prepare first drafts of the sections on biological effects. The complete drafts are then compiled by IARC staff and sent, prior to the meeting, to all participants of the Working Group for their comments.

The Working Group meets in Lyon for seven to eight days to discuss and finalize the texts of the monographs and to formulate the evaluations. After the meeting, the master copy of each monograph is verified by consulting the original literature, edited by a professional editor and prepared for reproduction. The aim is to publish monographs within nine months of the Working Group meeting. Each volume of monographs is printed in 4000 copies for distribution to governments, regulatory agencies and interested scientists. The monographs are also available *via* the WHO Distribution and Sales Service.

These procedures are followed for the preparation of most volumes of monographs, which cover chemicals and groups of chemicals; however, they may vary when the subject matter is an industry or life-style factor.

5. DATA FOR EVALUATIONS

With regard to biological data, only reports that have been published or accepted for publication are reviewed by the working groups, although a few exceptions have been made: in certain instances, reports from government agencies that have undergone peer review and are widely available are considered. The monographs do not cite all of the literature on a particular chemical or complex exposure: only those data considered by the Working Group to be relevant to the evaluation of carcinogenic risk to humans are included.

Anyone who is aware of data that have been published or are in press which are relevant to the evaluations of the carcinogenic risk to humans of chemicals or complex exposures for which monographs have appeared is asked to make them available to the Unit of Carcinogen Identification and Evaluation, Division of Environmental Carcinogenesis, International Agency for Research on Cancer, Lyon, France.

6. THE WORKING GROUP

The tasks of the Working Group are five-fold: (a) to ascertain that all data have been collected; (b) to select the data relevant for evaluation; (c) to ensure that the summaries of the data enable the reader to follow the reasoning of the Working Group; (d) to judge the significance of the results of experimental and epidemiological studies; and (e) to make an evaluation of the carcinogenicity of the chemical or complex exposure.

Working Group participants who contributed to the consideration and evaluation of chemicals or complex exposures within a particular volume are listed, with their addresses, at the beginning of each publication. Each member serves as an individual scientist and not as a representative of any organization or government. In addition, observers are often invited from national and international agencies and industrial associations.

7. GENERAL PRINCIPLES APPLIED BY THE WORKING GROUP IN EVALUATING CARCINOGENIC RISK OF CHEMICALS OR COMPLEX MIXTURES

The widely accepted meaning of the term 'chemical carcinogenesis', and that used in these monographs, is the induction by chemicals (or complex mixtures of chemicals) of neoplasms that are not usually observed, the earlier induction of neoplasms that are commonly observed, and/or the induction of more neoplasms than are usually found —although fundamentally different mechanisms may be involved in these three situations. Etymologically, the term 'carcinogenesis' means the induction of cancer, that is, of malignant neoplasms; however, the commonly accepted meaning is the induction of various types of neoplasms or of a combination of malignant and benign tumours. In the monographs, the words 'tumour' and 'neoplasm' are used interchangeably. (In the scientific literature, the terms 'tumorigen', 'oncogen' and 'blastomogen' have all been used synonymously with 'carcinogen', although occasionally 'tumorigen' has been used specifically to denote a substance that induces benign tumours.)

(a) Experimental Evidence

(i) *Evidence for carcinogenicity in experimental animals*

The Working Group considers various aspects of the experimental evidence reported in the literature and formulates an evaluation of that evidence.

Qualitative aspects: Both the interpretation and evaluation of a particular study as well as the overall assessment of the carcinogenic activity of a chemical (or complex mixture) involve several considerations of qualitative importance, including: (a) the experimental parameters under which the chemical was tested, including route of administration and exposure, species, strain, sex, age, etc.; (b) the consistency with which the chemical has been shown to be carcinogenic, e.g., in how many species and at which target organ(s); (c) the spectrum of neoplastic response, from benign neoplasm to multiple malignant tumours; (d) the stage of tumour formation in which a chemical may be involved: some chemicals act as complete carcinogens and have initiating and promoting activity, while others may have promoting activity only; and (e) the possible role of modifying factors.

There are problems not only of differential survival but of differential toxicity, which may be manifested by unequal growth and weight gain in treated and control animals. These complexities are also considered in the interpretation of data.

Many chemicals induce both benign and malignant tumours. Among chemicals that have been studied extensively, there are few instances in which the only neoplasms induced are benign. Benign tumours may represent a stage in the evolution of a malignant neoplasm or they may be 'end-points' that do not readily undergo transition to malignancy. If a substance is found to induce only benign tumours in experimental animals, it should nevertheless be suspected of being a carcinogen, and it requires further investigation.

Hormonal carcinogenesis: Hormonal carcinogenesis presents certain distinctive features: the chemicals involved occur both endogenously and exogenously; in many instances, long exposure is required; and tumours occur in the target tissue in association with a stimulation of non-neoplastic growth, although in some cases hormones promote the proliferation of tumour cells in a target organ. For hormones that occur in excessive amounts, for hormone-mimetic agents and for agents that cause hyperactivity or imbalance in the endocrine system, evaluative methods comparable with those used to identify chemical carcinogens may be required; particular emphasis must be laid on quantitative aspects and duration of exposure. Some chemical carcinogens have significant side effects on the endocrine system, which may also result in hormonal carcinogenesis. Synthetic hormones and anti-hormones can be expected to possess other pharmacological and toxicological actions in addition to those on the endocrine system, and in this respect they must be treated like any other chemical with regard to intrinsic carcinogenic potential.

Complex mixtures: There is an increasing amount of data from long-term carcinogenicity studies on complex mixtures and on crude materials obtained by sampling in occupational environments. The representativity of such samples must be considered carefully.

Quantitative aspects: Dose-response studies are important in the evaluation of carcinogenesis: the confidence with which a carcinogenic effect can be established is strengthened by the observation of an increasing incidence of neoplasms with increasing exposure.

The assessment of carcinogenicity in animals is frequently complicated by recognized differences among the test animals (species, strain, sex, age) and route and schedule of administration; often, the target organs at which a cancer occurs and its histological type may vary with these parameters. Nevertheless, indices of carcinogenic potency in particular experimental systems (for instance, the dose-rate required under continuous exposure to halve the probability of the animals remaining tumourless(9)) have been formulated in the hope that, at least among categories of fairly similar agents, such indices may be of some predictive value in other species, including humans.

Chemical carcinogens share many common biological properties, which include metabolism to reactive (electrophilic(10-11)) intermediates capable of interacting with DNA. However, they may differ widely in the dose required to produce a given level of tumour induction. The reason for this variation in dose-response is not understood, but it may be due to differences in metabolic activation and detoxification processes, in different DNA repair capacities among various organs and species or to the operation of qualitatively distinct mechanisms.

Statistical analysis of animal studies: It is possible that an animal may die prematurely from unrelated causes, so that tumours that would have arisen had the animal lived longer may not be observed; this possibility must be allowed for. Various analytical techniques have been developed which use the assumption of independence of competing risks to allow for the effects of intercurrent mortality on the final numbers of tumour-bearing animals in particular treatment groups.

For externally visible tumours and for neoplasms that cause death, methods such as Kaplan-Meier (i.e., 'life-table', 'product-limit' or 'actuarial') estimates(9), with associated significance tests(12), have been recommended. For internal neoplasms that are discovered 'incidentally'(12) at autopsy but that did not cause the death of the host, different estimates(13) and significance tests(12) may be necessary for the unbiased study of the numbers of tumour-bearing animals.

The design and statistical analysis of long-term carcinogenicity experiments were reviewed in Supplement 2 to the *Monographs* series(14). That review outlined the way in which the context of observation of a given tumour (fatal or incidental) could be included in an analysis yielding a single combined result. This method requires information on time to death for each animal and is therefore comparable to only a limited extent with analyses which include global proportions of tumour-bearing animals.

Evaluation of carcinogenicity studies in experimental animals: The evidence of carcinogenicity in experimental animals is assessed by the Working Group and judged to fall into one of four groups, defined as follows:

(1) *Sufficient evidence* of carcinogenicity is provided when there is an increased incidence of malignant tumours: (a) in multiple species or strains; or (b) in multiple experiments (preferably with different routes of administration or using different dose levels); or (c) to an unusual degree with regard to incidence, site or type of tumour, or age at onset. Additional evidence may be provided by data on dose-response effects.

(2) *Limited evidence* of carcinogenicity is available when the data suggest a carcinogenic effect but are limited because: (a) the studies involve a single species, strain or experiment; or (b) the experiments are restricted by inadequate dosage levels, inadequate duration of exposure to the agent, inadequate period of follow-up, poor survival, too few animals, or inadequate reporting; or (c) the neoplasms produced often occur spontaneously and, in the past, have been difficult to classify as malignant by histological criteria alone (e.g., lung adenomas and adenocarcinomas and liver tumours in certain strains of mice).

(3) *Inadequate evidence* of carcinogenicity is available when, because of major qualitative or quantitative limitations, the studies cannot be interpreted as showing either the presence or absence of a carcinogenic effect.

(4) *No evidence* of carcinogenicity applies when several adequate studies are available which show that, within the limits of the tests used, the chemical or complex mixture is not carcinogenic.

It should be noted that the categories *sufficient evidence* and *limited evidence* refer only to the strength of the experimental evidence that these chemicals or complex mixtures are carcinogenic and not to the extent of their carcinogenic activity nor to the mechanism involved. The classification of any chemical may change as new information becomes available.

(ii) *Evidence for activity in short-term tests*[1]

Many short-term tests bearing on postulated mechanisms of carcinogenesis or on the properties of known carcinogens have been developed in recent years. The induction of cancer is thought to proceed by a series of steps, some of which have been distinguished experimentally (15-19). The first step — initiation — is thought to involve damage to DNA, resulting in heritable alterations in or rearrangements of genetic information. Most short-term tests in common use today are designed to evaluate the genetic activity of a substance.

---
[1]Based on the recommendations of a working group which met in 1983(5).

Data from these assays are useful for identifying potential carcinogenic hazards, in identifying active metabolites of known carcinogens in human or animal body fluids, and in helping to elucidate mechanisms of carcinogenesis. Short-term tests to detect agents with tumour-promoting activity are, at this time, insufficiently developed.

Because of the large number of short-term tests, it is difficult to establish rigid criteria for adequacy that would be applicable to all studies. General considerations relevant to all tests, however, include (a) that the test system be valid with respect to known animal carcinogens and noncarcinogens; (b) that the experimental parameters under which the chemical (or complex mixture) is tested include a sufficiently wide dose range and duration of exposure to the agent and an appropriate metabolic system; (c) that appropriate controls be used; and (d) that the purity of the compound or, in the case of complex mixtures, that the source and representativity of the sample being tested be specified. Confidence in positive results is increased if a dose-response relationship is demonstrated and if this effect has been reported in two or more independent studies.

Most established short-term tests employ as end-points well-defined genetic markers in prokaryotes and lower eukaryotes and in mammalian cell lines. The tests can be grouped according to the end-point detected:

Tests of *DNA damage*. These include tests for covalent binding to DNA, induction of DNA breakage or repair, induction of prophage in bacteria and differential survival of DNA repair-proficient/-deficient strains of bacteria.

Tests of *mutation* (measurement of heritable alterations in phenotype and/or genotype). These include tests for detection of the loss or alteration of a gene product, and change of function through forward or reverse mutation, recombination and gene conversion; they may involve the nuclear genome, the mitochondrial genome and resident viral or plasmid genomes.

Tests of *chromosomal effects*. These include tests for detection of changes in chromosome number (aneuploidy), structural chromosomal aberrations, sister chromatid exchanges, micronuclei and dominant-lethal events. This classification does not imply that some chromosomal effects are not mutational events.

Tests for *cell transformation*, which monitor the production of preneoplastic or neoplastic cells in culture, are also of importance because they attempt to simulate essential steps in cellular carcinogenesis. These assays are not grouped with those listed above since the mechanisms by which chemicals induce cell transformation may not necessarily be the result of genetic change.

The selection of specific tests and end-points for consideration remains flexible and should reflect the most advanced state of knowledge in this field.

The data from short-term tests are summarized by the Working Group and the test results tabulated according to the end-points detected and the biological complexities of the test systems. The format of the table used is shown below. In these tables, a '+' indicates that the compound was judged by the Working Group to be significantly positive in one or more assays for the specific end-point and level of biological complexity; '-' indicates that it was judged to be negative in one or more assays; and '?' indicates that there were contradictory results from different laboratories or in different biological systems, or that the result was judged to be equivocal. These judgements reflect the assessment by the Working Group of the quality of the data (including such factors as the purity of the test compound, problems of metabolic activation and appropriateness of the test system) and the relative significance of the component tests.

Overall assessment of data from short-term tests

	Genetic activity			Cell transformation
	DNA damage	Mutation	Chromosomal effects	
Prokaryotes				
Fungi/ Green plants				
Insects				
Mammalian cells (*in vitro*)				
Mammals (*in vivo*)				
Humans (*in vivo*)				

An overall assessment of the evidence for *genetic activity* is then made on the basis of the entries in the table, and the evidence is judged to fall into one of four categories, defined as follows:

(1) *Sufficient evidence* is provided by at least three positive entries, one of which must involve mammalian cells *in vitro* or *in vivo* and which must include at least two of three end-points — DNA damage, mutation and chromosomal effects.

(2) *Limited evidence* is provided by at least two positive entries.

(3) *Inadequate evidence* is available when there is only one positive entry or when there are too few data to permit an evaluation of an absence of genetic activity or when there are unexplained, inconsistent findings in different test systems.

(4) *No evidence* applies when there are only negative entries; these must include entries for at least two end-points and two levels of biological complexity, one of which must involve mammalian cells *in vitro* or *in vivo*.

It is emphasized that the above definitions are operational, and that the assignment of a chemical or complex mixture into one of these categories is thus arbitrary.

In general, emphasis is placed on positive results; however, in view of the limitations of current knowledge about mechanisms of carcinogenesis, certain cautions should be respected: (i) At present, short-term tests should not be used by themselves to conclude whether or not an agent is carcinogenic nor can they predict reliably the relative potencies of compounds as carcinogens in intact animals. (ii) Since the currently available tests do not detect all classes of agents that are active in the carcinogenic process (e.g., hormones), one must be cautious in utilizing these tests as the sole criterion for setting priorities in carcinogenesis research and in selecting compounds for animal bioassays. (iii) Negative results from short-term tests cannot be considered as evidence to rule out carcinogenicity, nor does lack of demonstrable genetic activity attribute an epigenetic or any other property to a substance (5).

(b) Evaluation of Carcinogenicity in Humans

Evidence of carcinogenicity can be derived from case reports, descriptive epidemiological studies and analytical epidemiological studies.

An analytical study that shows a positive association between an exposure and a cancer may be interpreted as implying causality to a greater or lesser extent, on the basis of the following criteria: (a) There is no identifiable positive bias. (By 'positive bias' is meant the operation of factors in study design or execution that lead erroneously to a more strongly positive association between an exposure and disease than in fact exists. Examples of positive bias include, in case-control studies, better documentation of the exposure for cases than for controls, and, in cohort studies, the use of better means of detecting cancer in exposed individuals than in individuals not exposed.) (b) The possibility of positive confounding has been considered. (By 'positive confounding' is meant a situation in which the relationship between an exposure and a disease is rendered more strongly positive than it truly is as a result of an association between that exposure and another exposure which either causes or prevents the disease. An example of positive confounding is the association between coffee consumption and lung cancer, which results from their joint association with cigarette smoking.) (c) The association is unlikely to be due to chance alone. (d) The association is strong. (e) There is a dose-response relationship.

In some instances, a single epidemiological study may be strongly indicative of a cause-effect relationship; however, the most convincing evidence of causality comes when several independent studies done under different circumstances result in 'positive' findings.

Analytical epidemiological studies that show no association between an exposure and a cancer ('negative' studies) should be interpreted according to criteria analogous to those listed above: (a) there is no identifiable negative bias; (b) the possibility of negative confounding has been considered; and (c) the possible effects of misclassification of exposure or outcome have been weighed. In addition, it must be recognized that the probability that a given study can detect a certain effect is limited by its size. This can be perceived from the confidence limits around the estimate of association or relative risk. In a study regarded as 'negative', the upper confidence limit may indicate a relative risk substantially greater than unity; in that case, the study excludes only relative risks that are above the upper limit. This usually means that a 'negative' study must be large to be convincing. Confidence in a 'negative' result is increased when several independent studies carried out under different circumstances are in agreement. Finally, a 'negative' study may be considered to be relevant only to dose levels within or below the range of those observed in the study and is pertinent only if sufficient time has elapsed since first human exposure to the agent. Experience with human cancers of known etiology suggests that the period from first exposure to a chemical carcinogen to development of clinically observed cancer is usually measured in decades and may be in excess of 30 years.

The evidence for carcinogenicity from studies in humans is assessed by the Working Group and judged to fall into one of four groups, defined as follows:

(1) *Sufficient evidence* of carcinogenicity indicates that there is a causal relationship between the exposure and human cancer.

(2) *Limited evidence* of carcinogenicity indicates that a causal interpretation is credible, but that alternative explanations, such as chance, bias or confounding, could not adequately be excluded.

(3) *Inadequate evidence* of carcinogenicity, which applies to both positive and negative evidence, indicates that one of two conditions prevailed: (a) there are few pertinent data; or (b) the available studies, while showing evidence of association, do not exclude chance, bias or confounding.

(4) *No evidence* of carcinogenicity applies when several adequate studies are available which do not show evidence of carcinogenicity.

(c) Relevance of Experimental Data to the Evaluation of Carcinogenic Risk to Humans

Information compiled from the first 41 volumes of the *IARC Monographs* shows that, of the chemicals or groups of chemicals now generally accepted to cause or probably to cause cancer in humans, all of those that have been tested appropriately produce cancer in at least one animal species. For several of the chemicals (e.g., aflatoxins, 4-aminobiphenyl, diethylstilboestrol, melphalan, mustard gas and vinyl chloride), evidence of carcinogenicity in experimental animals preceded evidence obtained from epidemiological studies or case reports.

For many of the chemicals (or complex mixtures) evaluated in the *IARC Monographs* for which there is *sufficient evidence* of carcinogenicity in animals, data relating to carcinogenicity for humans are either insufficient or nonexistent. **In the absence of adequate data on humans, it is reasonable, for practical purposes, to regard chemicals or exposures for which there is sufficient evidence of carcinogenicity in animals as if they presented a carcinogenic risk to humans.** The use of the expressions 'for practical purposes' and 'as if they presented a carcinogenic risk' indicates that, at the present time, a correlation between carcinogenicity in animals and possible human risk cannot be made on a purely scientific basis, but only pragmatically. Such a pragmatic correlation may be useful to regulatory agencies in making decisions related to the primary prevention of cancer.

In the present state of knowledge, it would be difficult to define a predictable relationship between the dose (mg/kg bw per day) of a particular chemical required to produce cancer in test animals and the dose that would produce a similar incidence of cancer in humans. Some data, however, suggest that such a relationship may exist(20,21), at least for certain classes of carcinogenic chemicals, although no acceptable method is currently available for quantifying the possible errors that may be involved in such an extrapolation procedure.

8. EXPLANATORY NOTES ON THE CONTENTS OF MONOGRAPHS ON CHEMICALS AND COMPLEX MIXTURES

These notes apply to the format of most monographs, except for those that address industries or life-style factors. Thus, sections 1 and 2, as described below, are applicable in monographs on chemicals or groups of chemicals; in other monographs, they may be replaced by sections on the history of an industry or habit, a description of a process and other relevant information.

(a) Chemical and Physical Data (Section 1)

The Chemical Abstracts Services Registry Number, the Chemical Abstracts Primary Name (Ninth Collective Index)(22) and the IUPAC Systematic Name(23) are recorded in section 1. Other synonyms and trade names are given, but the list is not necessarily comprehensive. Some of the trade names may be those of mixtures in which the compound being evaluated is only one of the ingredients.

The structural and molecular formulae, molecular weight and chemical and physical properties are given. The properties listed refer to the pure substance, unless otherwise specified, and include, in particular, data that might be relevant to identification, environmental fate and human exposure, and biological effects, including carcinogenicity.

A separate description of the composition of technical products includes available information on impurities and formulated products.

(b) Production, Use, Occurrence and Analysis (Section 2)

The purpose of section 2 is to provide indications of the extent of past and present human exposure to the chemical.

Monographs on occupational exposures to complex mixtures or exposures to complex mixtures resulting from human habits include sections on: historical perspectives; description of the industry or habit; manufacturing processes and use patterns; exposures in the workplace; chemistry of the complex mixture.

(i) *Synthesis*

Since cancer is a delayed toxic effect, the dates of first synthesis and of first commercial production of the chemical are provided. This information allows a reasonable estimate to be made of the date before which no human exposure could have occurred. In addition, methods of synthesis used in past and present commercial production are described.

(ii) *Production*

Since Europe, Japan and the USA are reasonably representative industrialized areas of the world, most data on production, foreign trade and uses are obtained from those regions. It should not, however, be inferred that those areas or nations are the sole or even necessarily the major sources or users of any individual chemical.

Production and foreign trade data are obtained from both governmental and trade publications. In some cases, separate production data on organic chemicals manufactured in the USA are not available because their publication could disclose confidential information. In such cases, an indication of the minimum quantity produced can be inferred from the number of companies reporting commercial production. Each company is required to report on individual chemicals if the annual sales value or production volume exceeds a specified minimum level. These levels vary for chemicals classified for different uses, e.g., medicinals and plastics; in fact, the minimal reportable level for annual sales value ranges from $1000-$50 000, and the minimal reportable level for annual production volume ranges from 450-22 700 kg for different classes of use. Data on production are also obtained by means of general questionnaires sent to companies thought to produce the compounds being evaluated. Information from the completed questionnaires is compiled, by country, and the resulting estimates of production are included in the individual monographs.

(iii) *Use*

Information on uses is usually obtained from published sources but is often complemented by direct contact with manufacturers. Some uses identified may not be current or major applications, and the coverage is not necessarily comprehensive. In the case of drugs, mention of their therapeutic uses does not necessarily represent current practice nor does it imply judgement as to their clinical efficacy.

Statements concerning regulations, standards and guidelines (e.g., pesticide registrations, maximum levels permitted in foods, occupational standards and allowable limits) in specific countries may not reflect the most recent situation, since such standards are continuously reviewed and modified. The absence of information on regulatory status for a country should not be taken to imply that that country does not have regulations with regard to the chemical.

(iv) *Occurrence*

Information on the occurrence of a chemical in the environment is obtained from published data, including that derived from the monitoring and surveillance of levels of the chemical in occupational environments, air, water, soil, foods and tissues of animals and humans. When no published data are available to the Working Group, unpublished reports, deemed appropriate, may be considered. When available, data on the generation, persistence and bioaccumulation of a chemical are also included.

(v) *Analysis*

The purpose of the section on analysis is to give the reader an overview, rather than a complete list, of current methods cited in the literature. No critical evaluation or recommendation of any of the methods is meant or implied.

(c) Biological Data Relevant to the Evaluation of Carcinogenic Risk to Humans (Section 3)

In general, the data recorded in section 3 are summarized as given by the author; however, comments made by the Working Group on certain shortcomings of reporting, of statistical analysis or of experimental design are given in square brackets. The nature and extent of impurities/contaminants in the chemicals being tested are given when available.

(i) *Carcinogenicity studies in animals*

The monographs are not intended to cover all reported studies. A few studies are purposely omitted because they are inadequate (e.g., too short a duration, too few animals, poor survival) or because they are judged irrelevant for the purpose of the evaluation. In certain cases, however, such studies are mentioned briefly, particularly when the information is considered to be a useful supplement to other reports or when it is the only data available. Their inclusion does not, however, imply acceptance of the adequacy of their experimental design or of the analysis and interpretation of their results.

Mention is made of all routes of administration by which the test material has been adequately tested and of all species in which relevant tests have been done(24). In most cases, animal strains are given. Quantitative data are given to indicate the order of magnitude of the effective carcinogenic doses. In general, the doses and schedules are indicated as they appear in the original report; sometimes units have been converted for easier comparison. Experiments in which the compound was administered in conjunction with known carcinogens and experiments on factors that modify the carcinogenic effect are also reported. Experiments on the carcinogenicity of known metabolites and derivatives are also included.

(ii) *Other relevant biological data*

LD_{50} data are given when available, and other data on toxicity are included when considered relevant.

Data on effects on reproduction, on teratogenicity and embryo- and fetotoxicity and on placental transfer, from studies in experimental animals and from observations in humans, are included when considered relevant.

Information is given on absorption, distribution and excretion. Data on metabolism are usually restricted to studies that show the metabolic fate of the chemical in experimental animals and humans, and comparisons of data from animals and humans are made when possible.

Data from short-term tests are also included. In addition to the tests for genetic activity and cell transformation described previously (see pages 18-19), data from studies of related effects, but for which the relevance to the carcinogenic process is less well established, may also be mentioned.

The criteria used for considering short-term tests and for evaluating their results have been described (see pages 19-21). In general, the authors' results are given as reported. An assessment of the data by the Working Group which differs from that of the authors, and comments concerning aspects of the study that might affect its interpretation are given in square brackets. Reports of studies in which few or no experimental details are given, or in which the data on which a reported positive or negative result is based are not available for examination, are cited, but are identified as 'abstract' or 'details not given' and are not considered in the summary tables or in making the overall assessment of genetic activity.

For several recent reviews on short-term tests, see IARC(24), Montesano *et al.*(25), de Serres and Ashby(26), Sugimura *et al.*(27), Bartsch *et al.*(28) and Hollstein *et al.*(29).

(iii)   *Case reports and epidemiological studies of carcinogenicity to humans*

Observations in humans are summarized in this section. These include case reports, descriptive epidemiological studies (which correlate cancer incidence in space or time to an exposure) and analytical epidemiological studies of the case-control or cohort type. In principle, a comprehensive coverage is made of observations in humans; however, reports are excluded when judged to be clearly not pertinent. This applies in particular to case reports, in which either the clinico-pathological description of the tumours or the exposure history, or both, are poorly described; and to published routine statistics, for example, of cancer mortality by occupational category, when the categories are so broadly defined as to contribute virtually no specific information on the possible relation between cancer occurrence and a given exposure. Results of studies are assessed on the basis of the data and analyses that are presented in the published papers. Some additional analyses of the published data may be performed by the Working Group to gain better insight into the relation between cancer occurrence and the exposure under consideration. The Working Group may use these analyses in its assessment of the evidence or may actually include them in the text to summarize a study; in such cases, the results of the supplementary analyses are given in square brackets. Any comments by the Working Group are also reported in square brackets; however, these are kept to a minimum, being restricted to those instances in which it is felt that an important aspect of a study, directly impinging on its interpretation, should be brought to the attention of the reader.

(d)   Summary of Data Reported and Evaluation (Section 4)

Section 4 summarizes the relevant data from animals and humans and gives the critical views of the Working Group on those data.

(i)   *Exposures*

Human exposure to the chemical or complex mixture is summarized on the basis of data on production, use and occurrence.

(ii) *Experimental data*

Data relevant to the evaluation of the carcinogenicity of the test material in animals are summarized in this section. The animal species mentioned are those in which the carcinogenicity of the substance was clearly demonstrated. Tumour sites are also indicated. If the substance has produced tumours after prenatal exposure or in single-dose experiments, this is indicated. Dose-response data are given when available.

Significant findings on effects on reproduction and prenatal toxicity, and results from short-term tests for genetic activity and cell transformation assays are summarized, and the latter are presented in tables. An overall assessment is made of the degree of evidence for genetic activity in short-term tests.

(iii) *Human data*

Case reports and epidemiological studies that are considered to be pertinent to an assessment of human carcinogenicity are described. Other biological data that are considered to be relevant are also mentioned.

(iv) *Evaluation*

This section comprises evaluations by the Working Group of the degrees of evidence for carcinogenicity of the exposure to experimental animals and to humans. An overall evaluation is then made of the carcinogenic risk of the chemical or complex mixture to humans. This section should be read in conjunction with pages 18 and 22 of this Preamble for definitions of degrees of evidence.

When no data are available from epidemiological studies but there is *sufficient evidence* that the exposure is carcinogenic to animals, a footnote is included, reading: 'In the absence of adequate data on humans, it is reasonable, for practical purposes, to regard chemicals or exposures for which there is *sufficient evidence* of carcinogenicity in animals as if they presented a carcinogenic risk to humans' (see pp. 22-23 of this Preamble).

References

1. IARC (1979) Criteria to select chemicals for *IARC Monographs. IARC intern. tech. Rep. No. 79/003*

2. IARC (1977) IARC Monographs Programme on the Evaluation of the Carcinogenic Risk of Chemicals to Humans. Preamble. *IARC intern. tech. Rep. No. 77/002*

3. IARC (1978) Chemicals with *sufficient evidence* of carcinogenicity in experimental animals — *IARC Monographs* volumes 1-17. *IARC intern. tech. Rep. No. 78/003*

4. IARC (1982) *IARC Monographs on the Evaluation of the Carcinogenic Risk of Chemicals to Humans*, Supplement 4, *Chemicals, Industrial Processes and Industries Associated with Cancer in Humans (IARC Monographs Volumes 1 to 29)*, Lyon

5. IARC (1983) Approaches to classifying chemical carcinogens according to mechanism of action. *IARC intern. tech. Rep. No. 83/001*

6. IARC (1972-1986) *IARC Monographs on the Evaluation of the Carcinogenic Risk of Chemicals to Humans*, Volumes 1-41, Lyon

    Volume 1 (1972) Some Inorganic Substances, Chlorinated Hydrocarbons, Aromatic Amines, N-Nitroso Compounds and Natural Products (19 monographs), 184 pages

    Volume 2 (1973) Some Inorganic and Organometallic Compounds (7 monographs), 181 pages

    Volume 3 (1973) Certain Polycyclic Aromatic Hydrocarbons and Heterocyclic Compounds (17 monographs), 271 pages

    Volume 4 (1974) Some Aromatic Amines, Hydrazine and Related Substances, N-Nitroso Compounds and Miscellaneous Alkylating Agents (28 monographs), 286 pages

    Volume 5 (1974) Some Organochlorine Pesticides (12 monographs), 241 pages

    Volume 6 (1974) Sex Hormones (15 monographs), 243 pages

    Volume 7 (1974) Some Anti-thyroid and Related Substances, Nitrofurans and Industrial Chemicals (23 monographs), 326 pages

    Volume 8 (1975) Some Aromatic Azo Compounds (32 monographs), 357 pages

    Volume 9 (1975) Some Aziridines, N-, S- and O-Mustards and Selenium (24 monographs), 268 pages

    Volume 10 (1976) Some Naturally Occurring Substances (22 monographs), 353 pages

    Volume 11 (1976) Cadmium, Nickel, Some Epoxides, Miscellaneous Industrial Chemicals and General Considerations on Volatile Anaesthetics (24 monographs), 306 pages

    Volume 12 (1976) Some Carbamates, Thiocarbamates and Carbazides (24 monographs), 282 pages

    Volume 13 (1977) Some Miscellaneous Pharmaceutical Substances (17 monographs), 255 pages

    Volume 14 (1977) Asbestos (1 monograph), 106 pages

    Volume 15 (1977) Some Fumigants, the Herbicides, 2,4-D and 2,4,5-T, Chlorinated Dibenzodioxins and Miscellaneous Industrial Chemicals (18 monographs), 354 pages

    Volume 16 (1978) Some Aromatic Amines and Related Nitro Compounds — Hair Dyes, Colouring Agents, and Miscellaneous Industrial Chemicals (32 monographs), 400 pages

    Volume 17 (1978) Some N-Nitroso Compounds (17 monographs), 365 pages

    Volume 18 (1978) Polychlorinated Biphenyls and Polybrominated Biphenyls (2 monographs), 140 pages

Volume 19 (1979) Some Monomers, Plastics and Synthetic Elastomers, and Acrolein (17 monographs), 513 pages

Volume 20 (1979) Some Halogenated Hydrocarbons (25 monographs), 609 pages

Volume 21 (1979) Sex Hormones (II) (22 monographs), 583 pages

Volume 22 (1980) Some Non-nutritive Sweetening Agents (2 monographs), 208 pages

Volume 23 (1980) Some Metals and Metallic Compounds (4 monographs), 438 pages

Volume 24 (1980) Some Pharmaceutical Drugs (16 monographs), 337 pages

Volume 25 (1981) Wood, Leather and Some Associated Industries (7 monographs), 412 pages

Volume 26 (1981) Some Antineoplastic and Immunosuppressive Agents (18 monographs), 411 pages

Volume 27 (1981) Some Aromatic Amines, Anthraquinones and Nitroso Compounds, and Inorganic Fluorides Used in Drinking-Water and Dental Preparations (18 monographs), 344 pages

Volume 28 (1982) The Rubber Industry (1 monograph), 486 pages

Volume 29 (1982) Some Industrial Chemicals and Dyestuffs (18 monographs), 416 pages

Volume 30 (1982) Miscellaneous Pesticides (18 monographs), 424 pages

Volume 31 (1983) Some Food Additives, Feed Additives and Naturally Occurring Substances (21 monographs), 314 pages

Volume 32 (1983) Polynuclear Aromatic Compounds, Part 1, Chemical, Environmental and Experimental Data (42 monographs), 477 pages

Volume 33 (1984) Polynuclear Aromatic Compounds, Part 2, Carbon Blacks, Mineral Oils and Some Nitroarenes (8 monographs), 245 pages

Volume 34 (1984) Polynuclear Aromatic Compounds, Part 3, Industrial Exposures in Aluminium Production, Coal Gasification, Coke Production, and Iron and Steel Founding (4 monographs), 219 pages

Volume 35 (1984) Polynuclear Aromatic Compounds, Part 4, Bitumens, Coal-Tars and Derived Products, Shale-Oils and Soots (4 monographs), 271 pages

Volume 36 (1985) Allyl Compounds, Aldehydes, Epoxides and Peroxides (15 monographs), 369 pages

Volume 37 (1985) Tobacco Habits other than Smoking; Betel-quid and Areca-nut Chewing; and Some Nitroso Compounds (8 monographs), 291 pages

Volume 38 (1986) Tobacco Smoking (1 monograph), 421 pages

Volume 39 (1986) Some Chemicals Used in Plastics and Elastomers (19 monographs), 403 pages

Volume 40 (1986) Some Naturally Occurring and Synthetic Food Components, Furocoumarins and Ultraviolet Radiation (21 monographs), 444 pages

Volume 41 (1986) Some Halogenated Hydrocarbons and Pesticide Exposures (16 monographs), 434 pages

Supplement No. 1 (1979) Chemicals and Industrial Processes Associated with Cancer in Humans (IARC Monographs, Volumes 1 to 20), 71 pages

Supplement No. 2 (1980) *Long-term and Short-term Screening Assays for Carcinogens: A Critical Appraisal*, 426 pages

Supplement No. 3 (1982) *Cross Index of Synonyms and Trade Names in Volumes 1 to 26*, 199 pages

Supplement No. 4 (1982) *Chemicals, Industrial Processes and Industries Associated with Cancer in Humans, IARC Monographs, Volumes 1 to 29*, 292 pages

Supplement No. 5 (1985) *Cross Index of Synonyms and Trade Names in Volumes 1 to 36*, 259 pages

7. IARC (1973-1986) *Information Bulletin on the Survey of Chemicals Being Tested for Carcinogenicity*, Numbers 1-11, Lyon, France

   Number 1 (1973) 52 pages
   Number 2 (1973) 77 pages
   Number 3 (1974) 67 pages
   Number 4 (1974) 97 pages
   Number 5 (1975) 88 pages
   Number 6 (1976) 360 pages
   Number 7 (1978) 460 pages
   Number 8 (1979) 604 pages
   Number 9 (1981) 294 pages
   Number 10 (1983) 326 pages
   Number 11 (1984) 370 pages
   Number 12 (1986) 389 pages

8. PHS 149 (1951-1983) Public Health Service Publication No. 149, *Survey of Compounds which have been Tested for Carcinogenic Activity*, Washington DC, US Government Printing Office

   1951 Hartwell, J.L., 2nd ed., Literature up to 1947 on 1329 compounds, 583 pages

   1957 Shubik, P. & Hartwell, J.L., Supplement 1, Literature for the years 1948-1953 on 981 compounds, 388 pages

   1969 Shubik, P. & Hartwell, J.L., edited by Peters, J.A., Supplement 2, Literature for the years 1954-1960 on 1048 compounds, 655 pages

   1971 National Cancer Institute, Literature for the years 1968-1969 on 882 compounds, 653 pages

   1973 National Cancer Institute, Literature for the years 1961-1967 on 1632 compounds, 2343 pages

   1974 National Cancer Institute, Literature for the years 1970-1971 on 750 compounds, 1667 pages

   1976 National Cancer Institute, Literature for the years 1972-1973 on 966 compounds, 1638 pages

   1980 National Cancer Institute, Literature for the year 1978 on 664 compounds, 1331 pages

1983 National Cancer Institute, Literature for years 1974-1975 on 575 compounds, 1043 pages

9. Pike, M.C. & Roe, F.J.C. (1963) An actuarial method of analysis of an experiment in two-stage carcinogenesis. *Br. J. Cancer, 17*, 605-610

10. Miller, E.C. (1978) Some current perspectives on chemical carcinogenesis in humans and experimental animals: Presidential address. *Cancer Res., 38*, 1479-1496

11. Miller, E.C. & Miller, J.A. (1981) Searches for ultimate chemical carcinogens and their reactions with cellular macromolecules. *Cancer, 47*, 2327-2345

12. Peto, R. (1974) Guidelines on the analysis of tumour rates and death rates in experimental animals. *Br. J. Cancer, 29*, 101-105

13. Hoel, D.G. & Walburg, H.E., Jr (1972) Statistical analysis of survival experiments. *J. natl Cancer Inst., 49*, 361-372

14. Peto, R., Pike, M.C., Day, N.E., Gray, R.G., Lee, P.N., Parish, S., Peto, J., Richards, S. & Wahrendorf, J. (1980) *Guidelines for simple sensitive significance tests for carcinogenic effects in long-term animal experiments.* In: *IARC Monographs on the Evaluation of the Carcinogenic Risk of Chemicals to Humans, Supplement 2, Long-term and Short-term Screening Assays for Carcinogens: A Critical Appraisal,* Lyon, IARC, pp. 311-426

15. Berenblum, I. (1975) *Sequential aspects of chemical carcinogenesis: Skin.* In: Becker, F.F., ed., *Cancer. A Comprehensive Treatise,* Vol. 1, New York, Plenum Press, pp. 323-344

16. Foulds, L. (1969) *Neoplastic Development,* Vol. 2, London, Academic Press

17. Farber, E. & Cameron, R. (1980) The sequential analysis of cancer development. *Adv. Cancer Res., 31*, 125-126

18. Weinstein, I.B. (1981) The scientific basis for carcinogen detection and primary cancer prevention. *Cancer, 47*, 1133-1141

19. Slaga, T.J., Sivak, A. & Boutwell, R.K., eds (1978) *Mechanisms of Tumor Promotion and Cocarcinogenesis,* Vol. 2, New York, Raven Press

20. Rall, D.P. (1977) *Species differences in carcinogenesis testing.* In: Hiatt, H.H., Watson, J.D. & Winsten, J.A., eds, *Origins of Human Cancer,* Book C, Cold Spring Harbor, NY, Cold Spring Harbor Laboratory, pp. 1383-1390

21. National Academy of Sciences (NAS) (1975) *Contemporary Pest Control Practices and Prospects: The Report of the Executive Committee*, Washington DC

22. Chemical Abstracts Services (1978) *Chemical Abstracts Ninth Collective Index (9CI), 1972-1976*, Vols 76-85, Columbus, OH

23. International Union of Pure & Applied Chemistry (1965) *Nomenclature of Organic Chemistry*, Section C, London, Butterworths

24. IARC (1980) *IARC Monographs on the Evaluation of the Carcinogenic Risk of Chemicals to Humans, Supplement 2, Long-term and Short-term Screening Assays for Carcinogens: A Critical Appraisal*, Lyon

25. Montesano, R., Bartsch, H. & Tomatis, L., eds (1980) *Molecular and Cellular Aspects of Carcinogen Screening Tests (IARC Scientific Publications No. 27)*, Lyon, IARC

26. de Serres, F.J. & Ashby, J., eds (1981) *Evaluation of Short-Term Tests for Carcinogens. Report of the International Collaborative Program*, Amsterdam, Elsevier/North-Holland Biomedical Press

27. Sugimura, T., Sato, S., Nagao, M., Yahagi, T., Matsushima, T., Seino, Y., Takeuchi, M. & Kawachi, T. (1976) *Overlapping of carcinogens and mutagens*. In: Magee, P.N., Takayama, S., Sugimura, T. & Matsushima, T., eds, *Fundamentals in Cancer Prevention*, Tokyo/Baltimore, University of Tokyo/University Park Press, pp. 191-215

28. Bartsch, H., Tomatis, L. & Malaveille, C. (1982) *Qualitative and quantitative comparison between mutagenic and carcinogenic activities of chemicals*. In: Heddle, J.A., ed., *Mutagenicity: New Horizons in Genetic Toxicology*, New York, Academic Press, pp. 35-72

29. Hollstein, M., McCann, J., Angelosanto, F.A. & Nichols, W.W. (1979) Short-term tests for carcinogens and mutagens. *Mutat. Res.*, *65*, 133-226

GENERAL REMARKS ON THE SUBSTANCES AND EXPOSURES CONSIDERED

This forty-first volume of *IARC Monographs* is comprised of considerations of 11 halogenated aliphatic hydrocarbons, bis(2-chloro-1-methylethyl)ether, polybrominated biphenyls and amitrole. Several of these compounds (dichloromethane, methyl iodide, polybrominated biphenyls and amitrole) were evaluated by previous Working Groups (IARC, 1974, 1977a, 1978, 1979a) and in Supplement 4 of the *IARC Monographs* (IARC, 1982). In addition, the carcinogenic risks of occupational exposures to chlorophenols and chlorophenoxy (phenoxyacetic acid) herbicides have been re-evaluated; these exposures were first evaluated in Supplement 4 to the *Monographs* (IARC, 1982), but new data have recently become available and are incorporated here. Monographs on pentachlorophenol (IARC, 1979b), 2,4,5- and 2,4,6-trichlorophenol (IARC, 1979c), 2,4-D (IARC, 1977b), 2,4,5-T (IARC, 1977c), MCPA (IARC, 1983) and TCDD (IARC, 1977d) have not been updated.

Previous evaluations of the carcinogenicity of halogenated hydrocarbons and of their oxygenated derivatives, published in the first 40 volumes of *Monographs* are compiled in Table 1. Several compounds listed in the table are isomers or structural analogues of compounds under consideration here, and they may occur concomitantly as constituents or impurities in technical products.

Several of the compounds are very widely distributed in the atmosphere due either to their formation in the ocean (methyl chloride, methyl bromide, methyl iodide) or to losses during their use (dichloromethane and chlorodifluoromethane). Halogenated alkanes are produced in large quantities for use as solvents and intermediates. A major concern with this group of compounds is thus their widespread distribution and, in many cases, the persistence in the environment of the compounds themselves and of their impurities, in addition to occupational exposures.

Properties, production and use

Halogenated alkanes and alkenes of low molecular weight are produced by a number of basic chemical processes, including direct chlorination, hydrochlorination, oxychlorination, dehydrochlorination and chlorinolysis. World production of chlorine exceeds 30 million tonnes per year, and, for example, in the USA about 65% of the total production is

Table 1. Halogenated hydrocarbons and their oxygen-containing derivatives that have been evaluated in the *IARC Monographs*

Compound	Degree of evidence[a]		*Monographs* volume(s) (group in Suppl. 4)
	Humans	Animals	
Chloroform	I	S	*20*; Suppl. 4 (2B)
Carbon tetrachloride	I	S	*20*; Suppl. 4 (2B)
Dichloromethane*	I	I**	*20*; Suppl. 4 (3)
1,2-Dichloroethane	ND	S	*20*
1,1,1-Trichloroethane	ND	I	*20*
1,1,2-Trichloroethane	ND	L	*20*
1,1,2,2-Tetrachloroethane	ND	L	*20*
Hexachloroethane	ND	L	*20*
Vinyl chloride	S	S	*19*; Suppl. 4 (1)
Trichloroethylene	I	L	*20*; Suppl. 4 (3)
Tetrachloroethylene	I	L	*20*; Suppl. 4 (3)
Vinylidene chloride	I	L	*39*; Suppl. 4 (3)
Dichloroacetylene	ND	L	*39*
Allyl chloride	ND	I	*36*
Chloroprene	I	I	*19*; Suppl. 4 (3)
trans-1,4-Dichlorobutene	ND	I	*15*
Hexachlorobutadiene	ND	L	*20*
Hexachlorocyclohexane	I	L	*20*; Suppl. 4 (3)
ortho- and *para*-Dichlorobenzene	I	I	*29*; Suppl. 4 (3)
Hexachlorobenzene	ND	S	*20*
Benzotrichloride	I	S	*29*; Suppl. 4 (2B)
Benzyl chloride	I	L	*29*; Suppl. 4 (3)
Benzal chloride	I	L	*29*; Suppl. 4 (3)
Chlorinated camphenes (Toxaphene)	ND	S	*20*
Terpene polychlorinates	ND	L	*5*
Epichlorohydrin	I	S	*11*; Suppl. 4 (2B)
Bis(chloromethyl)ether	S	S	*4*; Suppl. 4 (1)
Chloromethyl methyl ether (technical-grade)	S	S	*4*; Suppl. 4 (1)
1,2-Bis(chloromethoxy)ethane	ND	L	*15*
1,4-Bis(chloromethoxymethyl)benzene	ND	L	*15*
Chlordane (Heptachlor)	I	L	*20*; Suppl. 4 (3)
Chlordecone (Kepone)	ND	S	*20*
Mirex	ND	S	*20*
Methoxychlor	ND	NE	*20*
Endrin	ND	I	*5*
Aldrin	I	L	*5*; Suppl. 4 (3)
Dieldrin	I	L	*5*; Suppl. 4 (3)
Chlorotrianisene	ND	I	*21*
Chlormadinone acetate	I	L	*21*; Suppl. 4 (3)
Chlorobenzilate	ND	L	*30*
Benzoyl chloride	I	I	*29*; Suppl. 4 (3)
Hexachlorophene	ND	I	*20*
DDT	I	S	*5*; Suppl. 4 (2B)

Table 1 (contd)

Compound	Degree of evidence[a]		Monographs volume(s) (group in Suppl. 4)
	Humans	Animals	
Clofibrate	I	L	24; Suppl. 4 (3)
2,4,5-Trichlorophenol	I	I	20; Suppl. 4 (3)
2,4,6-Trichlorophenol	I	S	20; Suppl. 4 (2B)
Pentachlorophenol	I	I	20; Suppl. 4 (3)
2,4-D and esters	I	I	15; Suppl. 4 (3)
2,4,5-T and esters	I	I	15; Suppl. 4 (3)
MCPA	I	I	30
Polychlorinated biphenyls	I	S	18; Suppl. 4 (2B)
Tetrachlorodibenzo-*para*-dioxin (TCDD)	I	S	15; Suppl. 4 (2B)
Chlorinated dibenzodioxins (other than TCDD)	ND	I	15
Chlorophenols* (occupational exposure to)	L	—	Suppl. 4 (2B)
Phenoxyacetic acid herbicides* (occupational exposure to)	L	—	Suppl. 4 (2B)
1,2-Dibromoethane (ethylene dibromide)	I	S	15; Suppl. 4 (2B)
Vinyl bromide	ND	S	39
1,2-Dibromo-3-chloropropane	ND	S	20
Polybrominated biphenyls*	ND	I**	18
Methyl iodide*	ND	S**	15
Vinylidene fluoride	ND	I	39
Tetrafluoroethylene	ND	I	19

[a]I, inadequate; S, sufficient; ND, no data; L, limited; NE, no evidence. For definitions of the degrees of evidence, see pp. 18 and 22 of the Preamble to this volume; for a description of the criteria for assigning an exposure to a particular group, see Supplement 4 (IARC, 1982).

*Re-evaluated in the present volume of *Monographs*

**The degree of evidence is that given to the compounds before their re-evaluation in the present volume.

used in plastics, solvents, pesticides and various organic chemicals, including many of the compounds considered here.

Halogenated hydrocarbons are widely used in products to which humans are exposed — in solvents, paints, glues, degreasing agents, dry-cleaning fluids, aerosol propellants, blowing agents, refrigerants, textile-processing chemicals, gasoline additives, flame retardants, fire extinguishers, cutting fluids and as intermediates in the production of other chemicals, synthetic fibres and plastics.

Production of dichloromethane has increased steadily since the 1950s, and, recently, it has substituted partially for trichloroethylene and tetrachloroethylene in metal-degreasing operations and food processing, and for fluorocarbons in aerosol products. In 1978, the estimated world production of chlorofluorocarbons was 850 thousand tonnes. Chlorodifluoromethane, the third most important compound of this chemical class, is used primarily in air conditioning and refrigeration applications.

Bis(2-chloro-1-methylethyl)ether is formed in substantial amounts as a byproduct of propylene oxide and propylene glycol manufacture by the chlorohydrin process and has been used to a limited extent as a soil fumigant. Similarly, pentachloroethane, 1,1,1,2-tetrachloroethane and chlorofluoromethane occur only as intermediates or byproducts in some industrial halogenation processes. 2-Chloro-1,1,1-trifluoroethane is an intermediate in the production of, and a metabolite of, halothane.

Methyl chloride is used principally in the production of silicone compounds, gasoline additives (e.g., tetramethyllead) and as a chemical intermediate and solvent. Methyl chloride, methyl bromide and methyl iodide are also used in the dyestuff and pharmaceutical industries. In the USA, agricultural soil fumigation accounts for about 65% of the total use of methyl bromide. 1,3-Dichloropropene and its mixtures with 1,2-dichloropropane have also been marketed mainly as soil fumigants.

Polybrominated biphenyls were once produced as flame retardants in plastics, textiles and cable coatings. Products containing mainly hexabromobiphenyls were made in large amounts in the late 1970s, and production of compounds containing mainly octabromo- and decabromobiphenyls has continued into the 1980s.

Several chlorinated phenols are of major commercial significance as wood preservatives and as intermediates in herbicide production. Others are used almost exclusively as intermediates in the chemical industry, and therefore give rise to less human exposure and have less impact on the environment. Chlorophenol production has been altered markedly since the 1970s due to the concern over the presence of polychlorinated dioxins and polychlorinated dibenzofurans as impurities.

Use of technical pentachlorophenol as a wood preservative began in the late 1930s. This compound and its salts now have a variety of industrial, agricultural and domestic uses. In Canada, the Nordic countries and the USA, about 95% of pentachlorophenol is used for commercial wood treatment and the rest for slime control in the pulp and paper industry, as well as for nonindustrial purposes, such as weed control and wood preservation in households. Significantly smaller quantities of tetrachlorophenol isomers have been produced for similar uses. 2,4,5-Trichlorophenol and 2,4-dichlorophenol have been commercially important chemical intermediates in the production of herbicides and disinfectants since the 1950s.

The chlorophenoxy herbicides, 2,4-D and 2,4,5-T, were first introduced for agricultural use in 1944; MCPA was introduced in 1945. The propionic acid derivatives, silvex, mecoprop and dichlorprop, have been used since 1953, 1957 and 1961, respectively. By the mid 1960s, chlorophenoxy herbicides represented the most important single class of herbicides. In the 1970s and 1980s, many countries cancelled pesticide registrations for 2,4,5-T formulations, but 2,4-D, MCPA products and amitrole are still used widely by agricultural, forestry, railroad, utility and municipal services.

Occurrence and exposure

Some algae and other marine organisms can introduce chlorine, bromine or iodine into organic molecules, and it has been estimated that the quantities of methyl chloride, methyl bromide and methyl iodide formed in the oceans exceed those from man-made sources. Most of the halocarbon, chlorophenol and chlorophenoxy compounds included in this volume are thus ubiquitous in the human environment, either from natural sources or through industrial and other pollution. Potential sources of human exposure include occupational exposures during manufacture, transportation, use and disposal; domestic exposures from consumer products, such as aerosols, paints and pesticides; environmental exposures from air, water and food contaminated with residues of chemicals, industrial effluents or waste-disposal emissions; intentional or accidental misuse; and industrial accidents. Incomplete combustion can result in emissions of compounds such as polychlorinated dibenzodioxins and polychlorinated dibenzofurans.

Because of the extreme toxicity of some polychlorinated dibenzodioxins and polychlorinated dibenzofurans, highly selective, sensitive, specific analytical methods are required to detect the low levels that occur in environmental and human samples. In recent years, mass spectrometric methods have been developed for the measurement of dioxins in commercial chlorophenol and chlorophenoxy products, incinerator emissions, aquatic organisms and human tissues. Progress has been rapid: for example, in 1971 the limit of detection of 2,3,7,8-TCDD in environmental samples was about 50 ng/g, whereas concentrations of 0.01 pg/g can now be measured, and specific isomers can be identified.

A volume of the *IARC Scientific Publications* series, *Environmental Carcinogens: Selected Methods of Analysis*, Vol. 7 (Fishbein & O'Neill, 1985), describes methods for the sampling and analysis of a number of the volatile halogenated hydrocarbons evaluated in the present volume.

Biological data

Monohalomethanes have a relatively high odour threshold, and this may be one reason for the numerous cases of acute intoxication that have been reported. Because they function as alkylating agents, they are also mutagenic in many test systems. The Working Group noted that there were few available long-term studies in experimental animals or studies *in vitro* or of acute toxicity. The Group was aware, however, of the existence of large volumes of relevant data on the effects of 2,4,5-T and chloromethanes in experimental systems and on humans that were not available in the open literature. Some of these data had been cited in passing, in reviews or in publications on related subjects and as abstracts; but the studies have not been finalized in published reports. Because only published studies are considered in the *IARC Monographs*, these data or their associated conclusions were not addressed. Publication of such studies is strongly encouraged, so that a thorough evaluation of the activity of these compounds can be made.

In arriving at an overall assessment of the strength of the evidence for carcinogenic activity in experimental animals, the Working Group had to consider the weight to give to different types of evidence. Cases in which several studies showed statistically significant increases in tumour incidence in more than one species could readily be regarded as providing 'sufficient evidence', particularly if malignant tumours were induced at more than one site and dose-response relationships were evident. However, the overall evaluation of carcinogenicity was more difficult for chemicals for which a single study reported the very rapid development of an unusually high incidence of uncommon cancers, and for those for which several studies each showed a modest tumour response over a full lifespan, largely affecting tissues in which a high background incidence of tumours had been observed in the species and/or strains in question. The Working Group considered that either of these cases could be regarded as providing 'limited evidence' of carcinogenicity. Other factors considered by the Working Group included the potential effects of known carcinogens present as contaminants in the chemical formulations under study, the weight to give to studies that failed to confirm earlier results, and how to judge the induction of benign tumours or increases in spontaneously occurring tumours, particularly when the induced rate fell within the range of occurrence in historical controls.

All short-term test systems were taken into consideration by the Working Group, with the exception of the BHK21 cell transformation assay. This assay was judged unacceptable because of the demonstrated lack of interlaboratory reproducibility, and a prior decision by a Working Group convened by the IARC (Griesemer et al., 1986) to exclude it. The Working Group recognized that some other test systems may also not be reproducible between laboratories; however, in the absence of controlled studies, these tests were given equal consideration with test procedures that have been validated.

The Group questioned the significance of results that were obtained using very high doses, e.g., with 1,2-dichloropropane, because low levels of contaminants, physical factors and kinetic considerations could affect the activity. Equally, the validity of negative results obtained with nontoxic doses is doubtful.

It was noted that the absence of mutagenicity in biological samples from individuals handling chemicals might not reflect the mutagenic potential of the chemical, since protective clothing might have been worn. Thus, adequate descriptions of the exposure situation are essential in order to interpret the results of such studies.

Epidemiology

The evaluation of epidemiological studies is comprised of a number of issues — particularly, exposure assessment, diagnostic criteria, confounding factors and study power. In this volume, these issues are particularly relevant to occupational exposures to chlorophenoxy herbicides and chlorophenols.

Groups exposed to chlorophenols have been studied mainly in manufacturing plants where trichlorophenols serve as intermediates in the production of chlorophenoxy herbicides. In these same plants, a number of other chemicals may have been produced, and exposure to TCDD may have occurred, particularly in association with chemical accidents. Strictly speaking, the possible carcinogenic risks of occupational exposures to chlorophenols could not be addressed in these cohort studies.

The other sources of exposure to chlorophenols that have been studied are those of cases and referent persons in case-control studies from Sweden and New Zealand, where exposure was reported in saw mills, in the pulp and paper industry, in fencing and in tanneries. In these working environments, also, other exposures have occurred. Difficulties related to variable diagnostic criteria are discussed in the context of soft-tissue sarcoma on p. 343 of the monograph on chlorophenols.

Case-control studies offer an alternative approach to cohort studies for assessing the relation between certain exposures and outcomes and are particularly suitable for rare diseases, although high population exposure frequencies are needed for them to be effective. Long-term occupational exposures to chlorophenoxy herbicides and chlorophenols occurred rarely in the epidemiological studies reported.

Occupational groups with potential exposure to chlorophenoxy herbicides include farmers, forestry workers, herbicide applicators and manufacturing groups. The first three categories of people have usually been exposed to a large number of herbicides as well as other chemicals. This has been taken into account in some case-control studies by inquiring about a number of chemicals — not only the herbicides. In those studies, the increased risks have been associated mainly with chlorophenoxy herbicides and chlorophenols and not with all chemicals, but synergistic effects could occur.

Some studies show statistically significant excess risks, while others do not. The reasons for these differences may be many, and the results are not necessarily contradictory, even in the case of well-designed studies; different levels or patterns of exposure, modifying factors, unknown concomitant exposures (e.g., impurities, solvents, additives) or chance may all play a part. The degree of imprecision in risk estimates that is attributable to random error is commonly measured by confidence intervals. The findings of conflicting studies should, in principle, be considered in terms of estimation of intervals rather than in terms of point estimations with their associated p values. In addition, possible systematic biases must be considered.

There has been considerable controversy about the possible carcinogenic hazards of chlorophenoxy herbicides and chlorophenols. This Working Group, like other IARC working groups, has come to its evaluation on the basis of information reported in scientific papers — published or in press. Expert testimonies and other information not documented in the scientific literature were not considered in making the evaluation.

References

Fishbein, L. & O'Neill, I.K., eds (1985) *Environmental Carcinogens. Selected Methods of Analysis*, Vol. 7, *Some Volatile Halogenated Hydrocarbons (IARC Scientific Publications No. 68)*, Lyon, International Agency for Research on Cancer

IARC (1974) *IARC Monographs on the Evaluation of Carcinogenic Risk of Chemicals to Man*, Vol. 7, *Some Anti-thyroid and Related Substances, Nitrofurans and Industrial Chemicals*, Lyon, pp. 31-43

IARC (1977a) *IARC Monographs on the Evaluation of the Carcinogenic Risk of Chemicals to Man*, Vol. 15, *Some Fumigants, the Herbicides 2,4-D and 2,4,5-T, Chlorinated Dibenzodioxins and Miscellaneous Industrial Chemicals*, Lyon, pp. 245-254

IARC (1977b) *IARC Monographs on the Evaluation of the Carcinogenic Risk of Chemicals to Man*, Vol. 15, *Some Fumigants, the Herbicides 2,4-D and 2,4,5-T, Chlorinated Dibenzodioxins and Miscellaneous Industrial Chemicals*, Lyon, pp. 111-138

IARC (1977c) *IARC Monographs on the Evaluation of the Carcinogenic Risk of Chemicals to Man*, Vol. 15, *Some Fumigants, the Herbicides 2,4-D and 2,4,5-T, Chlorinated Dibenzodioxins and Miscellaneous Industrial Chemicals*, Lyon, pp. 273-299

IARC (1977d) *IARC Monographs on the Evaluation of the Carcinogenic Risk of Chemicals to Man*, Vol. 15, *Some Fumigants, the Herbicides 2,4-D and 2,4,5-T, Chlorinated Dibenzodioxins and Miscellaneous Industrial Chemicals*, Lyon, pp. 41-102

IARC (1978) *IARC Monographs on the Evaluation of the Carcinogenic Risk of Chemicals to Humans*, Vol. 18, *Polychlorinated Biphenyls and Polybrominated Biphenyls*, Lyon, pp. 107-124

IARC (1979a) *IARC Monographs on the Evaluation of the Carcinogenic Risk of Chemicals to Humans*, Vol. 20, *Some Halogenated Hydrocarbons*, Lyon, pp. 449-465

IARC (1979b) *IARC Monographs on the Evaluation of the Carcinogenic Risk of Chemicals to Humans*, Vol. 20, *Some Halogenated Hydrocarbons*, Lyon, pp. 303-325

IARC (1979c) *IARC Monographs on the Evaluation of the Carcinogenic Risk of Chemicals to Humans*, Vol. 20, *Some Halogenated Hydrocarbons*, Lyon, pp. 349-367

IARC (1982) *IARC Monographs on the Evaluation of the Carcinogenic Risk of Chemicals to Humans*, Suppl. 4, *Chemicals, Industrial Processes and Industries Associated with Cancer in Humans, IARC Monographs, Volumes 1 to 29*, Lyon, pp. 38-40, 111-112, 88-89, 211-212

IARC (1983) *IARC Monographs on the Evaluation of the Carcinogenic Risk of Chemicals to Humans*, Vol. 30, *Miscellanous Pesticides*, Lyon, pp. 255-269

Montesano, R., Bartsch, H., Vainio, H., Wahrendorf, J., Wilbourn, J. & Yamasaki, H. (1986) *Long-term and Short-term Screening Assays for Carcinogens — A Critical Appraisal* (*IARC Scientific Publications No. 83*), Lyon, IARC (in press)

THE MONOGRAPHS

DICHLOROMETHANE

This substance was considered by previous Working Groups, in June 1978 (IARC, 1979a) and February 1982 (IARC, 1982a). Since that time, new data have become available, and these have been incorporated into the monograph and taken into consideration in the present evaluation.

1. Chemical and Physical Data

1.1 Synonyms and trade names

Chem. Abstr. Services Reg. No.: 75-09-2

Chem. Abstr. Name: Dichloromethane

IUPAC Systematic Name: Dichloromethane

Synonyms: Methane dichloride; methylene bichloride; methylene chloride; methylene dichloride

Trade Names: Aerothene MM; Narkotil; R 30; Solaesthin; Solmethine

1.2 Structural and molecular formulae and molecular weight

$$\text{Cl} - \underset{\underset{H}{|}}{\overset{\overset{H}{|}}{C}} - \text{Cl}$$

CH_2Cl_2 Mol. wt: 84.93

1.3 Chemical and physical properties of the pure substance

(a) *Description*: Colourless liquid with penetrating ether-like odour (Hawley, 1981; Verschueren, 1983; Windholz, 1983; Sax, 1984)

(b) *Boiling-point*: 40°C (Weast, 1985)

(c) *Melting-point*: -95.1°C (Weast, 1985)

(d) *Density*: d_4^{20} 1.3266 (Weast, 1985)

(e) *Spectroscopy data*[a]: Ultraviolet (Grasselli & Ritchey, 1975), infrared (Sadtler Research Laboratories, 1980; prism [6620 (gas), 1011], grating [28523]), nuclear magnetic resonance (Sadtler Research Laboratories, 1980; proton [6401], C-13 [167]) and mass spectral data (Grasselli & Ritchey, 1975) have been reported.

(f) *Solubility*: 1.38 g/100 ml in water at 20°C (Honeywill & Stein, undated); soluble in ethanol and diethyl ether (Windholz, 1983; Weast, 1985)

(g) *Volatility*: Vapour pressure, 400 mm Hg at 24.1°C (Weast, 1985); relative vapour density (air = 1), 2.93 (Verschueren, 1983; Sax, 1984)

(h) *Stability*: Vapour is noninflammable and is not explosive when mixed with air (Hawley, 1981; Windholz, 1983) but may form explosive mixtures in atmospheres with higher oxygen content (Sax, 1984)

(i) *Reactivity*: Reacts vigorously with active metals (lithium, sodium, potassium) and with strong bases (potassium *tert*-butoxide) (Sax, 1984)

(j) *Octanol/water partition coefficient (P)*: log P, 1.25 (Hansch & Leo, 1979)

(k) *Conversion factor*: mg/m³ = 3.47 × ppm[b]

1.4 Technical products and impurities

Dichloromethane is available as commercial/technical grade and grades intended specifically for vapour degreasing, aerosol use, food extraction, reagent use and spectrophotometry. Purity, when reported, ranges from 99-99.9% (reagent/high-performance liquid chromatography grade). Acidity (as hydrochloric acid) may be up to 5-10 mg/kg. The maximum concentration of water in commercial-grade dichloromethane is generally 100-200 mg/kg, but anhydrous dichloromethane (less than 50 mg/kg water) is also available (Anthony, 1979; Hays Chemicals, 1983; Aldrich Chemical Co., 1984; Burdick & Jackson Laboratories, 1984; Mannsville Chemical Products Corp., 1984; Eastman Kodak Co., 1985; Stauffer Chemical Co., undated a,b,c).

---

[a]In square brackets, spectrum number in compilation
[b]Calculated from: mg/m³ = (molecular weight/24.45) × ppm, assuming standard temperature (25°C) and pressure (760 mm Hg)

Small amounts of stabilizers are often added to dichloromethane at the time of manufacture (Anthony, 1979). Cyclohexane (50 mg/kg) and propylene oxide (see IARC, 1985) have been added to commercial aerosol and reagent grades of dichloromethane for this purpose (Burdick & Jackson Laboratories, 1984; Mannsville Chemical Products Corp., 1984). Other reported stabilizers include 2-methyl-2-butene at 50 mg/kg, ethanol or methanol at approximately 0.2%, and small quantities (1 mg/kg) of phenol, hydroquinone (see IARC, 1977a), *para*-cresol, resorcinol (see IARC, 1977b), thymol, 1-naphthol and various amines (US Environmental Protection Agency, 1978).

Commercial dichloromethane may contain methyl chloride (see monograph, p. 161), chloroform (see IARC, 1979b), 1,1-dichloroethane (vinylidene chloride; see IARC, 1986) and *trans*-1,2-dichloroethene as impurities (US Environmental Protection Agency, 1978; National Toxicology Program, 1986).

2. Production, Use, Occurrence and Analysis

2.1 Production and use

(a) Production

Dichloromethane was first prepared by Regnault in 1840 by the chlorination of methyl chloride in sunlight. Two commercial processes are currently used for the production of dichloromethane — hydrochlorination of methanol and direct chlorination of methane (Anthony, 1979).

Hydrochlorination of methanol represents the principal method of dichloromethane production. In this two-stage process, methyl chloride is produced initially by the reaction of methanol and hydrogen chloride in the presence of a catalyst and is then reacted with chlorine to form dichloromethane. Chloroform (see IARC, 1979b) and carbon tetrachloride (see IARC, 1979c, 1982b) are produced as coproducts of this reaction and are removed by fractionation (Anthony, 1979).

An older method for production of dichloromethane, used less frequently, is direct reaction of methane with chlorine at high temperature (485-510°C). Methyl chloride, chloroform and carbon tetrachloride are also produced as coproducts of this reaction (Anthony, 1979).

Production of dichloromethane from methyl chloride is conventionally performed in the gas phase; however, processes involving high-temperature gas-phase chlorination produce unwanted byproducts due to C-C bond formation and cleavage. These highly chlorinated products (up to C_6) comprise about 0.6% of the total products (Edwards *et al.*, 1982a). A liquid phase process has been described in which an aqueous mixture of methanol, hydrogen chloride and zinc chloride is reacted at 100-150°C, but this method is less efficient than the vapour phase processes and is not widely used (Anthony, 1979). A Japanese firm, however, has begun producing dichloromethane in the liquid phase; the lower temperature reportedly leads to higher conversion and lower byproduct formation (Akiyama *et al.*, 1981).

The 1981 world production capacity for dichloromethane was approximately 825 million kg/year (Edwards *et al.*, 1982b). Estimates of the world production of this compound are listed in Table 1.

Table 1. World production of dichloromethane (millions of kg)[a]

1960	1965	1970	1975	1980
93	175	331	402	570

[a]From Edwards *et al.* (1982b)

The USA has produced approximately 55% of the annual world output for many years (Edwards *et al.*, 1982b). US production of dichloromethane was approximately 276 million kg in 1984 (US International Trade Commission, 1985) and 265 million kg in 1983 (US International Trade Commission, 1984), with 33 million kg exported; imports were estimated at 20 million kg. Four major producers of dichloromethane have been identified in the USA, with a combined annual capacity of approximately 358 million kg for 1984 (Mannsville Chemical Products Corp., 1984). Mexico is also reported to be a major producer (201 million kg in 1982) (United Nations, 1985).

Eleven manufacturers of dichloromethane have been identified in western Europe with a combined annual capacity of 405 million kg. The second largest producer of this compound in the world is located in the UK (Edwards *et al.*, 1982b). Three dichloromethane manufacturers are located in the Federal Republic of Germany, two in Italy, and two in France; Belgium, Spain and the Netherlands have one producer each. Sweden produced dichloromethane until 1974 (United Nations, 1985).

There are currently four major manufacturers of dichloromethane in Japan (The Chemical Daily Co., 1984), which produced 44 million kg in 1982 (United Nations, 1985).

(*b*) *Use*

Historically, the most important use of dichloromethane has been as a paint and varnish remover (Anon., 1981). In 1974, paint removers and solvent degreasing accounted for the principal demand, aerosol formulations for 20%, food and drug processing for 13% and the plastics industries for 12% (Considine, 1974). Recently, new applications have arisen for this compound (Table 2), in particular, its use in a variety of aerosol products (hair sprays, insecticides, spray paints) as a cosolvent or vapour pressure depressant (Anon., 1981; Edwards *et al.*, 1982b; Anon., 1983; Mannsville Chemical Products Corp., 1984; Fishbein, 1985).

The importance of dichloromethane as a blowing agent for flexible polyurethane foams has grown in recent years, with its use to replace fluorocarbons (Anon., 1981; Mannsville

Table 2. Estimated worldwide dichloromethane pattern[a]

Use	Percentage
Aerosols	20-25
Paint remover	25
Process solvent[b]	35-40
Miscellaneous[c]	10-15

[a]From Anon. (1983); Mannsville Chemical Products Corp. (1984); Edwards et al. (1982b)

[b]Includes chemical and pharmaceutical manufacture, acetate film and fibre production, polycarbonate manufacture and other industrial processes

[c]Includes foam blowing, metal cleaning and food extraction

Chemical Products Corp., 1984). This application, which accounted for approximately 5% of US consumption in 1980, now represents as much as 15%.

Dichloromethane has also been introduced as a vapour degreasing solvent for metals and plastics as an alternative to trichloroethylene (Stauffer Chemical Co., 1976; Anon., 1981). Dichloromethane is used in the electronics industry as a cleaning solvent for circuit boards, and as a stripper solvent for photoresists (Mannsville Chemical Products Corp., 1984). In addition, dichloromethane has been used as a solvent for cellulose acetate fibre, plastic film, adhesives, protective coatings, in chemical processing (Anon., 1983) and as carrier solvent for herbicides and insecticides (Anthony, 1979). It is used in the pharmaceutical industry as a process solvent in the production of steroids, antibiotics and vitamins and as a solvent for tablet coatings (Anthony, 1979). Heat-sensitive, naturally occurring substances, such as cocoa, edible fats, spices and beer hops, can be extracted with dichloromethane. It has also been used since the late 1970s as a substitute for trichloroethylene for decaffeinating coffee, although it is estimated that less than 2% of all dichloromethane is used in this way (Anthony, 1979; Anon., 1981; Mannsville Chemical Products Corp., 1984).

Other minor miscellaneous uses of dichloromethane include use as a refrigerant, in oil dewaxing, as a dye and perfume intermediate, and as a carrier solvent in the textile industry (Anthony, 1979; Anon., 1981, 1983).

In agricultural applications, dichloromethane has been used as a post-harvest fumigant for strawberries, as a grain fumigant, and in combination with ethylene for degreening citrus fruits (Anon., 1985).

Exposure to dichloromethane has been reported through inhalant abuse of aerosol propellants (Garriott & Petty, 1980). Dichloromethane was used as an inhalation anaesthetic in the late 19th and early 20th centuries (Bourne & Stehle, 1923; Moskovitz & Shapiro, 1952). One clinical study of the effectiveness of dichloromethane as an anaesthetic agent was performed as late as 1950 (Grasset & Gauthier, 1950).

(c) *Regulatory status and guidelines*

Occupational exposure limits for dichloromethane in 19 countries are presented in Table 3.

In the USA, dichloromethane may be present as an extractant or process solvent residue in spice oleoresins at a level not to exceed 30 mg/kg (including all chlorinated solvents), in hops extract at less than or equal to 2.2%, and in coffee at a level not to exceed 10 mg/kg (US Food and Drug Administration, 1985).

The Joint FAO/WHO Expert Committee on Food Additives (WHO, 1983) withdrew the previously allocated temporary allowable daily intake (ADI) of 0-0.5 mg/kg bw and recommended that the use of dichloromethane as an extraction solvent be limited, in order to ensure that its residues in food are as low as practicable.

2.2 Occurrence

(a) *Natural occurrence*

Dichloromethane is not known to occur as a natural product.

(b) *Occupational exposure*

Occupational exposure levels of dichloromethane in locations in the USA are given in Table 4. High air concentrations have been measured in workplaces where dichloromethane is used as a solvent in paint stripping, cleaning and degreasing operations, in the manufacture of plastic film and synthetic fibres and in printing.

Concentrations of dichloromethane were measured in alveolar air and blood, and carboxyhaemoglobin (COHb) levels were determined in workers exposed in a shoe factory. Four hours after the beginning of a working shift, the levels were 26-421 mg/m^3, 0.22-4.71 mg/l and 2.5-15% COHb, respectively. The workers were exposed to workplace air concentrations ranging from 65 to 730 mg/m^3 (Perbellini *et al.*, 1977). In the production of film foils, formic acid was detected as a urinary metabolite in workers exposed to air concentrations of 100-17 000 mg/m^3 dichloromethane (Kuželová & Vlasák, 1966).

(c) *Air*

It has been suggested that most of the dichloromethane produced commercially is eventually released into the environment, especially the atmosphere (Derwent & Eggleton, 1978; Edwards *et al.*, 1982b; International Programme on Chemical Safety, 1984).

Average levels of dichloromethane in air at ten urban sites in the USA ranged from 390 to 3751 ppt (1353-13 000 ng/m^3) (Singh *et al.*, 1981, 1982). From these data, Singh *et al.* (1981) calculated that average human exposures to dichloromethane at three of the sites were up to 309 µg/person per day. The surface-level background concentration at 40°N latitude was estimated to be 173 ng/m^3 (50 ppt) (Singh *et al.*, 1982).

The primary mechanism for the removal of dichloromethane from the atmosphere is believed to be reaction with hydroxyl radicals. Based on this hypothesis, the residence time of dichloromethane in an urban atmosphere was estimated to be 77 days with a daily rate of loss (12 sunlit hours) of 1.3% (Singh *et al.*, 1981).

Table 3. Occupational exposure limits for dichloromethane[a]

Country	Year	Concentration (mg/m³)	Interpretation[b]
Australia	1978	720	TWA
Belgium	1978	720	TWA
Czechoslovakia	1976	500	TWA
		2500	Ceiling
Finland	1981	350	TWA
		525	STEL
France	1985	360	TWA
		1800	STEL
German Democratic Republic	1979	500	TWA
		1500	STEL
Germany, Federal Republic of	1985	360	TWA
Italy	1978	360	TWA
Japan	1978	1740	TWA
The Netherlands	1978	720	TWA
Norway	1981	245	TWA
Poland	1976	50	Ceiling
Romania	1975	500	TWA
		700	Ceiling
Sweden	1984	250	TWA
		500	STEL
Switzerland	1978	710	TWA
UK	1985	700	TWA
		870	STEL
USA	1985		
ACGIH		350	TWA
		1740	STEL
NIOSH		261	TWA
		1740	Ceiling
OSHA		1750	TWA
		3500	Ceiling
		7000	STEL
USSR	1977	50	Ceiling
Yugoslavia	1971	500	Ceiling

[a]From International Labour Office (1980); Direktoratet for Arbeidstilsynet (1981); Työsuojeluhallitus (1981); Arbetarskyddsstyrelsens Författningssamling (1984); American Conference of Governmental Industrial Hygienists (ACGIH) (1985); Deutsches Forschungsgemeinschaft (1985); Health and Safety Executive (1985); Institut National de Recherche et de Sécurité (1985); National Institute for Occupational Safety and Health (NIOSH) (1985); US Occupational Safety and Health Administration (OSHA) (1985)

[b]TWA, time-weighted average; STEL, short-term exposure limit

Table 4. Occupational exposures to dichloromethane

Location	Job classification	Concentration (mg/m³ air)[a]	Reference
Plastics industry (6 plants, 1972-1981)	Mixing, moulding preforming, pressing	18-35 (P) 0.4-48 (A) 17.5-130 (P) 66.5-108 (A)	Cohen & Vandervort (1972) Wagner (1974)
	Foam gun operator, waxer	690-1600 (P)	Burroughs & Moody (1982)
	Fabrication, assembly, finishing	≤0.4 (A) <0.4-40 (P) <0.3-35 (P) 59 (P) 16-325 (P)	Cohen & Vandervort (1972) Hollett (1977) Markel & Jannerfeldt (1981) Markel & Slovin (1981)
Plastic film industry (1 plant, 1968-1973)	Casting Filtration Winding Office work	458-2060 (A) 583-3350 (A) 652-659 (A) 160-1130 (A)	National Institute for Occupational Safety and Health (1976)
Synthetic fibres industry (2 plants, 1977-1979)	Pressman Extrusion area Bobbin stores Textile department	916-1300 (P) 239-1950 (P) 264-729 (P) 10.6-967 (P)	Cohen et al. (1980)
	Extrusion and preparation	486-1648 (P)	Ott et al. (1983a)
Footwear manufacture (4 plants, 1975-1982)	Four-part machine Crimping	118-597 (P)[b] 31 (P)[b]	Tharr et al. (1982)
	Assembly, moulding	<5-104 (P) 27-300 (A) 2-319 (P)	Gunter (1975) Hervin & Watanabe (1981)
General manufacturing (air conditioning and refrigeration equipment, vending machines, pipes, welding wire, toys, fibreglass boats, sporting goods, paints and coatings, drugs, medical equipment (14 plants, 1972-1981)	Degreasing, stripping, flushing, cleaning	7-1930 (P) 10-28 (A) 14-101 (P) <2-136 (A) <3.5-403 (P) 654-868 (A) 180-2190 (P) 27 (P) 52-141 (P) 1.3-467 (P) 2-240 (A)	Burton & Shmunes (1973) Markel & Shama (1974) Gunter & Lucas (1974) Hervin et al. (1974) Kronoveter (1977) Lee (1980) Ruhe (1981) Ruhe et al. (1981) Ruhe et al. (1982)
	Production, operations (fabrication, moulding, waxing, laminating)	1-74 (A) 100 (A) 4-38 (P)[b] <4-56 (A)[b] 22-85 (P)	Vandervort & Polakoff (1973) Gilles (1977) Rosensteel & Meyer (1977) Markel (1980)
	Solvent control	10-118 (A) 507 (P)	Tharr & Donohue (1980) Ruhe et al. (1982)

Table 4 (contd)

Location	Job classification	Concentration (mg/m³ air)[a]	Reference
Laboratories (2 laboratories, 1978-1980)	Laboratory technician	23-172 (P) 600 (A) 236-455 (P)	Ruhe (1978) Salisbury (1981)
Maintenance/repair (automotive, aircraft, furniture, general contracting, miscellaneous) (10 plants, 1976-1981)	Paint stripping, sanding	38-2820 (P) 81-1379 (A) 94-4882 (P) 1041 (A) 45-698 (P) 298-1461 (A) 55-930 (A)	Okawa & Keith (1977) Chrostek (1980) Hartle (1980) Cohen et al. (1980)
	Painting	25 (P) 30-503 (P) 22-233 (P) 0.06-0.48 (A)	Gunter (1976) Ruhe & Anderson (1977) Chrostek & Levine (1981) Pryor (1981)
	Other	2 (P) 30-412 (P) 2.61 (P) 1.1-70 (A) 0.01-2.4 (A) 3.4-31.1 (P)	Gunter (1976) Ruhe & Anderson (1977) White & Wegman (1978) Pryor (1981) Albrecht (1982)
Printing (5 plants, 1975-1981)	Printing operation	17 (P) 24-410 (P) 5-560 (P) 17-257 (A) 8.2-37 (P) 2.7-16 (A)	Ahrenholz (1980) Lewis & Thoburn (1981) Quinn (1981) Gorman (1982)
	Press checking and cleaning	360-1550 (P) 356 (A)	Quinn (1981)
	Tank cleaning	84-17 890 (P) 73-285 (A)	Rivera (1975)
	Darkroom, drafting, folding, collating, office work	<6-248 (P) <10-30 (A)	Quinn (1981)
Coffee decaffeination (1 plant, 1978)	Processing/extracting/evaporating	1.4-115 (P)	Cohen et al. (1980)
	Coffee handling, drying	1-86 (P)	

[a]P, personal samples collected in the breathing zone of workers; A, area samples, collected in general work area
[b]Time-weighted averages

Dichloromethane was detected at a level of 1.5 µg/m³ in the ambient air of a household basement. The sample was taken during February 1978 in a house that bordered a toxic chemical dump, the Old Love Canal, in Niagara Falls, NY, USA (Barkley et al., 1980).

Dichloromethane was detected at a mean concentration of 14.3 µg/m³ in indoor air at 13 of 16 residences in suburban Knoxville, TN, USA, monitored for volatile organic compounds. It was not detected in the outdoor air surrounding the houses (Gupta et al., 1984).

(d) *Water and sediments*

Samples of groundwater and surface-water used for potable supply were collected during 1977-1979 from every county and urban, suburban and rural area in New Jersey, USA, and analysed for dichloromethane. It was detected in 246 of 1047 groundwater samples and in 275 of 605 surface-water samples. The highest reported concentrations were 1900 and 743 µg/l, respectively (Page, 1981).

Although the concentrations were not reported, dichloromethane was identified in the drinking-water of all of five US cities surveyed (Coleman et al., 1976).

The levels of dichloromethane in raw and treated waters (both ground and surface) from 30 potable water treatment facilities serving about 5.5 million consumers across Canada were higher during summer than during winter. The mean concentrations of dichloromethane in raw water were 6 µg/l with a maximum of 50 µg/l in August-September, and <1 µg/l with a maximum of 12 µg/l in November-December. The mean concentrations in treated water were 10 µg/l with a maximum of 50 µg/l in August-September, and 3 µg/l with a maximum of 50 µg/l in November-December (Otson et al., 1982).

Dichloromethane was detected at an average concentration of 1.0 µg/l (range, 1.0-2.0 µg/l) in water samples taken in 1976 from the Rhine River near Lobith, Federal Republic of Germany, and at a maximum concentration of 1 µg/l in drinking-water samples from 100 cities in the country in 1977 (Bauer, 1981). Concentrations of dichloromethane in municipal drinking-water in Göteborg, Sweden, were reported to be <60 ng/l (Eklund et al., 1978a).

Dichloromethane was detected at levels between trace and 45 ng/l at six of 17 lower Niagara River sites and at an average level of 572 ng/l (traces to 4600 ng/l) in 51 samples from 83 sites in Lake Ontario (Kaiser et al., 1983). Although the levels found were not reported, a detection frequency of 7% for concentrations >1 µg/l was reported for dichloromethane in water samples from the Delaware River (DeWalle & Chian, 1978).

Levels of dichloromethane in water from a US sewage treatment plant which served a large industrial, as well as a municipal, area were 8.2 µg/l (influent before treatment), 2.9 µg/l (effluent before chlorination) and 3.4 µg/l (effluent after chlorination) (Bellar et al., 1974).

Water samples collected in February and May 1977 from the Back River estuary in Maryland, USA, which received effluent from an urban wastewater treatment plant, contained dichloromethane. The highest levels (66 µg/l) were found in samples taken in the treatment plant just before final chlorination, suggesting that dichloromethane was derived from commercial and industrial activities in the area (Helz & Hsu, 1978).

Dichloromethane was detected by purge-and-trap gas chromatography at levels ranging from 19-95 µg/l in six samples of raw sewage and effluent from Canadian sewage treatment plants (Lao et al., 1982).

The wastewater from a US specialty chemical plant manufacturing a broad range of chemicals contained 3-8 mg/l dichloromethane. It was not detected in the sediment but was identified (not quantified) in receiving water at the same facility (Jungclaus et al., 1978).

Dichloromethane was detected by gas chromatography/mass spectrometry at concentrations ranging from <0.01 to 1.0 mg/l in the volatile fraction of wastewater from the Oak Ridge Gaseous Diffusion Plant in Tennessee, USA (McMahon, 1983).

As part of the Swedish Drinking Water Project, dichloromethane was found at 640 µg/l in the effluent stream from a sulphate pulp mill. The amount of dichloromethane discharged from the mill was estimated to be 40 tonnes/year (Eklund et al., 1978b).

Levels of dichloromethane in sediment samples collected during May to June 1980 from Lake Pontchartrain in New Orleans, LA, USA, ranged from 1.5 to 3.2 µg/kg (Ferrario et al., 1985).

(e) Food and beverages

The average residual level of dichloromethane in coffee beans decaffeinated by dichloromethane extraction in the USA ranged from 0.32 to 0.42 mg/kg over a five-month period in 1978 (115-295 samples per month) (Cohen et al., 1980). In 12 samples of decaffeinated, ground, roasted coffee purchased in Canadian retail stores and analysed by gas chromatography (electrolytic conductivity detection) for solvent residues, dichloromethane was detected at 0.02-1.69 µg/g in four of eight samples of imported coffee, but was not detected in four domestically decaffeinated coffee samples (Page & Charbonneau, 1977).

(f) Animals

Oysters and clams collected during May-June 1980 from Lake Pontchartrain in New Orleans, LA, USA, contained mean levels of 7.8 and 27 µg/kg dichloromethane, respectively (Ferrario et al., 1985).

(g) Human tissues and secretions

Although levels were not reported, dichloromethane was identified by gas chromatography/mass spectrometry in all of eight samples of mother's milk collected in four US urban areas selected for the probability of emissions of various halogenated pollutants (Pellizzari et al., 1982).

2.3 Analysis

Selected methods for the analysis of dichloromethane in air, water, food and blood are identified in Table 5. The US Environmental Protection Agency methods for analysing water (8010 and 8240) have also been applied to liquid and solid wastes (US Environmental Protection Agency, 1982a,b). Volatile components of solid-waste samples are first extracted

Table 5. Methods for the analysis of dichloromethane

Sample matrix	Sample preparation	Assay procedure[a]	Limit of detection	Reference
Air	Adsorb (polymeric beads); desorb (heat, purge with helium); trap directly onto GC column	GC/MS	0.7 μg/m^3	Krost et al. (1982)
	Adsorb on Tenax-GC; desorb (heat, purge with helium); trap (cool with liquid nitrogen); vaporize onto GC column	GC/MS	0.7 μg/m^3	Pellizzari et al. (1978)
	Adsorb on activated charcoal; desorb (carbon disulphide); inject aliquot	GC/FID	10 mg/m^3	Eller (1984); Peers (1985)
Water	Purge (inert gas); trap (OV-1 on Chromosorb-W/Tenax/silica gel); desorb as vapour (heat to 180°C, backflush with inert gas) onto GC column	GC/ECD	0.25 μg/l	US Environmental Protection Agency (1984b)
		GC/MS	2.8 μg/l	US Environmental Protection Agency (1984a)
	Add internal standard (isotope-labelled dichloromethane); purge; trap and desorb as above	GC/MS	10 μg/l	US Environmental Protection Agency (1984c)
	Purge (80°C, nitrogen); trap (Tenax-GC); desorb (flash-heat) and trap in 'mini-trap' (Tenax-GC, -30°C); desorb (flash-heat) onto GC column	GC/EC	0.1 μg/l (tap-water)	Piet et al. (1985a)
		GC/MS	<0.05 μg/l (tap-water) 0.1 μg/l (surface-water)	
	Equilibrate sealed water sample at 30°C; inject aliquot of head-space vapour	GC/EC	3-15 μg/l	Piet et al. (1985b)
	Inject aqueous sample directly onto calcium carbide precolumn (to remove water)	GC/EC	400 μg/l	Boos et al. (1985)
Food	Dissolve (toluene); distill under vacuum	GC/EC GC/ECD	0.35 mg/kg 0.5 mg/kg	US Food and Drug Administration (1983)
Blood	Equilibrate with air in closed vessel, inject directly	GC/FID	–	Premel-Cabic et al. (1974) DiVincenzo et al. (1971)

Table 5 (contd)

Sample matrix	Sample preparation	Assay procedure[a]	Limit of detection	Reference
	Encapsulate sample; insert capsule into GC injector; allow to vaporize; pierce capsule; flush sample onto column with carrier gas	GC/FID	1 μg/ml	Laham & Potvin (1976)
Breath	Inject directly	GC/FID	0.7 mg/m³	DiVincenzo et al. (1971)

[a]Abbreviations: GC/EC, gas chromatography/electron capture detection; GC/ECD, gas chromatography/electrolytic conductivity detection; GC/FID, gas chromatography/flame ionization detection; GC/MS, gas chromatography/mass spectrometry

with polyethylene glycol or methanol prior to purge/trap concentration and analysis (US Environmental Protection Agency, 1982b). Method 624 (US Environmental Protection Agency, 1984a) has also been adapted to the analysis of dichloromethane in fish, with an estimated detection limit of 30 μg/kg (Easley et al., 1981).

Exposures to dichloromethane can also be monitored in air using a direct-reading infrared analyser, with minimum concentrations of 0.7 mg/m³ (Goelzer & O'Neill, 1985).

Biological monitoring of exposure to dichloromethane may be carried out by quantification of blood carboxyhaemoglobin saturation and dichloromethane concentrations (DiVincenzo et al., 1971; DiVincenzo & Kaplan, 1981).

3. Biological Data Relevant to the Evaluation of Carcinogenic Risk to Humans

3.1 Carcinogenicity studies in animals

(a) Oral administration

Mouse: Groups of male and female $B6C3F_1$ mice, seven weeks of age, received food-grade dichloromethane (containing <300 mg/kg cyclohexane, <20 mg/kg *trans*-1,2-dichloroethylene, <10 mg/kg chloroform (see IARC, 1979b), <2 mg/kg vinyl chloride (see IARC, 1979a, 1982c) and <1 mg/kg each methyl chloride, ethyl chloride, vinylidene chloride (see IARC, 1982d, 1986), carbon tetrachloride (see IARC, 1979c, 1982b) and trichloroethylene (see IARC, 1979e, 1982e)) in the drinking-water for 104 weeks according to the study design shown in Table 6. No significant treatment-related trend in survival was found in males; in females, a statistically significant trend towards longer survival in treated groups was reported. In male mice, the incidences of

Table 6. Design of studies of food-grade dichloromethane in drinking-water

Group	No. of animals		Dose (mg/kg bw/day)
	males	females	
Mice[a]			
Control 1	60	50	0 (deionized water)
Control 2	65	50	0 (deionized water)
Low-dose	200	100	60
Mid-dose 1	100	50	125
Mid-dose 2	100	50	185
High-dose	125	50	250
Rats[b]			
Control 1	85	85	0 (deionized water)
Control 2	50	50	0 (deionized water)
Low-dose	85	85	5
Mid-dose 1	85	85	50
Mid-dose 2	85	85	125
High-dose	85	85	250
High-dose (78 weeks + no treatment for 26 weeks)	25	25	250

[a]From Serota et al. (1986a)
[b]From Serota et al. (1986b)

hepatocellular adenoma were: 6/60, 4/65, 20/200, 14/100, 14/99 and 15/125; and the incidences of hepatocellular carcinomas were: 5/60, 9/65, 33/200, 18/100, 17/99 and 23/125 in the six groups, respectively. A slight but statistically significant [$p = 0.035$] dose-related increase in the incidence of hepatocellular adenomas and/or carcinomas (combined) was observed in male mice: 11/60, 13/65, 51/200, 30/100, 31/99, 35/125. However, the authors noted that tumour incidences in dosed groups were similar to those reported in historical controls (mean, 32.1%; range, 7-58%) (Serota et al., 1986a).

Rats: Groups of male and female Fischer 344 rats, seven weeks of age, were administered food-grade dichloromethane (containing <300 mg/kg cyclohexane, <20 mg/kg *trans*-1,2-dichloroethylene, 26 mg/kg chloroform, and <1 mg/kg each methyl chloride, vinyl chloride, ethyl chloride, vinylidene chloride and trichloroethylene) in the drinking-water for 104 weeks according to the study design shown in Table 6. Interim sacrifices were carried out at 26, 52 and 78 weeks in control group 1 and in the low-, mid-1 and -2, and high-dose groups, such that 50 males and 50 females per group received the treatment for 104 weeks. There was no statistically significant difference in survival between treated and control groups. Dose-related increases in 'altered foci/areas' [criteria not specified] in the liver were observed in animals of each sex. In females, the incidences of hepatocellular carcinomas were: 0/85, 0/50, 0/85, 2/83, 0/85, 2/85 and 0/25; those of neoplastic nodules were: 0/85, 0/50, 1/85, 2/83, 1/85, 4/85 and 2/25; and those of neoplastic nodules and/or hepatocellular carcinomas (combined) were: 0/85, 0/50, 1/85, 4/83, 1/85, 6/85 and 2/25 in the seven groups, respectively. This increasing trend was significant [$p < 0.01$]; however, tumour incidences in dosed groups were similar to those reported in historical controls in this laboratory (mean, 8%; range, 0-16%). In male rats, no increased incidence of liver tumours was observed (neoplastic nodules: 4/85, 5/50, 2/85, 3/84, 3/85, 1/85, 4/25; carcinomas: 2/85, 2/50, 0/85, 0/84, 0/85, 1/85, 0/25; neoplastic nodules and/or carcinomas combined: 6/85, 7/50, 2/85, 3/84, 3/85, 2/85 and 4/25). No other significant increase in tumour incidence was found (Serota et al., 1986b).

(b) *Inhalation exposure*

Mouse: Groups of 50 male and 50 female B6C3F$_1$ mice, eight to nine weeks of age, were exposed to 0, 2000 or 4000 ppm (0, 6940 or 13 880 mg/m³) dichloromethane (>99% pure) by inhalation for six hours per day on five days per week for 102 weeks and were killed after 104 weeks on study. Survival to the end of the study period in males was: control, 39/50; low-dose, 24/50; and high-dose, 11/50; and that in females was: 25/50, 25/49 and 8/49. Significant dose-related increases in the incidences of lung and liver tumours were observed in treated mice. The incidences of alveolar/bronchiolar adenomas were: males — 3/50, 19/50 and 24/50 ($p < 0.001$); and females — 2/50, 23/48 and 28/48 ($p < 0.001$). Those of alveolar/bronchiolar carcinomas were: males — 2/50, 10/50 and 28/50 ($p < 0.001$); and females — 1/50, 13/48 and 29/48 ($p < 0.001$). The incidences of hepatocellular adenomas were: males — 10/50, 14/49 and 14/49 ($p = 0.075$); and females — 2/50, 6/48 and 22/48 ($p < 0.001$). The incidences of hepatocellular carcinomas were: males — 13/50, 15/49 and 26/49 ($p = 0.016$); and females — 1/50, 11/48 and 32/48 ($p < 0.001$) (National Toxicology Program, 1986).

Rat: Groups of approximately 95 male and 95 female Sprague-Dawley rats, eight weeks of age, were exposed by inhalation to 0, 500, 1500 or 3500 ppm (1735, 5200 or 12 145 mg/m³) dichloromethane (99% pure, with ≤706 mg/kg *trans*-1,2-dichloroethylene, ≤467 mg/kg cyclohexane, ≤576 mg/kg chloroform, ≤90 mg/kg vinylidene chloride, ≤20 mg/kg carbon tetrachloride, ≤23 mg/kg methyl bromide (see monograph, p. 187), ≤11 mg/kg ethyl chloride, ≤4.5 mg/kg methyl chloride and ≤1 mg/kg vinyl chloride) for six hours per day on five days per week for two years. The numbers of animals still alive at the end of the study were 14, 14, six and seven males and 21, 24, 13 and four females, respectively. Mortality among high-dose females was statistically significantly increased from the 18th month. Non-neoplastic pathological changes in the liver and kidney were more frequently observed in treated animals. There was no significant increase in the proportion of animals with benign or malignant mammary tumours; however, the total number of benign mammary tumours showed a slight dose-related increase in males (control, 8/95; low-dose, 6/95; mid-dose, 11/95; and high-dose, 17/97; $p = 0.046$), and a dose-related increase [$p < 0.001$] in the total number of benign mammary tumours was observed in females (165/96; 218/95; 245/95;

287/97). The incidence of sarcomas located in the neck was increased in mid- and high-dose males (1/93, 0/94, 5/91 and 11/88; $p = 0.002$ [$p < 0.001$, trend test]) (Burek et al., 1984; US Environmental Protection Agency, 1985). [The Working Group noted the reported occurrence of sialoadenitis early in the treatment period.]

Groups of 50 male and 50 female Fischer 344/N rats, seven to eight weeks of age, were exposed by inhalation to 0, 1000, 2000 or 4000 ppm (0, 3470, 6940 or 13 880 mg/m^3) dichloromethane (>99% pure) for six hours per day on five days per week for 102 weeks and were killed after 104 weeks on study. Survival of treated males was comparable with that of controls. Survival at termination of the study was reduced in high-dose females as compared to controls: control, 30/50; low-dose, 22/50; mid-dose, 22/50; and high-dose, 15/50. Increased incidences of benign mammary-gland tumours (all fibroadenomas, except for one adenoma in the high-dose group) were observed in treated females (5/50, 11/50, 13/50 and 23/50; $p < 0.001$). There was a positive trend in the incidence of benign tumours in the mammary gland area in males (1/50, 1/50, 4/50 and 9/50; $p < 0.001$). No nasal tumour was observed, there was no increase in the incidence of respiratory-tract tumours and there was no difference considered biologically significant in the distribution of other types of tumour between the control and treated groups (National Toxicology Program, 1986).

Hamster: Groups of 95 male and 95 female Syrian golden hamsters, eight weeks of age, were exposed by inhalation to 0, 500, 1500 or 3500 ppm (0, 1735, 5200 or 12 145 mg/m^3) dichloromethane (99% pure, with ≤706 mg/kg *trans*-1,2-dichloroethylene, ≤467 mg/kg cyclohexane, ≤576 mg/kg chloroform, ≤90 mg/kg vinylidene chloride, ≤20 mg/kg carbon tetrachloride, ≤23 mg/kg methyl bromide, ≤11 mg/kg ethyl chloride, ≤4.5 mg/kg methyl chloride and ≤1 mg/kg vinyl chloride) for six hours per day on five days per week for two years. The numbers of animals surviving to the end of the study were 16, 20, 11 and 14 males and zero, four, ten and nine females. The incidence of lymphosarcomas was slightly higher in treated females than in controls: control, 1/91; low-dose, 6/92; mid-dose, 3/91; and high-dose, 7/91 ($p = 0.032$) (Burek et al., 1984; US Environmental Protection Agency, 1985). [The Working Group noted that the higher survival in treated animals may have contributed to this result.]

(c) *Intraperitoneal administration*

Mouse: In a screening assay based on the production of lung tumours in strain A mice, groups of 20 male mice, six to eight weeks of age, received thrice-weekly intraperitoneal injections of 0, 160, 400 or 800 mg/kg bw reagent-grade dichloromethane (purity, >95%; impurities unspecified) in tricaprylin for a total of 16-17 injections (total doses: 2720, 6800 and 12 800 mg/kg bw, in the treated groups respectively). After 24 weeks, 18, five and 12 animals were still alive in the three treated groups, respectively; these and 15/20 surviving vehicle controls were killed and their lungs examined for tumours. There was no significant increase in the number of lung tumours per mouse: vehicle-control, 0.27; low-dose, 0.94; mid-dose, 0.80; and high-dose, 0.50 (Theiss et al., 1977).

3.2 Other relevant biological data

(a) *Experimental systems*

Toxic effects

The toxicity of dichloromethane has been reviewed (National Institute for Occupational Safety and Health, 1976; US Environmental Protection Agency, 1982c; International Programme on Chemical Safety, 1984).

Intraperitoneal LD_{50} values for dichloromethane are approximately 1.5 ml/kg bw (2000 mg/kg bw) in mice (Klaassen & Plaa, 1966) and 0.95 ml/kg bw (1300 mg/kg bw) in dogs (Klaassen & Plaa, 1967); the oral LD_{50} in rats ranges from 1.6-2.3 ml/kg bw (2100-3000 mg/kg bw) (Kimura *et al.*, 1971); and the subcutaneous LD_{50} in mice is approximately 76 mmol/kg bw (6400 mg/kg bw) (Kutob & Plaa, 1962). The LC_{50} values in mice, rats and guinea-pigs are 16 000 ppm (56 000 mg/m^3; 7-h exposure plus 1-h observation), 5.7% (200 000 mg/m^3; 15-min exposure) and 11 600 ppm (40 000 mg/m^3; 6-h exposure plus 18-h observation), respectively (Svirbely *et al.*, 1947; Balmer *et al.*, 1976; Clark & Tinston, 1982).

The toxicity of dichloromethane is expressed mainly as disturbances of the central nervous system, involving sleep disturbance and reductions in spontaneous activity (Heppel & Neal, 1944; Schumacher & Grandjean, 1960; Fodor & Winneke, 1971; National Institute for Occupational Safety and Health, 1976).

Hepatotoxic effects are seen after exposure to near lethal concentrations of dichloromethane (Gehring, 1968). Inhalation exposure of guinea-pigs to 5200 ppm (18 000 mg/m^3) dichloromethane for 6 h increased hepatic triglyceride concentrations (Morris *et al.*, 1979). Exposure of guinea-pigs to approximately 11 000 ppm (40 000 mg/m^3) dichloromethane for 6 h also increased hepatic triglyceride concentrations, but concomitant exposure to 21 400-24 100 ppm (40 200-45 300 mg/m^3) ethanol blocked this effect (Balmer *et al.*, 1976). Continuous exposure of mice by inhalation to 5000 ppm (17 400 mg/m^3) dichloromethane caused swelling of the rough endoplasmic reticulum, fatty changes in the liver and necrosis of individual hepatocytes (Weinstein *et al.*, 1972). Slight liver damage was also observed following administration of dichloromethane by gavage (133-665 mg/kg bw) to mice, but no kidney damage was observed (Condie *et al.*, 1983). Following intraperitoneal administration of dichloromethane at near lethal doses, hydropic degeneration was observed in the kidneys of mice (Klaassen & Plaa, 1966) and slight calcification of the renal tubules was seen in dogs (Klaassen & Plaa, 1967). In rats, intraperitoneal administration of 1330 mg/kg bw dichloromethane produced renal proximal tubular swelling (Kluwe *et al.*, 1982).

In gerbils exposed continuously by inhalation to 350 ppm (1200 mg/m^3), but not in those exposed to 210 ppm (730 mg/m^3), dichloromethane for up to three months, increased brain concentrations of two astroglial proteins (S-100 and GFA) and decreased cerebellar DNA concentrations were observed. Decreased hippocampal DNA concentrations were observed at both exposure levels (Rosengren *et al.*, 1986).

In a long-term bioassay [see section 3.1] of dichloromethane, increased incidences of haemosiderosis, cytomegaly, cytoplasmic vacuolization, necrosis, granulomatous inflammation and bile-duct fibrosis were observed in the livers of treated male and female Fischer

344/N rats, and an increased incidence of testicular atrophy was observed in high-dose B6C3F$_1$ mice (National Toxicology Program, 1986).

Effects on reproduction and prenatal toxicity

Injection of dichloromethane into the air space of two-, three- and six-day-old White Leghorn chick embryos induced abnormalities and death. The LD$_{50}$ was estimated to be greater than 100 µmol/egg (Elovaara *et al.*, 1979). Dichloromethane was not teratogenic following injection into the yolk sac of 0-h-old White Leghorn chick embryos. The LD$_{50}$ was estimated to be 14 mg/egg (Verrett *et al.*, 1980).

In a classical teratology study, groups of Swiss Webster mice and Sprague-Dawley rats were exposed by inhalation to 0 or 1225 ppm (0 or 4000 mg/m^3) dichloromethane (purity, 97.86%) for 7 h per day on gestation days 6-15. Exposure of female mice resulted in a significant increase in body weight during and after exposure, while absolute but not relative liver weights were increased in both species. There was no statistically significant increase in visceral anomalies in the fetuses of either species, but evidence of skeletal anomalies was observed as decreased incidence of lumbar spurs and delayed ossification of the sternebrae in rats and increased incidence of a single extra sternal ossification centre in mice (Schwetz *et al.*, 1975).

Female Long-Evans rats were exposed to 0 or 4500 ppm (0 or 15 600 mg/m^3) dichloromethane (>97% pure) during either a three-week pregestational period or during the first 17 days of gestation or both. Ten females per group were allowed to give birth, and the offspring were examined for abnormal growth and behaviour. Dams exposed to dichloromethane during gestation had increased absolute and relative liver weights. There was no effect on litter size or viability, but fetal weight was reduced in both groups exposed during gestation. No treatment-related visceral or skeletal abnormality was detected in the fetuses of any exposure group, but a greater proportion of litters exposed during both the pregestational and gestational periods had fetuses with rudimentary lumbar ribs (Hardin & Manson, 1980). No difference in pup birth weight, viability or growth rate was observed, but alterations in spontaneous locomotor activities were seen in all exposure groups. No change was observed in running-wheel activity or acquisition of an avoidance response (Bornschein *et al.*, 1980).

Absorption, distribution, excretion and metabolism

The metabolism of halogenated methanes, including dichloromethane, has been reviewed (Ahmed *et al.*, 1980; Laib, 1982; International Programme on Chemical Safety, 1984; Anders & Jakobson, 1985; US Environmental Protection Agency, 1985; Kirschman *et al.*, 1986).

The highest levels of radioactivity in rats after 1 h exposure by inhalation to 1935 mg/m^3 ^{14}C-dichloromethane were found in the fat, with lower levels in the liver, kidney and adrenal glands. Two hours after exposure, the concentration in fat had decreased by more than 90% and that in the liver by 25% (Carlsson & Hultengren, 1975). Forty-eight hours after exposure of rats to dichloromethane either orally (1 or 50 mg/kg bw) or by inhalation (50,

500 or 1500 ppm; 174, 1735 or 5200 mg/m³), <10% or 7-23%, respectively, of the body burden was retained (McKenna & Zempel, 1981; McKenna et al., 1982).

In rats exposed by inhalation to 500 ppm (1735 mg/m³) dichloromethane for 1 h on day 21 of gestation, concentrations in maternal blood were higher than those in fetal blood (176 nmol versus 115 nmol/ml), while carbon monoxide concentrations were similar in the two compartments (approximately 160 nmol/ml) (Anders & Sunram, 1982).

Kinetic studies with dichloromethane administered by various routes to rats or mice revealed that the pharmacokinetics and metabolism are dose- and vehicle-dependent (Withey & Collins, 1980; McKenna & Zempel, 1981; McKenna et al., 1982; Withey et al., 1983; Angelo et al., 1986). When dichloromethane was given orally in corn oil to rats or mice, absorption and elimination were slower than when the vehicle was water (Withey et al., 1983; Angelo et al., 1986).

Dichloromethane is metabolized to carbon monoxide *in vivo* (Fodor et al., 1973; Kubic et al., 1974) and *in vitro* by hepatic microsomal cytochrome P450-dependent monooxygenases and by bacteria (*Salmonella typhimurium* TA100) (Kubic & Anders, 1975, 1978; Green, 1983). Dichloromethane is metabolized *in vitro* by rat hepatic cytosolic fractions to formaldehyde and inorganic chloride, apparently by glutathione-*S*-transferases (Heppel & Porterfield, 1948; Ahmed & Anders, 1976; Green, 1983).

Radioactivity from ¹⁴C-dichloromethane given orally (8.3 or 26 mmol/kg) to rats was associated preferentially with hepatic proteins and lipids. Trace amounts associated with nucleic acids were also reported (Reynolds & Yee, 1967).

After incubation of ¹⁴C-dichloromethane with rat hepatic microsomal fractions fortified with NADPH, metabolites were bound to both lipids and proteins (Anders et al., 1977).

¹⁴C-Dichloromethane metabolites did not bind to the RNA or DNA of rat hepatocytes exposed to 1, 14 or 42 μM dichloromethane, in the presence or absence of oxygen, although radioactivity was associated with lipid and protein (Cunningham et al., 1981).

Mutagenicity and other short-term tests

A number of reviews of the genetic effects of dichloromethane are available (International Programme on Chemical Safety, 1983, 1984; US Environmental Protection Agency, 1985).

Lambda phage were induced by 10 μl dichloromethane in a spot test with *Escherichia coli* K39 (λ) in a desiccator (Osterman-Golkar et al., 1983). It was reported in an abstract that dichloromethane gave negative results in the *Bacillus subtilis rec* assay (Kanada & Uyeta, 1978). [The Working Group noted that no indication was given as to whether the chemical was toxic.]

Dichloromethane was consistently mutagenic to *Salmonella typhimurium* when assayed in a desiccator or other closed chamber, both in the presence and absence of an Aroclor- or phenobarbital-induced rat-liver metabolic system (S9) (Simmon et al., 1977; Jongen et al., 1978; McGregor, 1979; Kirwin et al., 1980; Nestmann et al., 1980; Gocke et al., 1981; Nestmann et al., 1981; Jongen et al., 1982; Green, 1983; Osterman-Golkar et al., 1983;

Jongen, 1984). Positive responses were obtained with strains TA1535, TA1950, TA100 and TA98, although some laboratories found negative results with TA1535. A few studies found that dichloromethane was not mutagenic to *S. typhimurium* when assayed in the absence of S9 in standard plate incorporation tests (Simmon *et al.*, 1977; Nestmann *et al.*, 1980) and in a spot test (Buijs *et al.*, 1984).

Dichloromethane (10 µl) was mutagenic to *E. coli* WU361089 (mutation to tyrosine prototrophy), using a spot test in a desiccator; the same protocol did not lead to mutation of *E. coli* SD-4 to streptomycin independence (Osterman-Golkar *et al.*, 1983).

Dichloromethane (104, 157 and 209 mM) induced gene conversion, mitotic recombination and gene mutation in *Saccharomyces cerevisiae* D7 when the cells were grown under conditions that led to the synthesis of endogenous cytochrome P450. Strain D4 (which detects gene conversion only), grown under similar conditions, had one-fifth the P450 activity of strain D7 and responded only marginally to dichloromethane (Callen *et al.*, 1980). Simmon *et al.* (1977) reported that dichloromethane [concentration unspecified] did not increase the frequency of mitotic recombination in *S. cerevisiae* D3.

Liquid dichloromethane applied at concentrations of 1.17×10^{-3} M and above for 24 h was mutagenic in the *Tradescantia* stamen-hair somatic mutation assay (Schairer & Sautkulis, 1982; Schairer *et al.*, 1982). When tested in *Tradescantia* under conditions that allowed for the volatilization of dichloromethane, concentrations of 0.24-18% resulted in toxicity and a 'borderline' induction of micronuclei [details not given] (Ma *et al.*, 1984).

Dichloromethane was tested for induction of sex-linked recessive lethal mutations in *Drosophila melanogaster*. Following exposure by feeding (125 and 620 mM) [in which the flies would presumably also be exposed by inhalation], a weak positive response was observed in one of three broods (Gocke *et al.*, 1981).

Chinese hamster V79 cells and primary cultures of human fibroblasts were treated in closed vessels with 0.5-5% dichloromethane in the absence of S9. No increase in grain counts was seen, indicating that unscheduled DNA synthesis was not induced (Jongen *et al.*, 1981). Perocco and Prodi (1981) reported that concentrations of 2.5, 5 and 10 µl/ml dichloromethane did not induce unscheduled DNA synthesis in human lymphocytes in the presence or absence of phenobarbital-induced rat-liver S9. It was reported in an abstract that dichloromethane induced a marginal increase in unscheduled DNA synthesis in primary cultures of rat hepatocytes [details not given] (Thilagar *et al.*, 1984a). Dichloromethane (1-30 mM in glass-stoppered tubes) did not induce DNA repair, as measured by density-gradient analysis of bromodeoxyuridine incorporation, in primary cultures of rat hepatocytes. The two highest doses used (10 and 30 mM) inhibited replicative DNA synthesis (Andrae & Wolff, 1983).

Dichloromethane did not induce 6-thioguanine-resistant mutants in Chinese hamster V79 or Chinese hamster ovary cells exposed in sealed chambers (both liquid and gas exposure) in the absence of an exogenous metabolic system. The cells were exposed to concentrations ranging from 0.5-5%; the high doses led to a 25% reduction of survival (Jongen *et al.*, 1981). It was reported in an abstract that dichloromethane did not increase mutants at the thymidine kinase locus in L5178Y cells [details not given] (Thilagar et al., 1984a).

Dichloromethane (2-15 µl/ml) induced dose-related chromosomal damage, measured as chromatid gaps, chromatid breaks, isochromatid breaks and exchanges, in cultured Chinese hamster ovary cells. Damage was induced both with and without Aroclor-induced rat-liver S9, the S9-treated sample giving higher responses. Small but nonsignificant increases in the incidence of sister chromatid exchanges were produced, both with and without S9 (Thilagar & Kumaroo, 1983). It was reported in an abstract that dichloromethane induced 'extensive chromosome aberrations' in mouse lymphoma L5178Y cells both in the presence and absence of an undefined metabolic system. The authors concluded that the increase in sister chromatid exchange incidence in these cells was not significant because it was not double the background rate (Thilagar et al., 1984b). Jongen et al. (1981) reported a low level of induction of sister chromatid exchanges in Chinese hamster V79 cells without S9 in the dose range of 0.5-4% dichloromethane. In an abstract, McCarroll et al. (1983) reported that sister chromatid exchanges were induced in cultured Chinese hamster ovary cells exposed for 24 h to concentrations of 1.8-7% dichloromethane, but not in cells exposed for 2, 4 or 10 h.

It was reported in two abstracts that dichloromethane induced 'extensive chromosome aberrations' both with and without metabolic activation and a low but insignificant increase in sister chromatid exchanges in cultured human peripheral lymphocytes [details not given] (Thilagar et al., 1984a,b).

A dose-related increase in the number of SA7 adenovirus-transformed foci was observed following treatment of primary cultures of Syrian hamster embryo cells in sealed chambers with 0.63-2.5 ml (170-670 µg/cm^3) volatilized dichloromethane (Hatch et al., 1982, 1983). It was reported in an abstract that dichloromethane did not induce transformation in C3H 10T1/2 cells [details not given] (Thilagar et al., 1984a). Concentrations of 160 and 1600 µM dichloromethane transformed Fischer rat embryo cells; subcutaneous inoculation of cells transformed by 160 µM into newborn Fischer 344 rats led to the formation of undifferentiated fibrosarcomas at the site of injection (Price et al., 1978). [The Working Group noted that this transformation assay does not provide quantitative results.]

Dichloromethane did not induce micronuclei in polychromatic erythrocytes recovered from bone marrow of male and female NMRI mice given two intraperitoneal injections of 1700 mg/kg bw (Gocke et al., 1981).

In a long-term bioassay [see section 3.1] of dichloromethane, no exposure-related cytogenetic aberration was reported in bone marrow of male and female Sprague-Dawley rats [values not given] (Burek et al., 1984).

Dichloromethane induced sex-linked lethal mutations in the nematode *Panagrellus redivivus* maintained for 120 h in media containing 10^{-8}, 10^{-6} or 10^{-4} mol/l (Samoiloff et al., 1980).

(b) *Humans*

Toxic effects

The toxic effects of dichloromethane have been reviewed (International Programme on Chemical Safety, 1984). The odour threshold is 214 ppm (743 mg/m^3) (Stahl, 1973).

Fatalities have been associated with acute or prolonged exposure to dichloromethane (Moskowitz & Shapiro, 1952; Kuželová *et al.*, 1975; Stewart & Hake, 1976; Bakinson & Jones, 1985). This compound acts primarily on the central nervous system, causing narcosis at high doses (Fodor & Winneke, 1971; Winneke, 1974; Putz *et al.*, 1976). Temporary neurobehavioural effects have been reported after exposure to doses as low as 200 ppm (700 mg/m^3) by some authors (Winneke, 1974; Putz *et al.*, 1976) but not by others (Gamberale *et al.*, 1975).

An exposure-related increase in serum bilirubin was observed in workers exposed to dichloromethane, but no other sign of liver injury or haemolysis was reported (Ott *et al.*, 1983b). A cross-sectional study of 24 employees at a fibre production plant showed no excess of electrocardiographic abnormalities among those exposed to 60-475 ppm (208-1648 mg/m^3, time-weighted average [TWA]) dichloromethane and monitored for 24 h (Ott *et al.*, 1983c). No significant increase in overall mortality or deaths due to ischaemic heart disease was found among 1271 male and female employees exposed to 140-475 ppm (486-1648 mg/m^3, TWA) dichloromethane when compared to the mortality of the general US population (for further details, see p. 66). A statistically significant increased risk of ischaemic heart disease was found in comparison with an internal reference group, in which, however, there were less than half the observed deaths expected from US statistics [standardized mortality ratio, 39] (Ott *et al.*, 1983d). No such excess was found in another US cohort exposed to a time-weighted average range of 30-120 ppm (106-416 mg/m^3) dichloromethane, as compared to an internal reference group (see p. 66) (Hearne & Friedlander, 1981).

Effects on reproduction and prenatal toxicity

A case-control study on 44 women who had had a spontaneous abortion was performed within a cohort of female workers employed in Finnish pharmaceutical factories during 1973 or 1975 to 1980. Three controls matched for age at conception within 2.5 years were chosen for every case (except two). Information about pregnancy outcome was collected from hospital data, and data on exposures from health personnel at the factories. The odds ratio for dichloromethane exposure, based on 11 exposed cases, was 2.3 (95% confidence interval, 1.0-5.7; $p = 0.06$); the odds ratio was also increased for exposures to many other solvents. The odds ratio for those exposed once a week or more during the first trimester of pregnancy was 2.8, and that for those exposed less than once a week was 2.0 (Taskinen *et al.*, 1986).

Absorption, distribution, excretion and metabolism

The uptake and distribution of dichloromethane have been reviewed (Åstrand, 1975).

Dichloromethane is rapidly absorbed and eliminated by the lungs (Riley *et al.*, 1966; Morgan *et al.*, 1970; DiVincenzo *et al.*, 1972; Åstrand, 1975; Åstrand *et al.*, 1975) and is distributed to adipose tissues (Engström & Bjurström, 1977). The compound is absorbed through the skin (Stewart & Dodd, 1964).

Dichloromethane is metabolized to carbon monoxide (Stewart *et al.*, 1972a,b; Ratney *et al.*, 1974; Åstrand *et al.*, 1975; Stewart & Hake, 1976; Peterson, 1978; DiVincenzo & Kaplan, 1981; Ott *et al.*, 1983e). Elevated carboxyhaemoglobin saturation and increased urinary formic acid concentrations have been found in workers exposed to dichloromethane (Kuželová & Vlasák, 1966; DiVincenzo & Kaplan, 1981). [The Working Group noted that these observations indicate that both hepatic microsomal and cytosolic metabolism of dichloromethane occur in humans, as in experimental animals.]

Mutagenicity and chromosomal effects

No increase in the incidence of chromosomal aberrations or of sister chromatid exchanges was seen in peripheral lymphocytes of three factory workers exposed to 635-2421 mg/m^3 (median, 719 mg/m^3) dichloromethane (8-h time-weighted averages), in addition to other solvents, over those in three matched, unexposed controls (Haglund *et al.*, 1980). [The Working Group noted that no indication was given of the duration of exposure or of cytogenetic values.]

3.3 Case reports and epidemiological studies of carcinogenicity to humans

Friedlander *et al.* (1978) carried out a proportional mortality study of 334 deaths that occurred in 1956-1976 among active, disabled or retired men who had ever worked in dichloromethane (as a solvent) areas in one company, regardless of the duration of exposure. Expected numbers were calculated on the basis of age- and diagnosis-specific proportional mortality among nonexposed men in the same company in 1960-1976. The range of exposure levels in 1959, based on 188 samples, was reported to be 9-350 ppm (31-1215 mg/m^3), with a mean value of 79.4 ppm (276 mg/m^3). A total of 71 malignant neoplasms was observed *versus* 73.4 expected (standardized mortality ratio [SMR], 97; 95% confidence limits [CI], 76-122), and no significant difference was noted between observed and expected numbers for any specific cancer site.

In the same paper, Friedlander *et al.* (1978) reported a cohort mortality study of all 751 'hourly males' employed in 1964 in the same dichloromethane area. [The Working Group noted that the proportional mortality and cohort mortality studies concern partially overlapping populations.] Of this group, 252 had a minimum of 20 years' work exposure as at 1 July 1964 and were analysed separately. The study covers a maximum of 13 years' follow-up. The rates used for the internal industrial referent population were the age-group and cause-specific mortality rates for 1964-1976 of all other 'hourly males' in the same company. The rates used for the external referent population were the age-group and cause-specific mortality rates for 1966-1972 for men in New York State (exclusive of New

York City). In the total cohort, 14 malignant neoplasms were observed *versus* 19.5 expected [SMR, 72; 95% CI, 39-121] from the internal control rates and 25.0 from the external control rates [56; 31-94]. Among the employees with a minimum of 20 years of exposure, only seven malignancies were observed *versus* 12.5 expected from the internal control group [56; 23-115] and 16.2 expected from the New York State controls [43; 17-89]. No specific diagnostic group had a statistically significant excess.

Hearne and Friedlander (1981) updated the latter study through to 1980. Among the 252 men with 20 or more years of exposure, 13 malignant neoplasms were observed *versus* 17.8 expected from the internal control group [73; 39-125] and 24.7 from men in New York State [53; 28-90]. No diagnostic category of malignant neoplasm occurred significantly more frequently than in either set of controls.

Ott *et al.* (1983d) reported a cohort mortality study of 1271 white and non-white men and women employed in a fibre production plant in which dichloromethane was used as a general-purpose solvent. The range of exposure was a time-weighted average of approximately 140-475 ppm (486-1648 mg/m^3). Employees who had worked for at least three months subsequent to 1 January 1954 and prior to 1 January 1977 were included and followed up to 30 June 1977. The observed numbers were compared with the expected numbers from an internal referent cohort of 948 acetone-exposed employees and with mortality data for US white males, non-white males and white females. Vital status was not confirmed for 18% of the exposed cohort or 12% of the internal referent cohort. Among the exposed white men and women, seven deaths due to malignant neoplasms were observed compared to 10.1 expected in the US population [69; 28-143]. No specific cancer site was overrepresented. Seven malignant neoplasms were observed in the referent cohort, whereas 12.3 were expected for white men and women [57; 23-117]. [The Working Group noted that only 115 men and 154 women with more than five years' exposure had been first exposed before 1960, i.e., allowing for adequate follow-up time with respect to latency for malignant neoplasms.]

[The Working Group noted that, in the available studies, only limited numbers of persons had had long-term exposure and adequate follow-up time for identification of increased cancer rates.]

4. Summary of Data Reported and Evaluation

4.1 Exposure data

Dichloromethane has been produced on a large scale since the 1930s. It is used principally as a solvent, and in paint removers, degreasers, aerosol products, the manufacture of foam polymers and food extraction. Widespread exposures occur during the production and industrial use of dichloromethane and during the use of a variety of consumer products. Substantial losses to the environment lead to ubiquitous low-level exposure from ambient air and water.

4.2 Experimental data

Dichloromethane was tested by oral administration in drinking-water in mice and rats, by inhalation exposure in mice, rats and hamsters, and by intraperitoneal injection in a lung-adenoma assay in mice. Exposure by inhalation increased the incidence of benign and malignant lung and liver tumours in mice of each sex and the incidence or multiplicity of benign mammary tumours in rats of each sex; in male rats, an increased incidence of sarcomas located in the neck was also observed. The studies by oral administration in mice and in male rats and the inhalation study in male hamsters gave negative results. The study by oral administration in female rats and the inhalation study in female hamsters were inconclusive. In the lung-adenoma test in mice, negative results were obtained.

Exposure to dichloromethane by inhalation did not induce visceral malformations in fetal mice or rats. In rats, postnatal behavioural development is affected by prenatal exposure to dichloromethane.

Dichloromethane induces lambda prophage in *Escherichia coli*. It is mutagenic to bacteria, plants and nematodes and induces gene conversion, mitotic recombination and gene mutation in yeast. Results from one study for the induction of sex-linked recessive lethal mutations in *Drosophila melanogaster* were equivocal. In cultured mammalian cells, dichloromethane induces chromosomal aberrations, but it does not induce unscheduled DNA synthesis or mutations; equivocal results were reported for the induction of sister chromatid exchanges. Dichloromethane transforms cultured mammalian cells. It does not induce micronuclei in bone-marrow of mice.

Overall assessment of data from short-term tests: Dichloromethane[a]

	Genetic activity			Cell transformation
	DNA damage	Mutation	Chromosomal effects	
Prokaryotes	+	+		
Fungi/Green plants		+		
Insects		?		
Mammalian cells (*in vitro*)	−	−	+	+
Mammals (*in vivo*)			−	
Humans (*in vivo*)				
Degree of evidence in short-term tests for genetic activity: **Sufficient**				Cell transformation: Positive

[a]The groups into which the table is divided and the symbols '+' and '−' and '?' are defined on pp. 19-20 of the Preamble; the degrees of evidence are defined on pp. 20-21.

4.3 Human data

A study of possible causes of spontaneous abortions among women working in the pharmaceutical industry demonstrated an increased risk associated with exposure to several chemicals and solvents, including dichloromethane.

No excess risk of death from malignancies was observed in two cohort studies or in one proportional mortality study of workers exposed to dichloromethane, but the studies had limited power to detect excess risk.

4.4 Evaluation[1]

There is *sufficient evidence*[2] for the carcinogenicity of dichloromethane to experimental animals.

There is *inadequate evidence* for the carcinogenicity of dichloromethane to humans.

5. References

Ahmed, A.E. & Anders, M.W. (1976) Metabolism of dihalomethanes to formaldehyde and inorganic halide. I. In vitro studies. *Drug Metab. Disposition, 4,* 357-361

Ahmed, A.E., Kubic, V.L., Stevens, J.L. & Anders, M.W. (1980) Halogenated methanes: Metabolism and toxicity. *Fed. Proc., 39,* 3150-3155

Ahrenholz, S.H. (1980) *Health Hazard Evaluation Determination Report No. HE-80-18-691, Looart Press, Inc., Colorado Springs, CO,* Cincinnati, OH, National Institute for Occupational Safety and Health

Akiyama, S., Hisamoto, T. & Mochizuki, S. (1981) Chloromethanes from methanol. *Hydrocarbon Process., 60,* 76-78

Albrecht, W.N. (1982) *Health Hazard Evaluation Report No. HETA-81-468-1036, United Union of Roofers, Waterproofers, and Allied Workers, Baltimore, MD,* Cincinnati, OH, National Institute for Occupational Safety and Health

Aldrich Chemical Co. (1984) *1984-1985 Aldrich Catalog/Handbook of Fine Chemicals,* Milwaukee, WI, p. 371

American Conference of Governmental Industrial Hygienists (1985) *TLVs Threshold Limit Values and Biological Exposure Indices for 1985-86,* 2nd ed., Cincinnati, OH, p. 24

Anders, M.W. & Jakobson, I. (1985) Biotransformation of halogenated solvents. *Scand. J. Work Environ. Health, 11* (Suppl. 1), 23-32

---

[1]For definition of the italicized terms, see Preamble, pp. 18 and 22.
[2]In the absence of adequate data on humans, it is reasonable, for practical purposes, to regard chemicals or exposures for which there is *sufficient evidence* of carcinogenicity in animals as if they presented a carcinogenic risk to humans.

Anders, M.W. & Sunram, J.M. (1982) Transplacental passage of dichloromethane and carbon monoxide. *Toxicol. Lett.*, *12*, 231-234

Anders, M.W., Kubic, V.L. & Ahmed, A.E. (1977) Metabolism of halogenated methanes and macromolecular binding. *J. environ. Pathol. Toxicol.*, *1*, 117-124

Andrae, U. & Wolff, T. (1983) Dichloromethane is not genotoxic in isolated rat hepatocytes. *Arch. Toxicol.*, *52*, 287-290

Angelo, M., Pritchard, A.B., Hawkins, D.R., Waller, A.R. & Roberts, A. (1986) The pharmacokinetics of dichloromethane. I. Disposition in B6C3F1 mice following intravenous and oral administration. *Food Chem. Toxicol.*, *24*, 965-974

Anon. (1981) Chemical briefs. 2. Methylene chloride. *Chem. Purch.*, *17*, 57-58, 61

Anon. (1983) Chemical briefs. 1. Methylene chloride. *Chem. Purch.*, *19*, 17-18, 21

Anon. (1985) *Farm Chemicals Handbook*, Willoughby, OH, Meister Publishing Co., p. C-154

Anthony, T. (1979) *Methylene chloride*. In: Mark, H.F., Othmer, D.F., Overberger, C.G., Seaborg, G.T. & Grayson, M., eds, *Kirk-Othmer Encyclopedia of Chemical Technology*, 3rd ed., Vol. 5, New York, John Wiley & Sons, pp. 686-693

Arbetarskyddsstyrelsens Författningssamling (National Swedish Board of Occupational Safety and Health) (1984) *Occupational Exposure Limit Values (AFS 1984:5)* (Swed.), Solna, p. 20

Åstrand, I. (1975) Uptake of solvents in the blood and tissues of man: A review. *Scand. J. Work Environ. Health*, *1*, 199-218

Åstrand, I., Övrum, P. & Carlsson, A. (1975) Exposure to methylene chloride. I. Its concentration in alveolar air and blood during rest and exercise and its metabolism. *Scand. J. Work Environ. Health*, *1*, 78-94

Bakinson, M.A. & Jones, R.D. (1985) Gassings due to methylene chloride, xylene, toluene, and styrene reported to Her Majesty's Factory Inspectorate 1961-80. *Br. J. ind. Med.*, *42*, 184-190

Balmer, M.F., Smith, F.A., Leach, L.J. & Yuile, C.L. (1976) Effects in the liver of methylene chloride inhaled alone and with ethyl alcohol. *Am. ind. Hyg. Assoc. J.*, *37*, 345-352

Barkley, J., Bunch, J., Bursey, J.T., Castillo, N., Cooper, S.D., Davis, J.M., Erickson, M.D., Harris, B.S.H., III, Kirkpatrick, M., Michael, L.C., Parks, S.P., Pellizzari, E.D., Ray, M., Smith, D., Tomer, K.B., Wagner, R. & Zweidinger, R.A. (1980) Gas chromatography mass spectrometry computer analysis of volatile halogenated hydrocarbons in man and his environment. A multimedia environmental study. *Biomed. mass Spectrom.*, *7*, 139-147

Bauer, U. (1981) Human exposure to environmental chemicals — Investigations on volatile organic halogenated compounds in water, air, food, and human tissues. III. Communication: Results of investigations. *Zbl. Bakteriol. Mikrobiol. Hyg.*, *174*, 200-237

Bellar, T.A., Lichtenberg, J.J. & Kroner, R.C. (1974) The occurrence of organohalides in chlorinated drinking waters. *J. Am. Water Works Assoc.*, *66*, 703-706

Boos, R., Prey, T. & Begert, A. (1985) Determination of volatile chlorinated hydrocarbons by reaction gas chromatography (Ger.). *J. Chromatogr.*, *328*, 233-239

Bornschein, R.L., Hastings, L. & Manson, J.M. (1980) Behavioral toxicity in the offspring of rats following maternal exposure to dichloromethane. *Toxicol. appl. Pharmacol.*, *52*, 29-37

Bourne, W. & Stehle, R.L. (1923) Methylene chloride in anaesthesia. *Can. med. Assoc. J.*, *13*, 432-433

Buijs, W., van der Gen, A., Moh, G.R. & Breimer, D.D. (1984) The direct mutagenic activity of α,ω-dihalogenoalkanes in *Salmonella typhimurium*. Strong correlation between chemical properties and mutagenic activity. *Mutat. Res.*, *141*, 11-14

Burdick & Jackson Laboratories (1984) *Technical Data: Methylene Chloride*, Muskegon, MI

Burek, J.D., Nitschke, K.D., Bell, T.J., Wackerle, D.L., Childs, R.C., Beyer, J.E., Dittenber, D.A., Rampy, L.W. & McKenna, M.J. (1984) Methylene chloride: A two-year inhalation toxicity and oncogenicity study in rats and hamsters. *Fundam. appl. Toxicol.*, *4*, 30-47

Burroughs, G.E. & Moody, P.L. (1982) *Health Hazard Evaluation Report No. HETA-81-029-1088, Industrial Plastics, Valley City, OH*, Cincinnati, OH, National Institute for Occupational Safety and Health

Burton, D.J. & Shmunes, E. (1973) *Health Hazard Evaluation Report No. 71-20-49, Chemetron Chemical, Organics Division, Newport, TN*, Cincinnati, OH, National Institute for Occupational Safety and Health

Callen, D.F., Wolf, C.R. & Philpot, R.M. (1980) Cytochrome P-450 mediated genetic activity and cytotoxicity of seven halogenated aliphatic hydrocarbons in *Saccharomyces cerevisiae*. *Mutat. Res.*, *77*, 55-63

Carlsson, A. & Hultengren, M. (1975) Exposure to methylene chloride. III. Metabolism of ^{14}C-labelled methylene chloride in rat. *Scand. J. Work Environ. Health*, *1*, 104-108

The Chemical Daily Co. (1984) *JCW Chemicals Guide 1984/1985*, Tokyo, p. 269

Chrostek, W.J. (1980) *Health Hazard Evaluation Determination Report No. HE-80-108-705, Corporation of Veritas, Philadelphia, PA*, Cincinnati, OH, National Institute for Occupational Safety and Health

Chrostek, W.J. & Levine, M.S. (1981) *Health Hazard Evaluation Report No. HHE-80-154-1027, Bechtel Power Corporation, Berwick, PA*, Cincinnati, OH, National Institute for Occupational Safety and Health

Clark, D.G. & Tinston, D.J. (1982) Acute inhalation toxicity of some halogenated and non-halogenated hydrocarbons. *Human Toxicol.*, *1*, 239-247

Cohen, J.M., Dawson, R. & Koketsu, M. (1980) *Technical Report: Extent-of-Exposure Survey of Methylene Chloride (DHHS (NIOSH) Pub. No. 80-131)*, Washington DC, US Department of Health and Human Services

Cohen, S.R. & Vandervort, R. (1972) *Health Hazard Evaluation Report No. 72-68-25, North American Rockwell, Reinforced Plastic Operation, Ashtabula, OH*, Cincinnati, OH, National Institute for Occupational Safety and Health

Coleman, W.E., Lingg, R.D., Melton, R.G. & Kopfler, F.C. (1976) The occurrence of volatile organics in five drinking water supplies using gas chromatography/mass spectrometry. In: Keith, L.H. ed., *Identification and Analysis of Organic Pollutants in Water*, Ann Arbor, MI, Ann Arbor Science Publishers, pp. 305-327

Condie, L.W., Smallwood, C.L. & Laurie, R.D. (1983) Comparative renal and hepatotoxicity of halomethanes: bromodichloromethane, bromoform, chloroform, dibromochloromethane and methylene chloride. *Drug Chem. Toxicol.*, 6, 563-578

Considine, J.M., ed. (1974) *Chemical and Process Technology Encyclopedia*, New York, McGraw-Hill Book Co., p. 284

Cunningham, M.L., Gandolfi, A.J., Brendel, K. & Sipes, I.G. (1981) Covalent binding of halogenated volatile solvents to subcellular macromolecules in hepatocytes. *Life Sci.*, 29, 1207-1212

Derwent, R.G. & Eggleton, A.E.J. (1978) Halocarbon lifetimes and concentration distributions calculated using a two-dimensional tropospheric model. *Atmos. Environ.*, 12, 1261-1269

Deutsche Forschungsgemeinschaft (German Research Community) (1985) *Maximal Concentrations in the Workplace and Biological Tolerance Values for Substances in the Work Environment 1985* (Ger.), Vol. 21, Weinheim, Verlagsgesellschaft mbH, p. 29

DeWalle, F.B. & Chian, E.S.K. (1978) Presence of trace organics in the Delaware River and their discharge by municipal and industrial sources. *Proc. ind. Waste Conf.*, 32, 908-919

Direktoratet for Arbeidstilsynet (Directorate of Labour Inspection) (1981) *Administrative Norms for Pollution in Work Atmosphere, No. 36* (Norw.), Oslo, p. 10

DiVincenzo, G.D. & Kaplan, C.J. (1981) Uptake, metabolism, and elimination of methylene chloride vapor by humans. *Toxicol. appl. Pharmacol.*, 59, 130-140

DiVincenzo, G.D., Yanno, F.J. & Astill, B.D. (1971) The gas chromatographic analysis of methylene chloride in breath, blood, and urine. *Am. ind. Hyg. Assoc. J.*, 32, 387-391

DiVincenzo, G.D., Yanno, F.J. & Astill, B.D. (1972) Human and canine exposures to methylene chloride vapor. *Am. ind. Hyg. Assoc. J.*, 33, 125-135

Easley, D.M., Kleopfer, R.D. & Carasea, A.M. (1981) Gas chromatographic-mass spectrometric determination of volatile organic compounds in fish. *J. Assoc. off. anal. Chem.*, 64, 653-656

Eastman Kodak Co. (1985) *Kodak Laboratory Chemicals*, Rochester, NY, p. 168

Edwards, P.R., Campbell, I. & Milne, G.S. (1982a) The impact of chloromethanes in the environment. Part 1. The atmospheric chlorine cycle. *Chem. Ind.*, 16, 574-578

Edwards, P.R., Campbell, I. & Milne, G.S. (1982b) The impact of chloromethanes on the environment. Part 2. Methyl chloride and methylene chloride. *Chem. Ind., 17,* 619-622

Eklund, G., Josefsson, B. & Roos, C. (1978a) Trace analysis of volatile organic substances in Göteborg municipal drinking water. *Vatten, 34,* 195-206

Eklund, G., Josefsson, B. & Roos, C. (1978b) Determination of volatile halogenated hydrocarbons in tap water, seawater and industrial effluents by glass capillary gas chromatography and electron capture detection. *J. high Resolut. Chromatogr. chromatogr. Commun., 1,* 34-40

Eller, P.M. (1984) *NIOSH Manual of Analytical Methods,* 3rd ed., Vol. 2 *(DHHS (NIOSH) Publ. No. 84-100),* Washington DC, US Government Printing Office, pp. 1005-1 — 1005-3

Elovaara, E., Hemminki, K. & Vainio, H. (1979) Effects of methylene chloride, trichloroethane, trichloroethylene, tetrachloroethylene and toluene on the development of chick embryos. *Toxicology, 12,* 111-119

Engström, J. & Bjurström, R. (1977) Exposure to methylene chloride: content in subcutaneous adipose tissue. *Scand. J. Work Environ. Health, 3,* 215-224

Ferrario, J.B., Lawler, G.C., DeLeon, I.R. & Laseter, J.L. (1985) Volatile organic pollutants in biota and sediments of Lake Pontchartrain. *Bull. environ. Contam. Toxicol., 34,* 246-255

Fishbein, L. (1985) *Halogenated aliphatic hydrocarbons: uses and environmental occurrence.* In: Fishbein, L. & O'Neill, I.K., eds, *Environmental Carcinogens. Selected Methods of Analysis,* Vol. 7, *Some Volatile Halogenated Hydrocarbons (IARC Scientific Publications No. 68),* Lyon, International Agency for Research on Cancer, pp. 47-67

Fodor, G.G. & Winneke, H. (1971) *Nervous system disturbances in men and animals experimentally exposed to industrial solvent vapors.* In: Englund, H.M. & Beery, W.T., eds, *Proceedings of the 2nd International Clean Air Congress,* New York, Academic Press, pp. 238-243

Fodor, G.G., Prajsnar, D. & Schlipköter, H.-W. (1973) Endogenous formation of CO from incorporated halogenated hydrocarbons of the methane series (Ger.). *Staub-Reinhalt. Luft., 33,* 258-259

Friedlander, B.R., Hearne, T. & Hall, S. (1978) Epidemiologic investigation of employees chronically exposed to methylene chloride. Mortality analysis. *J. occup. Med., 20,* 657-666

Gamberale, F., Annwall, G. & Hultengren, M. (1975) Exposure to methylene chloride. II. Psychological functions. *Scand. J. Work Environ. Health, 1,* 95-103

Garriott, J.K. & Petty, C.S. (1980) Death from inhalant abuse: toxicological and pathological evaluation of 34 cases. *Clin. toxicol., 16,* 305-315

Gehring, P.J. (1968) Hepatotoxic potency of various chlorinated hydrocarbon vapors relative to their narcotic and lethal potencies in mice. *Toxicol. appl. Pharmacol., 13,* 287-298

Gilles, D. (1977) *Health Hazard Evaluation Determination Report No. 77-101-441, Dap Derusto, Inc., Tipp City, OH*, Cincinnati, OH, National Institute for Occupational Safety and Health

Gocke, E., King, M.-T., Eckhardt, K. & Wild, D. (1981) Mutagenicity of cosmetics ingredients licensed by the European Communities. *Mutat. Res., 90*, 91-109

Goelzer, B. & O'Neill, I.K. (1985) *Workplace air sampling*. In: Fishbein, L. & O'Neill, I.K., eds, *Environmental Carcinogens. Selected Methods of Analysis*, Vol. 7, *Some Volatile Halogenated Hydrocarbons (IARC Scientific Publications No. 68)*, Lyon, International Agency for Research on Cancer, pp. 107-140

Gorman, R. (1982) *Health Hazard Evaluation Report No. HETA-82-008-1226, Arts Consortium, Cincinnati, OH*, Cincinnati, OH, National Institute for Occupational Safety and Health

Grasselli, J.G. & Ritchey, W.M., eds (1975) *CRC Atlas of Spectral Data and Physical Constants for Organic Compounds*, Vol. 3, Cleveland, OH, CRC Press, p. 594

Grasset, J. & Gauthier, R. (1950) Clinical and graphical study of the obstetric analgesic action of methylene chloride (Fr.). *Sem. Hôp. Paris, 26*, 1280-1283

Green, T. (1983) The metabolic activation of dichloromethane and chlorofluoromethane in a bacterial mutation assay using *Salmonella typhimurium*. *Mutat. Res., 118*, 277-288

Gunter, B.J. (1975) *Health Hazard Evaluation Determination Report No. 74-148-239, Lange Company, Bloomfield, CO*, Cincinnati, OH, National Institute for Occupational Safety and Health

Gunter, B.J. (1976) *Health Hazard Evaluation Report No. 76-23-319, Western Gear Corp., Jamestown, ND*, Cincinnati, OH, National Institute for Occupational Safety and Health

Gunter, B.J. & Lucas, J.B. (1974) *Health Hazard Evaluation Determination Report No. 73-84-119, Head Ski Company, Boulder, CO*, Cincinnati, OH, National Institute for Occupational Safety and Health

Gupta, K.C., Ulsamer, A.G. & Gammage, R. (1984) *Volatile organic compounds in residential air: levels, sources and toxicity*. In: *Proceedings of the Air Pollution Control Association, 77th Annual Meeting, San Francisco, 1984*, Pittsburgh, PA, Air Pollution Control Association, pp. 1-9

Haglund, U., Lundberg, I. & Zech, L. (1980) Chromosome aberrations and sister chromatid exchanges in Swedish paint industry workers. *Scand. J. Work Environ. Health, 6*, 291-298

Hansch, C. & Leo, A. (1979) *Substituent Constants for Correlation Analysis in Chemistry and Biology*, New York, John Wiley & Sons, p. 173

Hardin, B.D. & Manson, J.M. (1980) Absence of dichloromethane teratogenicity with inhalation exposure in rats. *Toxicol. appl. Pharmacol., 52*, 22-28

Hartle, R.W. (1980) *Health Hazard Evaluation Report No. HE-80-057-781, Long Island Rail Road, Richmond Hill, NY*, Cincinnati, OH, National Institute for Occupational Safety and Health

Hatch, G.G., Mamay, P.D., Ayer, M.L., Casto, B.C. & Nesnow, S. (1982) Methods for detecting gaseous and volatile carcinogens using cell transformation assays. *Environ. Sci. Res.*, 25, 75-90

Hatch, G.G., Mamay, P.D., Ayer, M.L., Casto, B.C. & Nesnow, S. (1983) Chemical enhancement of viral transformation in Syrian hamster embryo cells by gaseous and volatile chlorinated methanes and ethanes. *Cancer Res.*, 43, 1945-1950

Hawley, G.G., ed. (1981) *The Condensed Chemical Dictionary*, 10th ed., New York, Van Nostrand Reinhold, p. 677

Hays Chemicals (1983) *Specification Sheet: Methylene Chloride*, London, Unalco Division

Health and Safety Executive (1985) *Occupational Exposure Limits 1985 (Guidance Note EH 40/85)*, London, Her Majesty's Stationery Office, p. 11

Hearne, F.T. & Friedlander, B.R. (1981) Follow-up of methylene chloride study. *J. occup. Med.*, 23, 660

Helz, G.R. & Hsu, R.Y. (1978) Volatile chloro- and bromocarbons in coastal waters. *Limnol. Oceanogr.*, 23, 858-869

Heppel, L.A. & Neal, P.A. (1944) Toxicology of dichloromethane (methylene chloride). II. Its effect upon running activity in the male rat. *J. ind. Hyg.*, 26, 17-21

Heppel, L.A. & Porterfield, V.T. (1948) Enzymatic dehalogenation of certain brominated and chlorinated compounds. *J. biol. Chem.*, 76, 763-769

Hervin, R.L. & Watanabe, A.S. (1981) *Health Hazard Evaluation Report No. HHE-80-014-920, Scott U.S.A., Clearfield, UT*, Cincinnati, OH, National Institute for Occupational Safety and Health

Hervin, R.L., Cromer, J.W., Jr & Butler, G.J. (1974) *Health Hazard Evaluation Determination Report No. 74-2&8-164, The Vendo Company, Kansas City, MO*, Cincinnati, OH, National Institute for Occupational Safety and Health

Hollett, B.A. (1977) *Health Hazard Evaluation Report No. 76-92-363, Jeffery Bigelow Design Group, Inc., Washington DC*, Cincinnati, OH, National Institute for Occupational Safety and Health

Honeywill & Stein (undated) *Methylene Chloride*, Wallington, Surrey, UK

IARC (1977a) *IARC Monographs on the Evaluation of the Carcinogenic Risk of Chemicals to Man*, Vol. 15, *Some Fumigants, the Herbicides 2,4-D and 2,4,5-T, Chlorinated Dibenzodioxins and Miscellaneous Industrial Chemicals*, Lyon, pp. 155-175

IARC (1977b) *IARC Monographs on the Evaluation of the Carcinogenic Risk of Chemicals to Man*, Vol. 15, *Some Fumigants, the Herbicides 2,4-D and 2,4,5-T, Chlorinated Dibenzodioxins and Miscellaneous Industrial Chemicals*, Lyon, pp. 155-175

IARC (1979a) *IARC Monographs on the Evaluation of the Carcinogenic Risk of Chemicals to Humans*, Vol. 20, *Some Halogenated Hydrocarbons*, Lyon, pp. 449-465

IARC (1979b) *IARC Monographs on the Evaluation of the Carcinogenic Risk of Chemicals to Humans*, Vol. 20, *Some Halogenated Hydrocarbons*, Lyon, pp. 401-427

IARC (1979c) *IARC Monographs on the Evaluation of the Carcinogenic Risk of Chemicals to Humans*, Vol. 20, *Some Halogenated Hydrocarbons*, Lyon, pp. 371-399

IARC (1979d) *IARC Monographs on the Evaluation of the Carcinogenic Risk of Chemicals to Humans*, Vol. 19, *Some Monomers, Plastics and Synthetic Elastomers, and Acrolein*, Lyon, pp. 377-438

IARC (1979e) *IARC Monographs on the Evaluation of the Carcinogenic Risk of Chemicals to Humans*, Vol. 20, *Some Halogenated Hydrocarbons*, Lyon, pp. 545-572

IARC (1982a) *IARC Monographs on the Evaluation of the Carcinogenic Risk of Chemicals to Humans*, Suppl. 4, *Chemicals, Industrial Processes and Industries Associated with Cancer in Humans, IARC Monographs, Volumes 1 to 29*, Lyon, pp. 111-112

IARC (1982b) *IARC Monographs on the Evaluation of the Carcinogenic Risk of Chemicals to Humans*, Suppl. 4, *Chemicals, Industrial Processes and Industries Associated with Cancer in Humans, IARC Monographs, Volumes 1 to 29*, Lyon, pp. 74-75

IARC (1982c) *IARC Monographs on the Evaluation of the Carcinogenic Risk of Chemicals to Humans*, Suppl. 4, *Chemicals, Industrial Processes and Industries Associated with Cancer in Humans, IARC Monographs, Volumes 1 to 29*, Lyon, pp. 260-262

IARC (1982d) *IARC Monographs on the Evaluation of the Carcinogenic Risk of Chemicals to Humans*, Suppl. 4, *Chemicals, Industrial Processes and Industries Associated with Cancer in Humans, IARC Monographs, Volumes 1 to 29*, Lyon, pp. 262-264

IARC (1982e) *IARC Monographs on the Evaluation of the Carcinogenic Risk of Chemicals to Humans*, Suppl. 4, *Chemicals, Industrial Processes and Industries Associated with Cancer in Humans, IARC Monographs, Volumes 1 to 29*, Lyon, pp. 247-249

IARC (1985) *IARC Monographs on the Evaluation of the Carcinogenic Risk of Chemicals to Humans*, Vol. 36, *Allyl Compounds, Aldehydes, Epoxides and Peroxides*, Lyon, pp. 227-243

IARC (1986) *IARC Monographs on the Evaluation of the Carcinogenic Risk of Chemicals to Humans*, Vol. 39, *Some Chemicals and Elastomers Used in the Plastics Industry*, Lyon, pp. 195-226

Institut National de Recherche et de Sécurité (National Institute for Research and Safety) (1985) *Limit Values for Concentrations of Dangerous Substances in Work Place Air (Notes Documentaires 1555-121-85)* (Fr.), Paris, p. 486

International Labour Office (1980) *Occupational Exposure Limits for Airborne Toxic Substances: A Tabular Compilation of Values from Selected Countries*, 2nd (rev.) ed. (*Occupational Safety and Health Series No. 37*), Geneva, pp. 146-147

International Programme on Chemical Safety (1983) *Toxicological Evaluation of Certain Food Additives and Contaminants (WHO Food Add. Ser. No. 18)*, Geneva, World Health Organization, pp. 65-66

International Programme on Chemical Safety (1984) *Methylene Chloride (Environmental Health Criteria 32)*, Geneva, World Health Organization

Jongen, W.M.F. (1984) Relationship between exposure time and metabolic activation of dichloromethane in *Salmonella typhimurium*. *Mutat. Res.*, *136*, 107-108

Jongen, W.M.F., Alink, G.M. & Koeman, J.H. (1978) Mutagenic effect of dichloromethane on *Salmonella typhimurium*. *Mutat. Res.*, *56*, 245-248

Jongen, W.M.F., Lohman, P.H.M., Kottenhagen, M.J., Alink, G.M., Berends, F. & Koeman, J.H. (1981) Mutagenicity testing of dichloromethane in short-term mammalian test systems. *Mutat. Res.*, *81*, 203-213

Jongen, W.M.F., Harmsen, E.G.M., Alink, G.M. & Koeman, J.H. (1982) The effect of glutathione conjugation and microsomal oxidation on the mutagenicity of dichloromethane in *S. typhimurium*. *Mutat. Res.*, *95*, 183-189

Jungclaus, G.A., Lopez-Avila, V. & Hites, R.A. (1978) Organic compounds in an industrial wastewater: A case study of their environmental impact. *Environ. Sci. Technol.*, *12*, 88-96

Kaiser, K.L.E., Comba, M.E. & Huneault, H. (1983) Volatile halocarbon contaminants in the Niagara River and in Lake Ontario. *J. Great Lakes Res.*, *9*, 212-223

Kanada, T. & Uyeta, M. (1978) Mutagenicity screening of organic solvents in microbial systems (Abstract). *Mutat. Res.*, *54*, 215

Kimura, E.T., Ebert, D.M. & Dodge, P.W. (1971) Acute toxicity and limits of solvent residue for sixteen organic solvents. *Toxicol. appl. Pharmacol.*, *19*, 699-704

Kirschman, J.C., Brown, N.M., Coots, R.H. & Morgareidge, K. (1986) Review of investigations of dichloromethane metabolism and subchronic oral toxicity as the basis for design of chronic oral studies in rats and mice. *Food Chem. Toxicol.*, *24*, 943-949

Kirwin, C.J., Thomas, W.C. & Simmon, V.F. (1980) In vitro microbiological mutagenicity of hydrocarbon propellants. *J. Soc. cosmet. Chem.*, *31*, 367-370

Klaassen, C.D. & Plaa, G.L. (1966) Relative effects of various chlorinated hydrocarbons on liver and kidney function in mice. *Toxicol. appl. Pharmacol.*, *9*, 139-151

Klaassen, C.D. & Plaa, G.L. (1967) Relative effects of various chlorinated hydrocarbons on liver and kidney function in dogs. *Toxicol. appl. Pharmacol.*, *10*, 119-131

Kluwe, W.M., Harrington, F.W. & Cooper, S.E. (1982) Toxic effects of organohalide compounds on renal tubular cells *in vivo* and *in vitro*. *J. Pharmacol. exp. Therap.*, *220*, 597-603

Kronoveter, K. (1977) *Health Hazard Evaluation Report No. 76-84-377, Kenner Products Company, Cincinnati, OH*, Cincinnati, OH, National Institute for Occupational Safety and Health

Krost, K.J., Pellizzari, E.D., Walburn, S.G. & Hubbard, S.A. (1982) Collection and analysis of hazardous organic emissions. *Anal. Chem.*, *54*, 810-817

Kubic, V.L. & Anders, M.W. (1975) Metabolism of dihalomethanes to carbon monoxide. II. *In vitro* studies. *Drug Metab. Disposition*, *3*, 104-112

Kubic, V.L. & Anders, M.W. (1978) Metabolism of dihalomethanes to carbon monoxide. III. Studies on the mechanism of the reaction. *Biochem. Pharmacol.*, *27*, 2349-2355

Kubic, V.L., Anders, M.W., Engel, R.R., Barlow, C.H. & Caughey, W.S. (1974) Metabolism of dihalomethanes to carbon monoxide. I. *In vivo* studies. *Drug Metab. Disposition, 2*, 53-57

Kutob, S.D. & Plaa, G.L. (1962) A procedure for estimating the hepatotoxic potential of certain industrial solvents. *Toxicol. appl. Pharmacol., 4*, 354-361

Kuželová, M. & Vlasák, R. (1966) The effect of methylene-dichloride on the health of workers in production of film-foils and investigation of formic acid as the methylene-dichloride metabolite (Czech.). *Pracov. Lek., 18*, 167-170

Kuželová, M., Černý, J., Hlavová, S., Hub, M., Kunor, V. & Popler, A. (1975) Lethal methylene chloride poisoning with severe chilblains (Czech.). *Pracov. Lek., 27*, 317-319

Laham, S. & Potvin, M. (1976) Microdetermination of dichloromethane in blood with a syringeless gas chromatographic injection system. *Chemosphere, 6*, 403-411

Laib, R.J. (1982) *Specific covalent binding and toxicity of aliphatic halogenated xenobiotics.* In: Beckett, A.J. & Gorrod, J.W., eds, *Reviews on Drug Metabolism and Drug Interactions*, Vol. IV, No. 1, London, Freund Publishing, pp. 1-48

Lao, R.C., Thomas, R.S., Bastien, P., Halman, R.A. & Lockwood, J.A. (1982) Analysis of organic priority and non-priority pollutants in environmental samples by GC/MS/computer systems. *Pergamon Ser. environ. Sci., 7*, 107-118

Lee, S.A. (1980) *Health Hazard Evaluation Determination Report No. HE-80-27-704, Airco Welding Products, Chester, WV*, Cincinnati, OH, National Institute for Occupational Safety and Health

Lewis, F.A. & Thoburn, T.W. (1981) *Health Hazard Evaluation Report No. HHE-79-020-839, Graphic Color Plate, Inc., Stamford, CT*, Cincinnati, OH, National Institute for Occupational Safety and Health

Ma, T.-H., Harris, M.M., Anderson, V.A., Ahmed, I., Mohammad, K., Bare, J.L. & Lin, G. (1984) *Tradescantia*-micronucleus (Trad-MCN) tests on 140 health-related agents. *Mutat. Res., 138*, 157-167

Mannsville Chemical Products Corp. (1984) *Chemical Products Synopsis: Methylene Chloride*, Cortland, NY

Markel, H.L., Jr (1980) *Health Hazard Evaluation Report No. HE-78-125-712, Owens-Corning Fiberglas Corporation, Conroe, TX*, Cincinnati, OH, National Institute for Occupational Safety and Health

Markel, H.L., Jr & Jannerfeldt, E. (1981) *Health Hazard Evaluation Report No. HHE-79-156-899, Gulf-Wandes Corp., Baton Rouge, LA*, Cincinnati, OH, National Institute for Occupational Safety and Health

Markel, H.L., Jr & Shama, S.K. (1974) *Health Hazard Evaluation Report No. 72-100-121, Whirlpool Corporation, Fort Smith, AK*, Cincinnati, OH, National Institute for Occupational Safety and Health

Markel, H.L., Jr & Slovin, D. (1981) *Health Hazard Evaluation Report No. HHE-79-158-819, Morrilton Plastics Corp., Morrilton, AK*, Cincinnati, OH, National Institute for Occupational Safety and Health

McCarroll, N.E., Cortina, T.A., Zito, M.J. & Farrow, M.G. (1983) Evaluation of methylene chloride and vinylidine chloride in mutational assays (Abstract). *Environ. Mutagenesis*, 5, 426-427

McGregor, D.B. (1979) *Practical experience in testing unknowns* in vitro. In: Paget, G.E., ed., *Mutagenesis in Sub-mammalian Systems, Status and Significance*, Lancaster, UK, MTP Press, pp. 53-71

McKenna, M.J. & Zempel, J.A. (1981) The dose-dependent metabolism of [^{14}C]methylene chloride following oral administration to rats. *Food Cosmet. Toxicol.*, 19, 73-78

McKenna, M.J., Zempel, J.A. & Braun, W.H. (1982) The pharmacokinetics of inhaled methylene chloride in rats. *Toxicol. appl. Pharmacol.*, 65, 1-10

McMahon, L.W. (1983) *Organic priority pollutants in wastewater*. In: Oakes, T.W., ed., *Proceedings of UCC-ND/GAT Environmental Protection Seminar (US NTIS CONF-820418)*, Washington DC, US Government Printing Office, pp. 220-250

Morgan, A., Black, A. & Belcher, D.R. (1970) The excretion in breath of some aliphatic halogenated hydrocarbons following administration by inhalation. *Ann. occup. Hyg.*, 13, 219-233

Morris, J.B., Smith, F.A. & Garman, R.H. (1979) Studies on methylene chloride-induced fatty liver. *Exp. mol. Pathol.*, 30, 386-393

Moskowitz, S. & Shapiro, H. (1952) Fatal exposure to methylene chloride vapor. *Arch. ind. Hyg. occup. Med.*, 6, 116-123

National Institute for Occupational Safety and Health (1976) *Criteria for a Recommended Standard — Occupational Exposure to Methylene Chloride (DHEW (NIOSH) Publ. No. 76-138)*, Washington DC, US Department of Health, Education, and Welfare, p. 165

National Institute for Occupational Safety and Health (1985) NIOSH recommendations for occupational safety and health standards. *Morbid. Mortal. wkly Rep. Suppl.*, 34, 22S

National Toxicology Program (1986) *Toxicology and Carcinogenesis Studies of Dichloromethane (Methylene Chloride) (CAS No. 75-09-2) in F344/N Rats and B6C3F$_1$ Mice (Inhalation Studies) (Technical Report No. 306)*, Research Triangle Park, NC, US Department of Health and Human Services

Nestmann, E.R., Lee, E.G.-H., Matula, T.I., Douglas, G.R. & Mueller, J.C. (1980) Mutagenicity of constituents identified in pulp and paper mill effluents using the Salmonella/mammalian-microsome assay. *Mutat. Res.*, 79, 203-212

Nestmann, E.R., Otson, R., Williams, D.T. & Kowbel, D.J. (1981) Mutagenicity of paint removers containing dichloromethane. *Cancer Lett.*, 11, 295-302

Okawa, M.T. & Keith, W. (1977) *Health Hazard Evaluation Determination Report No. 75-195-396, United Airlines Maintenance Base, San Francisco International Airport, Burlingame, CA*, Cincinnati, OH, National Institute for Occupational Safety and Health

Osterman-Golkar, S., Hussain, S., Walles, S., Anderstam, B. & Sigvardsson, K. (1983) Chemical reactivity and mutagenicity of some dihalomethanes. *Chem.-biol. Interactions*, *46*, 121-130

Otson, R., Williams, D.T. & Bothwell, P.D. (1982) Volatile organic compounds in water at thirty Canadian potable water treatment facilities. *J. Assoc. off. anal. Chem.*, *65*, 1370-1374

Ott, M.G., Skory, L.K., Holden, B.B., Bronson, J.M. & Williams, P.R. (1983a) Health evaluation of employees occupationally exposed to methylene chloride. General study design and environmental considerations. *Scand. J. Work Environ. Health*, *9* (Suppl. 1), 1-7

Ott, M.G., Skory, L.K., Holder, B.B., Bronson, J.M. & Williams, P.R. (1983b) Health evaluation of employees occupationally exposed to methylene chloride. Clinical laboratory evaluation. *Scand. J. Work Environ. Health*, *9* (Suppl. 1), 17-25

Ott, M.G., Skory, L.K., Holder, B.B., Bronson, J.M. & Williams, P.R. (1983c) Health evaluation of employees occupationally exposed to methylene chloride. Twenty-four hour electrocardiographic monitoring. *Scand. J. Work Environ. Health*, *9* (Suppl. 1), 26-30

Ott, M.G., Skory, L.K., Holder, B.B., Bronson, J.M. & Williams, P.R. (1983d) Health evaluation of employees occupationally exposed to methylene chloride. Mortality. *Scand. J. Work Environ. Health*, *9* (Suppl. 1), 8-16

Ott, M.G., Skory, L.K., Holder, B.B., Bronson, J.M. & Williams, P.R. (1983e) Health evaluation of employees occupationally exposed to methylene chloride. Metabolism data and oxygen half-saturation pressures. *Scand. J. Work Environ. Health*, *9* (Suppl. 1), 31-38

Page, B.D. & Charbonneau, C.F. (1977) Gas chromatographic determination of residual methylene chloride and trichloroethylene in decaffeinated instant and ground coffee with electrolytic conductivity and electron capture detection. *J. Assoc. off. anal. Chem.*, *60*, 710-715

Page, G.W. (1981) Comparison of groundwater and surface water for patterns and levels of contamination by toxic substances. *Environ. Sci. Technol.*, *15*, 1475-1481

Peers, A.M. (1985) *The determination of methylene chloride in air*. In: Fishbein, L. & O'Neill, I.K., eds, *Environmental Carcinogens. Selected Methods of Analysis*, Vol. 7, *Some Volatile Halogenated Hydrocarbons* (*IARC Scientific Publications No. 68*), Lyon, International Agency for Research on Cancer, pp. 191-196

Pellizzari, E.D., Zweidinger, R.A. & Erickson, M.D. (1978) *Environmental Monitoring Near Industrial Sites: Brominated Chemicals Part II: Appendix (EPA-560/6-78-002A; US NTIS PB-286483)*, Washington DC, Office of Toxic Substances, US Environmental Protection Agency,

Pellizzari, E.D., Hartwell, T.D., Harris, B.S.H., III, Waddell, R.D., Whitaker, D.A. & Erickson, M.D. (1982) Purgeable organic compounds in mother's milk. *Bull. environ. Contam. Toxicol.*, *28*, 322-328

Perbellini, L., Brugnone, F., Grigolini, L., Cunegatti, P. & Tacconi, A. (1977) Alveolar air and blood dichloromethane concentration in shoe sole factory workers. *Int. Arch. occup. environ. Health*, *40*, 241-247

Perocco, P. & Prodi, G. (1981) DNA damage by haloalkanes in human lymphocytes cultured *in vitro*. *Cancer Lett.*, *13*, 213-218

Peterson, J.E. (1978) Modeling the uptake, metabolism and excretion of dichloromethane by man. *Am. ind. Hyg. Assoc. J.*, *39*, 41-47

Piet, G.J., Luijten, W.C.M.M. & van Noort, P.C.M. (1985a) *Dynamic head-space determination of volatile organic halogen compounds in water*. In: Fishbein, L. & O'Neill, I.K., eds, *Environmental Carcinogens. Selected Methods of Analysis*, Vol. 7, *Some Volatile Halogenated Hydrocarbons (IARC Scientific Publications No. 68)*, Lyon, International Agency for Research on Cancer, pp. 331-343

Piet, G.J., Luijten, W.C.M.M. & van Noort, P.C.M. (1985b) *'Static' head-space determination of volatile organic halogen compounds in water*. In: Fishbein, L. & O'Neill, I.K., eds, *Environmental Carcinogens. Selected Methods of Analysis*, Vol. 7, *Some Volatile Halogenated Hydrocarbons (IARC Scientific Publications No. 68)*, Lyon, International Agency for Research on Cancer, pp. 321-330

Premel-Cabic, A., Cailleux, A. & Allain, P. (1974) Gas chromatographic assay of fifteen volatile organic solvents in blood (Fr.). *Clin. chim. Acta*, *56*, 5-11

Price, P.J., Hassett, C.M. & Mansfield, J.I. (1978) Transforming activities of trichloroethylene and proposed industrial alternatives. *In Vitro*, *14*, 290-293

Pryor, P.D. (1981) *Health Hazard Evaluation Report No. HHE-80-218-848, Ford Motor Company, San José, CA*, Cincinnati, OH, National Institute for Occupational Safety and Health

Putz, V.R., Johnson, B.L. & Setzer, J.V. (1976) A comparative study of the effects of carbon monoxide and methylene chloride on human performance. *J. environ. Pathol. Toxicol.*, *2*, 97-112

Quinn, M.M. (1981) *Health Hazard Evaluation Report No. HETA-81-106-1003, ABT Associates, Cambridge, MA*, Cincinnati, OH, National Institute for Occupational Safety and Health

Ratney, R.S., Wegman, D.H. & Elkins, H.B. (1974) In vivo conversion of methylene chloride to carbon monoxide. *Arch. environ. Health*, *28*, 223-226

Reynolds, E.S. & Yee, A.G. (1967) Liver parenchymal cell injury. V. Relationships between patterns of chloromethane-C^{14} incorporation into constituents of liver *in vivo* and cellular injury. *Lab. Invest.*, *16*, 591-603

Riley, E.C., Fassett, D.W. & Sutton, W.L. (1966) Methylene chloride vapor in expired air of human subjects. *Am. ind. Hyg. Assoc. J.*, *27*, 341-348

Rivera, R.O. (1975) *Health Hazard Evaluation Determination Report No. 74-135-226, GAF Corporation, Equipment Manufacturing Plant, Vestal, NY*, Cincinnati, OH, National Institute for Occupational Safety and Health

Rosengren, L.E., Kjellstrand, P., Aurell, A. & Haglid, K.G. (1986) Irreversible effects of dichloromethane on the brain after long term exposure: a quantitative study of DNA and the glial cell marker proteins S-100 and GFA. *Br. J. ind. Med.*, *43*, 291-299

Rosensteel, R.E. & Meyer, C.R. (1977) *Health Hazard Determination Report No. 75-150-378, Reinell Boats, Inc., Poplar Bluff, MO*, Cincinnati, OH, National Institute for Occupational Safety and Health

Ruhe, R.L. (1978) *Health Hazard Evaluation Determination Report No. HE-78-70-528, Hospal Medical Corporation, Littleton, CO*, Cincinnati, OH, National Institute for Occupational Safety and Health

Ruhe, R.L. (1981) *Health Hazard Evaluation Report No. HETA-81-378-1000, Keystone Diesel Engine Company, Wexford, PA*, Cincinnati, OH, National Institute for Occupational Safety and Health

Ruhe, R.L. & Anderson, L. (1977) *Health Hazard Evaluation Determination Report No. 76-17-395, The Hayes & Albion Company, Spencerville, OH*, Cincinnati, OH, National Institute for Occupational Safety and Health

Ruhe, R.L., Watanabe, A. & Stein, G. (1981) *Health Hazard Evaluation Report No. HHE-80-49-808, Superior Tube Company, Collegeville, PA*, Cincinnati, OH, National Institute for Occupational Safety and Health

Ruhe, R.L., Singal, M. & Hervin, R.L. (1982) *Health Hazard Evaluation Report No. HETA-80-79-1189, Rexall Drug Company, St Louis, MO*, Cincinnati, OH, National Institute for Occupational Safety and Health

Sadtler Research Laboratories (1980) *The Sadtler Standard Spectra, Cumulative Index*, Philadelphia, PA

Salisbury, S.A. (1981) *Health Hazard Evaluation Report No. HETA-81-053-876, Georgia Department of Human Resources, Drug Abuse Laboratory, Atlanta, GA*, Cincinnati, OH, National Institute for Occupational Safety and Health

Samoiloff, M.R., Schulz, S., Jordan, Y., Denich, K. & Arnott, E. (1980) A rapid simple long-term toxicity assay for aquatic contaminants using the nematode *Panagrellus redivivus*. *Can. J. Fish. aquat. Sci.*, *37*, 1167-1174

Sax, N.I. (1984) *Dangerous Properties of Industrial Materials*, 6th ed., New York, Van Nostrand Reinhold, p. 1763

Schairer, L.A. & Sautkulis, R.C. (1982) *Detection of ambient levels of mutagenic atmospheric pollutants with the higher plant* Tradescantia. In: Klekowski, E.J., Jr, ed., *Environmental Mutagenesis, Carcinogenesis, and Plant Biology*, Vol. II, New York, Praeger Scientific, pp. 154-194

Schairer, L.A., Sautkulis, R.C. & Tempel, N.R. (1982) Detection and identification of genotoxic agents in ambient air using the *Tradescantia* bioassay (Abstract). *Environ. Mutagenesis, 4,* 379

Schumacher, H. & Grandjean, E. (1960) Comparison of narcotic action and acute toxicity of new solvents (Ger.). *Arch. Gewerbepath. Gewerbehyg., 18,* 109-119

Schwetz, B.A., Leong, B.K.J. & Gehring, P.J. (1975) The effect of maternally inhaled trichloroethylene, perchloroethylene, methyl chloroform, and methylene chloride on embryonal and fetal development in mice and rats. *Toxicol. appl. Pharmacol., 32,* 84-96

Serota, D.G., Thakur, A.K., Ulland, B.M., Kirschman, J.C., Brown, N.M., Coots, R.G. & Morgareidge, K. (1986a) A two-year drinking water study of dichloromethane on rodents: II. Mice. *Food Chem. Toxicol., 24,* 959-963

Serota, D.G., Thakur, A.K., Ulland, B.M., Kirschman, J.C., Brown, N.M., Coots, R.G. & Morgareidge, K. (1986b) A two-year drinking water study of dichloromethane on rodents: I. Rats. *Food Chem. Toxicol., 24,* 951-958

Simmon, V.F., Kauhanen, K. & Tardiff, R.G. (1977) *Mutagenic activity of chemicals identified in drinking water*. In: Scott, D., Bridges, B.A. & Sobels, F.H., eds, *Progress in Genetic Toxicology*, Amsterdam, Elsevier/North-Holland Biomedical Press, pp. 249-258

Singh, H.B., Salas, L.J., Smith, A.J. & Shigeishi, H. (1981) Measurements of some potentially hazardous organic chemicals in urban environments. *Atmos. Environ., 15,* 601-612

Singh, H.B., Salas, L.J. & Stiles, R.E. (1982) Distribution of selected gaseous organic mutagens and suspect carcinogens in ambient air. *Environ. Sci. Technol., 16,* 872-880

Stahl, W.H., ed. (1973) *Compilation of Odor and Taste Threshold Values Data (ASTM Data Series DS48)*, Philadelphia, PA, American Society for Testing and Materials, p. 107

Stauffer Chemical Co. (1976) *MC+ Vapor Degreasing Grade of Methylene Chloride*, Westport, CT

Stauffer Chemical Co. (undated a) *Methylene Chloride, Vapor Degreasing Grade*, Westport, CT

Stauffer Chemical Co. (undated b) *Methylene Chloride, Reagent Grade*, Westport, CT

Stauffer Chemical Co. (undated c) *Methylene Chloride, Aerosol-Technical Grade*, Westport, CT

Stewart, R.D. & Dodd, H.C. (1964) Apsorption of carbon tetrachloride, trichloroethylene, tetrachloroethylene, methylene chloride, and 1,1,1-trichloroethane through the human skin. *Am. ind. Hyg. Assoc. J., 25,* 439-446

Stewart, R.D. & Hake, C.L. (1976) Paint-remover hazard. *J. Am. med. Assoc.*, *235*, 398-401

Stewart, R.D., Fisher, T.N., Hosko, M.J., Peterson, J.E., Baretta, E.D. & Dodd, H.C. (1972a) Carboxyhemoglobin elevation after exposure to dichloromethane. *Science*, *176*, 295-296

Stewart, R.D., Fisher, T.N., Hosko, M.J., Peterson, J.E., Baretta, E.D. & Dodd, H.C. (1972b) Experimental human exposure to methylene chloride. *Arch. environ. Health*, *25*, 342-348

Svirbely, J.L., Highman, B., Alford, W.C. & von Oettingen, W.F. (1947) The toxicity and narcotic action of mono-chloro-mono-bromo-methane with special reference to inorganic and volatile bromide in blood, urine and brain. *J. ind. Hyg. Toxicol.*, *29*, 382-389

Taskinen, H., Lindbohm, M.-L. & Hemminki, K. (1986) Spontaneous abortions among women working in the pharmaceutical industry. *Br. J. ind. Med.*, *43*, 199-205

Tharr, D.G. & Donohue, M. (1980) *Health Hazard Evaluation Report No. HE-79-80, 81-746, Cobe Laboratories, Inc., Lakewood and Arvada, CO*, Cincinnati, OH, National Institute for Occupational Safety and Health

Tharr, D.G., Murphy, D.C. & Mortimer, V. (1982) *Health Hazard Evaluation Report No. HETA-81-455-1229, Red Wing Shoe Company, Red Wing, MN*, Cincinnati, OH, National Institute for Occupational Safety and Health

Theiss, J.C., Stoner, G.D., Shimkin, M.B. & Weisburger, E.K. (1977) Test for carcinogenicity of organic contaminants of United States drinking waters by pulmonary tumor response in strain A mice. *Cancer Res.*, *37*, 2717-2720

Thilagar, A.K. & Kumaroo, V. (1983) Induction of chromosome damage by methylene chloride in CHO cells. *Mutat. Res.*, *116*, 361-367

Thilagar, A.K., Back, A.M., Kirby, P.E., Kumaroo, P.V., Pant, K.J., Clarke, J.J., Knight, R. & Haworth, S.R. (1984a) Evaluation of dichloromethane in short term in vitro genetic toxicity assays (Abstract). *Environ. Mutagenesis*, *6*, 418-419

Thilagar, A.K., Kumaroo, P.V., Clarke, J.J., Kott, S., Back, A.M. & Kirby, P.E. (1984b) Induction of chromosome damage by dichloromethane in cultures human peripheral lymphocytes, CHO cells and mouse lymphoma L5178Y cells (Abstract). *Environ. Mutagenesis*, *6*, 422

Työsuojeluhallitus (National Finnish Board of Occupational Safety and Health) (1981) *Airborne Contaminants in the Work Places* (*Safety Bull. 3*) (Finn.), Tampere, p. 10

United Nations (1985) *Industrial Statistics Yearbook 1982*, Vol. II, *Commodity Production Statistics 1973-1982*, New York, p. 363

US Environmental Protection Agency (1978) *Chemical Hazard Information Profile: Dichloromethane*, Washington DC, TSCA Interagency Testing Committee

US Environmental Protection Agency (1982a) *Method 8010. Halogenated volatile organics*. In: *Test Methods for Evaluating Solid Waste — Physical/Chemical Methods*, 2nd ed. (*US EPA No. SW-846*), Washington DC, Office of Solid Waste and Emergency Response

US Environmental Protection Agency (1982b) *Method 8240. GC/MS method for volatile organics*. In: *Test Methods for Evaluating Solid Waste — Physical/Chemical Methods*, 2nd ed. (*US EPA No. SW-846*), Washington DC, Office of Solid Waste and Emergency Response

US Environmental Protection Agency (1982c) *Health Assessment Document for Dichloromethane (Methylene Chloride)* (*Publ. No. PB83-135996*), Washington DC, US Department of Commerce, National Technical Information Service

US Environmental Protection Agency (1984a) Method 624. Guidelines establishing test procedures for the analysis of pollutants under the Clean Water Act (40 CFR 136). Purgeables. *Fed. Regist.*, *49*, 43373-43384

US Environmental Protection Agency (1984b) Method 601. Guidelines establishing test procedures for the analysis of pollutants under the Clean Water Act (40 CFR 136). Purgeable halocarbons. *Fed. Regist.*, *49*, 43261-43271

US Environmental Protection Agency (1984c) Method 1624, Revision B. Guidelines establishing test procedures for the analysis of pollutants under the Clean Water Act (40 CFR 136). Volatile organic compounds by isotope dilution GC/MS. *Fed. Regist.*, *49*, 43407-43415

US Environmental Protection Agency (1985) *Health Assessment Document for Dichloromethane (Methylene Chloride). Final Report* (*EPA/600/8-82/004F*), Washington DC, Office of Health and Environmental Assessment

US Food and Drug Administration (1983) *Methylene chloride*. In: Warner, C., Modderman, J., Fazio, T., Beroza, M., Schwartzman, G. & Fominaya, K., eds, *Food Additives Analytical Manual*, Vol. 1, Arlington, VA, Association of Official Analytical Chemists, pp. 224-232

US Food and Drug Administration (1985) Methylene chloride. *US Code fed. Regul.*, Title *21*, Part 173.255, p. 114

US International Trade Commission (1984) *Synthetic Organic Chemicals, US Production and Sales, 1983* (*USITC Publ. 1588*), Washington DC, US Government Printing Office, p. 259

US International Trade Commission (1985) *Synthetic Organic Chemicals, US Production and Sales, 1984* (*USITC Publ. 1745*), Washington DC, US Government Printing Office, p. 258

US Occupational Safety and Health Administration (1985) Labor. *US Code Fed. Regul.*, Title *29*, Part 1910.1000

Vandervort, R. & Polakoff, P.L. (1973) *Health Hazard Evaluation Report No. 72-84-31, Dunham-Bush, Inc., West Hartford, CT*, Cincinnati, OH, National Institute for Occupational Safety and Health

Verrett, M.J., Scott, W.F., Reynaldo, E.R., Alterman, E.K. & Thomas, C.A. (1980) Toxicity and teratogenicity of food additive chemicals in the developing chicken embryo. *Toxicol. appl. Pharmacol.*, *56*, 265-273

Verschueren, K. (1983) *Handbook of Environmental Data on Organic Chemicals*, 2nd ed., New York, Van Nostrand Reinhold, pp. 848-849

Wagner, W.L. (1974) *Health Hazard Evaluation Determination Report No. 73-124-127, Schnadig Corporation, Cornelia, GA*, Cincinnati, OH, National Institute for Occupational Safety and Health

Weast, R.C., ed. (1985) *CRC Handbook of Chemistry and Physics*, 66th ed., Boca Raton, FL, CRC Press, pp. C-349, D-196

Weinstein, R.S., Boyd, D.D. & Back, K.C. (1972) Effects of continuous inhalation of dichloromethane in the mouse: morphologic and functional observations. *Toxicol. appl. Pharmacol., 23*, 660-679

White, G.L. & Wegman, D.H. (1978) *Health Hazard Evaluation Determination Report No. HE-78-68-546, Lear Siegler, Inc., Marblehead, MA*, Cincinnati, OH, National Institute for Occupational Safety and Health

WHO (1983) *Evaluation of Certain Food Additives and Contaminants (Tech. Rep. Ser. No. 696)*, Geneva

Windholz, M., ed. (1983) *The Merck Index*, 10th ed., Rahway, NJ, Merck & Co., p. 869

Winneke, G. (1974) *Behavioral effects of methylene chloride and carbon monoxide as assessed by sensory and psychomotor performance*. In: Xintaras, C., Johnson, B. & de Groot, I., eds, *Behavioral Toxicology*, Washington DC, US Government Printing Office, pp. 130-144

Withey, J.R. & Collins, B.T. (1980) Chlorinated aliphatic hydrocarbons used in the foods industry: the comparative pharmacokinetics of methylene chloride, 1,2-dichloroethane, chloroform and trichloroethylene after i.v. administration in the rat. *J. environ. Pathol. Toxicol., 3*, 313-332

Withey, J.R., Collins, B.T. & Collins, P.G. (1983) Effect of vehicle on the pharmacokinetics and uptake of four halogenated hydrocarbons from the gastrointestinal tract of the rat. *J. appl. Toxicol., 3*, 249-253

1,1,1,2-TETRACHLOROETHANE

1. Chemical and Physical Data

1.1 Synonyms and trade names

Chem. Abstr. Services Reg. No.: 630-20-6
Chem. Abstr. Name: 1,1,1,2-Tetrachloroethane
IUPAC Systematic Name: 1,1,1,2-Tetrachloroethane

1.2 Structural and molecular formulae and molecular weight

$$\begin{array}{c} \text{Cl} \quad \text{H} \\ | \quad\quad | \\ \text{Cl} - \text{C} - \text{C} - \text{Cl} \\ | \quad\quad | \\ \text{Cl} \quad \text{H} \end{array}$$

$C_2H_2Cl_4$ Mol. wt: 167.85

1.3 Chemical and physical properties of the pure substance

(a) *Description*: Colourless liquid (Sittig, 1985)

(b) *Boiling-point*: 130.5°C (Weast, 1985)

(c) *Melting-point*: -70.2°C (Weast, 1985)

(d) *Density*: d_4^{20} 1.5406 (Weast, 1985)

(e) *Spectroscopy data*: Infrared (Pouchert, 1981), nuclear magnetic resonance (Pouchert, 1983) and mass spectral data (Grasselli & Ritchey, 1975) have been reported.

(f) *Solubility*: 0.056 g/100 g in water at 20°C (Canada Safety Council, 1981); soluble in ethanol, diethyl ether, acetone, benzene and chloroform (Weast, 1985)

(g) *Volatility*: Vapour pressure, 10 mm Hg at 19.3°C (Weast, 1985)

(h) *Stability*: Emits toxic fumes when heated to decomposition (Sax, 1984)

(i) *Octanol/water/partition coefficient (P)*: log P, 2.66 (ICIS Chemical Information System, 1985 [ISHOW])

(j) *Conversion factor*: mg/m^3 = 6.87 × ppm[a]

1.4 Technical products and impurities

1,1,1,2-Tetrachloroethane is not available in commercial quantities. Although the symmetrical isomer, 1,1,2,2-tetrachloroethane, has a number of industrial uses, the unsymmetrical tetrachloroethane is used only in research quantities (Chieruttini & Franklin, 1976). 1,1,1,2-Tetrachloroethane is available in 98.5% purity (Aldrich Chemical Co., 1984).

2. Production, Use, Occurrence and Analysis

2.1 Production and use

(a) Production

1,1,1,2-Tetrachloroethane is not produced for sale in bulk quantities (Chieruttini & Franklin, 1976; National Institute for Occupational Safety and Health, 1978; Kaveler, 1984). It is present as an unisolated intermediate in some processes for the manufacture of trichloroethylene and tetrachloroethylene from ethylene dichloride (Archer, 1979; National Toxicology Program, 1983).

(b) Use

1,1,1,2-Tetrachloroethane has no known commercial use other than as an unisolated process intermediate. Trichloroethylene and tetrachloroethylene, in which 1,1,1,2-tetrachloroethane may be present as an impurity (National Toxicology Program, 1983), are widely used as solvents in cleaning, degreasing and extraction processes (see IARC, 1979).

---

[a]Calculated from: mg/m^3 = (molecular weight/24.45) × ppm, assuming standard temperature (25°C) and pressure (760 mm Hg)

(c) Regulatory status and guidelines

No standard has been set for occupational exposure to 1,1,1,2-tetrachloroethane. In order to protect aquatic organisms, the US Environmental Protection Agency (1980) recommends that the maximum concentration of the chemical in fresh water not exceed 9.32 mg/l.

1,1,1,2-Tetrachloroethane is classified as a hazardous waste and as a priority toxic pollutant by the US Environmental Protection Agency (Sittig, 1985).

2.2 Occurrence

(a) Natural occurrence

1,1,1,2-Tetrachloroethane is not known to occur as a natural product.

(b) Occupational exposure

No data on exposure levels were available to the Working Group.

(c) Air

In the USA, mean urban air concentrations of 1,1,1,2-tetrachloroethane were 4-9 ppt (28-62 ng/m^3). The calculated average human exposure doses were 0.1-0.6 µg/day. The estimated residence time for 1,1,1,2-tetrachloroethane in a typical urban atmosphere was >1160 days, with a daily rate of loss of <0.1% (Singh *et al.*, 1981).

(d) Water and sediments

The concentration of 1,1,1,2-tetrachloroethane in water samples taken in 1976 from the Rhine River near Lobith ranged from <0.1 to 0.4 µg/l, with an average of 0.2 µg/l. Concentrations in drinking-water samples from 100 cities of the Federal Republic of Germany in 1977 ranged from nondetectable to 1.3 µg/l, with an average of <0.1 µg/l (Bauer, 1981).

(e) Human tissues

Bauer (1981) analysed samples taken from 15 autopsy cases in 1978 in the Ruhr district of the Federal Republic of Germany. The mean value of 1,1,1,2-tetrachloroethane was 1.6 µg/kg in kidney capsule fat, 0.6 µg/kg in hypodermis fat tissue, 2.2 µg/kg in lung, 0.4 µg/kg in liver and 0.2 µg/kg in muscle. The highest levels found in individual samples were 15.1 µg/kg in kidney capsule fat and 27.0 µg/kg in lung. The author did not correlate these levels with any specific exposure to 1,1,1,2-tetrachloroethane.

(f) Other

The compound has been reported as a trace impurity in trichloroethylene and tetrachloroethylene (National Toxicology Program, 1983).

1,1,1,2-Tetrachloroethane concentrations ranged from nondetectable to 0.2 μg/l in mouthwash and from nondetectable to 0.1 μg/l in seven cough mixtures purchased in shops in Bochum, Federal Republic of Germany (Bauer, 1981).

2.3 Analysis

A method for the quantitative analysis of 1,1,1,2-tetrachloroethane in groundwater and in liquid and solid wastes has been described (US Environmental Protection Agency, 1982) involving extraction (solid samples) with methanol, dilution with water and a standard purge/trap sequence for isolation of volatiles and analysis by gas chromatography with electrolytic conductivity detection. Gas chromatographic determination of 1,1,1,2-tetrachloroethane in drinking-water, using photoionization and electrolytic conductivity detectors in series, is reported to provide a sensitivity of 0.1-0.5 μg/l (Kingsley et al., 1983). A method using gas chromatography/Fourier-transform infrared detection is reported for the analysis of 1,1,1,2-tetrachloroethane in chemical plant still-bottom samples (Gurka et al., 1982).

3. Biological Data Relevant to the Evaluation of Carcinogenic Risk to Humans

3.1 Carcinogenicity studies in animals

Oral administration

Mouse: Groups of 50 male and 50 female B6C3F$_1$ mice, eight weeks of age, were administered 0, 250 or 500 mg/kg bw 1,1,1,2-tetrachloroethane (>99% pure with traces of chloroethane and ethylene derivatives) in corn oil by gavage on five days a week for 103 weeks (low dose) or 65 weeks (high dose) and then killed. Clinical signs of central nervous system toxicity occurred at week 51 in high-dose animals of each sex; all died or were killed when moribund after 65 weeks. Survival to the end of the study was 38/50 control males and 34/50 low-dose males and 41/50 control females and 31/50 low-dose females. There was a statistically significant dose-related increase in the incidence of hepatocellular adenomas in animals of each sex: 6/48, 14/46 and 21/50 ($p = 0.001$) in control, low-dose and high-dose males, and 4/49, 8/46 and 24/48 ($p < 0.001$) in females. In spite of low survival, there was a dose-related [$p < 0.05$] increase in the incidence of hepatocellular carcinomas in treated females: 1/49, 5/46 and 6/48 (National Toxicology Program, 1983).

Rat: Groups of 50 male and 50 female Fischer 344/N rats, seven weeks of age, were administered 0, 125 or 250 mg/kg bw 1,1,1,2-tetrachloroethane (>99% pure with traces of chloroethane and ethylene derivatives) in corn oil by gavage on five days a week for 103 weeks and killed at 104 weeks. Clinical signs of central nervous system involvement were observed from week 44 onwards. At the end of the study, survival was 29/50, 25/50 and

21/50 in control, low-dose and high-dose males, and 29/50, 27/50 and 24/50 in females. A statistically significant increase in the incidence of fibroadenomas of the mammary gland was observed in low-dose females: controls, 6/49; low-dose, 15/49 ($p = 0.024$); and high-dose, 7/46; however, no significant dose-response trend was observed (National Toxicology Program, 1983).

3.2 Other relevant biological data

(a) *Experimental systems*

Toxic effects

Oral LD_{50} values for 1,1,1,2-tetrachloroethane have been reported to be 1500 mg/kg bw in male mice, 670 mg/kg bw in male rats and 780 mg/kg bw in female rats. LC_{50} values for a four-hour exposure to 1,1,1,2-tetrachloroethane were 2100 ppm (14 400 mg/m³) in male and 2500 ppm (17 150 mg/m³) in female rats, and 2800 ppm (19 200 mg/m³) in male rabbits. The dermal LD_{50} for 1,1,1,2-tetrachloroethane in rabbits was reported to be approximately 20 g/kg (Truhaut *et al.*, 1974).

Acute oral toxicity studies with 1,1,1,2-tetrachloroethane (0.5 g/kg, 99.9% pure) in rabbits showed increases in the activities of several serum enzymes (serum glutamic-pyruvic transaminase, serum glutamic-oxaloacetic transaminase, creatinine phosphokinase, lactate dehydrogenase and α-hydroxybutyrate dehydrogenase) 24 h after treatment (Truhaut *et al.*, 1973). Female, but not male, rats given oral doses of 0.3 g/kg bw 1,1,1,2-tetrachloroethane on five days per week for two weeks developed hepatic steatosis (Truhaut *et al.*, 1975).

Male and female guinea-pigs and rabbits administered 0.5 g/kg bw 1,1,1,2-tetrachloroethane orally showed morphological evidence of liver damage 24 h to 11 days after treatment (Truhaut *et al.*, 1974).

Oral administration of 0.4 g/kg bw 1,1,1,2-tetrachloroethane on five days per week for two weeks to male and female rats or of 0.5 g/kg bw 1,1,1,2-tetrachloroethane every other day for two weeks to male rabbits produced 10 and 20% mortality in male and female rats, respectively; 4/7 rabbits died during the experiment. Following oral administration of 0.3 g/kg bw 1,1,1,2-tetrachloroethane to male and female rats on five days a week for 10 months, mortality was 25% in males and 40% in females (10% in controls). Morphological studies in both rats and rabbits showed evidence of liver damage (Truhaut *et al.*, 1974).

Exposure of male and female rats and rabbits by inhalation to 500 ml/m³ (3430 mg/m³) 1,1,1,2-tetrachloroethane for 4 h per day on five days per week for four weeks produced no mortality or change in growth. When male and female rats were exposed to the same regimen for 12 months, mortality was similar to that in controls (about 20%); no mortality was observed in male and female rabbits exposed to the same regimen for six months. Morphological examination revealed hepatic centrilobular necrosis and microvacuolation in both rats and rabbits (Truhaut *et al.*, 1974).

In a bioassay [see section 3.1], male and female Fischer 344/N rats or B6C3F$_1$ mice were given 1,1,1,2-tetrachloroethane by gavage for 103 weeks. Because of compound-related toxicity, the mice receiving the high dose were killed after 65 weeks. Central nervous system toxicity (weakness, inactivity, loss of coordination) was observed in the high-dose groups of rats after 44 weeks and in mice after 51 weeks. Mineralization of the kidney occurred in male rats. Non-neoplastic liver lesions (inflammation, necrosis, fatty metamorphosis and hepatomegaly) were observed in high-dose male and female mice (National Toxicology Program, 1983).

Effects on reproduction and prenatal toxicity

No reproductive disturbance was observed in rats exposed orally or by inhalation to 1,1,1,2-tetrachloroethane, although neonates born to exposed females died within two days of birth (Truhaut *et al.*, 1974). [The Working Group noted that the report did not specify whether or not the newborn animals had themselves been exposed to 1,1,1,2-tetrachloroethane, and control animals were not described.]

Absorption, distribution, excretion and metabolism

The metabolism of 1,1,1,2-tetrachloroethane and related chloroethanes has been reviewed (Loew *et al.*, 1984).

In mice given a subcutaneous dose of 1.2-2.0 g/kg bw 1,1,1,2-tetrachloroethane (<0.1% impurities, including 0.03% trichloroethylene), 21-62% was eliminated unchanged in exhaled air within 72 h. The major urinary metabolite (17-49% of the dose, although some faecal material was collected in the urine) was trichloroethanol and its glucuronide conjugate; trichloroacetic acid (1-7% of the dose) was also excreted in the urine (Yllner, 1971). Trichloroethanol has also been isolated as the major metabolite, with trichloroacetic acid, from the urine of rats, rabbits and guinea-pigs (Ikeda & Ohtsuji, 1972; Truhaut & Phu Lich, 1973). After intraperitoneal administration of 1,1,1,2-tetrachloroethane to phenobarbital-treated rats, 1,1-dichloroethylene and 1,1,2-trichloroethane were detected in the blood (Thompson *et al.*, 1984).

In the presence of oxygen, NADPH and rat liver microsomes, 1,1,1,2-tetrachloroethane undergoes little dechlorination (Van Dyke & Wineman, 1971). In contrast, NADPH-dependent reductive metabolism of 1,1,1,2-tetrachloroethane by hepatic microsomal fractions from rats yields 1,1-dichloroethylene as the major metabolite (Thompson *et al.*, 1984; Town & Leibman, 1984) and 1,1,2-trichloroethane as a minor metabolite (Thompson *et al.*, 1984, 1985).

Mutagenicity and other short-term tests

1,1,1,2-Tetrachloroethane (>99% pure) was not mutagenic to *Salmonella typhimurium* TA1535, TA1537, TA98 or TA100 when tested at up to toxic doses (1000 µg/plate) in a preincubation assay without an exogenous metabolic system (S9) or in the presence of Aroclor-induced rat or hamster liver S9 (Haworth *et al.*, 1983). It was also reported to be nonmutagenic when tested in *S. typhimurium* TA1535, TA1537, TA1538, TA98 and TA100 [using an unspecified protocol] in the presence or absence of Aroclor-induced rat-liver S9 [values not given] (Simmon *et al.*, 1977).

1,1,1,2-Tetrachloroethane did not induce transformation of BALB/c-3T3 clone 1-13 cells at concentrations of up to 250 µg/ml in the absence of S9; it was not tested in the presence of S9. Cell survival at the highest dose was 75% (Tu *et al.*, 1985).

(b) Humans

No data were available to the Working Group.

3.3 Case reports and epidemiological studies of carcinogenicity to humans

No data were available to the Working Group.

4. Summary of Data Reported and Evaluation

4.1 Exposure data

1,1,1,2-Tetrachloroethane is an intermediate in one process for the manufacture of trichloroethylene and tetrachloroethylene and has been reported to occur as an impurity in these widely used products. It has been detected at low levels in ambient air and in drinking-water.

4.2 Experimental data

1,1,1,2-Tetrachloroethane was tested for carcinogenicity by oral administration by gavage in one study in mice and one study in rats. An increased incidence of hepatocellular adenomas was observed in mice of each sex and of hepatocellular carcinomas in females. The experiment in male rats gave negative results and that in female rats was inconclusive.

No evaluation of the effects of 1,1,1,2-tetrachloroethane on reproduction or prenatal toxicity in experimental animals could be made on the basis of the available data.

1,1,1,2-Tetrachloroethane is not mutagenic to bacteria. It does not induce transformation in cultured mammalian cells.

4.3 Human data

No data were available to evaluate the reproductive effects or prenatal toxicity of 1,1,1,2-tetrachloroethane to humans.

No case report or epidemiological study of the carcinogenicity of 1,1,1,2-tetrachloroethane to humans was available to the Working Group.

Overall assessment of data from short-term tests: 1,1,1,2-Tetrachloroethane[a]

	Genetic activity			Cell transformation
	DNA damage	Mutation	Chromosomal effects	
Prokaryotes		–		
Fungi/Green plants				
Insects				
Mammalian cells (*in vitro*)				–
Mammals (*in vivo*)				
Humans (*in vivo*)				
Degree of evidence in short-term tests for genetic activity: **Inadequate**				Cell transformation: Negative

[a]The groups into which the table is divided and the symbol '–' are defined on pp. 19-20 of the Preamble; the degrees of evidence are defined on pp. 20-21.

4.4 Evaluation[1]

There is *limited evidence* for the carcinogenicity of 1,1,1,2-tetrachloroethane to experimental animals.

No evaluation could be made of the carcinogenicity of 1,1,1,2-tetrachloroethane to humans.

5. References

Aldrich Chemical Co. (1984) *1984-1985 Aldrich Catalog/Handbook of Fine Chemicals*, Milwaukee, WI, p. 1004

Archer, W.L. (1979) *Chlorocarbons, — hydrocarbons (other)*. In: Grayson, M. & Eckroth, D., eds, *Kirk-Othmer Encyclopedia of Chemical Technology*, 3rd ed., Vol. 5, New York, John Wiley & Sons, p. 734

---

[1]For definition of the italicized term, see Preamble, p. 18.

Bauer, U. (1981) Human exposure to environmental chemicals — investigations on volatile organic halogenated compounds in water, air, food and human tissues. III. Communication: Results of investigations (Ger.). *Zbl. Bakteriol. Mikrobiol. Hyg., Abt. 1, Orig. B, 174*, 200-237

Canada Safety Council (1981) *Chloroethanes (Bull. CIS 81-1609)*, Ottawa, p. 16

Chieruttini, M.E. & Franklin, C.S. (1976) The toxicology of tetrachloroethanes. *Br. J. Pharmacol., 57*, 421

Grasselli, J.G. & Ritchey, W.M., eds (1975) *CRC Atlas of Spectral Data and Physical Constants for Organic Compounds*, Vol. 3, Cleveland, OH, CRC Press, p. 256

Gurka, D.F., Laska, P.R. & Titus, R. (1982) The capability of GC/FT-IR to identify toxic substances in environmental sample extracts. *J. chromatogr. Sci., 20*, 145-154

Haworth, S., Lawlor, T., Mortelmans, K., Speck, W. & Zeiger, E. (1983) *Salmonella* mutagenicity test results for 250 chemicals. *Environ. Mutagenesis, 5* (Suppl. 1)

IARC (1979) *IARC Monographs on the Evaluation of the Carcinogenic Risk of Chemicals to Humans*, Vol. 20, *Some Halogenated Hydrocarbons*, Lyon, pp. 491-514, 545-572

ICIS Chemical Information System (1985) *Information System for Hazardous Organics in Water* (ISHOW), and *Environmental Fate* (ENVIROFATE), Washington DC, Information Consultants

Ikeda, M. & Ohtsuji, H. (1972) A comparative study of the excretion of Fujiwara reaction-positive substances in urine of humans and rodents given trichloro- or tetrachloro-derivatives of ethane and ethylene. *Br. J. ind. Med., 29*, 99-104

Kaveler, A.R., ed. (1984) *OPD Chemical Buyers 1985 Directory*, New York, Schnell Publishing Co., p. 695

Kingsley, B.A., Gin, C., Coulson, D.M. & Thomas, R.F. (1983) *Gas chromatographic analysis of purgeable halocarbon and aromatic compounds in drinking water using two detectors in series*. In: Jolley, R.L., Brungs, W.A., Cotruvo, J.A., Cumming, R.B., Mattice, J.S. & Jacobs, V.A., eds, *Water Chlorination: Environmental Impact and Health Effects*, Vol. 4 (Book 1), Ann Arbor, MI, Ann Arbor Science Publishers, pp. 593-608

Loew, G.H., Rebaghati, M. & Poulsen, M. (1984) Metabolism and relative carcinogenic potency of chloroethanes: a quantum chemical structure-activity study. *Cancer Biochem. Biophys., 7*, 109-132

National Institute for Occupational Safety and Health (1978) *NIOSH Current Intelligence Bulletin 27. Chloroethanes: Review of Toxicity (DHEW (NIOSH) Publ. No. 78-181)*, Cincinnati, OH

National Toxicology Program (1983) *Carcinogenesis Studies of 1,1,1,2-Tetrachloroethane (CAS No. 630-20-6) in F344/N Rats and B6C3F$_1$ Mice (Gavage Study) (Technical Report Series NO. 237)*, Research Triangle Park, NC, US Department of Health and Human Services

Pouchert, C.J., ed. (1981) *The Aldrich Library of Infrared Spectra*, 3rd ed., Milwaukee, WI, Aldrich Chemical Co., p. 53

Pouchert, C.J., ed. (1983) *The Aldrich Library of NMR Spectra*, 2nd ed., Vol. 1, Milwaukee, WI, Aldrich Chemical Co., p. 83

Sax, N.I. (1984) *Dangerous Properties of Industrial Materials*, 6th ed., New York, Van Nostrand Reinhold, p. 2517

Simmon, V.F., Kauhanen, K. & Tardiff, R.G. (1977) *Mutagenic activity of chemicals identified in drinking water*. In: Scott, D., Bridges, B.A. & Sobels, F.H., eds, *Progress in Genetic Toxicology*, Amsterdam, Elsevier/North-Holland Biomedical Press, pp. 249-258

Singh, H.B., Salas, L.J., Smith, A.J. & Shigeishi, H. (1981) Measurements of some potentially hazardous organic chemicals in urban environments. *Atmos. Environ.*, 15, 601-612

Sittig, M. (1985) *Handbook of Toxic and Hazardous Chemicals and Carcinogens*, 2nd ed., Park Ridge, NJ, Noyes Publications, p. 840

Thompson, J.A., Ho, B. & Mastovich, S.L. (1984) Reductive metabolism of 1,1,1,2-tetrachloroethane and related chloroethanes by rat liver microsomes. *Chem.-biol. Interactions*, 51, 321-333

Thompson, J.A., Ho, B. & Mastovich, S.L. (1985) Dynamic headspace analysis of volatile metabolites from the reductive dehalogenation of trichloro- and tetrachloroethanes by hepatic microsomes. *Anal. Biochem.*, 145, 376-384

Town, C. & Leibman, K.C. (1984) The in vitro dechlorination of some polychlorinated ethanes. *Drug. Metab. Disposition*, 12, 4-8

Truhaut, R. & Phu Lich, N. (1973) Metabolic transformation of 1,1,1,2-tetrachloroethane in the rat, guinea-pig and rabbit (Fr.). *J. Eur. Toxicol.*, 4-5, 211-217

Truhaut, R., Phu Lich, N., Le Squang Thuan, N.T. & Dutertre-Catella, H. (1973) Study of some enzymatic activities by subacute poisoning with 1,1,1,2-tetrachloroethane in the rabbit (Fr.). *J. Eur. Toxicol.*, 2, 81-84

Truhaut, R., Phu Lich, N., Dutertre-Catella, H., Molas, G. & Ngoc Huyen, V. (1974) Contribution to the toxicological study of 1,1,1,2-tetrachloroethane (Fr.). *Arch. Mal. prof.*, 35, 593-608

Truhaut, R., Thévenin, M., Warnet, J.-M., Claude, J.-R. & Phu Lich, N. (1975) Preliminary biochemical study on the hepatotoxicity of 1,1,1,2-tetrachloroethane in the Wistar rat. Influence of sex (Fr.). *Eur. J. Toxicol.*, 8, 175-179

Tu, A.S., Murray, T.A., Hatch, K.M., Sivak, A. & Milman, H.A. (1985) In vitro transformation of BALB/c-3T3 cells by chlorinated ethanes and ethylenes. *Cancer Lett.*, 28, 85-92

US Environmental Protection Agency (1980) *Ambient Water Quality Criteria for Chlorinated Ethanes* (*PB81-117400*), Washington DC, p. VI

US Environmental Protection Agency (1982) *Method 8010. Halogenated volatile organics*. In: *Test Methods for Evaluating Solid Waste — Physical/Chemical Methods*, 2nd ed. (*US EPA No. SW-846*), Washington DC, Office of Solid Waste and Emergency Response

Van Dyke, R.A. & Wineman, C.G. (1971) Enzymatic dechlorination: dechlorination of chloroethanes and propanes *in vitro*. *Biochem. Pharmacol.*, *20*, 463-470

Weast, R.C., ed. (1985) *CRC Handbook of Chemistry and Physics*, 66th ed., Boca Raton, FL, CRC Press, pp. C-265, D-197

Yllner, S. (1971) Metabolism of 1,1,1,2-tetrachloroethane in the mouse. *Acta pharmacol. toxicol.*, *29*, 471-480

PENTACHLOROETHANE

1. Chemical and Physical Data

1.1 Synonyms and trade names

Chem. Abstr. Services Reg. No.: 76-01-7
Chem. Abstr. Name: Pentachloroethane
IUPAC Systematic Name: Pentachloroethane
Synonyms: Pentalin; ethane pentachloride

1.2 Structural and molecular formulae and molecular weight

$$Cl-\underset{\underset{Cl}{|}}{\overset{\overset{Cl}{|}}{C}}-\underset{\underset{Cl}{|}}{\overset{\overset{H}{|}}{C}}-Cl$$

C_2HCl_5 Mol. wt: 202.29

1.3 Chemical and physical properties of the pure substance

(a) *Description*: Dense colourless liquid with chloroform-like odour (Hawley, 1981; Windholz, 1983; Sax, 1984)

(b) *Boiling-point*: 162°C (Weast, 1985)

(c) *Melting-point*: -29°C (Weast, 1985)

(d) *Density*: d_4^{20} 1.6796 (Weast, 1985)

(e) *Spectroscopy data*[a]: Infrared (Sadtler Research Laboratories, 1980, prism [178], grating [33331]), nuclear magnetic resonance (Sadtler Research Laboratories, 1980, proton [21132], C-13 [44]) and mass spectral data (Grasselli & Ritchey, 1975) have been reported.

---

[a] In square brackets, spectrum number in compilation

(f) *Solubility*: Insoluble in water (Windholz, 1983; Sax, 1984); soluble in ethanol and diethyl ether (Weast, 1985)

(g) *Volatility*: Vapour pressure, 3.4 mm Hg at 20°C (Verschueren, 1983), 10 mm at 39.8°C (Weast, 1985); relative vapour density (air = 1), 7.2 (Verschueren, 1983)

(h) *Stability*: Emits highly toxic fumes when heated to decomposition (Sax, 1984)

(i) *Reactivity*: Dehydrohalogenation under mild alkaline conditions produces tetrachloroethylene (US Environmental Protection Agency, 1983; see IARC, 1979a, 1982a); under reducing conditions, chloroacetylenes may be formed (Sax, 1984)

(j) *Octanol/water partition coefficient (P)*: log P, 3.67 (ICIS Chemical Information System, 1985 [ISHOW])

(k) *Conversion factor*: $mg/m^3 = 8.27 \times ppm$[a]

1.4 Technical products and impurities

Pentachloroethane is not available commercially in bulk quantities, but can be obtained in 95-96% purity for research purposes (Aldrich Chemical Co., 1984; Fluka Chemical Corp., 1984). Many chlorinated ethane and ethylene derivatives, including hexachloroethane (see IARC, 1979b), have been detected as impurities in various batches of pentachloroethane (National Toxicology Program, 1983).

2. Production, Use, Occurrence and Analysis

2.1 Production and use

(a) *Production*

Pentachloroethane is no longer produced other than for research purposes, but it may still be formed as an intermediate or byproduct in some processes. It can be produced by the chlorination of acetylene, and it is found as an intermediate product in the conversion of trichloroethylene (see IARC, 1979c, 1982b) to tetrachloroethylene (National Institute for Occupational Safety and Health, 1981).

Other production methods have been patented in France, the Federal Republic of Germany and Japan. The French patent provides for the oxychlorination of ethylene to

---

[a]Calculated from: mg/m^3 = (molecular weight/24.45) × ppm, assuming standard temperature (25°C) and pressure (760 mm Hg)

produce various chlorinated ethanes. The German patent describes a method to produce pentachloroethane through the ultraviolet-activated chlorination of 1,1,2,2-tetrachloroethane (see IARC, 1979d). Pentachloroethane is an intermediate in a Japanese process in which ethylene (see IARC, 1979e) is converted into tetrachloroethylene (National Institute for Occupational Safety and Health, 1981).

According to the TSCA (Toxic Substances Control Act) Inventory, more than 5-23 million kg pentachloroethane were produced by two US manufacturers in 1977 as an intermediate in the synthesis of tetrachloroethylene (US Environmental Protection Agency, 1983). The US International Trade Commission has not reported production statistics for pentachloroethane. The sole US plant producing tetrachloroethylene from acetylene feedstock (*via* pentachloroethane as an intermediate) ceased production in 1978 (National Institute for Occupational Safety and Health, 1981). No current European or Japanese producer of commercial quantities of pentachloroethane has been identified.

(*b*) *Use*

There appears to be no current commercial use for pentachloroethane. It may occur as an intermediate in the production of chlorinated ethylenes, and it remains as an unisolated component of production still bottoms (Gurka *et al.*, 1982), which are generally returned to feedstock. Pentachloroethane can be used as a solvent for cellulose plastics, natural gums and resins (JRB Associates, Inc., 1981) and for dry-cleaning (Browning, 1965), as a drying agent for timber (Sittig, 1985), and for coal purification (Hawley, 1981); however, there is no evidence that it is currently used in these ways.

Pentachloroethane was used in the past as an intermediate in the production of dichloroacetic acid (Freiter, 1978), as an oil and grease solvent in metal cleaning (Hawley, 1981), as a solvent for cellulose acetate, certain cellulose ethers, resins and gums (Sittig, 1985), and as a soil sterilizing agent (Browning, 1965).

(*c*) *Regulatory status and guidelines*

Occupational exposure limits for pentachloroethane in eight countries are presented in Table 1.

2.2 Occurrence

(*a*) *Natural occurrence*

Pentachloroethane is not known to occur as a natural product.

(*b*) *Occupational exposure*

No data on exposure levels were available to the Working Group.

Table 1. Occupational exposure limits for pentachloroethane[a]

Country	Year	Concentration (mg/m³)	Interpretation[b]
Finland	1981	40	TWA
		80	STEL
German Democratic Republic	1979	20	TWA
		40	STEL
Germany, Federal Republic of	1985	40	TWA
The Netherlands	1978	40	TWA
Norway	1981	40	TWA
Romania	1975	200	TWA
		250	Ceiling
Switzerland	1978	40	TWA
Yugoslavia	1971	40	Ceiling

[a]From International Labour Office (1980); Direktoratet for Arbeidstilsynet (1981); Työsuojeluhallitus (1981); Deutsche Forschungsgemeinschaft (1985)

[b]TWA, time-weighted average; STEL, short-term exposure limit

(c) *Air*

Atmospheric concentrations of pentachloroethane at Bochum University, Federal Republic of Germany, and its environs ranged from non-detectable to 0.1 µg/m³. Within the city of Bochum, the concentrations ranged from non-detectable to 0.6 µg/m³ (Bauer, 1981).

Pellizzari *et al.* (1979) identified pentachloroethane in ambient air in the USA, in Houston, TX (4 µg/m³) and in Baton Rouge, LA (0.01 µg/m³).

(d) *Water*

The concentration of pentachloroethane in water samples taken in 1976 from the Rhine river near Lobith, Federal Republic of Germany, ranged from non-detectable to 0.1 µg/l. Concentrations in drinking-water samples from 100 cities in the Federal Republic of Germany in 1977 ranged from non-detectable to 0.2 µg/l, with an average of <0.1 µg/l (Bauer, 1981).

Pentachloroethane has also been detected in trace quantities (generally less than 1 µg/l) in water supplies in the UK and the USA at an average level of 3.3 mg/l (US Environmental Protection Agency, 1983).

(e) Other

Pentachloroethane was detected in some cosmetic products purchased from shops in Bochum, Federal Republic of Germany, with maximum concentrations of 1.0 µg/l in mouthwash, 0.2 µg/l in after-shave lotion, and 0.3 µg/l in one of seven cough mixtures (Bauer, 1981).

2.3 Analysis

Selected methods for the analysis of pentachloroethane in air and water are identified in Table 2.

Table 2. Methods for the analysis of pentachloroethane

Sample matrix	Sample preparation	Assay procedure[a]	Limit of detection	Reference
Air	Adsorb (Porapak R); desorb (hexane); inject aliquot	GC/EC	1.3 µg/m^3	Eller (1985)
Water	Extract (*n*-pentane); dry (sodium sulphate); inject aliquot	GC/EC	0.02 µg/l	Deetman *et al.* (1976)

[a]GC/EC, gas chromatography/electron capture detection

A method has been reported for the analysis of pentachloroethane in chemical plant still-bottom samples (Gurka *et al.*, 1982), using gas chromatography/Fourier-transform infrared detection.

3. Biological Data Relevant to the Evaluation of Carcinogenic Risk to Humans

3.1 Carcinogenicity studies in animals

Oral administration

Mouse: Groups of 50 male and 50 female B6C3F$_1$ mice, eight weeks of age, were administered 0, 250 or 500 mg/kg bw technical-grade pentachloroethane (95.5% pure, with 4.2% hexachloroethane and traces of other chlorinated ethane and ethylene derivatives) in corn oil by gavage on five days a week for 103 weeks, when the survivors were killed. All high-dose males and females had died or were killed by weeks 41 and 74, respectively; 25 control males were killed at week 44 to be compared with the high-dose males. Of the low-dose groups, 22/50 males and 9/50 females survived to the end of the study. The incidence of hepatocellular carcinomas was significantly increased in low-dose males (26/44) as compared to controls (4/48) ($p < 0.001$). Early mortality of high-dose males precluded an evaluation of the life-time incidence of hepatocellular carcinomas, but there was a significantly increased incidence over that observed among the 25 controls at week 44 (7/45 versus 0/25, $p < 0.05$). A significantly increased incidence of hepatocellular carcinomas was observed in treated females as compared to controls (control, 1/46; low-dose, 28/42, $p < 0.001$; high-dose, 13/45, $p < 0.001$). There was also a significant dose-related increase in the incidence of hepatocellular adenomas in treated females (2/46, 8/42 and 19/45 in the three groups, respectively; $p < 0.001$, trend test) (Mennear et al., 1982; National Toxicology Program, 1983). [The Working Group noted the presence in the test compound of 4.2% hexachloroethane, which has been shown to produce liver tumours in mice (see IARC, 1979b); however, the possible contribution of this impurity to the effects observed could not be evaluated.]

Rat: Groups of 50 male and 50 female Fischer 344/N rats, seven weeks of age, were administered 0, 75 or 150 mg/kg bw technical-grade pentachloroethane (95.5% pure, with 4.2% hexachloroethane and traces of other chlorinated ethane and ethylene derivatives) in corn oil by gavage on five days a week for 103 weeks, when survivors were killed. Survival of males by the end of the study was: control, 82%; low-dose, 66%; high-dose, 52%; and that of females was: 76%, 72%, and 50%. The incidence of diffuse chronic inflammation of the kidney, distinguishable from old-age nephropathy, was significant and dose-related in treated male rats (4/50, 14/49 and 33/50; $p < 0.001$). There was also a dose-related trend in the incidence of tubular-cell adenomas of the kidney in males (0/50, 1/49 and 4/50; $p < 0.05$). In addition, one control and one low-dose male each had a tubular-cell adenocarcinoma and another low-dose male had a carcinoma of the kidney [type unspecified]. The trend is not statistically significant when the numbers of benign and malignant tubular-cell tumours are combined (1/50, 2/49 and 4/50). A tubular-cell tumour also occurred in 1/293 (0.3%) historical, male, corn-oil vehicle controls from this laboratory (Mennear et al., 1982; National Toxicology Program, 1983). [The Working Group noted the presence in the test compound of 4.2% hexachloroethane, which is suspected to produce renal tumours in rats (see IARC, 1979b); however the possible contribution of this impurity to the effects observed cannot be evaluated.]

3.2 Other relevant biological data

(a) Experimental systems

Toxic effects

The lowest observed lethal doses of pentachloroethane were reported to be 0.5 ml (840 mg)/kg bw after oral administration to dogs (Wright & Schaffer, 1932), 100 mg/kg bw and 1750 mg/kg bw after intravenous and oral administration to dogs, respectively, and 700 mg/kg bw after subcutaneous administration to rabbits (Barsoum & Saad, 1934). In inhalation studies with mice in a static exposure system, the lowest observed lethal concentration was 35 000 mg/m^3 for a 2-h exposure (Lazarew, 1929).

Oral administration of 2.6 mmol (525 mg)/kg bw pentachloroethane (practical grade; 95% pure) to male rats reduced hepatic cytochrome P450 content and microsomal epoxide hydrolase activities (Vainio et al., 1976).

Exposure of rabbits for 3 h per day, six times per week, for eight to ten months to 100 mg/m^3 pentachloroethane resulted in decreased total antibody titres, which were more pronounced at the highest dose (Shmuter, 1972, 1977).

In a bioassay [see section 3.1], in which 75 and 150 mg/kg bw pentachloroethane were given by gavage for 103 weeks, chronic, diffuse inflammation of the kidney and mineralization of the renal papilla were observed in both low- and high-dose male Fischer 344/N rats (Meenear et al., 1982; National Toxicology Program, 1983).

Effects on reproduction and prenatal toxicity

No data were available to the Working Group.

Absorption, distribution, excretion and metabolism

The metabolism of pentachloroethane and related chloroethanes has been reviewed (Loew et al., 1984).

Following subcutaneous administration of pentachloroethane (1.1-1.8 g/kg bw; 0.5% impurities) to female mice, 12-51% of the dose was expired as the parent compound, 2-16% as trichloroethylene and 3-9% as tetrachloroethylene; trichloroethanol (16-32% of the dose) and trichloroacetic acid (9-18% of the dose) were the major urinary metabolites (Urine and faeces were analysed together.) (Yllner, 1971).

In the presence of oxygen, NADPH and rat liver microsomes, 1.7% dechlorination of pentachloroethane was observed (Van Dyke & Wineman, 1971). In contrast, in the absence of oxygen, pentachloroethane was metabolized by NADPH-fortified hepatic microsomal fractions to trichloroethylene (96%) and 1,1,2,2-tetrachloroethane (4%) (Nastainczyk et al., 1982; Town & Leibman, 1984). The in-vitro reductive dechlorination of pentachloroethane was catalysed by a purified (rabbit liver) reconstituted cytochrome P450 system (Salmon et al., 1985).

Mutagenicity and other short-term tests

Pentachloroethane (96% pure) was not mutagenic to *Salmonella typhimurium* TA1535, TA1537, TA98 or TA100 when tested at up to toxic concentrations (333 µg/plate) in a preincubation assay either in the presence or absence of an exogenous metabolic system (S9) from the liver of Aroclor-induced rats or hamsters (Haworth et al., 1983). In an abstract, Douglas et al. (1985) reported that pentachloroethane was mutagenic to *S. typhimurium* in a plate assay, but provided no indication of the strains used, the need for an exogenous metabolic system or the magnitude of the response.

In an abstract, Douglas *et al.* (1985) reported that pentachloroethane failed to induce SCEs or micronuclei in Chinese hamster ovary cells; however, experimental details and data were not presented.

(*b*) *Humans*

No data were available to the Working Group.

3.3 Case reports and epidemiological studies of carcinogenicity to humans

No data were available to the Working Group.

4. Summary of Data Reported and Evaluation

4.1 Exposure data

Pentachloroethane was produced commercially as a chemical intermediate, and occupational exposure may have occurred. Trace levels have been reported in ambient air and water.

4.2 Experimental data

Technical-grade pentachloroethane (containing 4.2% hexachloroethane) was tested for carcinogenicity by oral administration by gavage in one experiment in mice and one experiment in rats. Hepatocellular carcinomas were induced in mice of each sex and hepatocellular adenomas in female mice; a marginally increased incidence of kidney tubular-cell adenomas was observed in male rats but not in female rats.

No data were available to the Working Group on the carcinogenicity of pure pentachloroethane to experimental animals.

No data were available to evaluate the reproductive effects or prenatal toxicity of pentachloroethane to experimental animals.

Pentachloroethane is not mutagenic to *Salmonella typhimurium*.

4.3 Human data

No data were available to evaluate the reproductive effects or prenatal toxicity of pentachloroethane to humans.

No case report or epidemiological study of the carcinogenicity of pentachloroethane to humans was available to the Working Group.

Overall assessment of data from short-term tests: Pentachloroethane[a]

	Genetic activity			Cell transformation
	DNA damage	Mutation	Chromosomal effects	
Prokaryotes		–		
Fungi/Green plants				
Insects				
Mammalian cells (*in vitro*)				
Mammals (*in vivo*)				
Humans (*in vivo*)				
Degree of evidence in short-term tests for genetic activity: **Inadequate**				Cell transformation: No data

[a]The groups into which the table is divided and the symbol '–' are defined on pp. 19-20 of the Preamble; the degrees of evidence are defined on pp. 20-21.

4.4 Evaluation[1]

There is *limited evidence* for the carcinogenicity of technical-grade pentachloroethane (containing hexachloroethane) to experimental animals.

No evaluation could be made of the carcinogenicity of pentachloroethane to humans.

5. References

Aldrich Chemical Co. (1984) *1984-1985 Aldrich Catalog/Handbook of Fine Chemicals*, Milwaukee, WI, p. 851

Barsoum, G.S. & Saad, K. (1934) Relative toxicity of certain chlorine derivatives of the aliphatic series. *Q. J. Pharm. Pharmacol.*, 7, 205-214

---

[1]For definition of the italicized term, see Preamble, p. 18.

Bauer, U. (1981) Human exposure to environmental chemicals — investigations on volatile organic halogenated compounds in water, air, food, and human tissues. III. Communication: results of investigations. *Zbl. Bakteriol. Mikrobiol. Hyg., Abt. 1, Orig. B, 174,* 200-237

Browning, E. (1965) *Toxicity and Metabolism of Industrial Solvents,* Amsterdam, Elsevier, p. 261

Deetman, A.A., Demeulemeester, P., Garcia, M., Hauck, G., Hollies, J.I., Krockenberger, D., Palin, D.E., Prigge, H., Rohrschneider, L. & Schmidhammer, L. (1976) Standardization of methods for the determination of traces of some volatile chlorinated aliphatic hydrocarbons in air and water by gas chromatography. *Anal. chim. Acta, 82,* 1-17

Deutsche Forschungsgemeinschaft (German Research Community) (1985) *Maximal Concentrations in the Workplace and Biological Tolerance Values for Substances in the Work Environment 1985* (Ger.), Vol. 21, Weinheim, Verlagsgesellschaft mbH, p. 47

Direktoratet for Arbeidstilsynet (Directorate of Labour Inspection) (1981) *Administrative Norms for Pollution in Workplaces, No. 361* (Norw.), Oslo, p. 18

Douglas, G.R., Nestmann, E.R., Lee, E., Marshall, R. & Heddle, J.A. (1985) How well do in vitro tests predict in vivo genotoxicity? (Abstract). *Environ. Mutagenesis, 7* (Suppl. 3), 31

Eller, P.M. (1985) *NIOSH Manual of Analytical Methods,* 3rd ed., 1st Suppl. (*DHHS (NIOSH) Pub. No. 84-100*), Washington DC, US Government Printing Office, pp. 2517-1 — 2517-4

Fluka Chemical Corp. (1984) *Biochemicals Reagents (Fluka Catalog 14, 1984/85),* Hauppauge, NY

Freiter, E.R. (1978) *Acetic acid derivatives (halogenated).* In: Grayson, M. & Eckroth, D., eds, *Kirk-Othmer Encyclopedia of Chemical Technology,* 3rd ed., Vol. 1, New York, John Wiley & Sons, pp. 171-178

Grasselli, J.G. & Ritchey, W.M., eds (1975) *CRC Atlas of Spectral Data and Physical Constants for Organic Compounds,* Vol. 3, Cleveland, OH, CRC Press, p. 255

Gurka, D.F., Laska, P.R. & Titus, R. (1982) The capability of GC/FT-IR to identify toxic substances in environmental sample extracts. *J. chromatogr. Sci., 20,* 145-154

Hawley, G.G., ed. (1981) *The Condensed Chemical Dictionary,* 10th ed., New York, Van Nostrand Reinhold, p. 782

Haworth, S., Lawlor, T., Mortelmans, K., Speck, W. & Zeiger, E. (1983) Salmonella mutagenicity test results for 250 chemicals. *Environ. Mutagenesis, 5* (Suppl. 1), 3-142

IARC (1979a) *IARC Monographs on the Evaluation of the Carcinogenic Risk of Chemicals to Humans,* Vol. 20, *Some Halogenated Hydrocarbons,* Lyon, pp. 491-514

IARC (1979b) *IARC Monographs on the Evaluation of the Carcinogenic Risk of Chemicals to Humans,* Vol. 20, *Some Halogenated Hydrocarbons,* Lyon, pp. 467-476

IARC (1979c) *IARC Monographs on the Evaluation of the Carcinogenic Risk of Chemicals to Humans*, Vol. 20, *Some Halogenated Hydrocarbons*, Lyon, pp. 545-572

IARC (1979d) *IARC Monographs on the Evaluation of the Carcinogenic Risk of Chemicals to Humans*, Vol. 20, *Some Halogenated Hydrocarbons*, Lyon, pp. 477-489

IARC (1979e) *IARC Monographs on the Evaluation of the Carcinogenic Risk of Chemicals to Humans*, Vol. 19, *Some Monomers, Plastics and Synthetic Elastomers, and Acrolein*, Lyon, pp. 157-186

IARC (1982a) *IARC Monographs on the Evaluation of the Carcinogenic Risk of Chemicals to Humans*, Suppl. 4, *Chemicals, Industrial Processes and Industries Associated with Cancer in Humans, IARC Monographs, Volumes 1 to 29*, Lyon, pp. 243-245

IARC (1982b) *IARC Monographs on the Evaluation of the Carcinogenic Risk of Chemicals to Humans*, Suppl. 4, *Chemicals, Industrial Processes and Industries Associated with Cancer in Humans, IARC Monographs, Volumes 1 to 29*, Lyon, pp. 247-249

ICIS Chemical Information System (1985) *Information System for Hazardous Organics in Water* (ISHOW), and *Environmental Fate* (ENVIROFATE), Washington DC, Information Consultants

International Labour Office (1980) *Occupational Exposure Limits for Airborne Toxic Substances: A Tabular Compilation of Values from Selected Countries*, 2nd (rev.) ed. (*Occupational Safety and Health Series No. 37*), Geneva, pp. 166-167

JRB Associates, Inc. (1981) *Pentachloroethane: Preliminary Report of Plants and Processes* (*US NTIS PB83-106344*), Cincinnati, OH, National Institute for Occupational Safety and Health, p. 6

Lazarew, N.W. (1929) Narcotic effectiveness of vapours of chlorinated derivatives of methane, ethane and ethylene (Ger.) *Naunyn-Schmiedeberg's Arch. exp. Pathol. Pharmacol.*, 141, 19-24

Loew, G.H., Rebagliati, M. & Poulsen, M. (1984) Metabolism and relative carcinogenic potency of chloroethanes: a quantum chemical structure-activity study. *Cancer Biochem. Biophys.*, 7, 109-132

Mennear, J.H., Haseman, J.K., Sullivan, D.J., Bernal, E. & Hildebrandt, P.K. (1982) Studies on the carcinogenicity of pentachloroethane in rats and mice. *Fundam. appl. Toxicol.*, 2, 82-87

Nastainczyk, W., Ahr, H.J. & Ullrich, V. (1982) The reductive metabolism of halogenated alkanes by liver microsomal cytochrome P-450. *Biochem. Pharmacol.*, 31, 391-396

National Institute for Occupational Safety and Health (1981) *Pentachloroethane: Preliminary Report of Plants and Processes* (*Report No. 210-79-0090; PB 83 106344*), Cincinnati, OH

National Toxicology Program (1983) *Carcinogenesis Bioassay of Pentachloroethane (CAS No. 76-01-7) in F344/N Rats and B6C3F$_1$ Mice (Gavage Study)* (*Technical Report Series No. 232*), Research Triangle Park, NC, US Department of Health and Human Services

Pellizzari, E.D., Erickson, M.D. & Zweidinger, R.A. (1979) *Formulation of a Preliminary Assessment of Halogenated Organic Compounds in Man and Environmental Media* (*EPA-560/13-79-006*), Washington DC, US Environmental Protection Agency, Office of Toxic Substances

Sadtler Research Laboratories (1980) *The Sadtler Standard Spectra, Cumulative Index*, Philadelphia, PA

Salmon, A.G., Nash, J.A., Walklin, C.M. & Freedman, R.B. (1985) Dechlorination of halocarbons by microsomes and vesicular reconstituted cytochrome P-450 systems under reductive conditions. *Br. J. ind. Med.*, *42*, 305-311

Sax, N.I. (1984) *Dangerous Properties of Industrial Materials*, 6th ed., New York, Van Nostrand Reinhold, p. 2126

Shmuter, L.M. (1972) Effect of chronic action of small concentrations of chlorinated hydrocarbons on the production of various classes of immunoglobulins (Russ.). *Gig. Sanit.*, *37*, 36-40

Shmuter, L.M. (1977) Effect of chronic exposure to low concentrations of the ethane series chlorinated hydrocarbons on the specific and nonspecific immunological reactivity in animal experiments (Russ.). *Gig. Tr. prof. Zabol.*, *8*, 38-43

Sittig, M. (1985) *Handbook of Toxic and Hazardous Chemicals and Carcinogens*, 2nd ed., Park Ridge, NJ, Noyes Publications, pp. 693-694

Town, C. & Leibman, K.C. (1984) The in vitro dechlorination of some polychlorinated ethanes. *Drug. Metab. Disposition*, *12*, 4-8

Työsuojeluhallitus (National Finnish Board of Occupational Safety and Health) (1981) *Airborne Contaminants in the Workplaces* (*Safety Bull. 3*) (Finn.), Tampere, p. 21

US Environmental Protection Agency (1983) *Chemical Hazard Information Profile, Draft Report: Pentachloroethane*, Washington DC, US Government Printing Office

Vainio, H., Parkki, M.G. & Marniemi, J. (1976) Effects of aliphatic chlorohydrocarbons on drug-metabolizing enzymes in rat liver *in vivo*. *Xenobiotica*, *6*, 599-604

Van Dyke, R.A. & Wineman, C.G. (1971) Enzymatic dechlorination: dechlorination of chloroethanes and propanes *in vitro*. *Biochem. Pharmacol.*, *20*, 463-470

Verschueren, K. (1983) *Handbook of Environmental Data on Organic Chemicals*, 2nd ed., New York, Van Nostrand Reinhold, p. 950

Weast, R.C., ed. (1985) *CRC Handbook of Chemistry and Physics*, 66th ed., Boca Raton, FL, CRC Press, pp. C-265, D-197

Windholz, M., ed. (1983) *The Merck Index*, 10th ed., Rahway, NJ, Merck & Co., p. 1021

Wright, W.H. & Schaffer, J.M. (1932) Critical anthelmintic tests of chlorinated alkyl hydrocarbons and a correlation between the anthelmintic efficacy, chemical structure and physical properties. *Am. J. Hyg.*, *16*, 325-426

Yllner, S. (1971) Metabolism of pentachloroethane in the mouse. *Acta pharmacol. toxicol.*, *29*, 481-489

1,3-DICHLOROPROPENE

1. Chemical and Physical Data

1.1 Synonyms and trade names

Chem. Abstr. Services Reg. No.: 542-75-6; *cis* isomer: 10061-01-5; *trans* isomer: 10061-02-6
Chem. Abstr. Name: 1,3-Dichloro-1-propene
IUPAC Systematic Name: 1,3-Dichloropropene
Synonyms: 3-Chloroallyl chloride; γ-chloroallyl chloride; 3-chloropropenyl chloride; DCP; 1,3-dichloro-2-propene; 1,3-dichloropropylene; α,γ-dichloropropylene
Trade Names: 1,3-D; Telone; Telone II

1.2 Structural and molecular formulae and molecular weight

$$\underset{\text{(cis)}}{\overset{H}{\underset{Cl}{>}}C=C\overset{H}{\underset{CH_2-Cl}{<}}} \qquad \underset{\text{(trans)}}{\overset{Cl}{\underset{H}{>}}C=C\overset{H}{\underset{CH_2-Cl}{<}}}$$

$C_3H_4Cl_2$ Mol. wt: 110.97

1.3 Chemical and physical properties of the pure substance (isomeric composition not specified, except when noted), unless otherwise specified

(*a*) *Description*: Liquid with chloroform-like odour (Windholz, 1983); clear, light straw-coloured liquid with sharp, sweet, penetrating and irritating odour (Verschueren, 1983)

(*b*) *Boiling-point*: 104.3°C (*cis* isomer) (Weast, 1985); 112°C (*trans* isomer) (Verschueren, 1983; Weast, 1985)

(c) *Density*: d_4^{20} 1.217 (*cis* isomer); 1.224 (*trans* isomer) (Weast, 1985)

(d) *Spectroscopy data*[a]: Infrared (Sadtler Research Laboratories, 1980; prism [23675; 23676 for *cis* isomer], grating [28450]), nuclear magnetic resonance (Sadtler Research Laboratories, 1980; proton [14414], C-13 [3063]) and mass spectral data (*cis* and *trans* isomers) (Grasselli & Ritchey, 1975) have been reported.

(e) *Solubility*: The technical product (92% pure) is slightly soluble in water (1 g/l) (Worthing, 1983); the pure product is soluble in acetone, toluene and octane (Hawley, 1981), diethyl ether, benzene and chloroform (Weast, 1985)

(f) *Volatility*: Vapour pressure, 43 (*cis* isomer); 34 (*trans* isomer) mm Hg at 25°C; relative vapour density (air = 1), 3.83 (Verschueren, 1983)

(g) *Stability*: Flash-point, 35°C (open-cup), inflammable (Hawley, 1981); emits toxic fumes when heated to decomposition (National Fire Protection Association, 1984; Sax, 1984)

(h) *Conversion factor*: mg/m^3 = 4.54 × ppm[b]

1.4 Technical products and impurities

Commercial 1,3-dichloropropene is produced as a mixture of *cis* and *trans* isomers, normally in an approximately 1:1 ratio. Impurities in 1,3-dichloropropene resulting from the production process include isomers of dichloropropane (see monograph, p. 131), other dichloropropene isomers, and epichlorohydrin (see IARC, 1976, 1982), used as a stabilizer (Worthing, 1983; Osterloh *et al.*, 1984; National Toxicology Program, 1985). 1,3-Dichloropropene products containing significant amounts of 1,2-dichloropropane and other dichloropropene isomers have been marketed in the USA and in Europe (Sittig, 1980; Tuazon *et al.*, 1984). One such product (D-D) contains 27.7% *cis*- and 27.4% *trans*-1,3-dichloropropene, in addition to 30% 1,2-dichloropropane (Hutson *et al.*, 1971) or 52% 1,3-dichloropropene, 29% 1,2-dichloropropane and minor components (Yang, 1986). Subsequent developments in distillation technology resulted in commercial products of higher purity (e.g., Telone II, 92% 1,3-dichloropropene) (Tuazon *et al.*, 1984; Miller, 1985; Yang, 1986). 1,3-Dichloropropene has also been marketed in mixtures with chloropicrin (trichloronitromethane) or methyl isothiocyanate (Tuazon *et al.*, 1984; Anon., 1985; Miller, 1985; Yang, 1986).

---

[a]In square brackets, spectrum number in compilation
[b]Calculated from: mg/m^3 = (molecular weight/24.45) × ppm, assuming standard temperature (25°C) and pressure (760 mm Hg)

2. Production, Use, Occurrence and Analysis

2.1 Production and use

(a) Production

1,3-Dichloropropene was first synthesized in 1872 by the reaction of 1,2,3-trichloropropane with sodium hydroxide (Windholz, 1983). The industrial preparation of 1,3-dichloropropene, described in 1936, involved dehydration of 1,3-dichloro-2-propanol (Hurd & Webb, 1936; Windholz, 1983).

1,3-Dichloropropene is currently synthesized by vapour-phase chlorination of propylene at high temperatures (>300°C). The initial product is a mixture of allyl chlorides (see IARC, 1985) and 1,3-dichloropropene, which are separated by fractional distillation (DeBenedictis, 1979).

Currently, only one US manufacturer produces 1,3-dichloropropene, but production statistics are not published by the US International Trade Commission. The US Environmental Protection Agency estimated US consumption at 16.3 million kg annually (Holtorf, 1984). Statistics for use in California indicate a dramatic increase from 1978 (2.5 million kg) to 1981 (7.5 million kg), with a drop to 5.6 million kg in 1982 (Tuazon *et al.*, 1984).

In Europe, the two primary producers of 1,3-dichloropropene are located in the Federal Republic of Germany. One of these firms, as well as one manufacturer in France, also produce the dichloropropene-dichloropropane mixture. In 1972, more than 2 million kg 1,3-dichloropropene were produced in Italy (De Lorenzo *et al.*, 1977).

One Japanese manufacturer of 1,3-dichloropropene has been identified (The Chemical Daily Co., 1984).

(b) Use

1,3-Dichloropropene was introduced in 1956 for use as a soil fumigant in vegetable and tobacco production (Hayes, 1982). All commercially available products, including the dichloropropene-dichloropropane mixture, have been used as soil fumigants, typically for the control of soil-borne nematodes (Anon., 1985). In the USA, 125 1,3-dichloropropene products have been registered for use on soil in the cultivation of a variety of crops, including cotton, grapes, sugar beets, tobacco, potatoes and other vegetables. Approximately 36% of US cropland devoted to potato production is treated with 1,3-dichloropropene (Holtorf, 1984). Use of 1,3-dichloropropene fumigants in the USA is expected to increase, since the US Environmental Protection Agency placed restrictions on use of two other fumigants, dibromochloropropane and ethylene dibromide (Holtorf, 1984; US Occupational Safety and Health Administration, 1985).

Registrations of 1,3-dichloropropene for pesticide use have also been reported in the Federal Republic of Germany, Sweden and the UK (International Register of Potentially Toxic Chemicals, 1984).

(c) *Regulatory status and guidelines*

Occupational exposure limits for 1,3-dichloropropene in five countries are presented in Table 1.

Table 1. Occupational exposure limits for 1,3-dichloropropene[a]

Country	Year	Concentration (mg/m^3)	Interpretation[b]
Bulgaria	1971	5	Ceiling
Norway	1981	25	TWA
UK	1985	5	TWA
		50	STEL
USA	1985		
ACGIH		5	TWA
		50	STEL
USSR	1977	5	Ceiling

[a]From International Labour Office (1980); Direktoratet for Arbeidstilsynet (1981); American Conference of Governmental Industrial Hygienists (ACGIH) (1985); Health and Safety Executive (1985)

[b]TWA, time-weighted average; STEL, short-term exposure limit

No tolerance limit or allowable daily intake (ADI) for residues of 1,3-dichloropropene in food commodities has been established by FAO/WHO (Vettorazzi, 1984).

2.2 Occurrence

(a) *Natural occurrence*

1,3-Dichloropropene is not known to occur as a natural product.

(b) *Occupational exposure*

During soil fumigation, concentrations in workers' breathing zone, measured in personal samples, ranged from 0.6 to 1.9 mg/m^3 (time-weighted average). In workers exposed to 1,3-dichloropropene, a good correlation was found between urinary *N*-acetyl-*S*-(3-chloroprop-2-enyl)cysteine concentration and the product of ambient air concentration (personal monitoring) and duration of exposure (Osterloh *et al.*, 1984).

(c) *Air*

Forty residences in suburban Knoxville, TN, USA, were monitored for volatile organic compounds. The mean concentration of 1,3-dichloropropene in indoor air in winter was

52.4 µg/m³; the mean concentration in the air surrounding the houses was 5.0 µg/m³ (Gupta et al., 1984).

(d) *Water*

1,3-Dichloropropene was detected in tap-water in New Orleans, LA, USA (Dowty et al., 1975).

(e) *Soil*

1,3-Dichloropropene does not persist in soil and is hydrolysed to the corresponding 3-chloroallyl alcohols (Worthing, 1983).

2.3 Analysis

Selected methods for the analysis of 1,3-dichloropropene in air and water are identified in Table 2. The methods of the US Environmental Protection Agency (1984a,b,c) for water have also been applied to liquid and solid wastes (US Environmental Protection Agency, 1982a,b). Volatile components in solid-waste samples are first extracted with polyethylene glycol or methanol prior to purge/trap concentration and analysis (limit of detection, 5 µg/g for the *trans* isomer).

Table 2. Methods for the analysis of 1,3-dichloropropene

Sample matrix	Sample preparation	Assay procedure[a]	Limit of detection	Reference
Air	Adsorb (Tenax-GC); desorb (isooctane); inject aliquot	GC/EC	2.3 mg/m³	Leiber & Berk (1984)
Water	Purge (inert gas); trap (OV-1 on Chromosorb-W/Tenax/silica gel); desorb as vapour (heat to 180°C, backflush with inert gas) onto GC column	GC/ECD	0.34 µg/l (*cis* isomer); 0.20 µg/l (*trans* isomer)	US Environmental Protection Agency (1984a)
		GC/MS	5.0 µg/l (*cis* isomer); ND (*trans* isomer)	US Environmental Protection Agency (1984b)
	Add internal standard (isotope-labelled *cis*- or *trans*-1,3-dichloropropene); purge, trap and desorb as above	GC/MS	ND (*cis* isomer); 10 µg/l (*trans* isomer)	US Environmental Protection Agency (1984c)
	Purge (40°C, nitrogen); trap directly on Tenax GC column (40°C); temperature-programme for analysis	GC/FID or GC/ECD	0.5 µg/l	Otson & Williams (1982)

[a]Abbreviations: GC/EC, gas chromatography/electron capture detection; GC/ECD, gas chromatography/electrolytic conductivity detection; GC/MS, gas chromatography/mass spectrometry; ND, not determined; GC/FID, gas chromatography/flame ionization detection

US EPA Method 624 (US Environmental Protection Agency, 1984b) has also been adapted for the analysis of 1,3-dichloropropene in fish, with an estimated limit of 10 μg/kg for the *trans* isomer (Easley *et al.*, 1981).

3. Biological Data Relevant to the Evaluation of Carcinogenic Risk to Humans

3.1 Carcinogenicity studies in animals

(a) *Oral administration*

Mouse: Groups of 50 male and 50 female B6C3F$_1$ mice, six to ten weeks of age, were administered 0, 50 or 100 mg/kg bw Telone II [1,3-dichloropropene, 88-90% (*cis* isomer, 41.6%; *trans* isomer, 45.9%); 1,2-dichloropropane, 2.5%; trichloropropene isomer, 1.5%; a second trichloropropene isomer; epichlorohydrin, 1.0%; and nine other impurities, 7.5%] in corn oil by gavage three times a week for 104 weeks (dose in volume, 5 ml/kg bw). Survival at 105-107 weeks in males was: vehicle-control, 8/50; low-dose, 28/50; high-dose, 31/50; and that in females was: 46/50, 45/50 and 36/50. A dose-related increase in the incidence of epithelial hyperplasia of the urinary bladder was observed: males — 0/50, 9/50 and 18/50 [$p < 0.001$]; females — 2/50, 15/50 and 19/48 [$p < 0.001$]. The incidence of transitional-cell carcinoma of the urinary bladder was: males — 0/50, 0/50 and 2/50; females — 0/50, 8/50 and 21/48 ($p < 0.001$). The incidence of alveolar/bronchial adenomas was: males — 1/50, 11/50 ($p = 0.026$, incidental tumour test) and 9/50; females — 0/50, 3/50 and 8/50 ($p < 0.001$). An increase in the incidence of epithelial hyperplasia of the forestomach was observed in high-dose males and females compared to controls (males: 0/50, 0/50 and 4/50 [$p = 0.011$]; females: 1/50, 1/50 and 21/50 [$p < 0.001$]). Squamous-cell papillomas of the forestomach were observed in males: 0/50, 2/50 and 3/50. In females, combined squamous-cell papillomas and carcinomas were observed in 0/50, 1/50 and 4/50 animals ($p = 0.021$) (National Toxicology Program, 1985; Yang *et al.*, 1986). [The Working Group noted the poor survival of the male vehicle-control animals and the presence in the test compound of 1% epichlorohydrin, which is carcinogenic to experimental animals (see IARC, 1982).]

Rat: Groups of 52 male and 52 female Fischer 344/N rats, six weeks of age, were administered 0, 25 or 50 mg/kg bw Telone II [1,3-dichloropropene, 88-90% (*cis* isomer, 41.6%; *trans* isomer, 45.9%); 1,2-dichloropropane, 2.5%; trichloropropene isomer, 1.5%; a second trichloropropene isomer; epichlorohydrin, 1.0%; and nine other impurities, 7.5%] in corn oil by gavage three times a week for 104 weeks (dose in volume, 5 ml/kg bw). Survival at 106-108 weeks in males was: vehicle-control, 43/52; low-dose, 38/52; and high-dose, 40/52; that in females was: 34/52, 35/52 and 38/52. An increased incidence of epithelial hyperplasia of the forestomach was observed in treated animals (males: 2/52, 5/52 and 13/52 [$p < 0.001$]; females: 1/52, 0/52 and 16/52 [$p < 0.001$]). In the forestomach, the incidences of squamous-cell papillomas, of carcinomas, and of papillomas and carcinomas combined showed a significant, positive trend in male rats: papillomas — 1/52, 1/52 and 9/52 ($p = 0.002$); carcinomas — 0/52, 0/52 and 4/52 ($p = 0.014$); papillomas and carcinomas combined — 1/52, 1/52 and 13/52 ($p < 0.001$). There was also a significant positive trend in the incidence of liver neoplastic nodules in males: 1/52, 6/52 and 7/52 ($p = 0.023$); additionally, one carcinoma of the liver was seen in the high-dose group. An ancillary study was conducted in which 28 rats of each sex were assigned to each dose group, and five male and five female rats were killed in each dose group after receiving 0, 25 or 50 mg/kg bw Telone II for nine, 16, 21, 24 or 27 months. The results of this study supported those of the two-year study. In addition, the combined incidences of squamous-cell papilloma of the forestomach in female rats in the two-year and ancillary studies showed a significant ($p = 0.002$) dose-related increase (0/75, 2/77, 8/77) (National Toxicology Program, 1985; Yang *et al.*, 1986). [The Working Group noted the presence in the test compound of 1% epichlorohydrin, which has been shown to produce forestomach tumours in rats (see IARC, 1982).]

(b) Skin application

Mouse: In an experiment involving repeated skin applications, groups of 30 female Ha:ICR mice, six to eight weeks of age, received thrice-weekly applications of 41 or 122 mg *cis*-1,3-dichloropropene ['compound distilled and purity checked by NMR (nuclear magnetic resonance) spectrometry'] in 0.2 ml acetone for 440-594 days. No skin tumour was observed in the low-dose group; three mice in the high-dose group developed skin papillomas and two had carcinomas. No tumour was observed in 30 acetone-treated controls; the difference was not, however, statistically significant (Van Duuren *et al.*, 1979).

(c) Subcutaneous application

Mouse: A group of 30 female Ha:ICR mice, six to eight weeks old, received weekly subcutaneous injections of 3 mg *cis*-1,3-dichloropropene ['compound was distilled and purity checked by NMR (nuclear magnetic resonance) spectrometry'] in 0.05 ml trioctanoin for 77 weeks. Six mice developed fibrosarcomas at the site of injection ($p < 0.0005$). No skin tumour was observed in 30 vehicle or 100 untreated controls (Van Duuren *et al.*, 1979).

(d) Initiation-promotion

cis-1,3-Dichloropropene did not have initiating activity in a two-stage mouse-skin assay with 12-*O*-tetradecanoylphorbol 13-acetate in female Ha:ICR mice (Van Duuren *et al.*, 1979).

3.2 Other relevant biological data

(a) *Experimental systems*

Toxic effects

The toxicity of 1,3-dichloropropene has been reviewed (Torkelson & Oyen, 1977; Torkelson & Rowe, 1981; Yang, 1986).

Oral LD_{50} values for 1,3-dichloropropene (technical product, 92% pure mixture of the two isomers) have been reported to be 710 mg/kg bw in male and 470 mg/kg bw in female rats (Torkelson & Oyen, 1977). In an abstract, an oral LD_{50} of 325 mg/kg bw and an intraperitoneal LD_{50} of 175 mg/kg bw were reported in male rats (Jeffrey *et al.*, 1985). The dermal LD_{50} for 1,3-dichloropropene (92% technical product) has been reported to be 504 mg/kg bw in rabbits. Exposure to 400 ppm (1800 mg/m³) 1,3-dichloropropene for 7 h was reported to be lethal to guinea-pigs (Torkelson & Oyen, 1977). In rats, the 4-h LC_{50} for D-D (see section 1.4) was 1000 ppm (4500 mg/m³); the dermal LD_{50} was 2.1 g/kg in rabbits (Hine *et al.*, 1953).

Liver and kidney injury have been reported following acute oral administration of 1,3-dichloropropene (92% technical product) in corn oil to rats (Torkelson & Oyen, 1977). In an abstract, depression of renal organic ion transport was reported after oral treatment of rats with 175-325 mg/kg bw 1,3-dichloropropene and after intraperitoneal treatment with 75-200 mg/kg bw (Jeffrey *et al.*, 1985).

Exposure of rats by inhalation to 1,3-dichloropropene caused injury to the nose ($\geqslant$1000 ppm; 4500 mg/m³), lung and liver (2700 ppm; 12 300 mg/m³) and death (1000 ppm; 4500 mg/m³) (Torkelson & Oyen, 1977).

1,3-Dichloropropene is reported to cause oedema and necrosis of the skin of rabbits (Torkelson & Rowe, 1981).

As reported in abstracts, depletion of glutathione by 1,3-dichloropropene has been observed in several organs in rats and mice (Dietz *et al.*, 1984a; Jeffrey *et al.*, 1985).

Rats and guinea-pigs exposed by inhalation to 50 ppm (230 mg/m³) 1,3-dichloropropene (92% technical product) for 7 h per exposure over a 28-day period (19 exposures) developed liver and kidney necrosis. Exposure of both species to 11 ppm (50 mg/m³) over a 39-day period (27 7-h exposures) induced changes in the liver and kidney in rats, but only in the kidney in guinea-pigs (Torkelson & Oyen, 1977).

When rats and mice were exposed by inhalation to 12, 32 or 93 ppm (55, 145 or 420 mg/m³) Telone II (47% *cis*- and 45% *trans*-1,3-dichloropropene) for 7 h per day on five days per week for 13 weeks, failure to gain weight was observed among high-dose animals; some high- and mid-dose animals developed alterations of the nasal epithelium (Torkelson & Rowe, 1981). Exposure of rats, rabbits, guinea-pigs and dogs by inhalation to 1 and 3 ppm (4.54 and 13.6 mg/m³) 1,3-dichloropropene (46% *cis* and 53% *trans* isomer) for 7 h per day on five days per week for six months induced slight changes in the kidneys of high-dose male rats (Torkelson & Oyen, 1977).

Rats administered oral doses of 10 and 30 mg/kg bw 1,3-dichloropropene on six days per week for 13 weeks had increased relative kidney weights; no such effect was observed with doses of 1 or 3 mg/kg bw (Torkelson & Rowe, 1981).

Mice exposed by inhalation to 50 ppm (230 mg/m³) D-D (25% cis- and 27% trans-1,3-dichloropropene, 29% 1,3-dichloropropane) for 6 h per day on five days per week for 12 weeks developed slight to moderate hepatocytic enlargement; rats receiving the same treatment had increased liver:body weight and kidney:body weight ratios. No treatment-related effect was reported following similar exposure to 5 or 15 ppm (23 or 68 mg/m³) D-D (Parker et al., 1982).

Effects on reproduction and prenatal toxicity
No data were available to the Working Group.

Absorption, distribution, excretion and metabolism
As reported in an abstract, when rats were exposed by inhalation to 30-900 ppm (135-4100 mg/m³) Telone II (91% cis- and trans-1,3-dichloropropene), an initial rapid elimination from blood, with a half-time of 2-3 min, was observed, followed by a second, slower phase with a half-time of about 40 min (Stott et al., 1985).

Following oral administration of 8.3-13.5 mg/kg bw cis- or trans-[2-^{14}C]-1,3-dichloropropene to rats, 80% of the radioactivity of the cis isomer and 57% of the radioactivity of the trans isomer was excreted in the urine within 24 h. Little further urinary excretion occurred over the next 72 h. After 96 h, 2-5% of the cis isomer and 23-24% of the trans isomer were expired as ^{14}C-CO_2; exhalation of other radioactive compounds was minor. Little faecal excretion was observed. Recovery of both isomers was about 90% (Hutson et al., 1971). In another experiment, reported in an abstract, rats and mice administered a mixture of cis- and trans-[^{14}C]-1,3-dichloropropene excreted 51-61% and 63-79% of the dose, respectively, in the urine within 48 h (Dietz et al., 1984b).

The major urinary metabolite (about 90% of the radioactivity) of cis-1,3-dichloropropene in rats is N-acetyl-S-(cis-3-chloroprop-2-enyl)cysteine. 1,3-Dichloropropene forms a conjugate with glutathione in the presence of rat-liver cytosol *in vitro*; the glutathione- and cytosol-dependent degradation of the cis isomer occurs four to five times faster than that of the trans isomer (Climie et al., 1979).

Studies with 4-(*para*-nitrobenzyl)pyridine have shown that 1,3-dichloropropene is an alkylating agent and that the trans isomer is less reactive than the cis isomer (Neudecker et al., 1980).

Mutagenicity and other short-term tests
The mutagenicity of 1,3-dichloropropene has been reviewed (Yang, 1986).

In a standard plate assay, commercial preparations of D-D soil fumigant (40% 1,3-dichloropropene, 40% 1,2-dichloropropane and 20% other unknown chemicals) and Telone (30% cis-1,3-dichloropropene, 30% trans-1,3-dichloropropene, 20% 1,2-dichloropropane, 5% 2,3-dichloro-1-propene, 2% allyl chloride and about 15% unknown compounds) were mutagenic to *Salmonella typhimurium* TA1978, TA1535 and TA100, but not to TA1537 or

TA98, both in the presence and absence of rat-liver homogenates. Mutagenicity of both preparations to strain TA1978 was significantly higher in the presence of the liver homogenate. Tested dose ranges were 0.5-25 mg/plate for D-D soil fumigant and 0.1-10 mg/plate for Telone. Purified *cis* and *trans* isomers of 1,3-dichloropropene were mutagenic to strains TA1535 and TA100 and only weakly mutagenic to strain TA1978, both in the presence and absence of rat-liver homogenates (tested dose range: 20-100 µg/plate). The authors considered that the mutagenicity of D-D soil fumigant and Telone was higher than could be accounted for by the mutagenicity of the identified components (De Lorenzo *et al.*, 1977).

The *cis* and *trans* isomers of 1,3-dichloropropene (99.97% and 97.46% pure, respectively) were mutagenic to *S. typhimurium* TA1535 both in the presence and absence of liver homogenates from Aroclor-treated rats when tested at 0.1-1 µl/ml top agar (Neudecker *et al.*, 1977).

In a standard plate assay, 1,3-dichloropropene [purity unspecified] was mutagenic to *S. typhimurium* TA100 when tested at 0.1-10 µmol/plate. The presence of Aroclor-induced rat liver S9 reduced the mutagenic effect (Stolzenberg & Hine, 1980).

In a preincubation assay, 1,3-dichloropropene [95.5% pure (Valencia *et al.*, 1985)] was mutagenic to *S. typhimurium* TA1535 and TA100, but not to strains TA1537 and TA98 (tested dose range, 3-3333 µg/plate). Addition of Aroclor-induced rat or hamster S9 reduced the mutagenic effect (Haworth *et al.*, 1983).

A single dose (100 µg) of 1,3-dichloropropene [purity unspecified] induced both his^+ revertants and forward mutations to rifampicin resistance in *S. typhimurium* TA98 in the absence of S9 (Vithayathil *et al.*, 1983).

Talcott and King (1984) reported that, after removal of polar impurities by silicic acid chromatography, none of four commercial preparations of 1,3-dichloropropene retained mutagenic activity when tested at the highest nontoxic doses in *S. typhimurium* TA100 in the absence of S9. The purities of the four preparations were reported to be 85%, 77%, 75% and 88% before, and 92%, 85%, 86% and 95% after purification, respectively. The first sample was not mutagenic either before or after purification. The isolated impurities from all four samples were mutagenic to strain TA100.

Both the *cis* (purity, 98%) and the *trans* (purity, 98%) isomers of 1,3-dichloropropene (tested dose range, 10-2000 µg/plate) were mutagenic to *S. typhimurium* TA100 in the presence and absence of Aroclor-induced rat-liver S9. Addition of a physiological concentration (5 mM) of glutathione markedly reduced the mutagenicity of both isomers (Creedy *et al.*, 1984).

Sex-linked recessive lethal mutations, but not reciprocal translocations, were induced in *Drosophila melanogaster* Canton-S stock flies fed 5750 µg/ml 1,3-dichloropropene (95.5% pure) (Valencia *et al.*, 1985). [The sample tested in this study was the same as that used by Haworth *et al.* (1983) cited above.]

It was reported in an abstract that 1,3-dichloropropene [purity unspecified] induced significant increases in the incidence of sister chromatid exchanges in cultured Chinese hamster ovary cells, both in the presence and absence of rat-liver homogenates. The maximum dose reduced the proportion of cells in their second division to 50% (Tomkins *et al.*, 1980).

(b) Humans

Toxic effects

Twenty-six cases of eye and skin injuries associated with accidental exposure to pesticides containing 1,3-dichloropropene were collected by the health authorities of the State of California in 1976-1977. The most common findings were conjunctivitis and burns; six patients developed systemic illness, characterized by weakness, difficulty in breathing, headache and nausea (Maddy & Edmiston, 1978).

In 46 people examined after accidental exposure to fumes leaking from a tank containing the pesticide Telone II (92% 1,3-dichloropropene and 8% other chlorinated hydrocarbons), the most common symptoms were headache, chest discomfort, mucous membrane irritation, dizziness, nausea and vomiting. Slightly elevated levels of serum transaminases were observed in 11 persons. One to two weeks after the accident, 28 persons interviewed still had headache, abdominal discomfort, chest discomfort and malaise. Of 21 persons examined two years after the accident, ten had headaches, ten had chest pain and 13 reported psychological changes (Flessel *et al.*, 1978).

Of an unspecified number of firemen exposed to 1,3-dichloropropene, nine were treated for headache, neck pain, nausea and difficulty in breathing; they were subsequently discharged from hospital (Markovitz & Crosby, 1984).

Effects on reproduction and prenatal toxicity

In a study of 64 men employed in the production of chlorinated compounds [time-weighted average exposures, <1 ppm (4.5 mg/m^3) 1,3-dichloropropene, 3.1 mg/m^3 allyl chloride and 3.8 mg/m^3 epichlorohydrin], sperm counts and percentages of normal sperm were similar in the study group and among 63 controls. The volunteer participation rate for the study group was 64% (Venable *et al.*, 1980).

Absorption, distribution, excretion and metabolism

N-Acetyl-*S*-(3-chloroprop-2-enyl)cysteine has been identified in the urine of workers exposed to 1,3-dichloropropene (Osterloh *et al.*, 1984).

Mutagenicity and chromosomal effects

No data were available to the Working Group.

3.3 Case reports and epidemiological studies of carcinogenicity to humans

Two cases of malignant histiocytic lymphoma were reported among the nine firemen accidentally exposed to 1,3-dichloropropene [see section 3.2(*b*)]. The two men were 32 and 38 years old at the time of diagnosis, six years after the accident. [The Working Group noted that firemen are exposed to a large number of chemicals.] A case of acute myelomonocytic leukaemia was reported in a farmer less than one year after accidental exposure to 1,3-dichloropropene employed as a nematocide (Markovitz & Crosby, 1984).

4. Summary of Data Reported and Evaluation

4.1 Exposure data

1,3-Dichloropropene has been used since the 1950s as a soil fumigant. Occupational exposures may occur during its production, formulation and application.

4.2 Experimental data

Technical-grade 1,3-dichloropropene (containing 1.0% epichlorohydrin) was tested by oral administration by gavage in one experiment in mice and in one experiment in rats. In mice, it produced dose-related increases in the incidences of benign and/or malignant tumours of the urinary bladder, lung and forestomach. In male rats, it produced dose-related increases in the incidences of benign and malignant forestomach tumours and benign liver tumours; in female rats, it produced benign forestomach tumours. In one experiment by subcutaneous administration in female mice, the *cis* isomer produced malignant tumours at the injection site. In a two-stage skin application study, the *cis* isomer was not active as an initiator. A study to evaluate *cis*-1,3-dichloropropene as a carcinogen by skin application was inconclusive.

No data were available to evaluate the reproductive effects or prenatal toxicity of 1,3-dichloropropene to experimental animals.

Both the *cis* and *trans* isomers of 1,3-dichloropropene and a mixture of the two are mutagenic to bacteria in the presence and absence of an exogenous metabolic system. It was found in one investigation that purification of 1,3-dichloropropene resulted in loss of mutagenic activity, although samples of higher purity were mutagenic in other studies. 1,3-Dichloropropene induces sex-linked recessive lethal mutations, but not reciprocal translocations, in *Drosophila melanogaster*.

4.3 Human data

No relevant data were available to evaluate the reproductive effects or prenatal toxicity of 1,3-dichloropropene to humans.

Two cases of malignant histiocytic lymphoma were reported in firemen previously treated for acute symptoms after exposure to 1,3-dichloropropene.

Overall assessment of data from short-term tests: 1,3-Dichloropropene[a]

	Genetic activity			Cell transformation
	DNA damage	Mutation	Chromosomal effects	
Prokaryotes		+	.	
Fungi/Green plants				
Insects		+	−	
Mammalian cells (*in vitro*)				
Mammals (*in vivo*)				
Humans (*in vivo*)				
Degree of evidence in short-term tests for genetic activity: **Limited**				Cell transformation: No data

[a]The groups into which the table is divided and the symbols '+' and '−' are defined on pp. 19-20 of the Preamble; the degrees of evidence are defined on pp. 20-21.

4.4 Evaluation[1]

There is *sufficient evidence*[2] for the carcinogenicity of technical-grade 1,3-dichloropropene to experimental animals.

There is *inadequate evidence* for the carcinogenicity of 1,3-dichloropropene to humans.

5. References

American Conference of Governmental Industrial Hygienists (1985) *TLVs Threshold Limit Values and Biological Exposure Indices for 1985-86*, 2nd ed., Cincinnati, OH, p. 16

Anon. (1985) *Farm Chemicals Handbook*, Willoughby, OH, Meister Publishing, pp. C79-C80

The Chemical Daily Co. (1984) *JCW Chemicals Guide 1984/1985*, Tokyo, p. 147

---

[1]For definitions of the italicized terms, see Preamble, pp. 18 and 22.
[2]In the absence of adequate data on humans, it is reasonable, for practical purposes, to regard chemicals or exposures for which there is *sufficient evidence* of carcinogenicity in animals as if they presented a carcinogenic risk to humans.

Climie, I.J.G., Hutson, D.H., Morrison, B.J. & Stoydin, G. (1979) Glutathione conjugation in the detoxication of (Z)-1,3-dichlorpropene (a component of the nematocide D-D) in the rat. *Xenobiotica, 9*, 149-156

Creedy, C.L., Brooks, T.M., Dean, B.J., Hutson, D.H. & Wright, A.S. (1984) The protective action of glutathione on the microbial mutagenicity of the Z- and E-isomers of 1,3-dichloropropene. *Chem.-biol. Interactions, 50*, 39-48

DeBenedictis, A. (1979) *Chlorocarbons, -hydrocarbons (allyl chloride)*. In: Grayson, M. & Eckroth, D., eds, *Kirk-Othmer Encyclopedia of Chemical Technology*, 3rd ed., Vol. 5, New York, John Wiley & Sons, pp. 766-767

De Lorenzo, F., Degl'Innocenti, S., Ruocco, A., Silengo, L. & Cortese, R. (1977) Mutagenicity of pesticides containing 1,3-dichloropropene. *Cancer Res., 37*, 1915-1917

Dietz, F.K., Dittenber, D.A., Kirk, H.D. & Ramsey, J.C. (1984a) Non-protein sulfhydryl content and macromolecular binding in rats and mice following oral administration of 1,3-dichloropropene (Abstract No. 586). *Toxicologist, 4*, 147

Dietz, F.K., Hermann, E.A. & Ramsey, J.C. (1984b) The pharmacokinetics of ^{14}C-1,3-dichloropropene in rats and mice following oral administration (Abstract No. 585). *Toxicologist, 4*, 147

Direktoratet for Arbeidstilsynet (Directorate of Labour Inspection) (1981) *Administrative Norms for Pollution in Work Atmosphere, No. 361* (Norw.), Oslo, p. 10

Dowty, B., Carlisle, D., Laseter, J.L. & Storer, J. (1975) Halogenated hydrocarbons in New Orleans drinking water and blood plasma. *Science, 187*, 75-77

Easley, D.M., Kleopfer, R.D. & Carasea, A.M. (1981) Gas chromatographic-mass spectrometric determination of volatile organic compounds in fish. *J. Assoc. off. anal. Chem., 64*, 653-656

Flessel, P., Goldsmith, J.R., Kahn, E., Wesolowski, J.J., Maddy, K.T. & Peoples, S.A. (1978) Acute and possible long-term effects of 1,3-dichloropropene — California. *Morb. Mortal. wkly Rep., 27*, 50, 55

Grasselli, J.G. & Ritchey, W.M., eds (1975) *CRC Atlas of Spectral Data and Physical Constants for Organic Compounds*, Vol. 4, Cleveland, OH, CRC Press, p. 301

Gupta, K.C., Ulsamer, A.G. & Gammage, R. (1984) *Volatile organic compounds in residential air: levels, sources and toxicity*. In: *Proceedings of the 77th Annual Meeting of the Air Pollution Control Association, San Francisco, CA*, Pittsburgh, PA, Air Pollution Control Association, pp. 1-9

Hawley, G.G., ed. (1981) *The Condensed Chemical Dictionary*, 10th ed., New York, Van Nostrand Reinhold, p. 338

Haworth, S., Lawlor, T., Mortelmans, K., Speck, W. & Zeiger, E. (1983) *Salmonella* mutagenicity test results for 250 chemicals. *Environ. Mutagenesis, 5* (Suppl. 1)

Hayes, W.J., Jr (1982) *Pesticides Studied in Man*, Baltimore, MD, Williams & Wilkins, p. 162

Health and Safety Executive (1985) *Occupational Exposure Limits 1985 (Guidance Note EH 40/85)*, London, Her Majesty's Stationery Office, p. 12

Hine, C.H., Anderson, H.H., Moon, H.D., Kodama, J.K., Morse, M. & Jacobsen, N.W. (1953) Toxicology and safe handling of CBP-55 (technical 1-chloro-3-bromopropene-1). *Arch. ind. Hyg. occup. Med.*, 7, 118-223

Holtorf, R.C. (1984) *Preliminary Quantitative Usage Analysis of Dichloropropene*, Washington DC, US Environmental Protection Agency

Hurd, C.D. & Webb, C.N. (1936) The effect of halogen substituents on the rearrangement of allyl aryl ethers. II. Ethers which behave normally. *J. Am. chem. Soc.*, 58, 2190-2193

Hutson, D.H., Moss, J.A. & Pickering, B.A. (1971) The excretion and retention of components of the soil fumigant D-D and their metabolites in the rat. *Food Cosmet. Toxicol.*, 9, 677-680

IARC (1976) *IARC Monographs on the Evaluation of Carcinogenic Risk of Chemicals to Man*, Vol. 11, *Cadmium, Nickel, Some Epoxides, Miscellaneous Industrial Chemicals and General Considerations on Volatile Anaesthetics*, Lyon, pp. 131-139

IARC (1982) *IARC Monographs on the Evaluation of the Carcinogenic Risk of Chemicals to Humans*, Suppl. 4, *Chemicals, Industrial Processes and Industries Associated with Cancer in Humans, IARC Monographs, Volumes 1 to 29*, Lyon, pp. 122-124

IARC (1985) *IARC Monographs on the Evaluation of the Carcinogenic Risk of Chemicals to Humans*, Vol. 36, *Allyl Compounds, Aldehydes, Epoxides and Peroxides*, Lyon, pp. 39-54

International Labour Office (1980) *Occupational Exposure Limits for Airborne Toxic Substances: A Tabular Compilation of Values from Selected Countries*, 2nd (rev.) ed. (*Occupational Safety and Health Series No. 37*), Geneva, pp. 94-95

International Register of Potentially Toxic Chemicals (1984) *IRPTC Data Profile on: 1,3-Dichloropropene*, Geneva, United Nations Environment Programme

Jeffrey, M.M., Baggett, J.M. & Berndt, W.O. (1985) Effects of 1,3-dichloropropene (DCP) on renal transport and glutathione (GSH) in the rat (Abstract No. 7). *Toxicologist*, 5, 2

Leiber, M.A. & Berk, H.C. (1984) Development and validation of an air monitoring method for 1,3-dichloropropene, *trans*-1,2,3-trichloropropene, *cis*-1,2,3-trichloropropene, 1,1,2,3-tetrachloropropene, 2,3,3-trichloro-3-propen-1-ol, and 1,1,2,2,3-pentachloropropane. *Anal. Chem.*, 56, 2134-2137

Maddy, K.T. & Edmiston, S. (1978) *Occupational Illnesses and Injuries Due to Pesticides Containing 1,3-Dichloropropene as Reported by Physicians in California in 1976 and 1977* (*Report No. 59-479*), Sacramento, CA, California Department of Food and Agriculture

Markovitz, A. & Crosby, W.H. (1984) A soil fumigant, 1,3-dichloropropene, as possible cause of hematologic malignancies. *Arch. intern. Med.*, 144, 1409-1411

Miller, D.M., ed. (1985) *Crop Protection Chemicals Reference*, New York, Chemical and Pharmaceutical Publishing, pp. 700-706

National Fire Protection Association (1984) *Fire Protection Guide on Hazardous Materials*, 8th ed., Quincy, MA, p. 325M-35

National Toxicology Program (1985) *Toxicology and Carcinogenesis Studies of Telone II®* *(Technical-Grade 1,3-Dichloropropene (CAS No. 542-75-6) Containing 1.0% Epichlorohydrin as a Stabilizer) in F344/N Rats and B6C3F$_1$ Mice (Gavage Studies (Technical Report No. 269)*, Research Triangle Park, NC, US Department of Health and Human Services

Neudecker, T., Stefani, A. & Henschler, D. (1977) In vitro mutagenicity of the soil nematicide 1,3-dichloropropene. *Experientia, 33*, 1084-1085

Neudecker, T., Lutz, D., Eder, E. & Henschler, D. (1980) Structure-activity relationship in halogen and alkyl substituted allyl and allylic compounds: correlation of alkylating and mutagenic properties. *Biochem. Pharmacol., 29*, 2611-2617

Osterloh, J.D., Cohen, B.-S., Popendorf, W. & Pond, S.M. (1984) Urinary excretion of the *N*-acetyl cysteine conjugate of *cis*-1,3-dichloropropene by exposed individuals. *Arch. environ. Health, 39*, 271-275

Otson, R. & Williams, D.T. (1982) Headspace chromatographic determination of water pollutants. *Anal. Chem., 54*, 942-946

Parker, C.M., Coate, W.B. & Voelker, R.W. (1982) Subchronic inhalation toxicity of 1,3-dichloropropene/1,3-dichloropropane (D-D) in mice and rats. *J. Toxicol. environ. Health, 9*, 899-910

Sadtler Research Laboratories (1980) *The Sadtler Standard Spectra, Cumulative Index*, Philadelphia, PA

Sax, N.I. (1984) *Dangerous Properties of Industrial Materials*, 6th ed., New York, Van Nostrand Reinhold, p. 963

Sittig, M., ed. (1980) *Pesticide Manufacturing and Toxic Materials Control Encyclopedia*, Park Ridge, NJ, Noyes Data Corp., pp. 290-292

Stolzenberg, S.J. & Hine, C.H. (1980) Mutagenicity of 2- and 3-carbon halogenated compounds in the *Salmonella*/mammalian-microsome test. *Environ. Mutagenesis, 2*, 59-66

Stott, W.T., Kastl, P.E. & McKenna, M.J. (1985) Inhalation pharmacokinetics of *cis*- and *trans*-1,3-dichloropropene in rats exposed to Telone* II vapors (Abstract No. 440). *Toxicologist, 5*, 110

Talcott, R.E. & King, J. (1984) Mutagenic impurities in 1,3-dichloropropene preparations. *J. natl Cancer Inst., 72*, 1113-1116

Tomkins, D., Kwok, E. & Douglas, G. (1980) Testing of pesticides for induction of sister chromatid exchange in Chinese hamster ovary cells (Abstract). *Can. J. Genet. Cytol., 22*, 681

Torkelson, T.R. & Oyen, F. (1977) The toxicity of 1,3-dichloropropene as determined by repeated exposure of laboratory animals. *Am. ind. Hyg. Assoc. J., 38*, 217-223

Torkelson, T.R. & Rowe, V.K. (1981) *Halogenated aliphatic hydrocarbons containing chlorine, bromide and iodine*. In: Clayton, G.D. & Clayton, F.E., eds, *Patty's Industrial Hygiene and Toxicology*, 3rd ed., Vol. 2B, New York, John Wiley & Sons, pp. 3573-3577

Tuazon, E.C., Atkinson, R., Winer, A.M. & Pitts, J.N., Jr (1984) A study of the atmospheric reactions of 1,3-dichloropropene and other selected organochlorine compounds. *Arch. environ. Contam. Toxicol.*, *13*, 691-700

US Environmental Protection Agency (1982a) *Method 8010. Halogenated volatile organics*. In: *Test Methods for Evaluating Solid Waste — Physical/Chemical Methods*, 2nd ed. (*US EPA No. SW-846*), Washington DC, Office of Solid Waste and Emergency Response

US Environmental Protection Agency (1982b) *Method 8240. GC/MS method for volatile organics*. In: *Test Methods for Evaluating Solid Waste — Physical/Chemical Methods*, 2nd ed. (*US EPA No. SW-846*), Washington DC, Office of Solid Waste and Emergency Response

US Environmental Protection Agency (1984a) Method 601. Guidelines establishing test procedures for the analysis of pollutants under the Clean Water Act (40 CFR 136). Purgeable halocarbons. *Fed. Regist.*, *49*, 43261-43271

US Environmental Protection Agency (1984b) Method 624. Guidelines establishing test procedures for the analysis of pollutants under the Clean Water Act (40 CFR 136). Purgeables. *Fed. Regist.*, *49*, 43373-43384

US Environmental Protection Agency (1984c) Method 1624, Revision B. Guidelines establishing test procedures for the analysis of pollutants under the Clean Water Act (40 CFR 136). Volatile organic compounds by isotope dilution GC/MS. *Fed. Regist.*, *49*, 43407-43415

US Occupational Safety and Health Administration (1985) Request for comments and secondary data on the use and health effects of 1,3-dichloropropene. *Fed. Regist.*, *50*, 4913-4914

Valencia, R., Mason, J.M., Woodruff, R.C. & Zimmering, S. (1985) Chemical mutagenesis testing in *Drosophila*. III. Results of 48 coded compounds tested for the National Toxicology Program. *Environ. Mutagenesis*, *7*, 325-348

Van Duuren, B.L., Goldschmidt, B.M., Loewengart, G., Smith, A.C., Melchionne, S., Seidman, I. & Roth, D. (1979) Carcinogenicity of halogenated olefinic and aliphatic hydrocarbons in mice. *J. natl Cancer Inst.*, *63*, 1433-1439

Venable, J.R., McClimans, C.D., Flake, R.E. & Dimick, D.B. (1980) A fertility study of male employees engaged in the manufacture of glycerine. *J. occup. Med.*, *22*, 87-91

Verschueren, K. (1983) *Handbook of Environmental Data on Organic Chemicals*, 2nd ed., New York, Van Nostrand Reinhold, pp. 507-508

Vettorazzi, G., ed. (1984) *Pesticides Reference Index, JMPR-IARC-IPCS-IRPTC-VBC —1961-1983*, Geneva, International Programme on Chemical Safety

Vithayathil, A.J., McClure, C. & Myers, J.W. (1983) *Salmonella*/microsome multiple indicator mutagenicity test. *Mutat. Res.*, *121*, 33-37

Weast, R.C., ed. (1985) *CRC Handbook of Chemistry and Physics*, 66th ed., Boca Raton, FL, CRC Press, p. C-456

Windholz, M., ed. (1983) *The Merck Index*, 10th ed., Rahway, NJ, Merck & Co., p. 446

Worthing, C.R. (1983) *The Pesticide Manual, A World Compendium*, 7th ed., Croydon, The British Crop Protection Council, pp. 182-183

Yang, R.S.H. (1986) 1,3-Dichloropropene. *Residue Rev.*, *97*, 19-35

Yang, R.S.H., Huff, J.E., Boorman, G.A., Haseman, J.K. & Kornreich, M. (1986) Chronic toxicology and carcinogenesis studies of Telone II by gavage in Fischer 344 rats and B6C3F$_1$ mice. *J. Toxicol. environ. Health* (in press)

1,2-DICHLOROPROPANE

1. Chemical and Physical Data

1.1 Synonyms and trade names

Chem. Abstr. Services Reg. No.: 78-87-5
Chem. Abstr. Name: 1,2-Dichloropropane
IUPAC Systematic Name: 1,2-Dichloropropane
Synonyms: ENT 15 406; propylene chloride; propylene dichloride

1.2 Structural and molecular formulae and molecular weight

$$Cl-\underset{\underset{H}{|}}{\overset{\overset{H}{|}}{C}}-\underset{\underset{H}{|}}{\overset{\overset{Cl}{|}}{C}}-\underset{\underset{H}{|}}{\overset{\overset{H}{|}}{C}}-H$$

$C_3H_6Cl_2$ Mol. wt: 112.99

1.3 Chemical and physical properties of the pure substance

(a) *Description*: Colourless liquid with chloroform-like odour (Hawley, 1981; Verschueren, 1983; Windholz, 1983)

(b) *Boiling-point*: 96.4°C (Weast, 1985)

(c) *Melting-point*: -100.4°C (Weast, 1985)

(d) *Density*: d_4^{20} 1.1560 (Weast, 1985)

(e) *Spectroscopy data*[a]: Infrared (Sadtler Research Laboratories, 1980; prism [3208], grating [10981]), nuclear magnetic resonance (Sadtler Research Laboratories, 1980; proton [10492], C-13 [541]) and mass spectral data (Grasselli & Ritchey, 1975) have been reported.

(f) *Solubility*: Slightly soluble in water (2.7 g/l at 20°C (Verschueren, 1983); soluble in ethanol, diethyl ether, benzene and chloroform (Weast, 1985)

(g) *Volatility*: Vapour pressure, 40 mm Hg at 19.4°C (Sax, 1984; Weast, 1985); relative vapour density (air = 1), 3.9 (Verschueren, 1983; National Fire Protection Association, 1984; Sax, 1984)

(h) *Stability*: Flash-point, 21°C (open-cup); inflammable (Windholz, 1983)

(i) *Octanol/water partition coefficient (P)*: log P, 2.00 (Hansch & Leo, 1979)

(j) *Conversion factor*: $mg/m^3 = 4.62 \times ppm$[b]

1.4 Technical products and impurities

Commercial 1,2-dichloropropane is marketed as a high-purity liquid (99-99.5%) for industrial use. Water and oxygenated organic impurities comprise a maximum of 0.05% and 0.1% of the product, respectively (Bayer AG, 1977; Dow Chemical Co., 1985). Trace levels of low molecular-weight chlorinated hydrocarbons, such as chloropropenes and chloropropanes, also occur. Fumigant products based on 1,3-dichloropropene also contain varying amounts of 1,2-dichloropropane. One such product (D-D) contains approximately 30% 1,2-dichloropropane, while another (Telone II) contains approximately 2% (Yang, 1986).

2. Production, Use, Occurrence and Analysis

2.1 Production and use

(a) *Production*

1,2-Dichloropropane, marketed as a solvent, is obtained as a byproduct of the synthesis of propylene oxide (see IARC, 1985) by the chlorohydrin reaction (Mannsville Chemical Products Corp., 1984). 1,2-Dichloropropane may also be produced by the reaction of

---

[a]In square brackets, spectrum number in compilation
[b]Calculated from: $mg/m^3 = (molecular\ weight/24.45) \times ppm$, assuming standard temperature (25°C) and pressure (760 mm Hg)

propylene and chlorine in the presence of an iron oxide catalyst at moderate temperature (45°C) and pressure (25-30 psi; 1.75-2.1 atm). The pesticide products that contain 1,2-dichloropropane (see monograph on 1,3-dichloropropene, p. 113) are distillates from the chlorination of propylene (Sittig, 1980).

Currently, there is only one US producer of 1,2-dichloropropane. Although the number of producers has decreased from five in 1975, the production volume in the USA remained relatively stable until 1984 (Table 1), when a major US manufacturer discontinued production (Mannsville Chemical Products Corp., 1984; Anon., 1985).

Table 1. US production of 1,2-dichloropropane (millions of kg)[a]

1975	1977	1978	1979	1980	1981	1982	1983	1984
38.2	26.5	33.7	31.7	34.9	31.8	25	27.2	27.2

[a]From US International Trade Commission (1976, 1978, 1979, 1980, 1981); Anon. (1984b)

In Europe, four manufacturers of 1,2-dichloropropane have been identified in the Federal Republic of Germany, as well as one each in France and the Netherlands. Five producers of 1,2-dichloropropane have been reported in Japan (The Chemical Daily Co., 1984).

(b) *Use*

1,2-Dichloropropane is a component of two insecticidal soil fumigants containing 1,3-dichloropropene, D-D mixture and Telone II (see monograph on 1,3-dichloropropene, p. 113) (Worthing, 1983). These fumigants have been used in the cultivation of a variety of crops, including citrus fruits, pineapple, soya beans, cotton, potatoes and tomatoes (Anon., 1985). This compound is registered for use as an insecticide in Japan.

1,2-Dichloropropane is used as an industrial solvent for oils, fats, resins, waxes and rubber. Because of its ability to dissolve bitumens and coal-tars, it is used as a solvent in impregnation technology (Bayer AG, 1977). It is also used as a chemical intermediate in the synthesis of tetrachloroethylene and carbon tetrachloride (Mannsville Chemical Products Corp., 1984), and as a lead scavenger in gasoline (Sittig, 1985).

Other uses of 1,2-dichloropropane are as a textile stain remover, oil and paraffin extractant, scouring compound, and as a metal degreaser, especially prior to electroplating (Bayer AG, 1977).

(c) *Regulatory status and guidelines*

Occupational exposure limits for 1,2-dichloropropane in 14 countries are shown in Table 2.

No tolerance limit or acceptable daily intake (ADI) has been established by FAO/WHO for residues of 1,2-dichloropropane in food commodities (Vettorazzi, 1984).

Table 2. Occupational exposure limits for 1,2-dichloropropane[a]

Country	Year	Concentration (mg/m^3)	Interpretation[b]
Australia	1978	350	TWA
Belgium	1978	350	TWA
Bulgaria	1971	10	Ceiling
Finland	1981	350	TWA
		515	STEL
German Democratic Republic	1979	50	TWA
		150	STEL
Germany, Federal Republic of	1985	350	TWA
The Netherlands	1978	350	TWA
Norway	1981	350	TWA
Poland	1976	50	Ceiling
Romania	1975	200	TWA
		300	Ceiling
Switzerland	1978	350	TWA
USA	1985		
ACGIH		350	TWA
		510	STEL
OSHA		350	TWA
USSR	1977	10	Ceiling
Yugoslavia	1971	350	Ceiling

[a]From International Labour Office (1980); Direktoratet for Arbeidstilsynet (1981); Työsuojeluhallitus (1981); American Conference of Governmental Industrial Hygienists (ACGIH) (1985); Deutsche Forschungsgemeinschaft (1985); US Occupational Safety and Health Administration (OSHA) (1985)

[b]TWA, time-weighted average; STEL, short-term exposure limit

2.2 Occurrence

(a) *Natural occurrence*

1,2-Dichloropropane is not known to occur as a natural product.

(b) *Occupational exposure*

No data on exposure levels were available to the Working Group.

(c) *Air*

The primary mechanism for the removal of 1,2-dichloropropane from the atmosphere is believed to be reaction with hydroxyl radicals at a daily rate (12 sunlit hours) of about 10.2%. Mean levels in air in selected US cities range from 0.11-0.37 $\mu g/m^3$ (Singh *et al.*, 1982). Estimated atmospheric levels ranged from trace to 2.2 $\mu g/m^3$ at seven of 11 sample sites in the USA where industrial activity included the production, use or storage of halogenated hydrocarbons (Pellizzari, 1982).

Exposures to 15 volatile organics in ambient air in a petrochemical manufacturing area in Texas and in a nonindustrial area in North Carolina were determined by personal air sampling with thermal desorption and gas chromatography/mass spectrometry analysis. The concentration of 1,2-dichloropropane in air averaged 0.10 $\mu g/m^3$ at both sites (Wallace *et al.*, 1982).

1,2-Dichloropropane was found at trace levels in ambient air outside two of nine houses at the Old Love Canal, Niagara Falls, NY, chemical dump site (Barkley *et al.*, 1980).

(d) *Water and sediments*

1,2-Dichloropropane was detected at levels up to 55 ng/l in 9 of 17 water samples taken from the lower Niagara River and at levels up to 440 ng/l in 20 of 82 sites in Lake Ontario (Kaiser *et al.*, 1983). The compound was detected at a level of <1 $\mu g/l$ in one drinking-water sample from 30 Canadian potable water-treatment facilities (Otson *et al.*, 1982).

In another Canadian study, the concentration of 1,2-dichloropropane in selected surface-, well- and treated water was determined in nine areas. Quantifiable levels (210 ng/l) were detected only in tap-water from one station, and traces (<30 ng/l) in three well-water samples (Comba & Kaiser, 1983).

Dichloropropane [isomer unspecified] was detected at 1200 ng/l in 1978 in drinking-water in one of eight houses that bordered a chemical dump site, the Old Love Canal in Niagara Falls, NY (Barkley *et al.*, 1980).

1,2-Dichloropropane was found at levels of >1 $\mu g/l$ in three of 30 water samples from unpolluted upstream reaches and polluted sections near wastewater discharges in the Delaware River Basin (1976) (DeWalle & Chian, 1978).

Mean levels of 1,2-dichloropropane in sediment samples collected during May-June 1980 from Lake Pontchartrain in New Orleans, LA, ranged from 0.2 to 0.4 $\mu g/l$ (Ferrario *et al.*, 1985).

2.3 Analysis

Selected methods for the analysis of 1,2-dichloropropane in air and water are identified in Table 3. The methods of the US Environmental Protection Agency (1984a,b,c) for water have also been applied to liquid and solid wastes (US Environmental Protection Agency, 1982a,b). Volatile components in solid-waste samples are first extracted with polyethylene glycol or methanol prior to purge/trap concentration and analysis.

Table 3. Methods for the analysis of 1,2-dichloropropane

Sample matrix	Sample preparation	Assay procedure[a]	Limit of detection	Reference
Air	Adsorb on charcoal; desorb (carbon disulphide); inject aliquot	GC/FID	1.7 mg/m^3	Eller (1984a)
	Adsorb on charcoal; desorb (15% acetone in cyclohexane); inject aliquot	GC/ECD	0.03 mg/m^3	Eller (1984b)
	Adsorb on Tenax-GC; desorb (heat, purge with helium or nitrogen); trap (liquid nitrogen cooled); vaporize (heat, purge with helium or nitrogen) onto GC column	GC/MS	0.1-1.0 μg/m^3	Riggin (1985)
Water	Purge (inert gas); trap (OV-1 on Chromosorb-W/Tenax/silica gel); desorb as vapour (heat to 180°C, backflush with inert gas) onto GC column	GC/ECD	0.04 μg/l	US Environmental Protection Agency (1984a)
		GC/MS	6.0 μg/l	US Environmental Protection Agency (1984b)
	Add internal standard (isotope-labelled 1,2-dichloropropane); purge, trap and desorb as above	GC/MS	10 μg/l	US Environmental Protection Agency (1984c)
	Purge (40°C, nitrogen); trap directly on Tenax GC column (40°C); temperature-programme for analysis	GC/FID GC/ECD	0.1 μg/l <0.1 μg/l	Otson & Williams (1982)

[a]Abbreviations: GC/FID, gas chromatography/flame ionization detection; GC/ECD, gas chromatography/electrolytic conductivity detection; GC/MS, gas chromatography/mass spectrometry

US EPA Method 624 (US Environmental Protection Agency, 1984b) has also been adapted for the analysis of 1,2-dichloropropane in fish, with an estimated limit of 10 μg/kg (Easley et al., 1981).

Exposures to 1,2-dichloropropane have also been monitored in air by a direct-reading infrared gas analyser, at minimum concentrations of 3.7 mg/m^3 (The Foxboro Co., 1985).

3. Biological Data Relevant to the Evaluation of Carcinogenic Risk to Humans

3.1 Carcinogenicity studies in animals

Oral administration

Mouse: Groups of 50 male and 50 female $B6C3F_1$ mice, seven to nine weeks old, were administered 0, 125 or 250 mg/kg bw 1,2-dichloropropane (99.4% pure; identified impurity, 0.24% toluene; dose in volume, 3 ml/kg bw) in corn oil by gavage on five days per week for 103 weeks. Survival at 105-107 weeks in males was: vehicle-control, 35/50; low-dose, 33/50; and high-dose, 35/50; that in females was 35/50, 29/50 and 26/50. The incidence of liver adenomas was increased in high-dose male mice (7/50, 10/50 and 17/50; $p = 0.017$ in the three groups, respectively). The incidences of liver carcinomas in males were 11/50, 17/50 and 16/50; a significantly increased incidence of combined liver adenomas and carcinomas was observed in the high-dose group (18/50, 26/50 and 33/50; $p = 0.002$). In females, the incidences of liver adenomas (1/50, 5/50 and 5/50) and of liver carcinomas (1/50, 3/50 and 4/50) were marginally increased in treated animals, and a significantly increased incidence of combined adenomas or carcinomas was observed (2/50, 8/50, $p = 0.046$; 9/50, $p = 0.026$) (National Toxicology Program, 1986).

Rat: Groups of 50 Fischer 344/N rats, seven to nine weeks old, were administered 0, 62 or 125 mg/kg bw (males) or 0, 125 or 250 mg/kg bw (females) 1,2-dichloropropane (99.4% pure; identified impurity, 0.24% toluene; dose in volume, 3 ml/kg bw) in corn oil by gavage on five days per week for 103 weeks. Survival in males at 105-108 weeks was: vehicle-control, 39/50; low-dose, 42/50; and high-dose, 41/50; that in females was 37/50, 43/50 and 16/50. There was a dose-related increase ($p < 0.05$, adjusted for survival) in females in the incidence of mammary-gland adenocarcinomas (1/50, 2/50 and 5/50), but not of mammary-gland fibroadenomas (15/50, 20/50, 7/50). There was no effect on tumour incidences in male rats (National Toxicology Program, 1986). [The Working Group noted the poor survival of high-dose females and that the authors reported that three of the five mammary-gland tumours were of low-grade malignancy that some pathologists would diagnose as a highly cellular variant of fibroadenoma or as adenofibroma.]

3.2 Other relevant biological data

(a) *Experimental systems*

Toxic effects

The toxicity of 1,2-dichloropropane has been reviewed (Torkelson & Rowe, 1981).

The oral and dermal LD_{50} values for 1,2-dichloropropane in rats are approximately 2 and 9 ml/kg bw (2 and 10 mg/kg bw), respectively (Smyth *et al.*, 1969). The LC_{50} in rats was 14 000 mg/m³ for an 8-h exposure (Pozzani *et al.*, 1959); two to four of six rats died following a 4-h exposure to 2000 ppm (9200 mg/m³) (Carpenter et al., 1949). An LC_{50} of 720 ppm (3325 mg/m³) was reported in mice after a 10-h exposure (Torkelson & Rowe, 1981).

Signs of acute 1,2-dichloropropane toxicity include central nervous system depression and irritation of the eyes and respiratory tract (Torkelson & Rowe, 1981).

Acute inhalation of 1,2-dichloropropane increases the levels of serum glutamate-oxalacetate transaminase and glutamate-pyruvate transaminase (Drew et al., 1978; Torkelson & Rowe, 1981).

When mice, rats, guinea-pigs, rabbits and dogs were exposed repeatedly to up to 2200 ppm (10 200 mg/m³) 1,2-dichloropropane for 7 h per day, deaths occurred among all species; mice were the most susceptible and rabbits the least susceptible (Heppel et al., 1946). When groups of mice, rats, guinea-pigs and dogs were exposed to 400 ppm (1850 mg/m³) 1,2-dichloropropane for 7 h per day on five days per week for 25-28 weeks, there was high mortality in mice and decreased weight gain in rats, but no compound-related toxic effect was observed in guinea-pigs or dogs (Heppel et al., 1948).

Liver damage was reported in rats following short-term exposure by inhalation to 1,2-dichloropropane (Sidorenko et al., 1976), and in female Fischer 344/N rats and male $B6C3F_1$ mice following administration by gavage for 103 weeks [see section 3.1] (National Toxicology Program, 1986).

Effects on reproduction and prenatal toxicity

No data were available to the Working Group.

Absorption, distribution, excretion and metabolism

Male and female rats were administered single oral doses of 3.5-5.3 mg/kg bw [1-^{14}C]-1,2-dichloropropane. After 96 h, approximately 5% of the radioactivity was present in the gut, skin and carcass. Within 24 h, approximately 50% of the radioactivity was excreted in the urine and 5% in the faeces. Little additional excretion was observed within the following 72 h. A total of 19% of the dose was expired as $^{14}C\text{-}CO_2$ and 23% as other volatile radioactivity (Hutson et al., 1971). The parent compound was detected in expired air of rats following intraperitoneal administration of 1,2-dichloropropane (100 mg/kg bw) (Jones & Gibson, 1980).

The major urinary metabolite (at least 25-35% of a single dose) following oral dosing of male rats with 1,2-dichloropropane (20 mg/kg bw per day for four days) was N-acetyl-S-(2-hydroxypropyl)cysteine; β-chlorolactate and N-acetyl-S-(2,3-dihydroxypropyl)cysteine were identified as minor metabolites. No parent compound was detected in the urine following daily oral treatment (Jones & Gibson, 1980).

Dechlorination of 1,2-dichloropropane has been demonstrated in rat-liver microsomes in the presence of oxygen and NADPH (Van Dyke & Wineman, 1971).

Mutagenicity and other short-term tests

1,2-Dichloropropane (practical grade) was tested in two laboratories in preincubation assays in *Salmonella typhimurium* in the presence and absence of a metabolic system (S9) from the livers of Aroclor-induced rats or hamsters at doses of 10-10 000 µg/plate. In both laboratories, a low, but reproducible increase in mutation frequency was obtained with strain TA100 in the absence of S9. Results with strain TA1535 were equivocal, and negative results were obtained with strains TA1537 and TA98 (Haworth *et al.*, 1983).

In a plate assay, 1,2-dichloropropane [purity unspecified] (tested at doses up to 100 µmol/plate) was not mutagenic to *S. typhimurium* TA100 in the presence or absence of S9 from Aroclor-induced rats (Stolzenberg & Hine, 1980).

In a standard plate assay, commercial preparations of D-D soil fumigant (40% 1,3-dichloropropene, 40% 1,2-dichloropropane and 20% other unknown chemicals) and Telone (30% *cis*-1,3-dichloroprene, 30% *trans*-1,3-dichloropropene, 20% 1,2-dichloropropane, 5% 2,3-dichloro-1-propene, 2% allyl chloride and about 15% unknown compounds) were mutagenic to *S. typhimurium* TA1978, TA1535 and TA100, but not to TA1537 or TA98, both in the presence and absence of rat-liver homogenates. Mutagenicity of both preparations to strain TA1978 was significantly higher in the presence of the liver homogenate. The doses tested were 0.5-25 mg/plate D-D soil fumigant and 0.1-10 mg/plate Telone. In the same study, purified 1,2-dichloropropane was mutagenic to *S. typhimurium* TA1535 and TA100 both in the presence and absence of rat-liver S9, when tested at 10-50 mg/plate (De Lorenzo *et al.*, 1977).

In a plate assay, 1,2-dichloropropane (analytical grade; tested at 1-10 µl/plate) was mutagenic to *S. typhimurium* TA1535 and TA100. The presence of Aroclor-induced rat-liver S9 reduced the mutagenic effect. Similar results were obtained with one dose (10 µl/plate) in strains TA1535 and TA100 in a spot test; strains TA1537, TA1538 and TA98 gave negative responses. The same sample (tested at 2-100 µl/plate) did not induce streptomycin-resistant mutants in *Streptomyces coelicolor* in either a plate assay or a spot test. 8-Azaguanine-resistant mutants were induced in *Aspergillus nidulans* in both a plate assay and a spot test at concentrations of 100-400 µl/plate (Principe *et al.*, 1981).

Crebelli *et al.* (1984) reported that 1,2-dichloropropane (>99% pure) did not induce crossing-over, mitotic nondisjunction or haploidization in *Aspergillus nidulans* in a plate-incorporation assay. [The Working Group noted that only a single, nontoxic dose (154 mM) was tested.]

[The Working Group noted that high concentrations of 1,2-dichloropropane were generally required to give positive responses in bacteria and *Aspergillus*.]

No sex-linked recessive lethal mutation was induced in adult *Drosophila melanogaster* treated by injection (4200 µg/ml) or inhalation (7200 ppm; 33 250 µg/l) of 1,2-dichloropropane (practical grade) (Woodruff *et al.*, 1985).

(b) Humans

Toxic effects

The toxicity of 1,2-dichloropropane has been investigated mainly with respect to accidental ingestion, self-poisoning and sniffing. Fatalities have been reported among subjects who were exposed occupationally by ingestion or inhalation of mixtures containing 1,2-dichloropropane (Gosselin *et al.*, 1976). The main target organs in cases exposed to solvents (mainly as domestic stain removers) containing 60-98% 1,2-dichloropropane were the liver and kidney. The severity of toxicity ranged from temporary elevation of serum liver enzymes to death from liver failure; tubular necrosis of the kidney and, in some cases, renal failure occurred. Disseminated intravascular coagulation and haemolytic anaemia were also reported (Chiappino & Secchi, 1968; Ponticelli *et al.*, 1968; Secchi *et al.*, 1968; Locatelli & Pozzi, 1983; Pozzi *et al.*, 1984; Perbellini *et al.*, 1985; Pozzi *et al.*, 1985).

Two cases of allergic dermatitis were attributed to 1,2-dichloropropane in workers in the plastics industry with exposure to commercial preparations containing 7.4-12.7% 1,2-dichloropropane (Grzywa & Rudzki, 1981).

Effects on reproduction and prenatal toxicity

No data were available to the Working Group.

Absorption, distribution, excretion and metabolism

No data were available to the Working Group.

Mutagenicity and chromosomal effects

No data were available to the Working Group.

3.3 Case reports and epidemiological studies of carcinogenicity to humans

No data were available to the Working Group.

4. Summary of Data Reported and Evaluation

4.1 Exposure data

1,2-Dichloropropane has been used as an industrial solvent, as a chemical intermediate and in soil fumigants. Human exposure may occur during its production and industrial and domestic use, and due to the presence of low levels in ambient air and in water.

4.2 Experimental data

1,2-Dichloropropane was tested for carcinogenicity by oral administration by gavage in one experiment in mice and one experiment in rats. A dose-related increase in the incidence of hepatocellular tumours was observed in male and female mice. Inconclusive results were obtained with regard to female rats, and no effect was seen in male rats.

No data were available to evaluate the reproductive effects or prenatal toxicity of 1,2-dichloropropane to experimental animals.

1,2-Dichloropropane is mutagenic to *Salmonella typhimurium* but not to *Streptomyces coelicolor*. It induces mutations but not chromosomal effects in *Aspergillus nidulans*. It does not induce sex-linked recessive lethal mutations in *Drosophila melanogaster*.

Overall assessment of data from short-term tests: 1,2-Dichloropropane[a]

	Genetic activity			Cell transformation
	DNA damage	Mutation	Chromosomal effects	
Prokaryotes		+		
Fungi/Green plants		+	−	
Insects		−		
Mammalian cells (*in vitro*)				
Mammals (*in vivo*)				
Humans (*in vivo*)				
Degree of evidence in short-term tests for genetic activity: **Limited**				Cell transformation: No data

[a]The groups into which the table is divided and the symbols '+' and '−' are defined on pp. 19-20 of the Preamble; the degrees of evidence are defined on pp. 20-21.

4.3 Human data

No data were available to evaluate the reproductive effects or prenatal toxicity of 1,2-dichloropropane to humans.

No case report or epidemiological study on the carcinogenicity of 1,2-dichloropropane to humans was available to the Working Group.

4.4 Evaluation[1]

There is *limited evidence* for the carcinogenicity of 1,2-dichloropropane to experimental animals.

No evaluation could be made of the carcinogenicity of 1,2-dichloropropane to humans.

---

[1]For definition of the italicized term, see Preamble, p. 18.

5. References

American Conference of Governmental Industrial Hygienists (1985) *TLVs Threshold Limit Values and Biological Exposure Indices for 1985-86*, 2nd ed., Cincinnati, OH, p. 28

Anon. (1985) *Farm Chemicals Handbook,* Willoughby, OH, Meister Publishing Co., pp. C79, C196

Barkley, J., Bunch, J., Bursey, J.T., Castillo, N., Cooper, S.D., Davis, J.M., Erickson, M.D., Harris, B.S.H., III, Kirkpatrick, M., Michael, L.C., Parks, S.P., Pellizzari, E.D., Ray, M., Smith, D., Tomer, K.B., Wagner, R. & Zweidinger, R.A. (1980) Gas chromatography mass spectrometry computer analysis of volatile halogenated hydrocarbons in man and his environment — a multimedia environmental study. *Biomed. mass Spectrom., 7*, 139-147

Bayer AG (1977) *1,2-Dichloropropane Distilled (DCP) (Technical Product Sheet)*, Cologne, OC Division — Sales E

Carpenter, C.P., Smyth, H.F. & Pozzani, U.C. (1949) The assay of acute toxicity, and the grading and interpretation of results on 96 chemical compounds. *J. ind. Hyg. Toxicol., 31*, 343-346

The Chemical Daily Co. (1984) *JCW Chemicals Guide 1984/1985*, Tokyo, p. 345

Chiappino, G. & Secchi, G.C. (1968) Description of a case of acute intoxication after accidental ingestion of 1,2-dichloropropane sold as trichloroethylene (Ital.). *Med. Lav., 59*, 334-341

Comba, M.E. & Kaiser, K.L.E. (1983) Determination of volatile contaminants at the ng.l^{-1} level in water by capillary gas chromatography with electron capture detection. *Int. J. environ. anal. Chem., 16*, 17-31

Crebelli, R., Conti, G., Conti, L. & Carere, A. (1984) Induction of somatic segregation by halogenated aliphatic hydrocarbons in *Aspergillus nidulans*. *Mutat. Res., 138*, 33-38

De Lorenzo, F., Degl'Innocenti, S., Ruocco, A., Silengo, L. & Cortese, R. (1977) Mutagenicity of pesticides containing 1,3-dichloropropene. *Cancer Res., 37*, 1915-1917

Deutsche Forschungsgemeinschaft (German Research Community) (1985) *Maximal Concentrations in the Workplace and Biological Tolerance Values for Substances in the Work Environment 1985* (Ger.), Vol. 21, Weinheim, Verlagsgesellschaft mbH, p. 30

DeWalle, F.B. & Chian, E.S.K. (1978) *Presence of trace organics in the Delaware River and their discharge by municipal and industrial sources.* In: Bell, J.M., ed., *Proceedings of the Industrial Waste Conference*, Vol. 32, Ann Arbor, MI, Ann Arbor Publishing Co., pp. 908-919

Direktoratet for Arbeidstilsynet (Directorate of Labour Inspection) (1981) *Administrative Norms for Pollution in Work Atmospheres, No. 361* (Norw.), Oslo, p. 19

Dow Chemical Co. (1985) *Material Safety Data Sheet for Propylene Chloride*, Midland, MI

Drew, R., Patel, J.M. & Lin, F.-N. (1978) Changes in serum enzymes in rats after inhalation of organic solvents singly and in combination. *Toxicol. appl. Pharmacol.*, *45*, 809-819

Easley, D.M., Kleopfer, R.D. & Carasea, A.M. (1981) Gas chromatographic-mass spectrometric determination of volatile organic compounds in fish. *J. Assoc. off. anal. Chem.*, *64*, 653-656

Eller, P.M. (1984a) *NIOSH Manual of Analytical Methods*, 3rd ed., Vol. 2 (*DHHS (NIOSH) Publ. No. 84-100*), Washington DC, US Government Printing Office, pp. 1003-1 — 1003-9

Eller, P.M. (1984b) *NIOSH Manual of Analytical Methods*, 3rd ed., Vol. 1 (*DHHS (NIOSH) Publ. No. 84-100*), Washington DC, US Government Printing Office, pp. 1013-1 — 1013-4

Ferrario, J.B., Lawler, G.C., DeLeon, I.R. & Laseter, J.L. (1985) Volatile organic pollutants in biota and sediments of Lake Pontchartrain. *Bull. environ. Contam. Toxicol.*, *34*, 246-255

The Foxboro Co. (1985) *Ambient Air Instrumentation Sheet*, Foxboro, MA

Gosselin, R.E., Hodge, H.C., Smith, R.P. & Gleason, M.N. (1976) *Clinical Toxicology of Commercial Products, Acute Poisoning*, 4th ed., Baltimore, MD, Williams & Wilkins Co., pp. 119-121

Grasselli, J.G. & Ritchey, W.M., eds (1975) *CRC Atlas of Spectral Data and Physical Constants for Organic Compounds*, Vol. 4, Cleveland, OH, CRC Press, p. 209

Grzywa, Z. & Rudzki, E. (1981) Dermatitis from dichloropropane. *Contact Dermatol.*, *7*, 151-152

Hansch, C. & Leo, A. (1979) *Substituent Constants for Correlation Analysis in Chemistry and Biology*, New York, John Wiley & Sons, p. 179

Hawley, G.G., ed. (1981) *The Condensed Chemical Dictionary*, 10th ed., New York, Van Nostrand Reinhold, p. 864

Haworth, S., Lawlor, T., Mortelmans, K., Speck, W. & Zeiger, E. (1983) *Salmonella* mutagenicity test results for 250 chemicals. *Environ. Mutagenesis*, *5* (Suppl. 1)

Heppel, L.A., Neal, P.A., Highman, B. & Porterfield, V.T. (1946) Toxicology of 1,2-dichloropropane (propylene dichloride). I. Studies on effects of daily inhalations. *J. ind. Hyg. Toxicol.*, *28*, 1-8

Heppel, L.A., Highman, B. & Peake, E.G. (1948) Toxicology of 1,2-dichloropropane (propylene dichloride). IV. Effects of repeated exposures to a low concentration of the vapor. *J. ind. Hyg. Toxicol.*, *30*, 189-191

Hutson, D.H., Moss, J.A. & Pickering, B.A. (1971) The excretion and retention of components of the soil fumigant D-D and their metabolites in the rat. *Food Cosmet. Toxicol.*, *9*, 677-680

IARC (1985) *IARC Monographs on the Evaluation of the Carcinogenic Risk of Chemicals to Humans*, Vol. 36, *Allyl Compounds, Aldehydes, Epoxides and Peroxides*, Lyon, pp. 227-243

International Labour Office (1980) *Occupational Exposure Limits for Airborne Toxic Substances: A Tabular Compilation of Values from Selected Countries*, 2nd (rev.) ed. (*Occupational Safety and Health Series No. 37*), Geneva, pp. 182-183

Jones, A.R. & Gibson, J. (1980) 1,2-Dichloropropane: metabolism and fate in the rat. *Xenobiotica, 10*, 835-846

Kaiser, K.L.E., Comba, M.E. & Huneault, H. (1983) Volatile halocarbon contaminants in the Niagara River and in Lake Ontario. *J. Great Lakes Res., 9*, 212-223

Locatelli, F. & Pozzi, C. (1983) Relapsing haemolytic-uraemic syndrome after organic solvent sniffing. *Lancet, ii*, 220

Mannsville Chemical Products Corp. (1984) *Propylene Chloride (Chemical Products Synopsis)*, Cortland, NY

National Fire Protection Association (1984) *Fire Protection Guide on Hazardous Materials*, 8th ed., Quincy, MA, p. 325M-82

National Toxicology Program (1986) *Carcinogenesis Studies of 1,2-Dichloropropane (Propylene Dichloride) (CAS No. 78-87-5) in F344/N Rats and B6C3F$_1$ Mice (Gavage Studies) (Technical Report No. 263)*, Research Triangle Park, NC, US Department of Health and Human Services

Otson, R. & Williams, D.T. (1982) Headspace chromatographic determination of water pollutants. *Anal. Chem., 54*, 942-946

Otson, R., Williams, D.T. & Bothwell, P.D. (1982) Volatile organic compounds in water at thirty Canadian potable water treatment facilities. *J. Assoc. off. anal. Chem., 65*, 1370-1374

Pellizzari, E.D. (1982) Analysis for organic vapor emissions near industrial and chemical waste disposal sites. *Environ. Sci. Technol., 16*, 781-785

Perbellini, L., Zedde, A., Schiavon, R. & Franchi, G.L. (1985) Disseminated vascular coagulation syndrome (DIC) from 1,2-dichloropropane (commercial trichloroethylene preparation). Description of two cases (Ital.). *Med. Lav., 76*, 412-417

Ponticelli, C., Imbasciati, E., Redaelli, B. & Salvadeo, A. (1968) Acute hepato-renal insufficiency after trichloroethylene poisoning (Ital.). *Lav. Um., 20*, 205-212

Pozzani, U.C., Weil, C.S. & Carpenter, C.P. (1959) The toxicological basis of threshold limit values: 5. The experimental inhalation of vapor mixtures by rats with notes upon the relationship between single dose inhalation and single dose oral data. *Am. ind. Hyg. Assoc. J., 20*, 364-369

Pozzi, C., Marai, P., Magni, M., Vernocchi, A., Zedda, S. & Locatelli, F. (1984) Renal, liver and haematological damage after 1,2-dichloropropane intoxication (Ital.). *G. ital. Nefrol., 1*, 105-110

Pozzi, C., Marai, P., Ponti, R., Cell'Oro, C., Sala, C., Zedda, S. & Locatelli, F. (1985) Toxicity in man due to stain removers containing 1,2-dichloropropane. *Br. J. ind. Med., 42*, 770-772

Principe, P., Dogliotti, E., Bignami, M., Crebelli, R., Falcone, E., Fabrizi, M., Conti, G. & Comba, P. (1981) Mutagenicity of chemicals of industrial and agricultural relevance in *Salmonella, Streptomyces* and *Aspergillus. J. Sci. Food Agric., 32*, 826-832

Riggin, R.M. (1985) *Determination of volatile organic compounds in ambient air using Tenax adsorption and gas chromatography/mass spectrometry.* In: Fishbein, L. & O'Neill, I.K., eds, *Environmental Carcinogens. Selected Methods of Analysis*, Vol. 7, *Some Volatile Halogenated Hydrocarbons (IARC Scientific Publications No. 68)*, Lyon, International Agency for Research on Cancer, pp. 269-289

Sadtler Research Laboratories (1980) *The Sadtler Standard Spectra, Cumulative Index*, Philadelphia, PA

Sax, N.I. (1984) *Dangerous Properties of Industrial Materials*, 6th ed., New York, Van Nostrand Reinhold, p. 962

Secchi, G.C., Chiappino, G., Lotto, A. & Zurlo, N. (1968) Actual chemical composition of commercial trichloroethylenes and their liver toxicity. Clinical and enzymological study (Ital.). *Med. Lav., 59*, 486-497

Sidorenko, G.I., Tsulaya, V.R., Korenevskaya, E.I. & Bonashevskaya, T.I. (1976) Methodological approaches to the study of the combined effect of atmospheric pollutants as illustrated by chlorinated hydrocarbons. *Environ. Health Perspect., 13*, 111-116

Singh, H.B., Salas, L.J. & Stiles, R.E. (1982) Distribution of selected gaseous organic mutagens and suspect carcinogens in ambient air. *Environ. Sci. Technol., 16*, 872-880

Sittig, M., ed. (1980) *Pesticide Manufacturing and Toxic Materials Control Encyclopedia*, Park Ridge, NJ, Noyes Data Corp., pp. 290-291

Sittig, M., ed. (1985) *Pesticide Manufacturing and Toxic Materials Control Encyclopedia*, 2nd ed., Park Ridge, NJ, Noyes Data Corp., p. 329

Smyth, H.F., Carpenter, C.P., Weil, C.S., Pozzani, U.C., Striegel, J.A. & Nycum, J.S. (1969) Range-finding toxicity data: list VII. *Am. ind Hyg. Assoc. J., 30*, 470-476

Stolzenberg, S.J. & Hine, C.H. (1980) Mutagenicity of 2- and 3-carbon halogenated compounds in the *Salmonella*/mammalian-microsome test. *Environ. Mutagenesis, 2*, 59-66

Torkelson, T.R. & Rowe, V.K. (1981) *Halogenated aliphatic hydrocarbons containing chlorine, bromine and iodine.* In: Clayton, G.D. & Clayton, F.E., eds, *Patty's Industrial Hygiene and Toxicology*, 3rd ed., Vol. 2B, New York, John Wiley & Sons, pp. 3529-3532

Työsuojeluhallitus (National Finnish Board of Occupational Safety and Health) (1981) *Airborne Contaminants in the Workplaces (Safety Bull. 3)* (Finn.), Tampere, p. 10

US Environmental Protection Agency (1982a) *Method 8010. Halogenated volatile organics.* In: *Test Methods for Evaluating Solid Waste — Physical/Chemical Methods*, 2nd ed. (*US EPA No. SW-846*), Washington DC, Office of Solid Waste and Emergency Response

US Environmental Protection Agency (1982b) *Method 8240. GC/MS method for volatile organics*. In: *Test Methods for Evaluating Solid Waste — Physical/Chemical Methods*, 2nd ed. (*US EPA No. SW-846*), Washington DC, Office of Solid Waste and Emergency Response

US Environmental Protection Agency (1984a) Method 601. Guidelines establishing test procedures for the analysis of pollutants under the Clean Water Act (40 CFR 136). Purgeable halocarbons. *Fed. Regist.*, *49*, 43261-43271

US Environmental Protection Agency (1984b) Method 624. Guidelines establishing test procedures for the analysis of pollutants under the Clean Water Act (40 CFR 136). Purgeables. *Fed. Regist.*, *49*, 43373-43384

US Environmental Protection Agency (1984c) Method 1624, Revision B. Guidelines establishing test procedures for the analysis of pollutants under the Clean Water Act (40 CFR 136). Volatile organic compounds by isotope dilution GC/MS. *Fed. Regist.*, *49*, 43407-43415

US International Trade Commission (1976) *Synthetic Organic Chemicals, US Production and Sales, 1975* (*USITC Publ. No. 804*), Washington DC, US Government Printing Office, p. 198

US International Trade Commission (1978) *Synthetic Organic Chemicals, US Production and Sales, 1977* (*USITC Publ. No. 920*), Washington DC, US Government Printing Office, p. 360

US International Trade Commission (1979) *Synthetic Organic Chemicals, US Production and Sales, 1978* (*USITC Publ. No. 1001*), Washington DC, US Government Printing Office, p. 313

US International Trade Commission (1980) *Synthetic Organic Chemicals, US Production and Sales, 1979* (*USITC Publ. No. 1099*), Washington DC, US Government Printing Office, p. 269

US International Trade Commission (1981) *Synthetic Organic Chemicals, US Production and Sales, 1980* (*USITC Publ. No. 1183*), Washington DC, US Government Printing Office, p. 265

US Occupational Safety and Health Administration (1985) Labor. *US Code fed. Regul.*, *Title 29*, Part 1910.1000

Van Dyke, R.A. & Wineman, C.G. (1971) Enzymatic dechlorination: dechlorination of chloroethanes and propanes *in vitro*. *Biochem. Pharmacol.*, *20*, 463-470

Verschueren, K. (1983) *Handbook of Environmental Data on Organic Chemicals*, 2nd ed., New York, Van Nostrand Reinhold, pp. 506-507

Vettorazzi, G., ed. (1984) *Pesticides Reference Index, JMPR-IARC-IPCS-IRPTC-VBC, 1961-1983*, Geneva, World Health Organization, International Programme on Chemical Safety

Wallace, L., Zweidinger, R., Erickson, M., Cooper, S., Whitaker, D. & Pellizzari, E. (1982) Monitoring individual exposure. Measurements of volatile organic compounds in breathing-zone air, drinking water, and exhaled breath. *Environ. int.*, *8*, 269-282

Weast, R.C., ed. (1985) *CRC Handbook of Chemistry and Physics*, 66th ed., Boca Raton, FL, CRC Press, pp. C-442, D-198

Windholz, M., ed. (1983) *The Merck Index*, 10th ed., Rahway, NJ, Merck & Co., p. 1130

Woodruff, R.C., Mason, J.M., Valencia, R. & Zimmering, S. (1985) Chemical mutagenesis testing in *Drosophila*. V. Results of 53 coded compounds tested for the National Toxicology Program. *Environ. Mutagenesis*, *7*, 677-702

Worthing, C.R. (1983) *The Pesticide Manual, A World Compendium*, 7th ed., Croydon, The British Crop Protection Council, pp. 181-183

Yang, R.S.H. (1986) 1,3-Dichloropropene. *Residue Rev.*, *97*, 19-35

BIS(2-CHLORO-1-METHYLETHYL)ETHER

1. Chemical and Physical Data

1.1 Synonyms and trade names

Chem. Abstr. Services Reg. No.: 108-60-1
Chem. Abstr. Name: 2,2'-Oxybis(1-chloropropane)
IUPAC Systematic Name: Bis(2-chloro-1-methylethyl) ether
Synonyms: BCMEE; bis(β-chloroisopropyl)ether; bis(2-chloroisopropyl)ether; bis(1-chloro-2-propyl) ether; DCIP; dichlorodiisopropyl ether; 2,2'-dichlorodiisopropyl ether; β,β'-dichlorodiisopropyl ether; dichloroisopropylether; 2,2'-oxybis(2-chloropropane)

1.2 Structural and molecular formulae and molecular weight

$$Cl-CH_2-\underset{\underset{H}{|}}{\overset{\overset{CH_3}{|}}{C}}-O-\underset{\underset{H}{|}}{\overset{\overset{CH_3}{|}}{C}}-CH_2-Cl$$

$C_6H_{12}Cl_2O$ \hfill Mol. wt: 171.07

1.3 Chemical and physical properties of the pure substance

(a) *Description*: Colourless liquid

(b) *Boiling-point*: 187°C (Weast, 1985)

(c) *Melting-point*: -100°C (Verschueren, 1983)

(d) *Density*: d_4^{20} 1.103 (Weast, 1985)

(e) *Spectroscopy data*[a]: Infrared spectral data (Sadtler Research Laboratories, 1980; prism [120, 13382], grating [10994]) have been reported.

(f) *Solubility*: Slightly soluble in water (1700 mg/l) (Verschueren, 1983); soluble in ethanol, diethyl ether, acetone and benzene (Weast, 1985)

(g) *Volatility*: Vapour pressure, 1 mm Hg at 29.6°C (Weast, 1985); relative vapour density (air = 1), 6.0 (Verschueren, 1983)

(h) *Octanol/water partition coefficient (P)*: log P, 1.76 (ISHOW); 2.58 (ENVIROFATE) (ICIS Chemical Information System, 1985)

(i) *Conversion factor*: $mg/m^3 = 7.00 \times ppm$[b]

1.4 Technical products and impurities

Bis(2-chloro-1-methylethyl)ether is available in commercial quantities only in Japan (The Chemical Daily Co., 1984). It has also been available in research quantities as a mixture containing approximately 30% of the isomeric 2-chloro-1-methylethyl(2-chloro-*n*-propyl)ether and smaller amounts of other isomers as impurities (National Toxicology Program, 1982).

2. Production, Use, Occurrence and Analysis

2.1 Production and use

(a) *Production*

Bis(2-chloro-1-methylethyl)ether is no longer commercially produced in the USA or western Europe. It is formed as a byproduct in the manufacture of propylene oxide (see IARC, 1985) and propylene glycol (National Toxicology Program, 1982) by the chlorohydrin process in amounts estimated at 1-2% (by weight) of propylene oxide. It can be isolated and purified by fractional distillation; however, it appears that most of this byproduct is incinerated on-site instead of being purified (US Environmental Protection Agency, 1983).

---

[a] In square brackets, spectrum number in compilation

[b] Calculated from: mg/m^3 = (molecular weight/24.45) × ppm, assuming standard temperature (25°C) and pressure (760 mm Hg)

In the USA, the total capacity for propylene oxide production in 1975 was approximately 1.2 million tonnes; production during that year was 680 million kg. Five companies produced propylene oxide; four of them, accounting for 65% of capacity, used the chlorohydrin process (Khan & Hughes, 1979). One of these plants closed down in 1981 (US Environmental Protection Agency, 1983), making the current total capacity of propylene oxide production by the chlorohydrin process in the USA 717 million kg per year. Sampling at propylene oxide plants showed that 2.03 g bis(2-chloro-2-methylethyl)ether were emitted for every kg of propylene oxide produced (Khan & Hughes, 1979). [The Working Group noted that this value is a calculated approximation. However, using this approximation, it can be estimated that 1.45 million kg bis(2-chloro-1-methylethyl)ether would be emitted each year in the USA if all plants were producing at full capacity. On the basis of the estimated production of 900 million kg propylene oxide in the USA in 1983 (US International Trade Commission, 1984), approximately 1.8 million kg of the ether would have been released as atmospheric emissions or in wastewater.]

In 1977, 324 000 kg bis(2-chloro-1-methylethyl)ether were used in Japan (US Environmental Protection Agency, 1983).

(b)   Use

Bis(2-chloro-1-methylethyl)ether has been used as a solvent for fats, waxes and greases; as an extractant; as an ingredient in paint and varnish removers; as an ingredient in spotting and cleaning solutions; and as a component in textile processing (US Environmental Protection Agency, 1983). There is no evidence that it is currently being used in any of these applications in the USA. It has also been marketed and used in Japan as a nematocide (Mitsumori *et al.*, 1979; US Environmental Protection Agency, 1983).

(c)   Regulatory status and guidelines

No specific regulation or guideline has been promulgated for the use of, or occupational exposure to, bis(2-chloro-1-methylethyl)ether.

In the USA, a provisional occupational limit of 15 ppm (105 mg/m^3) has been suggested. To protect aquatic life, the maximum permissible concentration of chloroalkyl ethers in fresh water is 238 mg/l. The ambient water criterion for bis(2-chloro-1-methylethyl)ether is 34.7 μg/l (Sittig, 1985).

2.2 Occurrence

(a)   Natural occurrence

Bis(2-chloro-1-methylethyl)ether is not known to occur as a natural product.

(b)   Occupational exposure

No data on exposure levels were available to the Working Group.

(c) *Water*

Bis(2-chloro-1-methylethyl)ether was identified in 1971 in effluents from industrial plants, in raw intake water downstream from these plants (2.0 µg/l), and in tap-water (0.8 µg/l) from the Ohio River, IN, USA, in which concentrations of 0.5-5 µg/l were detected (Kleopfer & Fairless, 1972). This compound has also been found in the Kanawha River at Nitro, WV, USA (Rosen *et al.*, 1963), in the Mississippi River at New Orleans, LA, USA, and in the Rhine and Scheldt Rivers in The Netherlands (National Toxicology Program, 1982).

2.3 Analysis

Selected methods for the analysis of bis(2-chloro-1-methylethyl)ether in water and in solid wastes are identified in Table 1.

Table 1. Methods for the analysis of bis(2-chloro-1-methylethyl)ether

Sample matrix	Sample preparation	Assay procedure[a]	Limit of detection	Reference
Water	Extract (dichloromethane); dry (anhydrous sodium sulphate); concentrate (60°C); add hexane and concentrate	GC/ECD	0.8 µg/l	US Environmental Protection Agency (1984a)
	Extract, dry and concentrate as above	GC/MS	5.7 µg/l	US Environmental Protection Agency (1984b)
	Add stable isotope-labelled bis(2-chloro-1-methylethyl)ether; extract (dichloromethane); dry (sodium sulphate); concentrate (heat)	GC/MS	10 µg/l	US Environmental Protection Agency (1984c)
Solid wastes	Extract (polyethylene glycol or methanol) and dilute (water); purge (nitrogen or helium); trap (OV-1 on Chromosorb-W/Tenax/silica gel); desorb as vapour (heat to 180°C, backflush with inert gas) onto GC column	GC/MS	5.7 µg/l	US Environmental Protection Agency (1982)

[a]Abbreviations: GC/ECD, gas chromatography/electrolytic conductivity detection; GC/MS, gas chromatography/mass spectrometry

3. Biological Data Relevant to the Evaluation of Carcinogenic Risk to Humans

3.1 Carcinogenicity studies in animals

Oral administration

Mouse: Groups of 56 male and 56 female SPF ICR mice, five weeks old, were fed diets containing 0, 80, 400, 2000 or 10 000 mg/kg bis(2-chloro-1-methylethyl)ether (purity, 98.5%) for 104 weeks. Seven males and females from each group were killed at weeks 13, 26 and 52; six males and females from each group were killed at week 78; and all surviving animals were killed at week 104. Food and water consumption decreased and body-weight gain was reduced in the highest-dose group. Survival at week 104 was: males — 8 controls; 5 fed 80 mg/kg; 8 fed 400 mg/kg; 5 fed 2000 mg/kg; and 6 fed 10 000 mg/kg; females — 5, 9, 9, 7 and 1, in the respective groups. No significant increase in the incidence of any tumour type was found in the treated groups as compared to controls (Mitsumori *et al.*, 1979). [The Working Group noted that the experiment was designed to test chronic toxicity and had limited sensitivity for detecting carcinogenicity.]

Groups of 50 male and 50 female B6C3F$_1$ mice, eight weeks old, were administered 0, 100 or 200 mg/kg bw per day technical-grade bis(2-chloro-1-methylethyl)ether [containing about 30% 2-chloro-1-methylethyl(2-chloro-*n*-propyl)ether] in corn oil at 5 ml/kg bw per day by stomach tube on five days per week for 103 weeks, and were killed at weeks 104-110. Two batches of test material were used: the first for the first 94 weeks and the second for the final nine weeks; both were a mixture of three isomers, bis(2-chloro-1-methylethyl)ether (69.4% and 71.5%, respectively), bis(2-chloro-*n*-propyl)ether (2.1 and 2.6%) and 2-chloro-1-methylethyl(2-chloro-*n*-propyl)ether (28.5 and 25.9%). The total percentages of test material were 98.1 and 97.6-98.2%, confirmed by gas chromatography; the numbers of unspecified impurities were 32 and 16-17. Mean body weights of treated and control mice were comparable. Survival at 104-110 weeks was: males —41/50 controls; 44/50 given the low dose; and 37/50 given the high dose; females — 31/50, 34/50 and 28/50, respectively. There was a positive dose-related trend in the incidence of alveolar/bronchiolar adenomas in males (vehicle-control, 5/50; low-dose, 13/50; high-dose, 11/50, $p < 0.05$) and females (1/50, 4/50 and 8/50, $p < 0.02$). Alveolar/bronchiolar carcinomas were observed in 1/50, 2/50 and 2/50 males and 0/50, 0/50 and 2/50 females. In males, significant ($p < 0.01$) dose-related increases in the incidences of hepatocellular carcinomas (5/50, 13/50 and 17/50) and in the combined incidences of hepatocellular adenomas and carcinomas (13/50, 23/50 and 27/50) were observed; no significant increase was observed in females (National Toxicology Program, 1982).

Rat: Groups of 50 male and 50 female Fischer 344 rats, five to seven weeks old, were administered 0, 100 or 200 mg/kg bw per day technical-grade bis(2-chloro-1-methylethyl)ether in corn oil at 1 ml/kg bw per day by stomach tube on five days per week for 103 weeks and observed for one to two weeks. A group of 50 male and 50 female rats served as untreated controls. The test material was obtained from three different sources, and each was a mixture of three isomers: bis(2-chloro-1-methylethyl)ether (~65%), 2-chloro-1-methylethyl(2-chloro-*n*-propyl)ether (~31%) and bis(2-chloro-*n*-propyl)ether (~4%) (confirmed by nuclear magnetic resonance spectrometry). The total percentage of the three isomers varied between 92.1 and 98.2%; the number of unspecified impurities ranged from 4-32 per batch per measurement. Survival at weeks 104-105 was: males — 32/50 untreated controls; 33/50 vehicle controls; 34/50 low-dose; and 3/50 high-dose; females — 37/50, 41/50, 28/50 and 1/50, respectively. Dose-related decreases in body weight were observed in males and females. The incidences of tumours in treated animals of either sex were not significantly higher than those in vehicle-control groups (National Cancer Institute, 1979).

3.2 Other relevant biological data

(*a*) *Experimental systems*

Toxic effects

The toxicity of bis(2-chloro-1-methylethyl)ether has been reviewed (Kirwin & Sandmeyer, 1981).

Oral LD_{50} values of 240 and 450 mg/kg bw have been reported for rats and guinea-pigs, respectively (Smyth *et al.*, 1951; Spector, 1956). The dermal LD_{50} in rabbits was estimated to be 3 ml/kg bw. Rats exposed by inhalation for 4 h to 1000 ppm (7000 mg/m³) bis(2-chloro-1-methylethyl)ether survived the exposure period but 1/6 died two weeks later (Smyth *et al.*, 1951). Dose- and time-dependent fatalities in rats have been reported following exposure by inhalation; the lowest dose at which deaths were observed has been reported to be 175 ppm (1225 mg/m³) for 8 h (Kirwin & Sandmeyer, 1981). Other effects reported following inhalation exposure include irritation of the eyes, nose and lung as well as hepatic and renal injury (Gage, 1970; Kirwin & Sandmeyer, 1981). Bis(2-chloro-1-methylethyl)ether is reported to be a relatively weak skin irritant (Kirwin & Sandmeyer, 1981).

Rats administered 22 oral doses of 200 mg/kg bw bis(2-chloro-1-methylethyl)ether over 31 days had decreased weight gain and increased weights of liver, kidney and spleen. Oral doses of 10 mg/kg bw caused reductions in weight gain only (Kirwin & Sandmeyer, 1981). Exposure of rats by inhalation to 350 ppm (2450 mg/m³) bis(2-chloro-1-methylethyl)ether for 5 h per day for eight exposures caused respiratory distress, reduced weight gain and congestion of the liver and kidneys. Twenty 6-h exposures to 20 or 70 ppm (140 or 490 mg/m³) resulted in reduced weight gain in the higher-dose group only (Gage, 1970).

All $B6C3F_1$ mice administered daily doses of 562 mg/kg bw bis(2-chloro-1-methylethyl)ether (~70% pure) by gavage for two weeks died, whereas administration of doses of 17.8-316 mg/kg bw caused few toxic effects. No death occurred in a group of mice administered 10-250 mg/kg bw bis(2-chloro-1-methylethyl)ether by gavage on seven days per week for 13 weeks, although morphological changes were observed in the lungs of some animals treated with doses over 50 mg/kg bw (National Toxicology Program, 1982).

In a bioassay [see section 3.1], administration of bis(2-chloro-1-methylethyl)ether (~70% pure) by gavage to Fischer 344 rats or B6C3F$_1$ mice for up to two years resulted in reduced weight gain in rats and in high-dose female mice (National Cancer Institute, 1979; National Toxicology Program, 1982). Following exposure of mice to bis(2-chloro-1-methylethyl)ether at doses of 2000-10000 mg/kg in the diet for up to two years, anaemia and reduced body-weight gain were observed. Mortality in the high-dose group was greater than among controls (Mitsumori *et al.*, 1979).

Effects on reproduction and prenatal toxicity

No data were available to the Working Group.

Absorption, distribution, excretion and metabolism

In rats administered a single oral dose of 0.2 µg/kg-300 mg/kg bw [2-^{14}C]-bis(2-chloro-1-methylethyl)ether (>95% pure), peak blood levels of radioactivity were reached 2-4 h after treatment. After administration of 300 mg/kg bw, the disappearance of radioactivity from the blood showed saturation kinetics. In monkeys receiving 30 mg/kg bw orally, two blood elimination phases were observed, with half-lives of 5 h and two days, whereas only one half-life of two days was observed in rats (Smith *et al.*, 1977).

Following administration to rats of ^{14}C-bis(2-chloro-1-methylethyl)ether orally (90 mg/kg bw; [1-^{14}C] label) (Lingg *et al.*, 1982) or intraperitoneally (30 mg/kg bw; [2-^{14}C] label) (Smith *et al.*, 1977), urinary excretion of radioactivity accounted for approximately 48% within 48 h and 60% within 24 h, respectively. After intravenous administration of 30 mg/kg bw [2-^{14}C]-bis(2-chloro-1-methylethyl)ether to a rhesus monkey, 30% of the radioactivity was excreted in the urine within 24 h (Smith *et al.*, 1977). Following parenteral administration to the rat and monkey, urinary excretion did not increase after 24 h within a seven-day observation period. Faecal excretion amounted to approximately 4% of the radioactivity within 48 h following oral administration to rats, and total faecal excretion amounted to 6% and 1%, respectively, in seven days, in rats treated intraperitoneally and in a monkey treated intravenously. Following oral administration of [1-^{14}C]-bis(2-chloro-1-methylethyl)ether to rats, approximately 20% of the radioactivity was exhaled as ^{14}C-CO$_2$ within 48 h, and approximately 2% of the radioactivity was retained. In rats, total recoveries of radioactivity were 75% following oral administration ([1-^{14}C] label) and 90% following intraperitoneal administration ([2-^{14}C] label) (Smith *et al.*, 1977; Lingg *et al.*, 1982). [The Working Group noted that two different ^{14}C-radiolabelled compounds were used in these studies, which may explain the differences in urinary excretion observed.]

In rats given an intraperitoneal injection of [2-^{14}C]-bis(2-chloro-1-methylethyl)ether, 1% of the radioactivity recovered in the urine was unchanged compound; the urinary metabolites were 1-chloropropan-2-ol, propylene oxide and 2-(2-chloro-1-methylethoxy)propanoic acid (Smith *et al.*, 1977). Following oral administration of [1-^{14}C]-bis(2-chloro-1-methylethyl)ether (90 mg/kg bw) to rats, 2-(2-chloro-1-methylethoxy)propanoic acid was recovered in the urine as approximately 17% of the dose, and *N*-acetyl-*S*-(2-hydroxypropyl)cysteine as approximately 9% of the dose (Lingg *et al.*, 1982).

Mutagenicity and other short-term tests

Bis(2-chloro-1-methylethyl)ether was mutagenic to *Salmonella typhimurium* TA100 treated with 1 ml in a dessicator for 3-9 h, both in the presence and absence of homogenates from uninduced mouse, hamster, human and rat liver. Mutation frequencies were higher when metabolic systems were present (Simmon, 1978; Simmon & Tardiff, 1978).

Bis(2-chloro-1-methylethyl)ether (purity, 34.3% [impurities unspecified]) was tested in two laboratories, in a preincubation assay with *S. typhimurium*, at doses of 10-3333 μg/plate. In the presence of Aroclor-induced hamster-liver S9, it induced a low but dose-related increase in mutation frequency in strains TA100 and TA1535 (Mortelmans *et al.*, 1986).

A commercial preparation of bis(2-chloro-1-methylethyl)ether was reported to be nonmutagenic in *S. typhimurium* TA1535, TA1537, TA1538, TA98 and TA100 and in *Escherichia coli* WP2 *hcr* (Moriya *et al.*, 1983). [The Working Group noted that values were not given and no precaution was taken to prevent volatilization of the compound.]

When *Drosophila melanogaster* Canton-S stock flies were treated with bis(2-chloro-1-methylethyl)ether (purity, 34.3% [impurities unspecified]) by feeding (250 or 283 μg/g) or injection (1600 μg/ml), no statistically significant increase in the incidence of sex-linked recessive lethal mutations was found (Valencia *et al.*, 1985).

It was reported in an abstract that bis(2-chloro-1-methylethyl)ether did not induce unscheduled DNA synthesis in hepatocytes of male mice treated *in vivo* [details not given] (Mirsalis *et al.*, 1985).

(*b*) *Humans*

No data were available to the Working Group.

3.3 Case reports and epidemiological studies of carcinogenicity to humans

No data were available to the Working Group.

4. Summary of Data Reported and Evaluation

4.1 Exposure data

Bis(2-chloro-1-methylethyl)ether has been produced as a solvent and soil fumigant and is also formed in large quantities as a by-product in some propylene oxide/propylene glycol production processes. Low levels have been found in water. Thus, both occupational and environmental exposures may occur.

4.2 Experimental data

Bis(2-chloro-1-methylethyl)ether, containing 2-chloro-1-methylethyl(2-chloro-*n*-propyl)ether and bis(2-chloro-*n*-propyl)ether, was tested for carcinogenicity by oral administration by gavage in one experiment in mice and in one experiment in rats. In mice, increased incidences of lung adenomas in males and females and of hepatocellular carcinomas in males were observed. In rats, no increase in tumour incidence was observed.

No data were available to evaluate the reproductive or prenatal effects of bis(2-chloro-1-methylethyl)ether to experimental animals.

Bis(2-chloro-1-methylethyl)ether is mutagenic to *Salmonella typhimurium*. It does not induce sex-linked recessive lethal mutations in *Drosophila melanogaster*.

Overall assessment of data from short-term tests: Bis(2-chloro-1-methylethyl)ether[a]

	Genetic activity			Cell transformation
	DNA damage	Mutation	Chromosomal effects	
Prokaryotes		+		
Fungi/Green plants				
Insects		−		
Mammalian cells (*in vitro*)				
Mammals (*in vivo*)				
Humans (*in vivo*)				
Degree of evidence in short-term tests for genetic activity: **Inadequate**				Cell transformation: No data

[a]The groups into which the table is divided and the symbols '+' and '−' are defined on pp. 19-20 of the Preamble; the degrees of evidence are defined on pp. 20-21.

4.3 Human data

No data were available to evaluate the reproductive effects or prenatal toxicity of bis(2-chloro-1-methylethyl)ether to humans.

No case report or epidemiological study of the carcinogenicity of bis(2-chloro-1-methylethyl)ether to humans was available to the Working Group.

4.4 Evaluation[1]

There is *limited evidence* for the carcinogenicity of bis(2-chloro-1-methylethyl)ether to experimental animals.

No evaluation could be made of the carcinogenicity of bis(2-chloro-1-methylethyl)ether to humans.

5. References

The Chemical Daily Co. (1984) *JCW Chemicals Guide 1984/1985*, Tokyo, p. 132

Gage, J.C. (1970) The subacute inhalation toxicity of 109 industrial chemicals. *Br. J. ind. Med., 27*, 1-18

IARC (1985) *IARC Monographs on the Evaluation of the Carcinogenic Risk of Chemicals to Humans*, Vol. 36, *Allyl Compounds, Aldehydes, Epoxides and Peroxides*, Lyon, pp. 227-243

ICIS Chemical Information System (1985) *Information System for Hazardous Organics in Water* (ISHOW), and *Environmental Fate* (ENVIROFATE), Washington DC, Information Consultants, Inc.

Khan, Z.S. & Hughes, T.W. (1979) *Source Assessment: Chlorinated Hydrocarbons Manufacture (EPA 600/2-79-019g)*, Washington DC, US Government Printing Office

Kirwin, C.J., Jr & Sandmeyer, E.E. (1981) *Ethers*. In: Clayton, G.D. & Clayton, F.E., eds, *Patty's Industrial Hygiene and Toxicology*, 3rd revised ed., Vol. 2A, *Toxicology*, New York, John Wiley & Sons, pp. 2505, 2519-2520

Kleopfer, R.D. & Fairless, B.J (1972) Characterization of organic components in a municipal water supply. *Environ. Sci. Technol., 6*, 1036-1037

Lingg, R.D., Kaylor, W.H., Pyle, S.M., Domino, M.M., Smith, C.C. & Wolfe, G.F. (1982) Metabolism of bis(2-chloroethyl)ether and bis(2-chloroisopropyl)ether in the rat. *Arch. environ. Contam. Toxicol., 11*, 173-183

Mirsalis, J., Tyson, K., Loh, E., Bakke, J., Hamilton, C., Spak, D., Steinmetz, K. & Spalding, J. (1985) Induction of unscheduled DNA synthesis (UDS) and cell proliferation in mouse and rat hepatocytes following *in vivo* treatment (Abstract). *Environ. Mutagenesis, 7*, 73

Mitsumori, K., Usui, T., Takahashi, K. & Shirasu, Y. (1979) Twenty-four month chronic toxicity studies of dichlorodiisopropyl ether in mice. *J. pestic. Sci., 4*, 323-335

---

[1]For definition of the italicized term, see Preamble, p. 18.

Moriya, M., Ohta, T., Watanabe, K., Miyazawa, T., Kato, K. & Shirasu, Y. (1983) Further mutagenicity studies on pesticides in bacterial reversion assay systems. *Mutat. Res.*, *116*, 185-216

Mortelmans, K., Haworth, S., Lawlor, T., Speck, W., Tainer, B. & Zeiger, E. (1986) *Salmonella* mutagenicity tests: II. Results from the testing of 270 chemicals. *Environ. Mutagenesis*, *8* (Suppl. 7), 1-119

National Cancer Institute (1979) *Bioassay of Technical-grade Bis(2-chloro-1-methylethyl)ether for Possible Carcinogenicity* (*Technical Report Series No. 191*), Bethesda, MD, US Department of Health, Education, and Welfare

National Toxicology Program (1982) *Carcinogenesis Bioassay of Bis(2-chloro-1-methylethyl)ether (~70%) (CAS No. 108-60-1) Containing 2-Chloro-1-methylethyl(2-chloropropyl)ether (~30%) (CAS No. 83270-31-9) in $B6C3F_1$ Mice (Gavage Study)* (*Technical Report No. 239*), Research Triangle Park, NC, US Department of Health and Human Services

Rosen, A., Skeel, R. & Ettinger, M. (1963) Relationship of river water odor to specific organic contaminants. *J. Water Pollut. Control Fed.*, *35*, 777-782

Sadtler Research Laboratories (1980) *The Sadtler Standard Spectra, Cumulative Index*, Philadelphia, PA

Simmon, V.F. (1978) *Structural correlations of carcinogenic and mutagenic alkyl halides*. In: Asher, I.M. & Zervos, C., eds, *Structural Correlates of Carcinogenesis and Mutagenesis. A Guide to Testing Priorities?* (*Proceedings of the Second FDA Office of Science Summer Symposium, US Naval Academy, August 31-September 2, 1977*), Washington DC, US Food and Drug Administration, Office of Science, pp. 163-171

Simmon, V.F. & Tardiff, R.G. (1978) *The mutagenic activity of halogenated compounds found in chlorinated drinking water*: In: Jolley, R.L., ed., *Water Chlorination. Environmental Impact and Health Effects*, Vol. 2, *Environment, Health and Risk*, Ann Arbor, MI, Ann Arbor Science Publishers, pp. 417-431

Sittig, M. (1985) *Handbook of Toxic and Hazardous Chemicals and Carcinogens*, 2nd ed., Park Ridge, NJ, Noyes Publications, pp. 131-133

Smith, C.C., Lingg, R.D. & Tardiff, R.G. (1977) Comparative metabolism of haloethers. *Ann. N.Y. Acad. Sci.*, *298*, 111-123

Smyth, H.F., Carpenter, C.P. & Weil, C.S. (1951) Range-finding toxicity data: list IV. *Arch. ind. Hyg. occup. Med.*, *4*, 119-122

Spector, W.S., ed. (1956) *Handbook of Toxicology*, Vol. 1, Philadelphia, PA, Saunders, pp. 94-95

US Environmental Protection Agency (1982) *Method 8250. GC/MS method for semivolatile organics: packed column technique*. In: *Test Methods for Evaluating Solid Waste — Physical/Chemical Methods*, 2nd ed. (*US EPA No. SW-846*), Washington DC, Office of Solid Waste and Emergency Response

US Environmental Protection Agency (1983) *Chemical Hazard Information Profile Draft Report: Bis(2-chloro-1-methylethyl) Ether*, Washington DC, US Government Printing Office

US Environmental Protection Agency (1984a) Method 611. Guidelines establishing test procedures for the analysis of pollutants under the Clean Water Act (40 CFR 136). Haloethers. *Fed. Regist.*, *49*, 43353-43359

US Environmental Protection Agency (1984b) Method 625. Guidelines establishing test procedures for the analysis of pollutants under the Clean Water Act (40 CFR 136). Base/neutrals and acids. *Fed. Regist.*, *49*, 43385-43406

US Environmental Protection Agency (1984c) Method 1625, Revision B. Guidelines establishing test procedures for the analysis of pollutants under the Clean Water Act (40 CFR 136). Semivolatile organic compounds by isotope dilution GC/MS. *Fed. Regist.*, *49*, 43416-43429

US International Trade Commission (1984) *Synthetic Organic Chemicals, US Production and Sales, 1983* (*USITC Publ. No. 1588*), Washington DC, US Government Printing Office, p. 260

Valencia, R., Mason, J.M., Woodruff, R.C. & Zimmering, S. (1985) Chemical mutagenesis testing in *Drosophila*. III. Results of 48 coded compounds tested for the National Toxicology Program. *Environ. Mutagenesis*, *7*, 325-348

Verschueren, K. (1983) *Handbook of Environmental Data on Organic Chemicals*, 2nd ed., New York, Van Nostrand Reinhold, p. 490

Weast, R.C., ed. (1985) *CRC Handbook of Chemistry and Physics*, 66th ed., Boca Raton, FL, CRC Press, pp. C-206, D-202

METHYL CHLORIDE

1. Chemical and Physical Data

1.1 Synonyms and trade names

Chem. Abstr. Services Reg. No.: 74-87-3
Chem. Abstr. Name: Chloromethane
IUPAC Systematic Name: Chloromethane
Synonym: Monochloromethane
Trade Names: Artic; R 40

1.2 Structural and molecular formulae and molecular weight

$$H-\underset{\underset{H}{|}}{\overset{\overset{H}{|}}{C}}-Cl$$

CH_3Cl Mol. wt: 50.49

1.3 Chemical and physical properties of the pure substance

(a) *Description*: Colourless gas with ethereal odour and sweet taste (Hawley, 1981; Windholz, 1983)

(b) *Boiling-point*: -24.2°C (Weast, 1985)

(c) *Melting-point*: -97.1°C (Weast, 1985)

(d) *Density*: d_4^{20} 0.9159 (Weast, 1985)

(e) *Spectroscopy data*[a]: Infrared (Sadtler Research Laboratories, 1980; prism [842], grating [15290]), nuclear magnetic resonance and mass spectral data (Grasselli & Ritchey, 1975) have been reported.

---

[a]In square brackets, spectrum number in compilation

(f) *Solubility*: Slightly soluble in water (303 ml/100 ml at 20°C) (Verschueren, 1983; Windholz, 1983); soluble in ethanol (3740 ml/100 ml at 20°C), diethyl ether, acetone, chloroform, benzene (4723 ml/100 ml at 20°C), carbon tetrachloride (3756 ml/100 ml at 20°C) and glacial acetic acid (3679 ml/100 ml at 20°C) (Hawley, 1981; Windholz, 1983; Sax, 1984; Weast, 1985)

(g) *Volatility*: Vapour pressure, 5.0 atm (3800 mm Hg) at 22°C (Weast, 1985); relative vapour density (air = 1), 1.8 (National Fire Protection Association, 1984; Sax, 1984)

(h) *Stability*: Flash-point, -45.5°C (National Fire Protection Association, 1984); decomposes in water with a half-time of 4.66 h at 100°C (Davis *et al.*, 1977)

(i) *Reactivity*: Reacts with active metals (aluminium, magnesium, potassium, sodium, zinc) (Hawley, 1981; Sax, 1984)

(j) *Octanol/water partition coefficient (P)*: log P, 0.91 (Hansch & Leo, 1979)

(k) *Conversion factor*: $mg/m^3 = 2.07 \times ppm$[a]

1.4 Technical products and impurities

Methyl chloride is marketed as a liquefied gas under pressure, generally in 99.5-99.95% purity. Impurities (maximum concentrations in $\mu g/g$) may include: water (40-100), acid (as hydrochloric acid, 5-10), dimethyl ether (20), methanol (50-100), acetone (25-50), acetylene (200), dimethyl ether plus dichloromethane (see monograph, p. 43) (25), ethyl chloride (10-125), and vinyl chloride (see IARC, 1979a, 1982a) (70-100). Higher chlorides may be present in trace amounts in methyl chloride produced by the chlorination of methane (Ahlstrom & Steele, 1979; Air Products & Chemicals, 1981; Linde, 1982; ICP Chemicals & Plastics, 1983; Matheson, 1985; ARC Chemical Corp., undated; Ethyl Corp., undated; Vista Chemical Co., undated).

2. Production, Use, Occurrence and Analysis

2.1 Production and use

(a) *Production*

Methyl chloride was synthesized as early as 1835. A purified form, prepared in 1874, was produced by reaction of hydrogen chloride and zinc chloride with methanol. The major commercial routes for the production of methyl chloride include chlorination of methane

---

[a]Calculated from: mg/m^3 = (molecular weight/24.45) × ppm, assuming standard temperature (25°C) and pressure (760 mm Hg)

and hydrochlorination of methanol. In the latter, more common reaction, equimolar portions of vaporized methanol and hydrogen chloride are reacted at approximately 350°C over alumina gel or other suitable catalysts. In 1974, 98% of methyl chloride produced in the USA was by methanol hydrochlorination (Ahlstrom & Steele, 1979; Walas, 1985). A low-temperature reaction method has also been described (Akiyama *et al.*, 1981).

The other, lesser used procedure, methane chlorination, may be accomplished thermally and through light-mitigated and catalytic reactions. In addition to methyl chloride, a typical reaction of this type yields substantial quantities of dichloromethane (see monograph, p. 43), chloroform (see IARC, 1979b, 1982b) and a small amount of carbon tetrachloride (see IARC, 1979c, 1982c) (Ahlstrom & Steele, 1979).

Methyl chloride may also be produced by reaction of methyl ether and hydrogen chloride with water; the reaction of dimethyl sulphate (see IARC, 1974, 1982d) with aluminium chloride or sodium chloride; decomposition of monochlorodimethyl ether with zinc; and the heating of methanol with chloroform or carbon tetrachloride (Ahlstrom & Steele, 1979).

Estimates for world production of methyl chloride are given in Table 1.

Table 1. Estimate of world production of methyl chloride (million kg)[a]

1960	1965	1970	1975	1980
63	142	318	232	360

[a]From Edwards *et al.* (1982)

The figure for 1980 reflects contributions of 150, 180 and 30 million kg per year by European countries, the USA and Japan, respectively (Edwards *et al.*, 1982). Production capacities in 1984 were estimated to be 320, 256 and 48 million kg per year in Europe, the USA and Japan, respectively (Matsushima, 1986).

In 1983, there were eight major US producers of methyl chloride (Anon., 1983). Five Japanese companies have also been reported to produce this compound (The Chemical Daily Co., 1984). In Europe, five manufacturers have been identified in the Federal Republic of Germany, as well as two each in France and the UK, and one each in Belgium, Greece, Italy, Spain and Switzerland.

(b) Use

Methyl chloride was first used as a refrigerant and dye intermediate in Europe in the late 1800s. It began to be widely used in the USA during the 1920s and in the UK in the 1930s, primarily as a refrigerant (Ahlstrom & Steele, 1979).

The current principal use of methyl chloride is in the production of methyl silicone polymers and resins, and in the manufacture of tetramethyl- and mixed tetramethylethyllead (see IARC, 1980, 1982e) antiknock compounds for gasoline. It is used to a lesser extent as a chemical intermediate and as a solvent. Methyl cellulose, a water-soluble resin used as a paint thickener, as a protective colloid and in paper processing, is produced by the etherification of cellulose with methyl chloride. Polymerization of isobutylene with isoprene to produce butyl rubber (see IARC, 1982f) is carried out in methyl chloride solvent. Methyl chloride or a combination of methyl chloride and fluorocarbons is used as a blowing agent in the production of polystyrene foams (see IARC, 1979d). In addition, methyl chloride has been used as an intermediate in the production of plastics, pharmaceuticals, herbicides (such as dimethyl arsenates), dyes, surfactants and disinfectants, methyl ethers, commercial Grignard reagents, methyl mercaptan and dichloromethane (Ahlstrom & Steele, 1979; Sittig, 1980; American Conference of Governmental Industrial Hygienists, 1984; Mannsville Chemical Products Corp., 1984).

Methyl chloride has been used to a limited extent as a local anaesthetic, alone or in combination with ethyl chloride. It was also used in the early 1900s as a general anaesthetic agent, but this application was discontinued due to its toxic side-effects (Browning, 1965; American Conference of Governmental Industrial Hygienists, 1984).

Regional use estimates for 1978-1979 are presented in Table 2. In a more recent analysis of US use patterns, 80% of the methyl chloride produced was used in the production of methyl silicone polymers and only 7% in lead antiknock compounds (Mannsville Chemical Products Corp., 1984). Use of methyl chloride in lead antiknock compounds will probably continue to decline with increasing restrictions on lead additives in gasoline (Anon., 1983; Mannsville Chemical Products Corp., 1984).

Table 2. Regional use estimates for methyl chloride (%)[a]

Use	USA	Europe	Japan
Silicones	50	45	71
Tetramethyllead	30	28	–
Butyl rubber	4	7	3
Methyl cellulose	4	4	14
Miscellaneous	12	16	12

[a]From Edwards et al. (1982)

(c) *Regulatory status and guidelines*

Occupational exposure limits for methyl chloride in 21 countries are presented in Table 3.

Table 3. Occupational exposure limits for methyl chloride[a]

Country	Year	Concentration (mg/m³)	Interpretation[b]
Australia	1978	210	TWA
Belgium	1978	210	TWA
Bulgaria	1971	5	Ceiling
Czechoslovakia	1976	100	TWA
		200	Ceiling
Finland	1981	105	TWA
		160	STEL
France	1985	105	TWA
		210	STEL
German Democratic Republic	1979	100	TWA
		200	STEL
Germany, Federal Republic of	1985	105	TWA
Hungary	1974	20	TWA
Italy	1978	200	TWA
Japan	1978	210	TWA
The Netherlands	1978	210	TWA
Norway	1981	105	TWA
Poland	1976	20	Ceiling
Romania	1975	100	TWA
		200	Ceiling
Sweden	1984	100	TWA
		200	STEL
Switzerland	1978	105	TWA
UK	1985	210	TWA
		260	STEL
USA	1985		
ACGIH		105	TWA
		205	STEL
OSHA		210	TWA
		420	Ceiling
		630	STEL
USSR	1977	5	Ceiling
Yugoslavia	1971	50	Ceiling

[a]From International Labour Office (1980); Direktoratet for Arbeidstilsynet (1981); Työsuojeluhallitus (1981); Arbetarskyddsstyrelsens Författningssamling (1984); American Conference of Governmental Industrial Hygienists (ACGIH) (1985); Deutsche Forschungsgemeinschaft (1985); Health and Safety Executive (1985); Institut National de Recherche et de Sécurité (1985); US Occupational Safety and Health Administration (OSHA) (1985)

[b]TWA, time-weighted average; STEL, short-term exposure limit

2.2 Occurrence

(a) Natural occurrence

Methyl chloride is produced by a variety of marine organisms (Lovelock, 1975).

(b) Occupational exposure

Four US chemical plants were selected by the National Institute for Occupational Safety and Health for surveys of worker exposure to methyl chloride in 1978-1979. Plant A was a multi-chemical production facility where methyl chloride was produced and also used in the production of tetramethyllead. Plant B was a major producer of methyl chloride. Plant C produced methyl chloride and used it in the production of methyl chlorosilanes. Plant D was a large chemical facility where methyl chloride was used as a blowing agent in the production of polystyrene foam. Personal, 8-h time-weighted average concentrations of methyl chloride in the methyl chloride production facility at Plant A ranged from 8.9 to 12.4 ppm (18.4-25.7 mg/m^3); in the tetramethyllead facility, the range was from <0.16 to 12.1 ppm (<0.33-25.1 mg/m^3); the average (geometric mean) for all workers surveyed was 1.3 ppm (2.7 mg/m^3). At Plant B, personal exposure levels ranged from <0.2 to 7.5 ppm (<0.4-15.5 mg/m^3); the average exposure for all workers in the survey was 1.1 ppm (2.3 mg/m^3). Personal exposures at Plant C varied from <0.1 (detectable limits) to 12.7 ppm (<0.2-26.3 mg/m^3), with a mean of 0.4 ppm (0.8 mg/m^3). At Plant D, personal exposures to methyl chloride ranged from 2.98 to 21.4 ppm (6.2-44.3 mg/m^3), with a mean of 6.75 ppm (14 mg/m^3) for all the jobs monitored (Cohen et al., 1980).

Workplace concentrations were measured by continuous monitoring at 12 sampling points in a Dutch methyl chloride production plant. The calculated 8-h average exposures of six workers ranged from 30 to 90 ppm (62-186 mg/m^3) during one working week (van Doorn et al., 1980).

(c) Air

Methyl chloride is probably the most abundant halocarbon in the atmosphere. The average global atmospheric concentration is estimated to be in the order of 1 ppb (2.1 μg/m^3), with lower concentrations (0.6-0.8 ppb [1.2-1.7 μg/m^3]) (Cronn et al., 1977; Singh et al., 1979; Rasmussen et al., 1980; Singh et al., 1982; Rasmussen & Khalil, 1983; Singh et al., 1983) in rural areas and higher concentrations in urban areas and near localized sources. The principal sources are formation in the oceans by seaweeds and marine microorganisms and by the combustion of organic matter. Combustion processes in which methyl chloride has been detected include forest fires (1-80 ppb [2.1-166 μg/m^3]); slash-and-burn agriculture; cigarettes (0.5-100 ppb [1-207 μg/m^3]); burning of fossil fuels; incineration of domestic, municipal and industrial rubbish; and accidental fires (Lovelock, 1977; Edwards et al., 1982).

The major source of atmospheric methyl chloride is believed to be the oceans, which release 1000-8000 million kg per year; urban sources and the combustion of vegetation are estimated to release 150-600 million kg per year (Tassios & Packham, 1985). Industrial emissions from the manufacture, processing and distribution of methyl chloride were estimated to contribute only about 20 million kg in 1980 (Edwards et al., 1982).

Average levels in air at ten urban sites in the USA ranged from 0.7-3.0 ppb (1.4-6.2 $\mu g/m^3$) (Singh et al., 1981, 1982). Singh et al. (1981) calculated that average human exposures to methyl chloride at these sites ranged up to 140 μg/person per day. The background concentration at surface level from natural sources at 40°N latitude was estimated to be 650 ppt (1.3 $\mu g/m^3$) (Singh et al., 1982).

The primary mechanism for the removal of methyl chloride from the atmosphere is believed to be reaction with hydroxyl radicals. On this hypothesis, the residence time of methyl chloride in an urban atmosphere was estimated to be 231 days with a daily rate of loss (12 sunlit hours) of 0.4% (Singh et al., 1981).

The average tropospheric background concentration of methyl chloride in March 1976 over the Pacific Northwest region of the USA was reported to be 569 ppt (1.2 $\mu g/m^3$) and accounted for about 24% of the total atmospheric burden of carbon-bound chlorine measured. The concentration in the lower stratosphere was only 5% lower than in the troposphere (Cronn et al., 1977).

Methyl chloride measurements from 1979 to 1981 showed fairly uniform atmospheric concentrations over the eastern Pacific, from 40°N to 32°S latitude, with a mean of 622 ppt (1.3 $\mu g/m^3$) (Singh et al., 1979, 1983). In other measurements over the Pacific, from about 65°N to 90°S latitude, Rasmussen et al. (1980) reported slightly higher average concentrations within the tropospheric layer of 815 ppt (1.7 $\mu g/m^3$) in 1977 and 755 ppt (1.6 $\mu g/m^3$) in 1978. Lower concentrations were found above the boundary layer, with average values of 629 ppt (1.3 $\mu g/m^3$) and 618 ppt (1.3 $\mu g/m^3$) in 1977 and 1978, respectively. In May 1982, the average concentration of methyl chloride in the lower troposphere over Point Barrow, Alaska (70°N), was reported to be 647 ppt (1.3 $\mu g/m^3$) (Rasmussen & Khalil, 1983).

Air samples collected in 1980 over the State of Washington in the vicinity of Mt St Helens during a period of volcanic activity were analysed for several trace gases. Methyl chloride concentrations in the volcanic plumes ranged from 1.1 to 3.3 ppb (2.3-6.8 $\mu g/m^3$) in the troposphere and from 0.5 to 2 ppb (1-4.1 $\mu g/m^3$) in the stratosphere (Cronn & Nutmagul, 1982; Hobbs et al., 1982). Leifer et al. (1981) reported non-plume concentrations of methyl chloride in the same vicinity during this time period of approximately 0.7 ppb (1.5 $\mu g/m^3$) in the troposphere and 0.4-0.6 ppb (0.8-1.2 $\mu g/m^3$) in the stratosphere.

Measurements of methyl chloride in the air of southern England between December 1974 and April 1975 showed a mean concentration of 1.1 ppb (2.3 $\mu g/m^3$) (Lovelock, 1975). The average concentration in the stratosphere over France in June 1978 was calculated to be approximately 0.7 ppb (1.5 $\mu g/m^3$), as determined from balloon-probe measurements (Penkett et al., 1980).

(d) Water

Methyl chloride concentrations in seawater have been reported as ranging from 0.01-0.05 $\mu g/l$. Traces (in the order of 1 $\mu g/l$) have been detected in some samples of drinking-water treated by chlorination (Edwards et al., 1982).

Concentrations of methyl chloride in seawater collected at Kimmeridge, Dorset, UK, in 1975 ranged from 5.9 to 21 µg/l (Lovelock, 1975). The near-shorewater concentration of methyl chloride at Point Reyes, CA, USA, was 1.2 µg/l (Singh et al., 1977). Seawater concentrations of methyl chloride in the Pacific Ocean in 1977, from 9° to 33°S latitude, were reported to range from 1.4 to 85.8 ng/l, with an average of 26.8 ng/l (0.0268 µg/l) (Singh et al., 1979).

In groundwater and surface-water in New Jersey, USA, sampled during 1977-1979, methyl chloride was detected in 3/1058 groundwater samples and in 24/605 surface-water samples. The highest reported concentrations were 6.0 µg/l and 222.4 µg/l, respectively (Page, 1981).

(e) Tobacco smoke

Tobacco smoke contains 150-840 µg methyl chloride per cigarette (Wynder & Hoffmann, 1967; see IARC, 1986). An average nonfilter US cigarette was found to contain 650 µg in the mainstream smoke, and the ratio in sidestream smoke:mainstream smoke was approximately 2.1 (US Department of Health and Human Services, 1982).

2.3 Analysis

Selected methods for the analysis of methyl chloride in air and water are identified in Table 4. The methods of the US Environmental Protection Agency (1984a,b) for analysing water have also been applied to liquid and solid wastes (US Environmental Protection Agency, 1982a,b). Volatile components of solid-waste samples are first extracted with polyethylene glycol or methanol prior to purge/trap concentration and analysis.

Method 624 (US Environmental Protection Agency, 1984c) has also been adapted to the analysis of methyl chloride in fish, with an estimated detection limit of 250 µg/kg (Easley et al., 1981).

Exposures to methyl chloride can be monitored in air by a direct-reading infrared analyser, at minimum detectable concentrations of 0.8-3.1 mg/m^3 (The Foxboro Co., 1985; Goelzer & O'Neill, 1985).

3. Biological Data Relevant to the Evaluation of Carcinogenic Risk to Humans

3.1 Carcinogenicity studies in animals

Inhalation exposure

Mouse: It was reported in an abstract on a two-year inhalation study that male and female B6C3F$_1$ mice [numbers and age unspecified] were exposed by inhalation to 0, 51, 224 or 997 ppm (0, 106, 464 or 2064 mg/m^3) methyl chloride for 6 h per day on five days per

Table 4. Methods for the analysis of methyl chloride

Sample matrix	Sample preparation	Assay procedure[a]	Limit of detection	Reference
Air	Adsorb on charcoal; desorb (dichloromethane); inject aliquot	GC/FID	122-455 mg/m^3 (range)	Taylor (1978); Peers (1985)
	Adsorb (porous glass beads, -173°C); desorb (heat, purge with nitrogen) onto GC column	GC/EC GC/PID GC/FID	6 ng/m^3 10 ng/m^3 103 ng/m^3	Rudolph & Jebsen (1983)
	Adsorb on charcoal; desorb (heat, purge with nitrogen) onto GC column	GC/MS	21 µg/m^3	Wilkes et al. (1982)
	Adsorb (polymeric beads); desorb (heat, purge with helium) onto GC column	GC/MS	2 mg/m^3	Krost et al. (1982)
	Inject air sample (5 ml) directly	GC/EC	21 µg/m^3	Pellizzari et al. (1978)
	Adsorb on charcoal; desorb (heat, purge with helium); dry (calcium sulphate); readsorb (Tenax GC); desorb as before; trap (liquid nitrogen cooled); vaporize onto GC	GC/MS	138-357 ng/m^3	Pellizzari et al. (1978)
Water	Purge (inert gas); trap (OV-1 on Chromosorb-W/Tenax/silica gel); desorb as vapour (heat to 180°C, backflush with inert gas) onto GC column	GC/ECD	0.08 µg/l	US Environmental Protection Agency (1984a)
	Add internal standard (isotope-labelled methyl chloride); purge, trap and desorb as above	GC/MS	50 µg/l	US Environmental Protection Agency (1984b)
	Purge (80°C, nitrogen); trap (Ambersorb or Porapak N); desorb (flash-heat) and trap in 'mini-trap' (Ambersorb or Porapak N, -30°C); desorb (flash-heat) onto GC column	GC/EC GC/MS GC/MS	0.05 µg/l (tap-water) 0.01 µg/l (tap-water) 0.05 µg/l (surface-water)	Piet et al. (1985a)
	Equilibrate sealed sample at 30°C; inject aliquot of head-space vapour	GC/EC	3-15 µg/l	Piet et al. (1985b)

[a]Abbreviations: GC/FID, gas chromatography/flame-ionization detection; GC/EC, gas chromatography/electron capture detection; GC/PID, gas chromatography/photoionization detection; GC/MS, gas chromatography/mass spectrometry; GC/ECD, gas chromatography/electrolytic conductivity detection

week. The reported analysis indicated an increase in the incidence of renal cortical adenomas and adenocarcinomas and cortical tubular cysts in males receiving the highest dose. Renal cortical adenomas were also seen in males exposed to 224 ppm (Pavkov *et al.*, 1982). [The Working Group noted the inadequate reporting of the study.]

Rat: It was reported in an abstract on a two-year inhalation study that male and female Fischer 344 rats [numbers and age unspecified] were exposed to 0, 51, 224 or 997 ppm (0, 106, 464 or 2064 mg/m^3) methyl chloride for 6 h per day on five days per week. No increase in tumour incidence was reported in the treated animals (Pavkov *et al.*, 1982). [The Working Group noted the inadequate reporting of the study.]

3.2 Other relevant biological data

(*a*) *Experimental systems*

Toxic effects

The toxicity of methyl chloride has been reviewed (von Oettingen, 1964; Repko & Lasley, 1979; Torkelson & Rowe, 1981).

The LC_{50} value for methyl chloride in mice is about 6300 mg/m^3 (7-h exposure) (von Oettingen *et al.*, 1950). In an abstract, 6-h LC_{50} values of 2250 ppm (4600 mg/m^3) and 8500 ppm (17 500 mg/m^3) were reported for male and female mice, respectively; the LC_{50} of methyl chloride is increased in mice treated with buthionine sulphoxime, an inhibitor of glutathione synthesis (White *et al.*, 1982). The average survival time of dogs exposed to 15 000 ppm (31 000 mg/m^3) was 6 h (von Oettingen *et al.*, 1950). Exposure of guinea-pigs to 3000 ppm (6200 mg/m^3) for 6 h was lethal to most animals (Smith & von Oettingen, 1947a). Mice exposed to 2500 ppm (5150 mg/m^3) methyl chloride exhaled ethane, an index of lipid peroxidation, 3 h after the start of exposure (Kornbrust & Bus, 1984).

Similar toxic effects have been reported following both short- and long-term exposure to methyl chloride and include apathy, drowsiness, staggering gait, paralysis of the hind limbs, anorexia, spasticity, convulsions, renal damage, hyperaemia and haemorrhages of the lung and death (Schwarz, 1926; Dunn & Smith, 1947; Smith & von Oettingen, 1947b; von Oettingen, 1964; Landry *et al.*, 1985).

Exposure to methyl chloride by inhalation decreased non-protein thiol concentrations in rodent liver, kidney, lung, brain, testis and epididymis (Dodd *et al.*, 1982; Landry *et al.*, 1983; Chapin *et al.*, 1984; Kornbrust & Bus, 1984). A similar effect was reported in tissues of dams and fetuses of mice and rats (as reported in an abstract) following exposure to methyl chloride during gestation (Bus *et al.*, 1980; Wolkowski-Tyl *et al.*, 1983a).

Following exposure to 2000 ppm (4100 mg/m^3) methyl chloride for 6 h per day on six days per week, 50% of guinea-pigs, mice and dogs died within three to four days, and 50% of monkeys, rats, rabbits, cats and chickens died after 10, 15, 23, 27 and 38 days, respectively (Smith & von Oettingen, 1947a).

The major morphological finding in mice exposed to 500-2000 ppm (1030-4100 mg/m^3) methyl chloride for 6 h per day for up to 12 days was degeneration of the liver, kidney and cerebellum. Rats exposed to 2000-5000 ppm (4100-10 300 mg/m³) for up to nine days out of 11 developed similar lesions and, in addition, degeneration of the adrenal glands, seminiferous tubules and the epididymis (Morgan et al., 1982). In several other studies, testicular and epididymal damage have been reported in rats following exposure to methyl chloride by inhalation (Chapin et al., 1984; Hamm et al., 1985; Working et al., 1985a,b).

Lesions in the cerebellum have been reported in mice following continuous or intermittent exposure to methyl chloride (Jiang et al., 1985; Landry et al., 1985). Degenerative changes in the cerebellum have also been found following exposure of guinea-pigs to 20 000 ppm (41 200 mg/m³) methyl chloride for 10 min per day on six days per week for up to 70 days (Kolkmann & Volk, 1975).

As reported in an abstract, mice (B6C3F$_1$) and rats (Fischer 344) were exposed to 51, 224 and 997 ppm (106, 464 and 2064 mg/m³) methyl chloride for 6 h per day on five days per week for 24 months. Survival and growth rates were reduced in all animals exposed to the highest dose, and nearly all mice in this exposure group showed neurofunctional impairment, cerebellar degeneration and hepatocellular necrosis. Male mice also developed renal and testicular damage. Male rats exposed to 997 ppm showed only testicular changes; no lesion was observed in female rats (Pavkov et al., 1982).

Effects on reproduction and prenatal toxicity

Exposure of male Fischer 344 rats to methyl chloride (purity, 99.98%) by inhalation of 1500 ppm (3090 mg/m³) for 6 h per day on five days per week for ten weeks, and for 6 h per day on seven days per week for a further two weeks, resulted in severe atrophy of the seminiferous tubules in all animals necropsied and in epididymal granulomas in 3/10 animals. Fewer litters were born to females bred to males similarly exposed to 475 ppm (980 mg/m³), and no litter resulted from breeding of males exposed to 1500 ppm (3090 mg/m³). No such effect was observed following exposures to 150 ppm (310 mg/m³) (Hamm et al., 1985).

In a classical teratology study, groups of 25 Fischer 344 rats were exposed by inhalation to 0, 100, 500 or 1500 ppm (0, 206, 1030 or 3090 mg/m³) methyl chloride (purity, >99.99%) for 6 h per day on gestation days 7-19. Maternal food consumption and body weight were significantly lower following exposure to 1500 ppm, and fetal body weights and skeletal maturity were also reduced. No exposure-related skeletal or visceral abnormality was seen (Wolkowski-Tyl et al., 1983a).

In a classical teratology study, groups of 33 female C57BL/6 mice were mated with C3H males, and the dams were then exposed by inhalation to 0, 100, 500 or 1500 ppm (0, 206, 1030 or 3090 mg/m³) methyl chloride (purity, >99.9%) for 6 h per day on gestation days 6-17. After five days of exposure, vaginal bleeding and an exudate from the eyes were observed in a high-dose female. After six to nine days of treatment, signs of overt neurotoxicity were observed in the group receiving the highest dose, and treatment was stopped. No fetal skeletal abnormality was detected, but a low, significant incidence of heart defects (reduction in size or absence of the atrioventricular valves, chordae tendinae and papillary muscles) was observed in the 500-ppm group (Wolkowski-Tyl et al., 1983a).

To examine further these cardiac defects, groups of 74-77 female C57BL/6 mice that had been mated to C3H males were exposed by inhalation to 0, 250, 500 or 750 ppm (0, 515, 1030 or 1545 mg/m³) methyl chloride (purity, 99.97%) for 6 h per day on gestation days 6-17. Some of the dams exposed to 500 and 750 ppm became ataxic, and seven of the dams exposed to 750 ppm died or were killed during the exposure period. Dams exposed to 750 ppm had a greater number of affected (defined as nonlive implantation sites plus malformed embryos) fetuses per litter and a greater percentage of litters thus affected than did controls. The incidence of cardiac defects was 0.7% of the fetuses in control litters, 1.3% at 250 ppm, 2.5% at 500 ppm and 4.3% at 750 ppm (significant at the two higher exposures) (Wolkowski-Tyl et al., 1983b).

Absorption, distribution, excretion and metabolism

Following inhalation of methyl chloride by rats, biphasic elimination kinetics were observed, with both saturable and first-order components (Andersen et al., 1980). After exposure of rats (6 h) and dogs (3 h) to 50 and 1000 ppm (103 and 2060 mg/m³) methyl chloride, the steady-state levels of the parent compound in blood were rapidly reached and were similar in both species. In rats exposed by inhalation to ^{14}C-methyl chloride (50-1000 ppm; 103-2060 mg/m³ for 6 h), total (nonvolatile) radioactivity was highest in liver and kidney and lower in testes (Landry et al., 1983). Mice cleared ^{14}C-methyl chloride from a closed system two to three times faster than rats (Peter et al., 1985). It was reported in an abstract that, following exposure of rats to 1500 ppm (3090 mg/m³) ^{14}C-methyl chloride for 6 h, about 64% of the radioactivity was recovered within 24 h in the exhaled air, 32% in urine and approximately 4% in faeces (Bus, 1978).

After exposure of rats by inhalation to ^{14}C-methyl chloride, about 50% of the radiolabel was expired as ^{14}C-CO_2 (Kornbrust et al., 1982; Landry et al., 1983).

In rats exposed by inhalation, the reported urinary metabolites of methyl chloride were S-methylthioacetic acid sulphoxide, N-acetyl-S-methylcysteine and N-(methylthioacetyl)glycine (Landry et al., 1983), which are known metabolites of S-methylcysteine (Sklan & Barnsley, 1968).

^{14}C-Methyl-S-methylglutathione and ^{14}C-methyl-S-methylcysteine were found after incubation of ^{14}C-methyl chloride with rat and guinea-pig liver, brain and kidney homogenates; heat inactivation of the homogenates reduced tissue binding. Little methylation of protein thiols occurred (Redford-Ellis & Gowenlock, 1971a).

The methyl group of methyl chloride is metabolized via S-methylcysteine to formate. Elevated formate levels were found in blood and urine of rats exposed to methyl chloride by inhalation and treated either with nitrous oxide or methotrexate, compounds known to inhibit folate-dependent formate metabolism (Kornbrust & Bus, 1982). Formaldehyde has also been detected in the blood of mice and rabbits exposed intraperitoneally or by inhalation to methyl chloride (Evtushenko, 1967; Sujbert, 1967) and in rat-liver microsomes incubated with methyl chloride and NADPH (Kornbrust & Bus, 1983).

Immediately following exposure of rats by inhalation to ^{14}C-methyl chloride, up to 20% of the total radioactivity was incorporated into tissue proteins, lipids and nucleic acids (Kornbrust et al., 1982). Following oral dosing of rats with ^{14}C-methyl chloride, radioactivity in hepatic protein was associated with methionine and serine (Reynolds & Yee, 1967). Methylated DNA bases were not found after exposure of rats and mice by inhalation to ^{14}C-methyl chloride (Kornbrust *et al.*, 1982; Peter *et al.*, 1985), although, as reported in an abstract, methyl chloride has been shown to methylate 4-(*para*-nitrobenzyl)pyridine *in vitro* (Kornbrust & Bus, 1980).

Mutagenicity and other short-term tests

Methyl chloride (2.5-20%) was mutagenic to *Salmonella typhimurium* TA100 both in the presence and absence of a metabolic system (S9) from the livers of Aroclor-induced rats (Simmon *et al.*, 1977). It was also mutagenic (at concentrations of 0.5-20.7%) to strain TA1535 both in the presence and absence of S9 (Andrews *et al.*, 1976). A dose-dependent increase in the number of 8-azaguanine-resistant mutants was found after exposure of *S. typhimurium* TM677 to 5-30% methyl chloride for 3 h (Fostel *et al.*, 1985). All the above studies were performed in closed vessels.

Chromatid breaks were induced in pollen grains of *Tradescantia paludosa* following exposure to gaseous methyl chloride for 5 min (Smith & Lofty, 1954).

In studies with TK6 human lymphoblasts, no increase in the incidence of DNA damage (as measured by alkaline elution) was found following treatment with 1-5% methyl chloride for 3 h; dose-dependent increases in the numbers of trifluorothymidine-resistant mutants and sister chromatid exchanges were observed after exposure to 1-5% and 0.3-3% methyl chloride, respectively, for 3 h (Fostel *et al.*, 1985).

It was reported in an abstract (Working & Butterworth, 1984) that exposure of male Fischer 344 rats to 3500 ppm (7245 mg/m^3) methyl chloride by inhalation for 6 h per day for one to nine days or to 15 000 ppm (31 050 mg/m^3) for 3 h did not induce unscheduled DNA synthesis in pachytene spermatocytes. However, a dose-related increase in unscheduled DNA synthesis (measured by autoradiography) was seen following in-vitro exposure of pachytene spermatocytes to 30-100 000 ppm (62-207 000 mg/m^3) methyl chloride for 3 h.

Male Fischer 344 rats exposed to 0, 1000 or 3000 ppm (0, 2070 or 6200 mg/m^3) methyl chloride for 6 h per day on five consecutive days were used in an assay for dominant lethal mutations. Treatment with the high dose resulted in increased preimplantation loss throughout the eight weeks following exposure (i.e., after treatment of spermatogonia, spermatocytes, spermatids and spermatozoa). Also at the high dose, a slight increase in postimplantation loss was found only during the first week after exposure (i.e., after treatment of spermatozoa), indicating induction of dominant lethal mutations at the highest concentration tested (Working *et al.*, 1985a). The same authors, in a subsequent abstract, suggested that the postimplantation loss was a secondary effect due to methyl chloride-induced inflammation of the epididymis (Chellman *et al.*, 1986).

Transformation of Syrian hamster embryo cells by SA7 adenovirus was enhanced after exposure of cells to 3-50 parts per thousand (6.2-103.5 g/m^3) methyl chloride in sealed chambers for 20 h (Hatch *et al.*, 1983).

(b) Humans

Toxic effects

The odour threshold for methyl chloride in humans has been reported to be 10 ppm (21 mg/m^3) (Stahl, 1973). Methyl chloride poisonings have occurred both in industry and as a consequence of leakages from domestic refrigerators. A review reported 263 severe cases of poisoning, observed mainly between the 1940s and the 1960s, 21 of which were fatal (von Oettingen, 1964).

The symptoms associated with fatal poisoning include nausea, vomiting and abdominal pain (Kegel et al., 1929; McNally, 1946), followed by headache, mental confusion, loss of balance and, eventually, loss of consciousness (Kegel et al., 1929; McNally, 1946; Baird, 1954). Pathological examination of cases has revealed congestion of the lungs and kidneys and fatty degeneration of the liver (Kegel et al., 1929; McNally, 1946).

Non-fatal cases develop similar but milder symptoms, which include nausea and vomiting, vertigo, tremor, visual disturbances and lack of coordination (Baker, 1927; Weinstein, 1937; Morgan Jones, 1942; Verrière & Vachez, 1949; Hansen et al., 1953; Hartman et al., 1955; MacDonald, 1964; Borovská et al., 1976; Spevak et al., 1976; Lanham, 1982).

Renal damage (e.g., nephritis), detected by proteinuria, oliguria and uraemia, has also been reported (Kegel et al., 1929; Verrière & Vachez, 1949; Mackie, 1961; Spevak et al., 1976). Cirrhosis of the liver was described in one case after long-term exposure to methyl chloride fumes (Wood, 1951).

Neurological damage (Kegel et al., 1929; Morgan Jones, 1942) and severe personality changes have been reported as late effects of toxic exposure (MacDonald, 1964; Thordarson et al., 1965; Borovská et al., 1976). Mild neurological or psychiatric sequelae were still evident several years after poisoning (Gudmundsson, 1977).

Effects on reproduction and prenatal toxicity

In a tabulation of case reports of agenesis of the sacrum, one concerned a baby who died at birth whose mother had been exposed to 'vapours of methyl chloride and ammonia'. Further details concerning the exposure were not given (Kučera, 1968).

Absorption, distribution, excretion and metabolism

Following inhalation of ^{38}Cl-methyl chloride by volunteers as a single breath, 29% of the total inhaled radioactivity was excreted in expired air within 1 h (Morgan et al., 1970).

Following exposure of six male volunteers (25-41 years old) to 50 ppm (103 mg/m^3) or 10 ppm (20.6 mg/m^3) methyl chloride for 6 h on two days separated by two weeks, the concentration in blood and expired air was proportional to the exposure concentration. The concentration in two of the volunteers was two to three times higher than in the others, and, following the exposures, both the clearance from the blood and the decrease in expired air concentrations were slower in these individuals. A five-fold difference in the rate constant for methyl chloride metabolism was calculated between the two groups (Nolan et al., 1985). Similar findings were reported in an abstract (Hake et al., 1977). [The Working Group noted the small number of subjects reported.]

S-Methylglutathione was identified in human erythrocytes following incubation with ^{14}C-methyl chloride; heating decreased the binding of radiolabel in whole blood. No methylation of haemoglobin was detected. After incubation of human plasma with ^{14}C-methyl chloride, the small amounts of radioactivity associated with albumin were identified as *S*-methylcysteine (Redford-Ellis & Gowenlock, 1971b).

Increased excretion of urinary *S*-methylcysteine has been reported in persons occupationally exposed to methyl chloride (up to 90 ppm; 185 mg/m^3) (van Doorn *et al.*, 1980). No correlation was found between exposure of volunteers to 10 and 50 ppm (20.6 and 103 mg/m^3) methyl chloride and urinary excretion of *S*-methylcysteine (Nolan *et al.*, 1985).

Mutagenicity and chromosomal effects

No data were available to the Working Group.

3.3 Case reports and epidemiological studies of carcinogenicity to humans

A cohort study was conducted of 852 male workers employed for at least one month between 1943 and 1978 in a butyl rubber manufacturing plant using methyl chloride (Holmes *et al.*, 1986). There was a total of 30 deaths from cancer. In comparison with the US population, the standardized mortality ratios for white men were 82 (120 observed) for all causes of death, 66 (19 observed) for deaths from all cancers and 70 (seven observed) for deaths from respiratory cancers; among non-white men, the corresponding figures were 59 (59 observed), 63 (11 observed) and 120 (statistically nonsignificant; six observed). Further analyses, by time of first employment, duration of employment and level of exposure to methyl chloride, provided no indication of an increase in standardized mortality ratio or a dose-response relationship for all cancers taken together. [The Working Group noted that the small number of deaths provides an insufficient basis for assessing site-specific cancer risk.]

4. Summary of Data Reported and Evaluation

4.1 Exposure data

Methyl chloride is formed extensively in the oceans and during combustion processes. It is also produced commercially in large quantities for use as an intermediate in the production of methyl silicone polymers, tetramethyllead and methyl cellulose, and as a solvent in the preparation of butyl rubber. It has been used as a refrigerant. Human exposure occurs in production and process plants, from cigarette smoke and from ubiquitous low-level concentrations in the ambient air and water.

4.2 Experimental data

A study in which methyl chloride was tested for carcinogenicity in mice and rats by inhalation exposure was reported only in an abstract. Although an excess of kidney tumours was reported in male mice exposed to the highest dose, the incomplete reporting precluded an evaluation of this finding. The studies in female mice and in male and female rats were reported to give negative results.

Inhalation exposure to methyl chloride causes fetal growth retardation and impaired male reproductive capacity in rats and malformations of the heart in fetal mice.

Methyl chloride is mutagenic to bacteria and induces chromosomal aberrations in plants. It induces DNA damage in mammalian cells *in vitro* but not *in vivo*. In cultured mammalian cells, it induces mutations and sister chromatid exchanges and enhances viral cell transformation. It induces dominant lethal mutations in rats.

Overall assessment of data from short-term tests: Methyl chloride[a]

	Genetic activity			Cell transformation
	DNA damage	Mutation	Chromosomal effects	
Prokaryotes		+		
Fungi/Green plants			+	
Insects				
Mammalian cells (*in vitro*)	+	+	+	+
Mammals (*in vivo*)	−		+	
Humans (*in vivo*)				
Degree of evidence in short-term tests for genetic activity: **Sufficient**				Cell transformation: Positive

[a]The groups into which the table is divided and the symbols '+' and '−' are defined on pp. 19-20 of the Preamble; the degrees of evidence are defined on pp. 20-21.

4.3 Human data

No evaluation of the effects of methyl chloride on reproduction or prenatal toxicity could be made on the basis of the available data.

A small cohort study of butyl rubber manufacturing workers exposed to methyl chloride was uninformative with regard to the carcinogenic hazard of this chemical.

4.4 Evaluation[1]

There is *inadequate evidence* for the carcinogenicity of methyl chloride to experimental animals.

There is *inadequate evidence* for the carcinogenicity of methyl chloride to humans.

5. References

Ahlstrom, R.C., Jr & Steele, J.M. (1979) *Methyl chloride.* In: Grayson, M. & Eckroth, D., eds, *Kirk-Othmer Encyclopedia of Chemical Technology*, 3rd ed., Vol. 5, New York, John Wiley & Sons, pp. 677-685

Air Products & Chemicals (1981) *Methyl Chloride Data Sheet*, Allentown, PA

Akiyama, S., Hisamoto, T. & Mochizuki, S. (1981) Chloromethanes from methanol. *Hydrocarbon Process., 60*, 76-78

American Conference of Governmental Industrial Hygienists (1984) *Documentation of the Threshold Limit Values*, 4th ed., Cincinnati, OH, pp. 268-269

American Conference of Governmental Industrial Hygienists (1985) *TLVs Threshold Limit Values and Biological Exposure Indices for 1985-86*, 2nd ed., Cincinnati, OH, p. 23

Andersen, M.E., Gargas, M.L., Jones, R.A. & Jenkins, L.J. (1980) Determination of the kinetic constants for metabolism of inhaled toxicants *in vivo* using gas uptake measurements. *Toxicol. appl. Pharmacol., 54*, 100-116

Andrews, A.W., Zawistowski, E.S. & Valentine, C.R. (1976) A comparison of the mutagenic properties of vinyl chloride and methyl chloride. *Mutat. Res., 40*, 273-276

Anon. (1983) Chemical profile: methyl chloride. *Chem. Mark. Rep., 223*, 54

Arbetarskyddsstyrelsens Författningssamling (National Swedish Board of Occupational Safety and Health) (1984) *Occupational Exposure Limit Values (AFS 1984:5)* (Swed.), Solna, p. 20

ARC Chemical Corp. (undated) *Methyl Chloride, Technical Data and Product Information*, Slate Hill, NY

Baird, T.T. (1954) Methyl chloride poisoning. *Br. med. J., ii*, 1353

Baker, H.M. (1927) Intoxication with commercial methyl chloride. Report of a series of cases. *J. Am. med. Assoc., 88*, 1137-1138

Borovská, V.D., Jindřichová, J. & Klima, M. (1976) Methyl chloride poisoning in East Bohemia (Ger.). *Z. ges. Hyg., 22*, 241-245

Browning, E. (1965) *Toxicity and Metabolism of Industrial Solvents*, Amsterdam, Elsevier, pp. 230-232

---

[1]For definitions of the italicized terms, see Preamble, pp. 18 and 22.

Bus, J.S. (1978) Disposition of ^{14}C-methyl chloride in Fischer 344 rats after inhalation exposure (Abstract). *Pharmacologist*, *20*, 214

Bus, J.S., Wolkowski-Tyl, R. & Barrow, C. (1980) Alterations in maternal and fetal non-protein sulfhydryl (NSPH) concentrations in pregnant Fisher-344 rats after acute inhalation exposure to methyl chloride. *Teratology*, *21*, 32A

Chapin, R.E., White, R.D., Morgan, K.T. & Bus, J.S. (1984) Studies of lesions induced in the testis and epididymis of F-344 rats by inhaled methyl chloride. *Toxicol. appl. Pharmacol.*, *76*, 328-343

Chellman, G.J., Bus, J.S. & Working, P.K. (1986) Role of epididymal inflammation in the induction of postimplantation loss in F-344 rats by methyl chloride (Abstract No. 391). *Toxicologist*, *6*, 98

The Chemical Daily Co. (1984) *JCW Chemicals Guide 1984/1985*, Tokyo, p. 267

Cohen, J.M., Dawson, R. & Koketsu, M. (1980) *A Technical Report: Extent-of-Exposure Survey of Methyl Chloride (DHHS (NIOSH) Publ. No. 80-134)*, Washington DC, US Department of Health and Human Services, National Institute for Occupational Safety and Health

Cronn, D.R. & Nutmagul, W. (1982) Characterization of trace gases in 1980 volcanic plumes of Mt St Helens. *J. geophys. Res.*, *87*, 11.153-11.160

Cronn, D.R., Rasmussen, R.A., Robinson, E. & Harsch, D.E. (1977) Halogenated compound identification and measurement in the troposphere and lower stratosphere. *J. geophys. Res.*, *82*, 5935-5944

Davis, L.N., Strange, J.R., Hoecker, J.E., Howard, P.H. & Santodonato, J. (1977) *Investigation of Selected Potential Environmental Contaminants: Monohalomethanes (EPA-560/2-77-007)*, Springfield, VA, National Technical Information Service, p. 16

Deutsche Forschungsgemeinschaft (German Research Community) (1985) *Maximal Concentrations in the Workplace and Biological Tolerance Values for Substances in the Work Environment 1985* (Ger.), Vol. 21, Weinheim, Verlagsgesellschaft mbH, p. 25

Direktoratet for Arbeidstilsynet (Directorate of Labour Inspection) (1981) *Administrative Norms for Pollution in Work Atmosphere, No. 361* (Norw.), Oslo, p. 16

Dodd, D.E., Bus, J.S. & Barrow, C.S. (1982) Nonprotein sulfhydryl alterations in F-344 rats following acute methyl chloride inhalation. *Toxicol. appl. Pharmacol.*, *62*, 228-236

van Doorn, R., Borm, P.J.A., Leijdekkers, C.-M., Henderson, P.T., Reuvers, J. & van Bergen, T.J. (1980) Detection and identification of S-methylcysteine in urine or workers exposed to methyl chloride. *Int. Arch. occup. environ. Health*, *46*, 99-109

Dunn, R.C. & Smith, W.W. (1947) Acute and chronic toxicity of methyl chloride. *Arch. Pathol.*, *43*, 296

Easley, D.M., Kleopfer, R.D. & Carasea, A.M. (1981) Gas chromatographic-mass spectrometric determination of volatile organic compounds in fish. *J. Assoc. off. anal. Chem.*, *64*, 653-656

Edwards, P.R., Campbell, I. & Milne, G.S. (1982) The impact of chloromethanes on the environment. Part 2. Methyl chloride and methylene chloride. *Chem. Ind., 17*, 619-622

Ethyl Corp. (undated) *Methyl Chloride Technical Information Brochure*, Baton Rouge, LA

Evtushenko, G.Y. (1967) Concerning methyl chloride metabolism. *Farmakol. Toksikol., 30*, 239-240

Fostel, J., Allen, P.F., Bermudez, E., Kligerman, A.D., Wilmer, J.L. & Skopek, T.R. (1985) Assessment of the genotoxic effects of methyl chloride in human lymphoblasts. *Mutat. Res., 155*, 75-81

The Foxboro Co. (1985) *Ambient Air Instrumentation Sheet*, Foxboro, MA

Goelzer, B. & O'Neill, I.K. (1985) *Workplace air-sampling for gases and vapours: strategy, equipment, procedure and exposure limits*. In: Fishbein, L. & O'Neill, I.K., eds, *Environmental Carcinogenesis. Selected Methods of Analysis*, Vol. 7, *Some Volatile Halogenated Hydrocarbons* (*IARC Scientific Publications No. 68*), Lyon, International Agency for Research on Cancer, pp. 107-136

Grasselli, J.G. & Ritchey, W.M., eds (1975) *CRC Atlas of Spectral Data and Physical Constants for Organic Compounds*, Vol. 3, Cleveland, OH, CRC Press, p. 592

Gudmundsson, G. (1977) Methyl chloride poisoning 13 years later. *Arch. environ. Health, 32*, 236-237

Hake, C.L., Stewart, R.D., Wu, A., Forster, H.V. & Newton, P.E. (1977) Experimental human exposures to methyl chloride at industrial environmental levels. *Toxicol. appl. Pharmacol., 41*, 198

Hamm, T.E., Jr, Raynor, T.H., Phelps, M.C., Auman, C.D., Adams, W.T., Proctor, J.E. & Wolkowski-Tyi, R. (1985) Reproduction in Fischer-344 rats exposed to methyl chloride by inhalation for two generations. *Fundam. appl. Toxicol., 5*, 568-577

Hansch, C. & Leo, A. (1979) *Substituent Constants for Correlation Analysis in Chemistry and Biology*, New York, John Wiley & Sons, p. 173

Hansen, H., Weaver, N.K. & Venable, F.S. (1953) Methyl chloride intoxication. Report of fifteen cases. *Arch. ind. Hyg. occup. Med., 8*, 328-334

Hartman, T.L., Wacker, W. & Roll, R.M. (1955) Methyl chloride intoxication. Report of two cases, one complicating pregnancy. *New Engl. J. Med., 253*, 552-554

Hatch, G.G., Mamay, P.D., Ayer, M.L., Casto, B.C. & Nesnow, S. (1983) Chemical enhancement of viral transformation in Syrian hamster embryo cells by gaseous and volatile chlorinated methanes and ethanes. *Cancer Res., 43*, 1945-1950

Hawley, G.G., ed. (1981) *The Condensed Chemical Dictionary*, 10th ed., New York, Van Nostrand Reinhold, p. 673

Health and Safety Executive (1985) *Occupational Exposure Limits 1985* (*Guidance Note EH 40/85*), London, Her Majesty's Stationery Office, p. 10

Hobbs, P.V., Tuell, J.P., Hegg, D.A., Radke, L.F. & Eltgroth, M.W. (1982) Particles and gases in the emissions from the 1980-1981 volcanic eruptions of Mt St Helens. *J. geophys. Res., 87*, 11062-11086

Holmes, T.M., Buffler, P.A., Holguin, A.H. & Hsi, B.P. (1986) A mortality study of employees at a synthetic rubber manufacturing plant. *Am. J. ind. Med., 9*, 355-362

IARC (1974) *IARC Monographs on the Evaluation of Carcinogenic Risk of Chemicals to Man*, Vol. 4, *Some Aromatic Amines, Hydrazine and Related Substances, N-Nitroso Compounds and Miscellaneous Alkylating Agents*, Lyon, pp. 271-276

IARC (1979a) *IARC Monographs on the Evaluation of the Carcinogenic Risk of Chemicals to Humans*, Vol. 19, *Some Monomers, Plastics and Synthetic Elastomers, and Acrolein*, Lyon, pp. 377-438

IARC (1979b) *IARC Monographs on the Evaluation of the Carcinogenic Risk of Chemicals to Humans*, Vol. 20, *Some Halogenated Hydrocarbons*, Lyon, pp. 401-417

IARC (1979c) *IARC Monographs on the Evaluation of the Carcinogenic Risk of Chemicals to Humans*, Vol. 20, *Some Halogenated Hydrocarbons*, Lyon, pp. 371-399

IARC (1979d) *IARC Monographs on the Evaluation of the Carcinogenic Risk of Chemicals to Humans*, Vol. 19, *Some Monomers, Plastics and Synthetic Elastomers, and Acrolein*, Lyon, pp. 231-274

IARC (1980) *IARC Monographs on the Evaluation of the Carcinogenic Risk of Chemicals to Humans*, Vol. 23, *Some Metals and Metallic Compounds*, Lyon, pp. 325-415

IARC (1982a) *IARC Monographs on the Evaluation of the Carcinogenic Risk of Chemicals to Humans*, Suppl. 4, *Chemicals, Industrial Processes and Industries Associated with Cancer in Humans, IARC Monographs, Volumes 1 to 29*, Lyon, pp. 260-262

IARC (1982b) *IARC Monographs on the Evaluation of the Carcinogenic Risk of Chemicals to Humans*, Suppl. 4, *Chemicals, Industrial Processes and Industries Associated with Cancer in Humans, IARC Monographs, Volumes 1 to 29*, Lyon, pp. 87-88

IARC (1982c) *IARC Monographs on the Evaluation of the Carcinogenic Risk of Chemicals to Humans*, Suppl. 4, *Chemicals, Industrial Processes and Industries Associated with Cancer in Humans, IARC Monographs, Volumes 1 to 29*, Lyon, pp. 74-75

IARC (1982d) *IARC Monographs on the Evaluation of the Carcinogenic Risk of Chemicals to Humans*, Suppl. 4, *Chemicals, Industrial Processes and Industries Associated with Cancer in Humans, IARC Monographs, Volumes 1 to 29*, Lyon, pp. 119-120

IARC (1982e) *IARC Monographs on the Evaluation of the Carcinogenic Risk of Chemicals to Humans*, Suppl. 4, *Chemicals, Industrial Processes and Industries Associated with Cancer in Humans, IARC Monographs, Volumes 1 to 29*, Lyon, pp. 149-150

IARC (1982f) *IARC Monographs on the Evaluation of the Carcinogenic Risk of Chemicals to Humans*, Vol. 28, *The Rubber Industry*, Lyon

IARC (1986) *IARC Monographs on the Evaluation of the Carcinogenic Risk of Chemicals to Humans*, Vol. 38, *Tobacco Smoking*, Lyon

ICP Chemicals & Plastics (1983) *Product Sheet, Methyl Chloride*, Edison, NJ

Institut National de Recherche et de Sécurité (National Institute for Research and Safety) (1985) *Limit Values for Concentrations of Dangerous Substances in Air of Work Places (Cahiers de Notes Documentaires No. 1555-121-85)* (Fr.), Paris, p. 483

International Labour Office (1980) *Occupational Exposure Limits for Airborne Toxic Substances: A Tabular Compilation of Values from Selected Countries*, 2nd rev. ed. (*Occupational Safety and Health Series No. 37*), Geneva, pp. 144-145

Jiang, X.Z., White, R. & Morgan, K.T. (1985) An ultrastructural study of lesions induced in the cerebellum of mice by inhalation exposure to methyl chloride. *Neurotoxicology, 6*, 93-104

Kegel, A.H., McNally, W.D. & Pope, A.S. (1929) Methyl chloride poisoning from domestic refrigerators. *J. Am. med. Assoc., 93*, 353-358

Kolkmann, F.W. & Volk, B. (1975) Necrosis in the granular cell layer of the cerebellum due to methyl chloride intoxication in guinea pig. *Exp. Pathol., 10*, 298-308

Kornbrust, D.J. & Buss, J.S. (1980) Association of inhaled (14)C-methyl chloride with macromolecules from various rat tissues. *Pharmacologist, 22*, 247

Kornbrust, D.J. & Bus, J.S. (1982) Metabolism of methyl chloride to formate in rats. *Toxicol. appl. Pharmacol., 65*, 135-143

Kornbrust, D.J. & Bus, J.S. (1983) The role of glutathione and cytochrome P-450 in the metabolism of methyl chloride. *Toxicol. appl. Pharmacol., 67*, 246-256

Kornbrust, D.J. & Bus, J.S. (1984) Glutathione depletion by methyl chloride and association with lipid peroxidation in mice and rats. *Toxicol. appl. Pharmacol., 72*, 388-399

Kornbrust, D.J., Bus, J.S., Doerjer, G. & Swenberg, J.A. (1982) Association of inhaled (^{14}C)methyl chloride with macromolecules from various rat tissues. *Toxicol. appl. Pharmacol., 65*, 122-134

Krost, K.J., Pellizzari, E.D., Walburn, S.G. & Hubbard, S.A. (1982) Collection and analysis of hazardous organic emissions. *Anal. Chem., 54*, 810-817

Kučera, J. (1968) Exposure to fat solvents: a possible cause of sacral agenesis in man. *J. Pediatr., 72*, 857-859

Landry, T.D., Gushow, T.S., Langvardt, P.W., Wall, J.M. & McKenna, M.J. (1983) Pharmacokinetics and metabolism of inhaled methyl chloride in the rat and dog. *Toxicol. appl. Pharmacol., 68*, 473-486

Landry, T.D., Quast, J.F., Gushow, T.S. & Mattsson, J.L. (1985) Neurotoxicity of methyl chloride in continuously versus intermittently exposed female C57BL/6 mice. *Fundam. appl. Toxicol., 5*, 87-98

Lanham, J.M. (1982) Methyl chloride: an unusual incident of intoxication. *Can. med. Assoc. J., 126*, 593

Leifer, R., Sommers, K. & Guggenheim, S.F. (1981) Atmospheric trace gas measurements with a new clean air sampling system. *Geophys. Res. Lett., 8*, 1079-1082

Linde Co. (1982) *Specialty Gases and Related Products 1982, Methyl Chloride*, Somerset, NJ, p. 35

Lovelock, J.E. (1975) Natural halocarbons in the air and in the sea. *Nature*, *256*, 193-194

Lovelock, J.E. (1977) Halogenated hydrocarbons in the atmosphere. *Ecotoxicol. environ. Saf.*, *1*, 399-406

MacDonald, J.D.C. (1964) Methyl chloride intoxication — Report of 8 cases. *J. occup. Med.*, *6*, 81-84

Mackie, I.J. (1961) Methyl chloride intoxication. *Med. J. Aust.*, *i*, 203-205

Mannsville Chemical Products Corp. (1984) *Chemical Products Synopsis: Methyl Chloride*, Cortland, NY

Matheson Co. (1985) *Gas Products Catalog, Methyl Chloride*, Bridgeport, NJ

Matsushima, K. (1986) Chlorine solvent. *Jpn. chem. Week*, January 16, 8

McNally, W.D. (1946) Eight cases of methyl chloride poisoning with three deaths. *J. ind. Hyg. Toxicol.*, *28*, 94-97

Morgan, A., Black, A. & Belcher, D.R. (1970) The excretion in breath of some aliphatic halogenated hydrocarbons following administration by inhalation. *Ann. occup. Hyg.*, *13*, 219-233

Morgan, K.T., Swenberg, J.A., Hamm, T.E., Jr, Wolkoski-Tyl, R. & Phelps, M. (1982) Histopathology of acute toxic response in rats and mice exposed to methyl chloride by inhalation. *Fundam. appl. Toxicol.*, *2*, 293-299

Morgan Jones, A. (1942) Methyl chloride poisoning. *Q. J. Med.*, *11*, 29-43

National Fire Protection Association (1984) *Fire Protection Guide on Hazardous Materials*, 8th ed., Quincy, MA, p. 325M-67

National Institute for Occupational Safety and Health (1985) NIOSH recommendations for occupational safety and health standards. *Morbid. Mortal. wkly Rep. Suppl.*, *34*, 6S-31S

Nolan, R.J., Rick, D.L., Landry, T.D., McCarty, L.P., Agin, G.L. & Saunders, J.H. (1985) Pharmacokinetics of inhaled methyl chloride (CH_3Cl) in male volunteers. *Fundam. appl. Toxicol.*, *5*, 361-369

von Oettingen, W.F. (1964) *The Halogenated Hydrocarbons of Industrial and Toxicological Importance*, Amsterdam, Elsevier, pp. 5-25

von Oettingen, W.F., Powell, C.C., Sharpless, N.E., Alford, W.C. & Pecora, L.J. (1950) Comparative studies of the toxicity and pharmacodynamic action of chlorinated methanes with special reference to their physical and chemical characteristics. *Arch. int. Pharmacodyn.*, *81*, 17-34

Page, G.W. (1981) Comparison of groundwater and surface water for patterns and levels of contamination by toxic substances. *Environ. Sci. Technol.*, *15*, 1475-1481

Pavkov, K.L., Kerns, W.D., Chrisp, C.E., Thake, D.C., Persing, R.L. & Harroff, H.H. (1982) Major findings in a twenty-four month inhalation toxicity study of methyl chloride in mice and rats (Abstract No. 566). *Toxicologist*, *2*, 161

Peers, A.M. (1985) *The determination of methyl chloride in air*. In: Fishbein, L. & O'Neill, I.K., eds, *Environmental Carcinogens. Selected Methods of Analysis*, Vol. 7, *Some Volatile Halogenated Hydrocarbons* (*IARC Scientific Publications No. 68*), Lyon, International Agency for Research on Cancer, pp. 219-226

Pellizzari, E.D., Zweidinger, R.A. & Erickson, M.D. (1978) *Environmental Monitoring Near Industrial Sites: Brominated Chemicals Part II: Appendix* (*US EPA Report No. EPA-560/6-78-002A; US NTIS PB-286483*), Washington DC, Office of Toxic Substances, US Environmental Protection Agency

Penkett, S.A., Derwent, R.G., Fabian, P., Borchers, R. & Schmidt, U. (1980) Methyl chloride in the stratosphere. *Nature, 283*, 58-60

Peter, H., Laib, R.J., Ottenwälder, H., Topp, H., Rupprich, N. & Bolt, H.M. (1985) DNA-binding assay of methyl chloride. *Arch. Toxicol., 57*, 84-87

Piet, G.J., Luijten, W.C.M.M. & van Noort, P.C.M. (1985a) *Dynamic head-space determination of volatile organic halogen compounds in water*. In: Fishbein, L. & O'Neill, I.K., eds, *Environmental Carcinogens. Selected Methods of Analysis*, Vol. 7, *Some Volatile Halogenated Hydrocarbons* (*IARC Scientific Publications No. 68*), Lyon, International Agency for Research on Cancer, pp. 331-343

Piet, G.J., Luijten, W.C.M.M. & van Noort, P.C.M. (1985b) *Static head-space determination of volatile organic halogen compounds in water*. In: Fishbein, L. & O'Neill, I.K., eds, *Environmental Carcinogens. Selected Methods of Analysis*, Vol. 7, *Some Volatile Halogenated Hydrocarbons* (*IARC Scientific Publications No. 68*), Lyon, International Agency for Research on Cancer, pp. 321-330

Rasmussen, R.A. & Khalil, M.A.K. (1983) Natural and anthropogenic trace gases in the lower troposphere of the Arctic. *Chemosphere, 12*, 371-375

Rasmussen, R.A., Rasmussen, L.E., Khalil, M.A.K. & Dalluge, R.W. (1980) Concentration distribution of methyl chloride in the atmosphere. *J. geophys. Res., 85*, 7350-7356

Redford-Ellis, M. & Gowenlock, A.H. (1971a) Studies on the reaction of chloromethane with preparations of liver, brain and kidney. *Acta pharmacol. toxicol., 30*, 49-58

Redford-Ellis, M. & Gowenlock, A.H. (1971b) Studies on the reaction of chloromethane with human blood. *Acta pharmacol. toxicol., 30*, 36-48

Repko, J.D. & Lasley, S.M. (1979) Behavioral, neurological, and toxic effects of methyl chloride: a review of the literature. *Crit. Rev. Toxicol., 6*, 283-302

Reynolds, E.S. & Yee, A.G. (1967) Liver parenchymal cell injury. V. Relationships between patterns of chloromethane-C^{14} incorporation into constituents of liver in vivo and cellular injury. *Lab. Invest., 16*, 591-603

Rudolph, J. & Jebsen, C. (1983) The use of photoionization, flame ionization and electron capture detectors in series for the determination of low molecular weight trace components in the non urban atmosphere. *Int. J. environ. anal. Chem., 13*, 129-139

Sadtler Research Laboratories (1980) *The Sadtler Standard Spectra, Cumulative Index*, Philadelphia, PA

Sax, N.I. (1984) *Dangerous Properties of Industrial Materials*, 6th ed., New York, Van Nostrand Reinhold, pp. 730-731

Schwarz, F. (1926) Cases of poisoning and animal experiments with methyl chloride (Ger.). *Dtsch. Z. ges. gerichtl. Med.*, 7, 278-288

Simmon, V.F., Kauhanen, K. & Tardiff, R.G. (1977) *Mutagenic activity of chemicals identified in drinking water*. In: Scott, D., Bridges, B.A. & Sobels, F.H., eds, *Progress in Genetic Toxicology*, Amsterdam, Elsevier/North-Holland Biomedical Press, pp. 249-258

Singh, H.B., Salas, L.J. & Cavanagh, L.A. (1977) Distribution, sources and sinks of atmospheric halogenated compounds. *J. Air Pollut. Control Assoc.*, 27, 332-336

Singh, H.B., Salas, L.J., Shigeishi, H. & Scribner, E. (1979) Atmospheric halocarbons, hydrocarbons, and sulfur hexafluoride: global distributions, sources, and sinks. *Science*, 203, 899-903

Singh, H.B., Salas, L.J., Smith, A.J. & Shigeishi, H. (1981) Measurements of some potentially hazardous organic chemicals in urban environments. *Atmos. Environ.*, 15, 601-612

Singh, H.B., Salas, L.J. & Stiles, R.E. (1982) Distribution of selected gaseous organic mutagens and suspect carcinogens in ambient air. *Environ. Sci. Technol.*, 16, 872-880

Singh, H.B., Salas, L.J. & Stiles, R.E. (1983) Methyl halides in and over the Eastern Pacific (40°N-32°S). *J. geophys. Res.*, 88, 3684-3690

Sittig, M., ed. (1980) *Pesticide Manufacturing and Toxic Materials Control Encyclopedia*, Park Ridge, NJ, Noyes Data Corp., pp. 356-358

Sklan, N.M. & Barnsley, E.A. (1968) The metabolism of S-methyl-L-cysteine. *Biochem. J.*, 107, 217-223

Smith, H.H. & Lofty, T.A. (1954) Comparative effects of certain chemicals on *Tradescantia* chromosomes as observed at pollen tube mitosis. *Am. J. Bot.*, 41, 589-593

Smith, W.W. & von Oettingen, W.F. (1947a) The acute and chronic toxicity of methyl chloride. I. Mortality resulting from exposures to methyl chloride in concentrations of 4,000 to 300 parts per million. *J. ind. Hyg. Toxicol.*, 29, 47-52

Smith, W.W. & von Oettingen, W.F. (1947b) The acute and chronic toxicity of methyl chloride. II. Symptomatology of animals poisoned by methyl chloride. *J. ind. Hyg. Toxicol.*, 29, 123-128

Spevak, L., Nadj, V. & Fellé, D. (1976) Methyl chloride poisoning in four members of a family. *Br. J. ind. Med.*, 33, 272-278

Stahl, W.H., ed. (1973) *Compilation of Odor and Taste Threshold Values Data (ASTM Data Series DS 48)*, Philadelphia, PA, American Society for Testing and Materials, p. 107

Sujbert, L. (1967) Study on the decomposition of methyl chloride in the mouse (Ger.). *Arch. Toxikol.*, 22, 233-235

Tassios, S. & Packham, D.R. (1985) The release of methyl chloride from biomass burning in Australia. *J. Air Pollut. Control Assoc.*, 35, 41-42

Taylor, D.G. (1978) *NIOSH Manual of Analytical Methods*, 2nd ed., Vol. 2 (*DHEW (NIOSH) Pub. No. 77-157-B*), Washington DC, US Government Printing Office, pp. S99-1 — S99-9

Thordarson, O., Gudmundsson, G., Bjarnason, O. & Jóhannesson, T. (1965) Methyl chloride poisonings (Norw.). *Nord. Med.*, *18*, 150-154

Torkelson, T.R. & Rowe, V.K. (1981) *Halogenated aliphatic hydrocarbons containing chlorine, bromine and iodine.* In: Clayton, G.D. & Clayton, F.E., eds, *Patty's Industrial Hygiene and Toxicology*, 3rd ed., Vol. 2B, New York, John Wiley & Sons, pp. 3436-3442

Työsuojeluhallitus (National Finnish Board of Occupational Safety and Health) (1981) *Airborne Contaminants in the Workplaces (Safety Bull. 3)* (Finn.), Tampere, p. 19

US Department of Health and Human Services (1982) *The Health Consequences of Smoking. Cancer. A Report of the Surgeon General*, Washington DC, US Government Printing Office, p. 240

US Environmental Protection Agency (1982a) *Method 8010. Halogenated volatile organics.* In: *Test Methods for Evaluating Solid Waste — Physical/Chemical Methods*, 2nd ed. (*US EPA No. SW-846*), Washington DC, Office of Solid Waste and Emergency Response

US Environmental Protection Agency (1982b) *Method 8240. GC/MS method for volatile organics.* In: *Test Methods for Evaluating Solid Waste — Physical/Chemical Methods*, 2nd ed. (*US EPA No. SW-846*), Washington DC, Office of Solid Waste and Emergency Response

US Environmental Protection Agency (1984a) Method 601. Guidelines establishing test procedures for the analysis of pollutants under the Clean Water Act (40 CFR 136). Purgeable halocarbons. *Fed. Regist.*, *49*, 43261-43271

US Environmental Protection Agency (1984b) Method 1624, Revision B. Guidelines establishing test procedures for the analysis of pollutants under the Clean Water Act (40 CFR 136). Volatile organic compounds by isotope dilution GC/MS. *Fed. Regist.*, *49*, 43407-43415

US Environmental Protection Agency (1984c) Method 624. Guidelines establishing test procedures for the analysis of pollutants under the Clean Water Act (40 CFR 136). Purgeables. *Fed. Regist.*, *49*, 43373-43384

US Occupational Safety and Health Administration (1985) Labor. *US Code fed. Regul.*, Title 29, Part 1910.1000

Verrière, M.P. & Vachez, M. (1949) Severe acute nephritis following methyl chloride poisoning (Fr.). *Lyon. méd.*, *1*, 296-297

Verschueren, K. (1983) *Handbook of Environmental Data on Organic Chemicals*, 2nd ed., New York, Van Nostrand Reinhold, pp. 839-840

Vista Chemical Co. (undated) *Methyl Chloride Specification Sheet*, Houston, TX

Walas, S.M. (1985) Chemical reactor data. *Chem. Eng. News*, *63*, 79-83

Weast, R.C., ed. (1985) *CRC Handbook of Chemistry and Physics*, 66th ed., Boca Raton, FL, CRC Press, pp. C-349, D-211

Weinstein, A. (1937) Methyl chloride (refrigerator) gas poisoning. An industrial hazard. *J. Am. med. Assoc.*, 108, 1603-1605

White, R.D., Norton, R. & Bus, J.S. (1982) Evidence for S-methyl glutathione metabolism in mediating the acute toxicity of methyl chloride (MeCl). *Pharmacologist*, 24, 172

Wilkes, B.E., Priestley, L.J., Jr & Scholl, L.K. (1982) An improved thermal desorption GC/MS method for the determination of low ppb concentrations of chloromethane in ambient air. *Microchem. J.*, 27, 420-424

Windholz, M., ed. (1983) *The Merck Index*, 10th ed., Rahway, NJ, Merck & Co., p. 866

Wolkowski-Tyl, R., Phelps, M. & Davis, J.K. (1983a) Structural teratogenicity evaluation of methyl chloride in rats and mice after inhalation exposure. *Teratology*, 27, 181-195

Wolkowski-Tyl, R., Lawton, A.D., Phelps, M. & Hamm, T.E., Jr (1983b) Evaluation of heart malformations in $B_6C_3F_1$ mouse foetuses induced by in utero exposure to methyl chloride. *Teratology*, 27, 197-206

Wood, M.W.W. (1951) Cirrhosis of the liver in a refrigeration engineer, attributed to methyl chloride. *Lancet*, i, 508-509

Working, P.K. & Butterworth, B.E. (1984) Induction of unscheduled DNA synthesis (UDS) in rat spermatocytes by exposure to methyl chloride *in vitro* and *in vivo* (Abstract). *Environ. Mutagenesis*, 6, 392

Working, P.K., Bus, J.S. & Hamm, T.E., Jr (1985a) Reproductive effects of inhaled methyl chloride in the male Fischer 344 rat. I. Mating performance and dominant lethal assay. *Toxicol. appl. Pharmacol.*, 77, 133-143

Working, P.K., Bus, J.S. & Hamm, T.E., Jr (1985b) Reproductive effects of inhaled methyl chloride in the male Fischer 344 rat. II. Spermatogonial toxicity and sperm quality. *Toxicol. appl. Pharmacol.*, 77, 144-157

Wynder, E.L. & Hoffmann, D. (1967) *Tobacco and Tobacco Smoke. Studies in Experimental Carcinogenesis*, New York, Academic Press, p. 455

METHYL BROMIDE

1. Chemical and Physical Data

1.1 Synonyms and trade names

Chem. Abstr. Services Reg. No.: 74-83-9
Chem. Abstr. Name: Bromomethane
IUPAC Systematic Name: Bromomethane
Synonyms: Monobromomethane
Trade Names: Brom-O-gas; Curafume; Dowfume MC-2; Dowfume MC-33; Embafume; Halon 1001; Haltox; Iscobrome; Meth-O-gas; Methyl Fume; R 40B1; Terabol; Terr-O-gas 67

1.2 Structural and molecular formulae and molecular weight

$$\begin{array}{c} H \\ | \\ H-C-Br \\ | \\ H \end{array}$$

CH_3Br Mol. wt: 94.94

1.3 Chemical and physical properties of the pure substance

(a) *Description*: Colourless gas with chloroform-like odour and burning taste (Hawley, 1981; Verschueren, 1983; Windholz, 1983)

(b) *Boiling-point*: 3.6°C (Weast, 1985)

(c) *Melting-point*: -93.6°C (Weast, 1985)

(d) *Density*: d_4^{20} 1.6755 (Weast, 1985)

(e) *Spectroscopy data*[a]: Ultraviolet (Grasselli & Ritchey, 1975), infrared (Sadtler Research Laboratories, 1980; prism [1098], grating [10972]), C-13 nuclear magnetic resonance and mass spectral data (Grasselli & Ritchey, 1975) have been reported.

(f) *Solubility*: Slightly soluble in water (0.9 g/l at 20°C) (Verschueren, 1983); soluble in ethanol, benzene, carbon disulphide, carbon tetrachloride, chloroform and diethyl ether (Windholz, 1983; Weast, 1985)

(g) *Volatility*: Vapour pressure, 2 atm (1520 mm Hg) at 23.3°C (Weast, 1985); relative vapour density (air = 1), 3.3 (Verschueren, 1983; National Fire Protection Association, 1984; Sax, 1984)

(h) *Stability*: Noninflammable in air but burns in oxygen (Hawley, 1981; Windholz, 1983); decomposes in water with a half-time of 67.8 h at 100°C (Davis *et al.*, 1977)

(i) *Reactivity*: Reacts with some metals (aluminium); alkylating agent (Stenger, 1978; Sax, 1984)

(j) *Octanol/water partition coefficient (P)*: log P, 1.19 (Hansch & Leo, 1979)

(k) *Conversion factor*: mg/m^3 = 3.88 × ppm[b]

1.4 Technical products and impurities

Methyl bromide is typically available as a liquefied gas with a minimum purity of 99.5% (Union Carbide Corp., 1982; Matheson Gas Products, 1985; Ethyl Corp., undated a). Water content and acidity (as hydrobromic acid) are less than 0.01% and 0.001%, respectively (Ethyl Corp., undated a,b). Methyl bromide is also available in formulations containing 2% (Ethyl Corp., undated b) or 32-33% chloropicrin, added at low levels as a sensory warning and at higher levels as a fumigant (Gosselin *et al.*, 1984).

2. Production, Use, Occurrence and Analysis

2.1 Production and use

(a) *Production*

Commercial and laboratory methods for methyl bromide production involve the reaction of methanol with hydrobromic acid, which is generated by addition of sulphuric acid to a solution of sodium bromide and methanol in a reactor. The methyl bromide

---

[a]In square brackets, spectrum number in compilation

[b]Calculated from: mg/m^3 = (molecular weight/24.45) × ppm, assuming standard temperature (25°C) and pressure (760 mm Hg)

product is removed through distillation, and is then dried, condensed and fractionated to remove impurities (Stenger, 1978).

Methyl bromide can also be produced by the action of bromine on methanol in the presence of phosphorus (Hawley, 1981) or sulphur (Sittig, 1980), the direct addition of hydrobromic acid to methanol, the reaction of hydrogen bromide with excess methyl chloride at 400-500°C, or the thermal/photochemical reaction of bromine chloride with methane (Stenger, 1978). Methyl bromide has been produced by the direct catalytic bromination of methane at elevated temperatures, but this reaction produced polybrominated methanes as undesirable byproducts (Considine, 1974).

Two manufacturers in the USA currently produce methyl bromide, with a combined annual capacity of 19.5 million kg. One other producer, with a capacity of 9.5 million kg, ceased operation in 1983. The smaller of the two current producers opened a plant with a capacity of 3 million kg in 1984 (Anon., 1985a). Current US demand is estimated to be 17-22 million kg annually (Holtorf, 1985). Data on the production, imports and exports of methyl bromide in the USA are given in Table 1.

Table 1. US production, imports and exports of methyl bromide (millions of kg)[a]

	1977	1978	1979	1980	1981	1982
Production	15.8	19.3	22.3	ND	21.0	ND
Imports	0.59	0.45	0.71	0.96	0.97	0.74
Exports	ND	3.18	4.96	5.48	4.25	2.78

[a]From Opresko and Daugherty (1985)
ND, no data

There are currently seven major manufacturers of methyl bromide in Japan (The Chemical Daily Co., 1984a). Japanese production in 1982 was approximately 5.3 million kg (ATOCHEM, 1984).

Several European producers of methyl bromide have been identified, including one each in Spain and Italy, and three in France (ATOCHEM, 1984; Anon., 1985b).

(b) Use

Methyl bromide, used primarily as a soil fumigant for the control of nematodes, fungi and weeds, was introduced for use as a pesticide in 1932 (Hayes, 1982) and 65% of the methyl bromide used in the USA is for this purpose (Anon., 1985a). As a soil fumigant, methyl bromide is applied prior to planting, often by injection into the soil followed by tarping (Mannsville Chemical Products Corp., 1979). It is used to fumigate plant beds primarily for tomatoes and strawberries and to sterilize soil in tobacco seed beds and plant and tree nurseries; only a a small percentage is used to fumigate soil for other fruits and ornamental plants (Holtorf, 1985).

Methyl bromide is also used as a space fumigant in warehouses, mills, ships, freight cars, vaults and grain elevators to control insects and rodents. It has been used as a fumigant for stored tobacco, peanuts, tree nuts and some fruits, and for structural pest control against wood-damaging insects such as termites (Mannsville Chemical Products Corp., 1979). Approximately 15% of the methyl bromide currently used in the USA is for this purpose (Anon., 1985a). It is used in the poultry industry as a fumigant, both on chicken feed and in animal living areas (Tucker *et al.*, 1974; Davis *et al.*, 1977).

Other miscellaneous applications account for approximately 20-25% of methyl bromide use. These include use as a solvent for the extraction of oils from nuts, seeds and flowers; as a wool-degreasing agent; and as a methylating agent, especially for the preparation of aniline dyes and antipyrine and other pharmaceuticals (Hawley, 1981; Windholz, 1983; Opresko & Daugherty, 1985; Sittig, 1985).

Methyl bromide was used in fire extinguishers in Europe and the USA from the 1920s through the 1940s and may still be found in some aircraft fire-extinguishing systems (Alexeeff & Kilgore, 1983; Opresko & Daugherty, 1985).

(*c*) *Regulatory status and guidelines*

Occupational exposure limits for methyl bromide in 18 countries are presented in Table 2.

In 1979, methyl bromide was re-evaluated by the Joint Meeting of the Food and Agriculture Organization Panel of Experts on Pesticide Residues in Food and the Environment and the WHO Expert Group on Pesticide Residues (Food and Agriculture Organization, 1980), but no admissible dietary intake (ADI) was set. Levels of methyl bromide that are not to be exceeded in products that have been fumigated are as follows: (1) in raw agricultural products at point of entry into a country — nuts and peanuts, 10 mg/kg; raw cereals and cocoa beans, 5 mg/kg; and dried fruits, 2 mg/kg; (2) in milled cereal products that will be subject to cooking — 1 mg/kg; (3) in foods when offered for consumption — nuts, peanuts, bread, other cooked cereal products, cocoa products, dried fruit, 0.01 mg/kg. Levels of bromide ion from all sources should not exceed 50 mg/kg in cereal grains and wholemeal flour and 100 mg/kg in lettuce and cabbage.

The tolerances set by the US Environmental Protection Agency for methyl bromide in or on raw agricultural commodities vary between 5 and 240 mg/kg, according to the products. Tolerances for bromide ion in animal feed resulting from fumigation with methyl bromide vary from 25 to 400 mg/kg (Sittig, 1980).

2.2 Occurrence

(*a*) *Natural occurrence*

Methyl bromide is produced by a variety of marine organisms (Lovelock, 1975).

Table 2. Occupational exposure limits for methyl bromide[a]

Country	Year	Concentration (mg/m³)	Interpretation
Australia	1978	60	TWA
Belgium	1978	60	TWA
Finland	1981	60	TWA
		90	STEL
France	1985	20	TWA
German Democratic Republic	1979	50	TWA
		50	STEL
Germany, Federal Republic of	1985	20	TWA
Hungary	1974	10	TWA
Italy	1978	60	TWA
The Netherlands	1978	60	TWA
Norway	1981	60	TWA
Poland	1976	5	Ceiling
Romania	1975	50	TWA
		80	Ceiling
Sweden	1984	60	TWA
		80	STEL
Switzerland	1978	60	TWA
UK	1985	60	TWA
USA	1985		
ACGIH		20	TWA
		60	STEL
OSHA		80	TWA
USSR	1977	1	Ceiling
Yugoslavia	1971	80	Ceiling
Council of Europe	1974	80	Not given

[a]From International Labour Office (1980); Direktoratet for Arbeidstilsynet (1981); Työsuojeluhallitus (1981); Arbetarskyddsstyrelsens Författningssamling (1984); American Conference of Governmental Industrial Hygienists (ACGIH) (1985); Deutsche Forschungsgemeinschaft (1985); Health and Safety Executive (1985); Institut National de Recherche et de Sécurité (1985); National Institute for Occupational Safety and Health (1985); US Occupational Safety and Health Administration (OSHA) (1985)

[b]TWA, time-weighted average; STEL, short-term exposure limit

(b) *Occupational exposure*

In a methyl bromide plant, workplace air concentrations of 20-30 ppm (78-116 mg/m^3) have been recorded (Evans *et al.*, 1979).

In a fruit processing and packing plant, methyl bromide values ranging up to 100 ppm (388 mg/m^3) were measured in the general workroom air and up to 1000 ppm (3880 mg/m^3) at the breathing zone of workers entering a fumigation chamber (Ingram, 1951).

During soil disinfection in greenhouses, methyl bromide concentrations varied between 30 and 3000 ppm (116.4-11640 mg/m^3). The exposure levels of applicators were greatly dependent on the type of fumigation techniques and equipment (Roosels *et al.*, 1981; Van Den Oever *et al.*, 1982).

Blood bromine content of 33 workers engaged in soil fumigation inside greenhouses varied from 4 to 23 mg/l. The samples were taken during the fumigation season (Verberk *et al.*, 1979). A serum sample from a worker suspected of methyl bromide poisoning contained 412 mg/l of bromine 13 days after exposure (Ohmori & Hirata, 1982).

Serum or plasma bromide ion concentrations ranged from 40 to 656 mg/l in four persons who died and from 17.9 to 321 mg/l in three persons who survived following accidental exposure to methyl bromide used in fumigation (Marraccini *et al.*, 1983).

(c) *Air*

Both man-made and natural sources contribute to the concentrations of methyl bromide in ambient air. Average levels in air at ten urban sites in the USA ranged from 41 to 259 ppt (159-1005 ng/m^3); average human exposures to methyl bromide at these sites were calculated to be up to 25 μg/person per day. The background concentration from natural sources at surface level at 40°N latitude is estimated to be 78 ng/m^3 (20 ppt) (Singh, H.B. *et al.*, 1981, 1982).

The primary mechanism for the removal of methyl bromide from the atmosphere is believed to be reaction with hydroxyl radicals. On this hypothesis, the residence time of methyl bromide in an urban atmosphere was estimated to be 289 days, with a daily rate of loss (12 sunlit hours) of 0.4% (Singh, H.B. *et al.*, 1981).

Measurements over the eastern Pacific (40°N and 32°S latitude) showed mean air concentrations of 23 ppt (89 ng/m^3) (Singh, H.B. *et al.*, 1983). Measurements of five bromine-containing trace gases in Point Barrow, AK, showed that the monthly average concentration of methyl bromide was higher in the summer (15 ppt; 58.2 ng/m^3) than in the winter (9 ppt; 35 ng/m^3) (Rasmussen & Khalil, 1984).

(d) *Water*

Concentrations of methyl bromide in seawater collected off Dorset, UK, in 1975 ranged from 1.5 to 3.9 μg/l (Lovelock, 1975). The near-shorewater concentration at Point Reyes, CA, USA, was 0.14 μg/l (Singh, H.B. *et al.*, 1977). Mean concentrations in surface seawater in the eastern Pacific at 40°N to 32°S latitude were approximately 1.2 ng/l (Singh, H.B. *et al.*, 1983).

Surface-water in a greenhouse crop-growing region of Malines-Antwerp, Belgium, was sampled for levels of bromide ion before, during and after soil treatment with methyl bromide. Accumulation of bromide ion was noted downstream of some brooks; however, the maximum concentration was 9.6 mg/l. In rivers, only a very slight increase of bromide ion was noted during periods of intensive leaching, and concentrations did not exceed 1 mg/l except at one site where levels ranging from 1.1 to 4.4 mg/l were thought to be due to the presence of seawater (Vanachter et al., 1981).

(e) *Soil*

Decomposition of methyl bromide in soil results in the production of bromide ion. The rate of bromide production is influenced by soil type: it is greatest in peaty manure, intermediate in loam (clay soil), and least in sand. Bromide ion in soil can be absorbed and concentrated by plants and may be ingested subsequently by humans and animals (Brown & Rolston, 1980).

Retention of bromide ion in soil after fumigation with methyl bromide is related to the organic content of the soil. The total bromide ion content of soil containing 2.81 and 0.93% organic carbon was 9 and 5 mg/kg when untreated and 63 and 25 mg/kg following fumigation with 500 mg/kg methyl bromide for 24 h, respectively (Brown & Jenkinson, 1971).

(f) *Food products*

Surveys of bromide ion residues in commercial vegetable crops in the UK following soil sterilization with methyl bromide showed levels ranging from 1 to 109 mg/kg in 29 late-season cucumbers sampled, 5 to 326 mg/kg in 242 tomato samples, and 2 to 521 mg/kg in 38 samples of celery (Roughan & Roughan, 1984).

Vegetables grown at Metaponto, Italy, in an experimental field treated with 60 g/m^2 methyl bromide, contained bromide ion at levels of 20-51 mg/kg (tomatoes), 8-44 mg/kg (string beans), 25-149 mg/kg (radishes), 18-60 mg/kg (aubergines), 6-16 mg/kg (cucumbers), 13-46 mg/kg (courgettes) and 3-27 mg/kg (peppers) (Basile & Lamberti, 1981).

Fruit imported into Australia from the USA is treated upon arrival with a mandatory fumigation of 32 g/m^3 methyl bromide for 2 or 4 h at 20°C. In a study duplicating this procedure, methyl bromide and inorganic bromide residues were determined in green and ripe avocados. Methyl bromide was detected in all fumigated fruit immediately after 30 min aeration and after one day's storage, but after two days only 0.1 mg/kg was detected in green fruit. Bromide ion residues were within the maximum residue level of 75 mg/kg recommended by the Australian National Health and Medical Research Council at all times tested (Singh, G. et al., 1982).

In an earlier study, bromide ion residues over seven days in avocados fumigated with 43 g/m^3 methyl bromide ranged from 39 to 68 mg/kg (mature green fruit) and 74 to 79 mg/kg (full ripe fruit). These values were close to, and in one case exceeded, the maximum residue level recommended in Australia (Hargreaves et al., 1978).

Dried fruit, cereals, nuts and spices imported into New Zealand in 1977 and 1978 were analysed for total bromide as residues of the fumigants methyl bromide and ethylene dibromide. About one-half of nut and spice samples contained levels above the permissible limit of 50 mg/kg, and occasional high bromide levels were found in cereals (Love *et al.*, 1979).

Methyl bromide levels in dry cocoa beans and cocoa products from a chocolate factory in Takoradi, Ghana, ranged from 3 to 257 mg/kg (as total bromide on a wet-weight basis) (Owusu-Manu, 1977).

Three methyl bromide fumigation treatments effective in controlling the Mediterranean fruit fly were tested on fruits obtained from production areas of the San Joaquin Valley in California, USA. Pears, plums, cherries, nectarines and peaches were fumigated at a concentration of 48 g/m^3 for 2 or 3 h, or 32 g/m^3 for 4 h. Pears and plums had relatively higher methyl bromide residues and retained the residues longer (30-60 mg/kg at 2 h post-fumigation; 0.6-3.5 mg/kg at 48 h) than did cherries, nectarines or peaches (7-31 mg/kg at 2 h post-fumigation; <0.01-0.03 mg/kg at 48 h). Desorption of methyl bromide was rapid in the first 24-48 h, even when the fruit was in cold storage. Bromide ion residues ranged from 5 to 9 mg/kg (Tebbets *et al.*, 1983). Similar results were obtained after the aeration of grapefruit with methyl bromide (King *et al.*, 1981).

Patches of discolouration (scorching) developed on winter wheat grown in a field treated with methyl bromide (980 kg/ha) during late September 1967. Scorching developed in wheat sown near methyl bromide injection sites and appeared to be associated with bromine uptake by the plants. The total bromide ion content in parts of plants above ground, sampled in July 1968, August 1968, July 1969, and July 1970, was 6100, 4200, 2500, and 940 mg/kg, respectively, for scorched plants grown in treated soil, and 14, 5, 10, and 14 mg/kg for plants grown in untreated soil (Brown & Jenkinson, 1971).

2.3 Analysis

Selected methods for the analysis of methyl bromide in air, water and various foods are identified in Table 3. The methods of the US Environmental Protection Agency (1984a,b) for analysing water have also been applied to liquid and solid wastes (US Environmental Protection Agency, 1982a,b). Volatile components of solid waste samples are first extracted with polyethylene glycol or methanol prior to purge/trap concentration and analysis. Method 624 of the US Environmental Protection Agency (1984c) has also been adapted to the analysis of methyl bromide in fish, with an estimated detection limit of 200 μg/kg (Easley *et al.*, 1981).

Exposures to methyl bromide can be monitored in air by a direct-reading infrared analyser, at minimum concentrations of 9 mg/m^3 (The Foxboro Co., 1985).

Table 3. Methods for the analysis of methyl bromide

Sample matrix	Sample preparation	Assay procedure[a]	Limit of detection	Reference
Air	Adsorb on charcoal; desorb (carbon disulphide); inject aliquot	GC/FID	1 mg/m^3	Eller (1985); Peers (1985)
	Adsorb (polymeric beads); desorb (heat, purge with helium); trap directly on GC column	GC/MS	500 ng/m^3	Krost et al. (1982)
	Inject sample directly (5 ml)	GC/EC	2 µg/m^3	Pellizzari et al. (1978)
	Adsorb on charcoal; desorb (heat, purge with helium); dry (calcium sulphate); readsorb (Tenax GC); desorb as before; trap (liquid nitrogen cooled); vaporize onto GC	GC/MS	8-24 ng/m^3	Pellizzari et al. (1978)
Water	Purge (inert gas); trap (OV-1 on Chromosorb-W/Tenax/silica gel); desorb as vapour (heat to 180°C, backflush with inert gas) onto GC column	GC/ECD	1.18 µg/l	US Environmental Protection Agency (1984a)
	Add internal standard (isotope-labelled methyl bromide); purge, trap and desorb as above	GC/MS	50 µg/l	US Environmental Protection Agency (1984b)
	Purge (80°C, nitrogen); trap (Ambersorb or Porapak N); desorb (flash-heat) and trap in 'mini-trap' (Ambersorb or Porapak N, -30°C); desorb (flash-heat) onto GC column	GC/EC GC/MS	0.05 µg/l (tap-water) <0.05 µg/l (tap-water) 0.1 µg/l (surface-water)	Piet et al. (1985)
Cereal grains and other foods	Extract with acetone:water; add sodium chloride; separate layers; dry acetone solution over anhydrous calcium chloride; inject aliquot	GC/EC	150 µg/kg	Scudamore (1985a)
	Extract with acetone:water; inject aliquot of head-space vapour	GC/EC	10 µg/kg	Scudamore (1985b)
Serum, plasma and blood	Analyse 5-ml sample directly	NAA	<5 µg/g	Ohmori & Hirata (1982)

[a]Abbreviations: GC/FID, gas chromatography/flame-ionization detection; GC/MS, gas chromatography/mass spectrometry; GC/EC, gas chromatography/electron capture detection; GC/ECD, gas chromatography/electrolytic conductivity detection; NAA, neutron activation analysis

3. Biological Data Relevant to the Evaluation of Carcinogenic Risk to Humans

3.1 Carcinogenicity studies in animals[1]

Oral administration

Rat: Groups of 10 male and 10 female weanling Wistar (Riv:Tox/M) rats, weighing 40-60 g, were administered 0, 0.4, 2, 10 or 50 mg/kg bw methyl bromide (purity, >98%) dissolved in arachis oil by gavage five times per week for 13 weeks. All animals were killed on day 90 except for one female receiving 2 mg/kg that died after eight days of treatment and one high-dose male that was killed in week 10. Incidences of hyperplasia and hyperkeratosis of the forestomach squamous epithelium were: males — control, 0/10; 0.4-mg, 0/10; 2-mg, 2/10; 10-mg, 6/10; 50-mg, 10/10; females — control, 1/10; 0.4-mg, 1/10; 2-mg, 1/10; 10-mg, 9/10; 50-mg, 10/10. Papillomas of the forestomach were observed in 2/10 50-mg males; no such tumour was found in the other groups. In the high-dose groups, 7/10 males and 6/10 females developed squamous-cell carcinomas of the forestomach. [These differences are statistically significant: $p = 0.002$ for males and $p = 0.005$ for females (Fisher exact test)]. No such tumour was found in the other groups (Danse *et al.*, 1984). The incidence of forestomach tumours in animals of this strain is very low even in aged rats (Kroes *et al.*, 1981).

It was reported in an abstract that oral administration of 25 and 50 mg/kg bw methyl bromide [purity unspecified] in peanut oil, administered by gavage for periods of 30, 60, 90 and 120 days, caused ulceration and epithelial hyperplasia of the forestomach in rats [number, strain, age, weight and sex unspecified] without evidence of malignancy (Hubbs & Harrington, 1986). [The Working Group noted that the level of detail provided was inadequate for an evaluation of carcinogenicity.]

3.2 Other relevant biological data

(a) Experimental systems

Toxic effects

The toxicity of methyl bromide has been reviewed (von Oettingen, 1964; Torkelson & Rowe, 1981; Alexeeff & Kilgore, 1983).

The oral LD_{50} for methyl bromide in rats was reported to be 214 mg/kg bw (Danse *et al.*, 1984). An 8-h LC_{50} of 300 ppm (1160 mg/m³) has been estimated in rats (Honma *et al.*, 1985), and a 1-h LC_{50} of 4680 mg/m³ has been estimated in mice (Alexeeff *et al.*, 1985). Previous results (summarized by Alexeeff & Kilgore, 1983) indicated a 30-min LC_{50} of

---

[1]The Working Group was aware of one study in progress in rats and one in mice by inhalation exposure (IARC, 1984).

11 000 mg/m^3 and a 24-h LC$_{50}$ of 50 mg/m^3 methyl bromide for rats. Rabbits were less sensitive than rats following single exposures by inhalation. Signs of toxicity in rats following acute exposure included drowsiness, dyspnoea, lachrymation, irritation, congestion and oedema of the lung, pneumonia and kidney damage. Although similar symptoms were not observed in rabbits, most animals that survived exposure to 1000 mg/m^3 became paralysed (Irish et al., 1940).

Other symptoms reported in various species following acute exposure to methyl bromide include tremor, incoordination, staggering gait, depression of the central nervous system and convulsions, depending on the exposure concentrations applied (von Oettingen, 1964).

Death following exposure to methyl bromide may occur hours or days later (Irish et al., 1940; Alexeeff & Kilgore, 1983; Alexeeff et al., 1985).

A decrease of 70% or more in liver glutathione concentrations has been found in mice exposed by inhalation to 4700 and 5930 mg/m^3 methyl bromide for 1 h (Alexeeff et al., 1985).

Rats and guinea-pigs were exposed to 130-850 mg/m^3 methyl bromide, rabbits to 65-850 mg/m^3 and monkeys to 130-420 mg/m^3 for 7.5-8 h per day on five days per week for up to six months. All rats and most guinea-pigs died after very few exposures to 850 mg/m^3, whereas some survived up to approximately 100 exposures to 420 mg/m^3. In rats, the signs of toxicity were poor general appearance, body-weight loss, pulmonary congestion and renal and hepatic lesions. No toxic effect was observed in rats or guinea-pigs following repeated exposures to 130 or 250 mg/m^3; but rabbits exposed to 130 or 250 mg/m^3 developed characteristic paralysis of the legs and died after several exposures, although they tolerated repeated exposures to 65 mg/m^3. In monkeys, exposure to 250 mg/m^3, but not to 130 mg/m^3, led to hyperexcitability, loss of equilibrium, inability to stand, convulsions and paralysis (Irish et al., 1940).

As reported in an abstract, high mortality was observed among mice exposed to 200 ppm (800 mg/m^3) methyl bromide for 6 h per day for 10 days (Haber et al., 1985). Rabbits exposed to 65 ppm (250 mg/m^3) methyl bromide for 7.5 h per day on four days per week for one month (total exposure time, 100 h) lost weight and had reduced eyeblink responses and reduced nerve-conduction velocity; no effect was observed in rats (Anger et al., 1981).

Effects on reproduction and prenatal toxicity

Groups of about 36 Wistar rats were exposed by inhalation to 0, 20 or 70 ppm (0, 78 or 272 mg/m^3) methyl bromide (purity, >99.5%) during either pregestational or gestational periods, or both. Pregestational exposure lasted 7 h per day, five days per week for three weeks; gestational exposures began immediately after breeding and continued daily through day 19 of gestation. Maternal body weights were reduced during gestation in the group receiving pre- and gestational exposure to 70 ppm (272 mg/m^3). Fetal growth was not affected by any exposure regimen, and no significant treatment-related visceral or skeletal abnormality was found (Sikov et al., 1980; Hardin et al., 1981).

Groups of 24 New Zealand rabbits were exposed by inhalation to 0, 20 or 70 ppm (0, 78 or 272 mg/m^3) methyl bromide (purity, >99.5%) for 7 h per day for a period planned to consist of days 1-24 of gestation; however, exposure of all groups was stopped after 15 days because of toxicity, and excessive maternal mortality prevented examination of fetuses in the high-dose group. No maternal or fetal toxicity was observed in rabbits exposed to 20 ppm (Sikov et al., 1980; Hardin et al., 1981).

Absorption, distribution, excretion and metabolism

Absorption of methyl bromide by the lung was rapid and exhibited first-order kinetics without a saturable component following exposure of rats to 100-3000 ppm (390-11 640 mg/m^3) (Andersen et al., 1980). During a 6-h exposure of rats to 50-10 400 mmol/l (4.75-9874 mg/m^3) ^{14}C-methyl bromide, approximately 27-50% of the compound inhaled was absorbed (Medinsky et al., 1985). After oral administration of 100 mg/kg bw methyl bromide to rats, approximately 4% was exhaled within 2 h. Following intraperitoneal administration of 120-180 mg/kg bw in hourly divided doses, a total of 24-45% was exhaled (Miller & Haggard, 1943). Following either oral or intraperitoneal administration of 250 μmol/kg bw (24 mg/kg bw) ^{14}C-methyl bromide to rats, 14-17% of the radioactivity remained in the body after 72 h. Expiration of ^{14}C-carbon dioxide amounted to 32% and 45%, respectively, and the respective urinary excretion of ^{14}C-radioactivity was 43% and 16%. Less than 3% of the radioactivity was excreted in the faeces. In animals with bile-duct cannulations, 46% of the radioactivity of an oral dose appeared in the bile over a 24-h period (Medinsky et al., 1984). Similar excretion patterns were seen after inhalation exposure (Bond et al., 1985; Medinsky et al., 1985).

Bromide ion has been identified as a metabolite of methyl bromide in rabbits (Irish et al., 1941). Following exposure of rats by inhalation, similar first-order rate constants have been reported for methyl bromide uptake and plasma bromide formation (Gargas & Andersen, 1982), and a linear relationship has been observed between exposure concentrations and blood bromine levels (Honma et al., 1985).

Methyl bromide-induced cytotoxicity to HeLa cells was reduced by the addition of glutathione (Nishimura et al., 1980).

Following in-vitro treatment with radiolabelled methyl bromide, methylation of cysteine-S and histidine-N^π and -N^τ residues of haemoglobin in suspended erythrocytes, of guanine-N-7 of DNA in solution and of DNA in CBA mouse spleen-cell suspensions was seen. After CBA mice were exposed by inhalation, alkylation of cysteine-S residues in haemoglobin and liver proteins, and of guanine-N-7 in DNA from liver and spleen cells was reported (see below) (Djalali-Behzad et al., 1981).

Mutagenicity and other short-term tests

Methyl bromide was mutagenic to *Salmonella typhimurium* TA100 when tested at concentrations of 0.02-0.2% in desiccators in the absence of an exogenous metabolic system (Simmon et al., 1977). Positive results were also obtained in strain TA 100 in a liquid assay (tested at 10-1000 mg/l) and a plate assay (tested in closed containers at concentrations of 500-50 000 mg/m^3); activity in the plate assay was unaffected by the presence of liver homogenates from Aroclor-induced rats (Kramers et al., 1985). A commercial preparation of methyl bromide (tested at 0.5-5 g/m^3 in a closed container) was mutagenic to

S. typhimurium TA1535 and TA100 (but not to TA1537, TA1538 or TA98) and to *Escherichia coli* WP2 *hcr* in the absence of an exogenous metabolic system (Moriya *et al.*, 1983). An aqueous solution of methyl bromide (tested at 0.5-6 mM) induced mutations to streptomycin independence in *E. coli* Sd-4 (Djalali-Behzad *et al.*, 1981). In a fluctuation test, methyl bromide (tested at 950-19 000 mg/m^3) induced mutations to streptomycin resistance in *Klebsiella pneumoniae* (Kramers *et al.*, 1985).

A few chlorophyll mutations were induced in barley after treatment of kernels with 1.4 mM methyl bromide for 24 h in closed vessels (Ehrenberg *et al.*, 1974).

In a sex-linked recessive lethal test in *Drosophila melanogaster*, flies of the Berlin K strain were exposed to methyl bromide at concentrations of 70-750 mg/m^3 for increasing periods; mutation frequencies were significantly increased at the highest nontoxic concentrations (Kramers *et al.*, 1985). It was reported in an abstract that the incidences of wing twin spots (indicating somatic recombination) and wing single spots (indicating a variety of somatic genetic alterations, including mutation, deletion, recombination, nondisjunction and chromosome loss) were increased after exposure of *D. melanogaster* larvae to 0-20 mg/l for 1 h (Katz, 1985).

As measured by autoradiography, methyl bromide (tested at 10-30 mg/l) did not induce unscheduled DNA synthesis in primary cultures of rat hepatocytes treated in air-tight bottles (Kramers *et al.*, 1985).

Treatment of L5178Y mouse lymphoma cells with 0.03-30 mg/l methyl bromide in air-tight bottles resulted in a dose-related increase in 6-thioguanine- and bromodeoxyuridine-resistant mutants (Kramers *et al.*, 1985).

It was reported in an abstract that exposure of human lymphocyte cultures to 4.3% methyl bromide for 100 sec increased the frequency of sister chromatid exchanges from 10.0 to 16.8 per cell (Tucker *et al.*, 1985).

Alkylation of guanine-N-7 in DNA of liver and spleen was observed after treatment of male CBA mice with ^{14}C-methyl bromide (4.9-5.0 mCi/mmol) by inhalation (80 μCi/11-l box for 4 h, corresponding to 340 μCi/kg bw) or intraperitoneal injection (4.4 μmol/kg bw) (Djalali-Behzad *et al.*, 1981).

It was reported in an abstract that micronuclei were induced in bone-marrow cells of Fischer 344 rats and BDF$_1$ mice and in peripheral blood cells of BDF$_1$ mice exposed to methyl bromide by inhalation for 6 h per day on five days per week for two weeks. In mice, the incidence of polychromatic erythrocytes with micronuclei in the bone marrow increased by ten fold in males (200 ppm; 776 mg/m^3) and by six fold in females (154 ppm; 600 mg/m^3), and those in peripheral blood increased by 32 fold in males (200 ppm) and by three fold in females (154 ppm). In rats, the increases were ten fold in males (338 ppm; 1311 mg/m^3) and three fold in females (338 ppm) (Ikawa *et al.*, 1986).

Transformation of Syrian hamster embryo cells by SA7 adenovirus was not enhanced by exposure of cells to 1000-8000 ppm (4-16 g/m^3) methyl bromide in sealed chambers (Hatch *et al.*, 1983). [The Working Group noted that the compound was tested only over a nontoxic dose range.]

(b) *Humans*

Toxic effects

An odour threshold of 65 mg/m^3 has been reported for methyl bromide (Worthing, 1983).

The first cases of methyl bromide poisoning were reported in the late 1800s. Since then, over 950 cases, involving fatalities, systemic poisoning, skin and eye injury, and damage to the central nervous system, have been reported. In the first half of this century, most poisonings were due to accidents that occurred during chemical manufacture and filling operations, or were associated with the use of methyl bromide as a fire extinguisher; in the last 30 years, the use of this agent as a fumigant has been the main source of poisoning (von Oettingen, 1946, 1964; Gosselin *et al.*, 1976; Torkelson & Rowe, 1981; Alexeeff & Kilgore, 1983),

After inhalation exposure to methyl bromide, lachrymation due to eye irritation, blurred vision, diplopia and transient blindness have been reported, and retinal haemorrhages were found in several instances both clinically and at autopsy (Morton Grant, 1962).

Deaths from inhalation of methyl bromide have generally occurred after short exposures to very high concentrations or after an exposure that was higher than usual after a history of long-term exposure to low doses (Alexeeff & Kilgore, 1983). Wyers (1945) reported a fatal case after exposure to over 60 000 ppm (233 000 mg/m^3) for 2 h; Astley Clarke *et al.* (1945) stated that inhalation of 10 000 ppm (38 800 mg/m^3) for more than a few minutes could be fatal.

Acute respiratory-tract symptoms include chest pain and difficulty in breathing, and are consistent with post-mortem findings of pulmonary oedema, bronchopneumonia, congestion and haemorrhage (Holling & Clarke, 1944; Astley Clarke *et al.*, 1945; Viner, 1945; Wyers, 1945; Mazel *et al.*, 1946; Ravault *et al.*, 1946; Miller, 1948; Prain & Harvey Smith, 1952; Hine, 1969; Marraccini *et al.*, 1983).

Acute gastrointestinal-tract symptoms include gastric and abdominal pain, nausea and vomiting; pathological examinations revealed stomach congestion and submucous haemorrhages (Wyers, 1945; MacDonald *et al.*, 1950).

Other pathological findings in fatal cases were congestion of the kidneys and glomeruli, degenerative changes in the tubuli and congestion of the liver (Holling & Clarke, 1944; Astley Clarke *et al.*, 1945; Mazel *et al.*, 1946; Wyers, 1945; MacDonald *et al.*, 1950; Prain & Harvey Smith, 1952; Hine, 1969; Davay, 1972; Marraccini *et al.*, 1983). Reported neurological symptoms include headache and dizziness, followed by muscular pain, numbness and, in the most severe case, convulsions and unconsciousness. Autopsy revealed congestion of the brain with multiple haemorrhages associated with degenerative changes, such as necrosis (Holling & Clarke, 1944; Astley Clarke *et al.*, 1945; Viner, 1945; Wyers, 1945; Mazel *et al.*, 1946; MacDonald *et al.*, 1950; Prain & Harvey Smith, 1952; Davay, 1972; Goulon *et al.*, 1975; Marraccini *et al.*, 1983).

Non-fatal exposure to methyl bromide has given rise to respiratory symptoms, such as chest pains and dyspnoea (Hine, 1969), and gastrointestinal symptoms, including pain, nausea and vomiting (Drawneek *et al.*, 1964; Longley & Jones, 1965; Hine, 1969; Zatuchni & Hong, 1981). The neurological symptoms reported in these cases include headache, weakness, speech impairment, visual difficulties, tremors and convulsions; permament brain damage can follow (Johnstone, 1945; Rathus & Landy, 1961; Drawneek *et al.*, 1964; Longley & Jones, 1965; Hine, 1969; Greenberg, 1971; Shield *et al.*, 1977; Zatuchni & Hong, 1981). Slight electroencephalographic alterations were observed in 10/33 fumigators using methyl bromide, even in the absence of subjective symptoms (Verberk *et al.*, 1979).

Watrous (1942) studied the prevalence of skin lesions among 90 workers filling and sealing glass ampoules containing methyl bromide. As a consequence of accidental contact, blisters, dermatitis and vesicles similar to those in second-degree burns appeared. Butler *et al.* (1945) described similar lesions. The systemic symptoms reported by a proportion of these patients were similar to those observed in subjects poisoned after inhalation; dermal absorption of methyl bromide has been associated with severe poisoning and even fatality (Schifferli, 1942).

Effects on reproduction and prenatal toxicity

No data were available to the Working Group.

Absorption, distribution, excretion and metabolism

Methyl bromide is absorbed through the lungs and the skin (Alexeeff & Kilgore, 1983).

Bromide ion has been identified in human blood following methyl bromide poisoning (for review see Alexeeff & Kilgore, 1983). See also the section on occupational exposures, p. 192.

Mutagenicity and chromosomal effects

No data were available to the Working Group.

3.3 Case reports and epidemiological studies of carcinogenicity to humans

Wong *et al.* (1984) studied the mortality of a cohort of 3579 white male workers with potential exposure to brominated compounds at three chemical manufacturing plants and at a research establishment between 1935 and 1976. The exposures included 1,2-dibromo-3-chloropropane (DBCP; see IARC, 1979a), tris(2,3-dibromopropyl)phosphate (Tris; see IARC, 1979b), polybrominated biphenyls (PBB; see this volume, p. 261), various organic and inorganic bromides and DDT (see IARC, 1974, 1982). Among a subgroup of 655 men exposed to organic brominated compounds other than DBCP, Tris and PBB, and with potential exposure to methyl bromide, 51 deaths occurred *versus* 44.77 expected. Ten deaths from cancer were observed *versus* 7.86 expected, yielding a standardized mortality ratio (SMR) of 127 (95% confidence interval, 61-234). In this group of workers, there were two deaths from testicular cancer *versus* 0.11 expected (SMR, 1799; $p < 0.05$). An investigation of the work histories showed that methyl bromide was the only common potential exposure of these two cases. However, these men died at the ages of 17 and 33 years, respectively. [The Working Group noted that no information is available on duration of exposure or on time between first exposure and death from testicular cancer.]

In a mortality study on 4411 pesticide applicators licensed in Florida in 1965-1966, 2210 subjects had worked at some time as fumigators, and therefore had potential exposure to methyl bromide: 17 lung cancer deaths were observed *versus* 12.7 expected (Blair *et al.*, 1983).

[The Working Group noted that in both these studies, methyl bromide was only one of many potential exposures.]

4. Summary of Data Reported and Evaluation

4.1 Exposure data

Methyl bromide is formed in the oceans and is produced industrially as a soil, fruit and space fumigant and as a chemical intermediate. Its use as a fumigant is the major potential source of high-level exposure to this compound. There is ubiquitous low-level exposure in ambient air and in water.

4.2 Experimental data

In one 90-day study, methyl bromide was tested in rats by oral administration by gavage. An increased incidence of squamous-cell carcinomas of the forestomach was observed in animals of each sex.

Methyl bromide did not cause prenatal toxic effects in rats or rabbits exposed by inhalation. No data were available to evaluate the reproductive effects of methyl bromide.

Methyl bromide is mutagenic to bacteria and plants. It induces sex-linked recessive lethal mutations in *Drosophila melanogaster* and mutations in cultured mammalian cells. It does not induce unscheduled DNA synthesis in cultured mammalian cells. Methyl bromide alkylates DNA in liver and spleen of mice treated by various routes. It induces micronuclei in bone-marrow and peripheral blood cells of rats and mice.

4.3 Human data

No data were available to evaluate the reproductive effects or prenatal toxicity of methyl bromide to humans.

The two available cohort studies mentioning exposure to methyl bromide were uninformative with regard to the carcinogenic hazard of this chemical.

Overall assessment of data from short-term tests: Methyl bromide[a]

	Genetic activity			Cell transformation
	DNA damage	Mutation	Chromosomal effects	
Prokaryotes		+		
Fungi/Green plants		+		
Insects		+		
Mammalian cells (*in vitro*)	–	+		
Mammals (*in vivo*)	+		+	
Humans (*in vivo*)				
Degree of evidence in short-term tests for genetic activity: **Sufficient**				Cell transformation: No data

[a]The groups into which the table is divided and the symbols '+' and '–' are defined on pp. 19-20 of the Preamble; the degrees of evidence are defined on pp. 20-21.

4.4 Evaluation[1]

There is *limited evidence* for the carcinogenicity of methyl bromide to experimental animals.

There is *inadequate evidence* for the carcinogenicity of methyl bromide to humans.

5. References

Alexeeff, G.V. & Kilgore, W.W. (1983) Methyl bromide. *Residue Rev.*, *88*, 101-153

Alexeeff, G.V., Kilgore, W.W., Munoz, P. & Watt, D. (1985) Determination of acute toxic effects in mice following exposure to methyl bromide. *J. Toxicol. environ. Health*, *15*, 109-123

American Conference of Governmental Industrial Hygienists (1985) *TLVs Threshold Limit Values and Biological Exposure Indices for 1985-86*, 2nd ed., Cincinnati, OH, p. 23

Andersen, M.E., Gargas, M.L., Jones, R.A. & Jenkins, L.J., Jr (1980) Determination of the kinetic constants for metabolism of inhaled toxicants *in vivo* using gas uptake measurements. *Toxicol. appl. Pharmacol.*, *54*, 100-116

---

[1]For definitions of the italicized terms, see Preamble, pp. 18 and 22.

Anger, W.K., Setzer, J.V., Russo, J.M., Brightwell, W.S., Wait, R.G. & Johnson, B.L. (1981) Neurobehavioral effects of methyl bromide in inhalation exposures. *Scand. J. Work Environ. Health*, 7 (Suppl. 4), 40-47

Anon. (1985a) Chemical profile: methyl bromide. *Chem. Mark. Rep.*, 227, 54

Anon. (1985b) *Farm Chemicals Handbook*, Willoughby, OH, Meister Publishing Co., p. C-154

Arbetskyadsstyrelsens Författningssamling (National Swedish Board of Occupational Safety and Health) (1984) *Occupational Exposure Limit Values (AFS 1984:5)* (Swed.), Solna, p. 20

Astley Clarke, C., Roworth, C.G. & Holling, H.E. (1945) Methyl bromide poisoning. An account of four recent cases met within one of H.M. ships. *Br. J. ind. Med.*, 2, 17-23

ATOCHEM (1984) *A Brief Summary*, Paris

Basile, M. & Lamberti, F. (1981) Bromide residues in edible organs of plants grown in soil treated with methyl bromide. *Med. Fac. Landbouwwet. Rijksuniv. Gent*, 46, 337-341

Blair, A., Grauman, D.J., Lubin, J.H. & Fraumeni, J.F., Jr (1983) Lung cancer and other causes of death among licensed pesticide applicators. *J. natl Cancer Inst.*, 71, 31-37

Bond, J.A., Dutcher, J.S., Medinsky, M.A., Henderson, R.F. & Birnbaum, L.S. (1985) Disposition of (^{14}C)methyl bromide in rats after inhalation. *Toxicol. appl. Pharmacol.*, 78, 259-267

Brown, D. & Rolston, D.E. (1980) Transport and transformation of methyl bromide in soils. *Soil Sci.*, 130, 68-75

Brown, G. & Jenkinson, D.S. (1971) Bromine in wheat grown on soil fumigated with methyl bromide. *Soil Sci. Plant Anal.*, 2, 45-54

Butler, E.C.B., Perry, K.M.A. & Williams, J.R.F. (1945) Methyl bromide burns. *Br. J. ind. Med.*, 2, 30-31

The Chemical Daily Co. (1984a) *JCW Chemicals Guide 1984/1985*, Tokyo, p. 267

The Chemical Daily Co. (1984b) *Specialty Chemicals Handbook*, Tokyo, p. 74

Considine, D.M., ed. (1974) *Chemical and Process Technology Encyclopedia*, New York, McGraw-Hill Book Co., p. 183

Danse, L.H.J.C., van Velsen, F.L. & van der Heijden, C.A. (1984) Methyl bromide: carcinogenic effects in the rat forestomach. *Toxicol. appl. Pharmacol.*, 72, 262-271

Davay, G.G. (1972) Methyl bromide poisoning. *Indian J. ind. Med.*, 18, 78-85

Davis, L.N., Strange, J.R., Hoecker, J.E., Howard, P.H. & Santodonato, J. (1977) *Investigation of Selected Potential Environmental Contaminants: Monohalomethanes (EPA-560/2-77-007)*, Washington DC, US Environmental Protection Agency, pp. 16, 46-51

Deutsche Forschungsgemeinschaft (German Research Community) (1985) *Maximal Concentrations in the Workplace and Biological Tolerance Values for Substances in the Work Environment 1985* (Ger.), Vol. 21, Weinheim, Verlagsgesellschaft, p. 21

Direktoratet for Arbeidstilsynet (Directorate of Labour Inspection) (1981) *Administrative Norms for Pollution in Work Atmosphere, No. 361* (Norw.), Oslo, p. 16

Djalali-Behzad, G., Hussain, S., Osterman-Golkar, S. & Segerbäck, D. (1981) Estimation of genetic risks of alkylating agents. VI. Exposure of mice and bacteria to methyl bromide. *Mutat. Res., 84*, 1-9

Drawneek, W., O'Brian, M.J., Goldsmith, H.J. & Bourdillon, R.E. (1964) Industrial methyl bromide poisoning in fumigators. *Lancet, ii*, 855-856

Easley, D.M., Kleopfer, R.D. & Carasea, A.M. (1981) Gas chromatographic-mass spectrometric determination of volatile organic compounds in fish. *J. Assoc. off. anal. Chem., 64*, 653-656

Ehrenberg, L., Osterman-Golkar, S., Singh, D. & Lundqvist, U. (1974) On the reaction kinetics and mutagenic activity of methylating and *beta*-halogenoethylating gasoline additives. *Radiat. Bot., 15*, 185-194

Eller, P.M. (1985) *NIOSH Manual of Analytical Methods*, 3rd ed., 1st Suppl. (*DHHS (NIOSH) Pub. No. 84-100*), Washington DC, US Government Printing Office, pp. 2520-1 — 2520-4

Ethyl Corp. (undated a) *M-B-R 100 Methyl Bromide*, Baton Rouge, LA

Ethyl Corp. (undated b) *M-B-R 98*, Baton Rouge, LA

Evans, W.A., Wilcox, T., Flesch, J.P., Hollett, B., Grote, A., Eller, P. & Carnow, B. (1979) *Health Hazard Evaluation Determination Report* (*HE 77-73-610*), Cincinnati, OH, National Institute for Occupational Safety and Health

Food and Agriculture Organization (1980) *Pesticide Residues in Food — 1979* (*FAO Plant Production and Protection Paper, 20 sup.*), Rome, pp. 89-97

The Foxboro Co. (1985) *Ambient Air Instrumentation Sheet*, Foxboro, MA

Gargas, M.L. & Andersen, M.E. (1982) Metabolism of inhaled brominated hydrocarbons: validation of gas uptake results by determination of a stable metabolite. *Toxicol. appl. Pharmacol., 66*, 55-68

Gosselin, R.E., Hodge, H.C., Smith, R.P. & Gleason, M.N. (1976) *Clinical Toxicology of Commercial Products — Acute Poisoning*, 4th ed., Baltimore, Williams & Wilkins, pp. 233-237

Gosselin, R.E., Smith, R.P., Hodge, H.C. & Braddock, J.E. (1984) *Clinical Toxicology of Commercial Products*, 5th ed., Baltimore, Williams & Wilkins, pp. V-101 — V-102, V-400, V-646

Goulon, M., Nouailhat, F., Escourolle, R., Zarranz-Imirizaldu, J.J., Grosbuis, S. & Lévy-Alcover, M.A. (1975) Methyl bromide intoxication — three cases, with one death. Neuropathological study of one case of stupor and myoclonus followed for five years (Fr.). *Rev. Neurol., 131*, 445-468

Grasselli, J.G. & Ritchey, W.M., eds (1975) *CRC Atlas of Spectral Data and Physical Constants for Organic Compounds*, Vol. 3, Cleveland, OH, CRC Press, p. 590

Greenberg, J.O. (1971) The neurological effects of methyl bromide poisoning. *Ind. Med., 40*, 27-29

Haber, S.B., Drew, R.T., Eustis, S. & Yang, R.S.H. (1985) Methyl bromide toxicity: a target organ? (Abstract No. 518). *Toxicologist, 5*, 130

Hansch, C. & Leo, A. (1979) *Substituent Constants for Correlation Analysis in Chemistry and Biology*, New York, John Wiley & Sons, p. 173

Hardin, B.D., Bond, G.P., Sikov, M.R., Andrew, F.D., Beliles, R.P. & Niemeier, R.W. (1981) Testing of selected workplace chemicals for teratogenic potential. *Scand. J. Work Environ. Health, 7* (Suppl. 4), 66-75

Hargreaves, P.A., Wainwright, D.H., Swaine, G. & Corcoran, R.J. (1978) Residual ethylene dibromide and inorganic bromide levels in some fruit and vegetables after fumigation with ethylene dibromide or methyl bromide. *Aust. J. exp. Agric. anim. Husb., 18*, 586-590

Hatch, G.G., Mamay, P.D., Ayer, M.L., Casto, B.C. & Nesnow, S. (1983) Chemical enhancement of viral transformation in Syrian hamster embryo cells by gaseous and volatile chlorinated methanes and ethanes. *Cancer Res., 43*, 1945-1950

Hawley, G.G., ed. (1981) *The Condensed Chemical Dictionary*, 10th ed., New York, Van Nostrand Reinhold, p. 670

Hayes, W.J., Jr (1982) *Pesticides Studied in Man*, Baltimore, MD, Williams & Wilkins, p. 140

Health and Safety Executive (1985) *Occupational Exposure Limits 1985 (Guidance Note EH 40/85)*, London, Her Majesty's Stationery Office, p. 9

Hine, C.H. (1969) Methyl bromide poisoning — a review of ten cases. *J. occup. Med., 11*, 1-10

Holling, H.E. & Clarke, C.A. (1944) Methyl bromide intoxication. *J. R. Nav. med. Serv., 30*, 218-224

Holtorf, R.C. (1985) *Preliminary Quantitative Usage Analysis of Methyl Bromide*, Washington DC, US Environmental Protection Agency

Honma, T., Miyagawa, M., Sato, M. & Hasegawa, H. (1985) Neurotoxicity and metabolism of methyl bromide in rats. *Toxicol. appl. Pharmacol., 81*, 183-191

Hubbs, A.F. & Harrington, D.D. (1986) *Further evaluation of the potential gastric carcinogenic effects of subchronic methyl bromide administration.* In: Proceedings of the 36th Annual Meeting of the American College of Veterinary Pathologists and the Annual Meeting of the American Society for Veterinary and Clinical Pathology, December 1985, Denver, CO, p. 92

IARC (1974) *IARC Monographs on the Evaluation of Carcinogenic Risk of Chemicals to Man*, Vol. 5, *Some Organochlorine Pesticides*, Lyon, pp. 83-124

IARC (1979a) *IARC Monographs on the Evaluation of the Carcinogenic Risk of Chemicals to Humans*, Vol. 20, *Some Halogenated Hydrocarbons*, Lyon, pp. 83-96

IARC (1979b) *IARC Monographs on the Evaluation of the Carcinogenic Risk of Chemicals to Humans*, Vol. 20, *Some Halogenated Hydrocarbons*, Lyon, pp. 575-588

IARC (1982) *IARC Monographs on the Evaluation of the Carcinogenic Risk of Chemicals to Humans*, Suppl. 4, *Chemicals, Industrial Processes and Industries Associated with Cancer in Humans, IARC Monographs, Volumes 1 to 29*, Lyon, pp. 105-108

IARC (1984) *Information Bulletin on the Survey of Chemicals Being Tested for Carcinogenicity*, No. 11, Lyon, pp. 111, 243

Ikawa, N., Araki, A., Nozaki, K. & Matsuchima, T. (1986) Micronucleus test of methyl bromide by the sub-chronic inhalation test (Abstract). *Mutat. Res.* (in press)

Ingram, F.R. (1951) Methyl bromide fumigation control in the date-packing industry. *Arch. ind Hyg. occup. Med.*, *4*, 193-198

Institut National de Recherche et de Sécurité (National Institute for Research and Safety) (1985) *Limit Values for Concentrations of Dangerous Substances in Work Place Air (Notes Documentaires 1555-21-85)* (Fr.), Paris, p. 481

International Labour Office (1980) *Occupational Exposure Limits for Airborne Toxic Substances: A Tabular Compilation of Values from Selected Countries*, 2nd (rev.) ed. (*Occupational Safety and Health Series No. 37*), Geneva, pp. 144-145

Irish, D.D., Adams, E.M., Spencer, H.C. & Rowe, V.K. (1940) The response attending exposure of laboratory animals to vapors of methyl bromide. *J. ind. Hyg. Toxicol.*, *22*, 218-230

Irish, D.D., Adams, E.M., Spencer, H.C. & Rowe, V.K. (1941) Chemical changes of methyl bromide in the animal body in relation to its physiological effects. *J. ind. Hyg. Toxicol.*, *23*, 408-411

Johnstone, R.T. (1945) Methyl bromide intoxication of a large group of workers. *Ind. Med.*, *14*, 495-497

Katz, A.J. (1985) Genotoxicity of methyl bromide in somatic cells of *Drosophila* larvae (Abstract). *Environ. Mutagenesis*, *7* (Suppl. 3), 13

King, J.R., Benschoter, C.A. & Burditt, A.K., Jr. (1981) Residues of methyl bromide in fumigated grapefruit determined by a rapid, headspace assay. *J. agric. Food Chem.*, *29*, 1003-1005

Kramers, P.G.N., Voogd, C.E., Knaap, A.G.A.C & van der Heijden, C.A. (1985) Mutagenicity of methyl bromide in a series of short-term tests. *Mutat. Res.*, *155*, 41-47

Kroes, R., Garbis-Berkvens, J.M., de Vries, T. & van Nesselrooy, J.H.J. (1981) Histopathological profile of a Wistar rat stock including a survey of the literature. *J. Gerontol.*, *36*, 259-279

Krost, K.J., Pellizzari, E.D., Walburn, S.G. & Hubbard, S.A. (1982) Collection and analysis of hazardous organic emissions. *Anal. Chem.*, *54*, 810-817

Longley, E.O. & Jones, A.T. (1965) Methyl bromide poisoning in man. *Ind. Med. Surg.*, *34*, 499-502

Love, J.L., Winchester, R.V. & Keating, D.L. (1979) Fumigant and contact insecticide residues on dried fruits, cereals, nuts, and spices imported into New Zealand. *N.Z. J. Sci.*, *22*, 95-98

Lovelock, J.E. (1975) Natural halocarbons in the air and in the sea. *Nature*, *256*, 193-194

MacDonald, A.C., Monro, I.C. & Scott, G.I. (1950) Fatal case of poisoning due to inhalation of methyl bromide. *Br. med. J., ii*, 441-442

Mannsville Chemical Products Corp. (1979) *Chemical Products Synopsis: Methyl Bromide*, Cortland, NY

Marraccini, J.V., Thomas, G.E., Ongley, J.P., Pfaffenberger, C.D., Davis, J.H. & Bednarczyk, L.R. (1983) Death and injury caused by methyl bromide, an insecticide fumigant. *J. forens. Sci., 28*, 601-607

Matheson Gas Products (1985) *Matheson Gas Products Catalog 85*, Secaucus, NJ, p. 46

Mazel, P., Bourret, J. & Roche, L. (1946) Familial intoxication of five persons after fumigation with methyl bromide — three deaths (Fr.). *Arch. Mal. prof., 7*, 38-42

Medinsky, M.A., Bond, J.A., Dutcher, J.S. & Birnbaum, L.S. (1984) Disposition of [^{14}C]methyl bromide in Fischer-344 rats after oral or intraperitoneal administration. *Toxicology, 32*, 187-196

Medinsky, M.A., Dutcher, J.S., Bond, J.A., Henderson, R.F., Mauderly, J.L., Snipes, M.B., Mewhinney, J.A., Cheng, Y.S. & Birnbaum, L.S. (1985) Uptake and excretion of [^{14}C]methyl bromide as influenced by exposure concentration. *Toxicol. appl. Pharmacol., 78*, 215-225

Miller, J.W. (1948) Fatal methyl bromide poisoning. *Arch. Pathol., 30*, 505-507

Miller, D.P. & Haggard, H.W. (1943) Intracellular penetration of bromide as a feature in the toxicity of alkyl bromides. *J. ind. Hyg. Toxicol., 25*, 423-433

Moriya, M., Ohta, T., Watanabe, K., Miyazawa, T., Kato, K. & Shirasu, T. (1983) Further mutagenicity studies on pesticides in bacterial reversion assay systems. *Mutat. Res., 116*, 185-216

Morton Grant, W. (1962) *Toxicology of the Eye*, Springfield, VA, Charles C. Thomas, pp. 349-353

National Fire Protection Association (1984) *Fire Protection Guide on Hazardous Materials*, 8th ed., Quincy, MA, p. 325M-67

National Institute for Occupational Safety and Health (1985) NIOSH recommendations for occupational safety and health standards. *Morbid. Mortal. wkly Rep., Suppl., 34*, 22S

Nishimura, M., Umeda, M., Ishizu, S. & Sato, M. (1980) Effect of methyl bromide on cultured mammalian cells. *J. toxicol. Sci., 5*, 321-330

von Oettingen, W.F. (1946) *The Toxicity and Potential Dangers of Methyl Bromide with Special Reference to its Use in the Chemical Industry, in Fire Extinguishers and in Fumigation (Natl Inst. Health Bull. No. 185)*, Washington DC, US Government Printing Office, pp. 1-41

von Oettingen, W.F. (1964) *The Halogenated Hydrocarbons of Industrial and Toxicological Importance*, Amsterdam, Elsevier, pp. 25-55

Ohmori, S. & Hirata, M. (1982) Determination of bromine contents in blood and hair of workers exposed to methyl[bromide] by radioactivity analysis method bromide (Jpn.). *Jpn. J. ind. Health, 24*, 119-125

Opresko, D. & Daugherty, M.L. (1985) *Chemical Hazard Information Profile Draft Report: Methyl Bromide*, Washington DC, US Environmental Protection Agency

Owusu-Manu, E. (1977) *Insecticide residues and tainting in cocoa*. In: Watson, D.L. & Brown, A.W.A., eds, *Pesticide Management and Insecticide Resistance [Papers of the Section on Pesticide Development, Management, and Regulation at the 15th International Congress of Entomology], Washington, DC*, New York, Academic Press, pp. 555-564

Peers, A.M.K. (1985) *The determination of methyl bromide in air*. In: Fishbein, L. & O'Neill, I.K., eds, *Environmental Carcinogens. Selected Methods of Analysis*, Vol. 7, *Some Volatile Halogenated Hydrocarbons (IARC Scientific Publications No. 68)*, Lyon, International Agency for Research on Cancer, pp. 227-233

Pellizzari, E.D., Zweidinger, R.A. & Erickson, M.D. (1978) *Environmental Monitoring Near Industrial Sites: Brominated Chemicals Part II: Appendix (EPA-560/6-78-002A; US NTIS PB-286483)*, Washington DC, Office of Toxic Substances, US Environmental Protection Agency

Piet, G.J., Luijten, W.C.M.M. & van Noost, P.C.M. (1985) *'Dynamic' head-space determination of volatile organic halogen compounds in water*. In: Fishbein, L. & O'Neill, I.K., eds, *Environmental Carcinogens. Selected Methods of Analysis*, Vol. 7, *Some Volatile Halogenated Hydrocarbons (IARC Scientific Publications No. 68)*, Lyon, International Agency for Research on Cancer, pp. 331-343

Prain, J.H. & Harvey Smith, G. (1952) A clinical-pathological report of eight cases of methyl bromide poisoning. *Br. J. ind. Med., 9*, 44-49

Rasmussen, R.A. & Khalil, M.A.K. (1984) Gaseous bromine in the Arctic and Arctic haze. *Geophys. Res. Lett., 11*, 433-436

Rathus, E.M. & Landy, P.J. (1961) Methyl bromide poisoning. *Br. J. ind. Med., 18*, 53-57

Ravault, P., Bourret, J. & Vignon, G. (1946) Acute intoxication with death with methyl bromide (Fr.). *Arch. Mal. prof., 7*, 43-44

Roosels, D., Van Den Oever, R. & Lahaye, D. (1981) Dangerous concentrations of methyl bromide used as a fumigant in Belgian greenhouses. *Int. Arch. occup. environ. Health, 48*, 243-250

Roughan, J.A. & Roughan, P.A. (1984) Pesticide residues in foodstuffs in England and Wales. Part II: Inorganic bromide ion in cucumber, tomato and self-blanching celery grown in soil fumigated with bromomethane, and the 'natural' bromide ion content in a range of fresh fruit and vegetables. *Pestic. Sci., 15*, 630-636

Sadtler Research Laboratories (1980) *The Sadtler Standard Spectra, Cumulative Index*, Philadelphia, PA

Saltzman, B.E. (1983) *Direct reading colorimetric indicators*. In: Lioy, P.J. & Lioy, M.J.Y., eds, *Air Sampling Instruments for Evaluation of Atmospheric Contaminants*, 6th ed., Cincinnati, OH, American Conference of Governmental Industrial Hygienists, pp. T-1 — T-29

Sax, N.I. (1984) *Dangerous Properties of Industrial Materials*, 6th ed., New York, Van Nostrand Reinhold, p. 531

Schifferli, E. (1942) Methyl bromide intoxications (Fr.). *Rev. méd. Suisse romande, 62*, 244-250

Scudamore, K.A. (1985a) *Multi-residue gas chromatographic method for determination of fumigant residues in cereal grains and other foods*. In: Fishbein, L. & O'Neill, I.K., eds, *Environmental Carcinogens. Selected Methods of Analysis*, Vol. 7, *Some Volatile Halogenated Hydrocarbons (IARC Scientific Publications No. 68)*, Lyon, International Agency for Research on Cancer, pp. 351-359

Scudamore, K.A. (1985b) *Determination of methyl bromide in grain using head-space analysis*. In: Fishbein, L. & O'Neill, I.K., eds, *Environmental Carcinogens. Selected Methods of Analysis*, Vol. 7, *Some Volatile Halogenated Hydrocarbons (IARC Scientific Publications No. 68)*, Lyon, International Agency for Research on Cancer, pp. 375-380

Shield, L.K., Coleman, T.L. & Markesbery, W.R. (1977) Methyl bromide intoxication: neurologic features, including simulation of Reye syndrome. *Neurology, 27*, 959-962

Sikov, M.R., Cannon, W.C., Carr, D.B., Miller, R.A., Montgomery, L.F. & Phelps, D.W. (1980) *Teratologic Assessment of Butylene Oxide, Styrene Oxide and Methyl Bromide (Contract No. 210-78-0025)*, Cincinnati, OH, US Department of Health, Education, and Welfare

Simmon, V.F., Kauhanen, K. & Tardiff, R.G. (1977) *Mutagenic activity of chemicals identified in drinking water*. In: Scott, D., Bridges, B.A. & Sobels, F.H., eds, *Progress in Genetic Toxicology*, Amsterdam, Elsevier/North-Holland Biomedical Press, pp. 249-258

Singh, G., Rippon, L.E. & Gilbert, W.S. (1982) Methyl bromide and inorganic bromide residues in avocados after fumigation and storage. *Aust. J. exp. Agric. anim. Husb., 22*, 343-347

Singh, H.B., Salas, L.J. & Cavanagh, L.A. (1977) Distribution, sources and sinks of atmospheric halogenated compounds. *J. Air Pollut. Control Assoc., 27*, 332-336

Singh, H.B., Salas, L.J., Smith, A.J. & Shigeishi, H. (1981) Measurements of some potentially hazardous organic chemicals in urban environments. *Atmos. Environ., 15*, 601-612

Singh, H.B., Salas, L.J. & Stiles, R.E. (1982) Distribution of selected gaseous organic mutagens and suspect carcinogens in ambient air. *Environ. Sci. Technol., 16*, 872-880

Singh, H.B., Salas, L.J. & Stiles, R.E. (1983) Methyl halides in and over the Eastern Pacific (40°N-32°S). *J. geophys. Res., 88*, 3684-3690

Sittig, M., ed. (1980) *Pesticide Manufacturing and Toxic Materials Control Encyclopedia*, Park Ridge, NJ, Noyes Data Corp., pp. 517-521

Sittig, M., ed. (1985) *Handbook of Toxic and Hazardous Chemicals and Carcinogens*, 2nd ed., Park Ridge, NJ, Noyes Publications, pp. 587-588

Stenger, V.A. (1978) *Bromine compounds*. In: Grayson, M. & Eckroth, eds, *Kirk-Othmer Encyclopedia of Chemical Technology*, 3rd ed., Vol. 4, New York, John Wiley & Sons, pp. 243-263

Tebbets, J.S., Hartsell, P.L., Nelson, H.D. & Tebbets, J.C. (1983) Methyl bromide fumigation of tree fruits for control of the Mediterranean fruit fly: concentrations, sorption, and residues. *J. agric. Food Chem.*, *31*, 247-249

Torkelson, T.R. & Rowe, V.K. (1981) *Halogenated aliphatic hydrocarbons containing chlorine, bromine and iodine*. In: Clayton, G.D. & Clayton, F.E., eds, *Patty's Industrial Hygiene and Toxicology*, Vol. 2B, *Toxicology*, 3rd rev. ed., New York, John Wiley & Sons, pp. 3442-3446

Tucker, J.D., Xu, J., Stewart, J. & Ong, T. (1985) Development of a method to detect volatile genotoxins using sister chromatid exchanges (Abstract). *Environ. Mutagenesis*, *7* (Suppl. 3), 48

Tucker, J.F., Brown, W.B. & Goodship, G. (1974) Fumigation with methyl bromide of poultry foods artificially contaminated with *Salmonella*. *Br. poult. Sci.*, *15*, 587-595

Työsuojeluhallitus (National Finnish Board of Occupational Safety and Health) (1981) *Airborne Contaminants in the Work Places* (*Safety Bull. 3*) (Finn.), Tampere, p. 19

Union Carbide Corp. (1982) *Specialty Gases and Related Products, 1982*, Somerset, NJ

US Environmental Protection Agency (1982a) *Method 8010. Halogenated volatile organics*. In: *Test Methods for Evaluating Solid Waste — Physical/Chemical Methods*, 2nd ed. (*US EPA No. SW-846*), Washington DC, Office of Solid Waste and Emergency Response

US Environmental Protection Agency (1982b) *Method 8240. GC/MS method for volatile organics*. In: *Test Methods for Evaluating Solid Waste — Physical/Chemical Methods*, 2nd ed. (*US EPA No. SW-846*), Washington DC, Office of Solid Waste and Emergency Response

US Environmental Protection Agency (1984a) Method 601. Guidelines establishing test procedures for the analysis of pollutants under the Clean Water Act (40 CFR 136). Purgeable halocarbons. *Fed. Regist.*, *49*, 43261-43271

US Environmental Protection Agency (1984b) Method 1624, Revision B. Guidelines establishing test procedures for the analysis of pollutants under the Clean Water Act (40 CFR 136). Volatile organic compounds by isotope dilution GC/MS. *Fed. Regist.*, *49*, 43407-43415

US Environmental Protection Agency (1984c) Method 624. Guidelines establishing test procedures for the analysis of pollutants under the Clean Water Act (40 CFR 136). Purgeables. *Fed. Regist.*, *49*, 43373-43384

US Occupational Safety and Health Administration (1985) Labor. *US Code fed. Regul.*, Title 29, Part 1910.1000

Vanachter, A., Feyaerts, J., Van Wambeke, E. & Van Assche, C. (1981) Bromide concentrations in water after methyl bromide soil disinfestation. II. Relation between leaching of methyl bromide fumigated greenhouse soils and bromide concentrations in the surrounding surface waters. *Med. Fac. Landbouwwet. Rijksuniv. Gent*, *46*, 351-358

Van Den Oever, R., Roosels, D. & Lahaye, D. (1982) Actual hazard of methyl bromide fumigation in soil disinfection. *Br. J. ind. Med.*, *39*, 140-144

Verberk, M.M., Rooyakkers-Beemster, T., De Vlieger, M. & van Vliet, A.G.M. (1979) Bromine in blood, EEG and transaminases in methyl bromide workers. *Br. J. ind. Med.*, *36*, 59-62

Verschueren, K. (1983) *Handbook of Environmental Data on Organic Chemicals*, 2nd ed., New York, Van Nostrand Reinhold, pp. 835-836

Viner, N. (1945) Methyl bromide poisoning: a new industrial hazard. *Can. med. Assoc. J.*, *53*, 43-45

Watrous, R.M. (1942) Methyl bromide — local and mild systemic toxic effects. *Ind. Med.*, *11*, 575-579

Weast, R.C., ed. (1985) *CRC Handbook of Chemistry and Physics*, 66th ed., Boca Raton, FL, CRC Press, pp. C-348, D-211

Windholz, M., ed. (1983) *The Merck Index*, 10th ed., Rahway, NJ, Merck & Co., p. 865

Wong, O., Brocker, W., Davis, H.V. & Nagle, G.S. (1984) Mortality of workers potentially exposed to organic and inorganic brominated chemicals, DBCP, TRIS, PBB, and DDT. *Br. J. ind. Med.*, *41*, 15-24

Worthing, C.R. (1983) *The Pesticide Manual. A World Compendium*, 7th ed., Croydon, The British Crop Protection Council, p. 372

Wyers, H. (1945) Methyl bromide intoxication. *Br. J. ind. Med.*, *2*, 24-29

Zatuchni, J. & Hong, K. (1981) Methyl bromide poisoning seen initially as psychosis. *Arch. Neurol.*, *38*, 529-530

METHYL IODIDE

This substance was considered by a previous Working Group, in February 1977 (IARC, 1977). Since that time, new data have become available and these are incorporated in the present monograph and taken into consideration in the evaluation.

1. Chemical and Physical Data

1.1 Synonyms and trade names

Chem. Abstr. Services Reg. No.: 74-88-4
Chem. Abstr. Name: Iodomethane
IUPAC Systematic Name: Iodomethane

1.2 Structural and molecular formulae and molecular weight

$$\begin{array}{c} H \\ | \\ H-C-I \\ | \\ H \end{array}$$

CH_3I 	 Mol. wt: 141.94

1.3 Chemical and physical properties of the pure substance

(a) *Description*: Colourless, transparent liquid (Windholz, 1983) with a pungent odour (Mazac, 1981; National Institute for Occupational Safety and Health, 1984)

(b) *Boiling-point*: 42.4°C (Weast, 1985)

(c) *Melting-point*: -66.4°C (Weast, 1985)

(d) *Density*: d_4^{20} 2.279 (Weast, 1985)

(e) *Spectroscopy data*[a]: Infrared (Sadtler Research Laboratories, 1980; prism [59], grating [28502]), nuclear magnetic resonance (Sadtler Research Laboratories, 1980; proton [7872], C-13 [13]) and mass spectral data (Grasselli & Ritchey, 1975) have been reported.

(f) *Solubility*: Soluble in water (14 g/l at 20°C) (Verschueren, 1983); very soluble in acetone, ethanol, benzene, diethyl ether (Weast, 1985) and carbon tetrachloride (Mazac, 1981)

(g) *Volatility*: Vapour pressure, 400 mm Hg at 25.3°C (Weast, 1985); relative vapour density (air = 1), 4.9 (Verschueren, 1983)

(h) *Stability*: Turns brown on exposure to light (Hawley, 1981; Windholz, 1983); decomposes in water with a half-time of 23.9 h at 100°C (Davis *et al.*, 1977)

(i) *Reactivity*: Noninflammable (Hawley, 1981); alkylating agent (Mazac, 1981)

(j) *Octanol/water partition coefficient (P)*: log P, 1.51-1.69 (Hansch & Leo, 1979)

(k) *Conversion factor*: mg/m^3 = 5.81 × ppm[b]

1.4 Technical products and impurities

Technical-grade methyl iodide is generally ⩾99% pure. Nonvolatile matter, or evaporation residue, is limited to less than 0.01% and water content to less than 0.1% (Fairmount Chemical Co., 1984; ACF Chemiefarma NV, undated; RSA Corporation, undated).

2. Production, Use, Occurrence and Analysis

2.1 Production and use

(a) *Production*

Methyl iodide was prepared in 1909 by electrolysis of an aqueous solution of potassium acetate with iodine or potassium iodide (Prager & Jacobson, 1918). Methyl iodide is currently produced by the reaction of methanol, iodine and phosphorus. Another production method involves use of dimethyl sulphate with an aqueous slurry of iodine in the presence of a reducing agent such as powdered iron or sodium bisulphite (Mazac, 1981; Windholz, 1983).

---

[a]In square brackets, spectrum number in compilation

[b]Calculated from: mg/m^3 = (molecular weight/24.45) × ppm, assuming standard temperature (25°C) and pressure (760 mm Hg)

Methyl iodide is produced in the USA, western Europe and Japan. The USA has four companies, three of which produced a total of 46 800 kg in 1983 and less than 23 000 kg in 1979 (Mazac, 1981; US International Trade Commission, 1984). There are two manufacturers of methyl iodide in France, one each in the Netherlands and Spain, and three in the UK. There are two producers of methyl iodide in Japan (The Chemical Daily Co., 1984).

(*b*) *Use*

Methyl iodide is used primarily as a methylating agent in pharmaceutical (e.g, quarternary ammonium compounds) and chemical synthesis, although its cost in comparison with that of methyl chloride (see monograph, p. 161) or methyl bromide (see monograph, p. 187) limits its use in other large-scale industrial methylations (Mazac, 1981; ACF Chemiefarma NV, undated).

Methyl iodide is used in microscopy, because of its high refractive index, as an embedding material for diatom examination and as a reagent in testing for pyridine (American Conference of Governmental Industrial Hygienists, 1984).

Other reported uses are as a catalyst in the production of tetramethyl- and tetraethyllead, boron and magnesium hydrides, and some rubber antioxidants and ion-exchange materials; as a light-sensitive etching agent for micronized electronic circuits; and as a component in fire extinguishers (ACF Chemiefarma NV, undated).

Methyl iodide has been used as a soil disinfectant for tobacco (International Technical Information Institute, 1979) and as a fumigant to control fungi in grain sorghum (Ragunathan *et al.*, 1974), but is reported to be no longer used in agricultural applications (Meister, 1985).

(*c*) *Regulatory status and guidelines*

Occupational exposure limits for methyl iodide in 12 countries are presented in Table 1.

2.2 Occurrence

(*a*) *Natural occurrence*

Methyl iodide is produced by a variety of marine organisms (Lovelock, 1975).

(*b*) *Occupational exposure*

No data on exposure levels were available to the Working Group.

(*c*) *Air*

The major sources of methyl iodide in the environment are considered to be the oceans. Global measurements of methyl iodide showed concentrations of 5.8-17.4 ng/m^3 in the tropospheric boundary layer and half of these values above the boundary layer. Near three oceanic regions characterized by high biomass productivity, atmospheric levels of methyl iodide ranged from 40-128 ng/m^3. The global flux from the oceans was calculated to be about 1.3 million tonnes per year (Rasmussen *et al.*, 1982).

Table 1. Occupational exposure limits for methyl iodide[a]

Country	Year	Concentration (mg/m³)	Interpretation[b]
Australia	1978	28	TWA
Belgium	1978	28	TWA
Finland	1981	28	TWA
		56	STEL
The Netherlands	1978	28	TWA
Norway	1981	28	TWA
Poland	1976	10	Ceiling
Romania	1975	15	TWA
		25	Ceiling
Sweden	1984	6	TWA
		30	STEL
Switzerland	1978	28	TWA
UK	1985	28	TWA
		56	STEL
USA	1985		
ACGIH		10	TWA
		30	STEL
OSHA		28	TWA
Yugoslavia	1971	28	Ceiling

[a]From International Labour Office (1980); Direktoratet for Arbeidstilsynet (1981); Työsuojeluhallitus (1981); Arbetarskyddsstyrelsens Författningssamling (1984); American Conference of Governmental Industrial Hygienists (ACGIH) (1985); Health and Safety Executive (1985); National Institute for Occupational Safety and Health (1985); US Occupational Safety and Health Administration (OSHA) (1985)

[b]TWA, time-weighted average; STEL, short-term exposure limit

The primary mechanism for removal of methyl iodide from the atmosphere is photolysis in the troposphere, at a daily rate (12 sunlit hours) of about 12.2% (Singh *et al.*, 1982). The atmospheric residence time of methyl iodide has been estimated as 50 h (Lovelock *et al.*, 1973). At 40°N latitude, the surface level background concentration is approximately 12 ng/m³ (2 ppt) (Singh *et al.*, 1983).

Mean levels in air at seven urban sites in the USA ranged from 6-21 ng/m³ (Singh *et al.*, 1983).

Radioactive methyl iodide ($CH_3{}^{131}I$) is produced as an off-gas by some nuclear reactors (boiling-water reactors, light-water reactors, pressurized-water reactors). Hypoiodous acid and organic iodides, including methyl iodide, constitute at least 50% of the radioactive iodine released from boiling-water reactors (Voilleque, 1979).

(d) Water and sediments

Methyl iodide has been found in all ocean waters examined and is a common product of marine algae. Surface seawater concentrations near land have been reported to be 0.04 in California, USA, and 3.4 µg/l in Ireland, in head-space gas equilibrated with seawater, with one report of 120 µg/l near kelp beds in Ireland (Lovelock, 1975; Singh *et al.*, 1977). Concentrations of methyl iodide in the open ocean have been reported to average 0.14 ng/l (as vapour) in the Atlantic, Antarctic and Caribbean Oceans (Lovelock, 1975) and 1.6 ng/l in the eastern Pacific Ocean (Singh *et al.*, 1983).

(e) Animals

Marine organisms collected from the Irish Sea in the vicinity of Port Erin, Isle of Man, contained methyl iodide (Table 2). Levels were highest in the brain of coalfish (*Pollachius birens*) and lowest (0.4 µg/kg) in the skeletal tissue of cod (*Gadus morhua*) (Dickson & Riley, 1976).

Table 2. Methyl iodide concentrations in marine organisms (µg/kg)[a]

Species	Brain	Liver	Muscle
Eel (*Conger conger*)	32	42	11
Cod (*Gadus morhua*)	54	17	4
Coalfish (*Pollachius birens*)	166	100	4
Dogfish (*Scylliorhinus canicula*)	103	24	31
Bib (*Trisopterus luscus*)	137	4	12
Mollusc (*Modiolus modolus*)	–	–	15
Mollusc (*Pecten maximus*)	–	–	9

[a] From Dickson and Riley (1976)

2.3 Analysis

Selected methods for the analysis of methyl iodide in air and cereal grains are identified in Table 3.

Gas chromatography with microwave emission detection, in conjunction with standard purge-and-trap techniques, has also been used for the analysis of methyl iodide in water (Quimby *et al.*, 1979). This detection method increased selectivity and provided good sensitivity (0.1 µg/l for methyl iodide).

Minimum concentrations of 10.4 mg/m^3 methyl iodide can be monitored in air by a direct-reading infrared analyser (The Foxboro Co., 1985).

Table 3. Methods for the analysis of methyl iodide

Sample matrix	Sample preparation	Assay procedure[a]	Limit of detection	Reference
Air	Adsorb on charcoal; desorb (toluene); inject aliquot	GC/FID	0.2 mg/m^3	Eller (1985)
Cereal grains	Extract with acetone:water; add sodium chloride; separate layers; dry acetone solution over anhydrous calcium chloride; inject aliquot	GC/EC	0.002 mg/kg	(Scudamore 1985)

[a]Abbreviations: GC/FID, gas chromatography/flame ionization detection; GC/EC, gas chromatography/electron capture detection

3. Biological Data Relevant to the Evaluation of Carcinogenic Risk to Humans

3.1 Carcinogenicity studies in animals

(a) Subcutaneous administration

Rat: Groups of BD rats [substrain and sex unspecified], about 100 days old, received weekly subcutaneous injections of 10 (16 animals) or 20 mg/kg bw (eight animals) methyl iodide [purity unspecified] in arachis oil for about one year (total dose, 500 or 900 mg/kg bw), or a single subcutaneous injection of 50 mg/kg bw (14 animals), and were observed for life. Four and two animals in the first two groups, respectively, died of pneumonia. Subcutaneous sarcomas occurred in 9/12 rats injected with 10 mg/kg bw, in 6/6 rats injected with 20 mg/kg bw and in 4/14 rats given a single injection of 50 mg/kg bw. No subcutaneous tumour was reported to have occurred in control rats [number unspecified] injected with arachis oil alone. Local tumours occurred more than one year after the first injection; histologically, these were fibrosarcomas and spindle-cell and round-cell sarcomas. In most cases [number unspecified], pulmonary and lymph-node metastases were observed (Druckrey *et al.*, 1970).

(b) Intraperitoneal administration

Mouse: In a screening assay based on the production of lung tumours in strain A mice (Shimkin & Stoner, 1975), groups of 10 male and 10 female A/He mice, six to eight weeks old, were injected intraperitoneally thrice weekly with three dose levels (the highest being the maximum tolerated dose) of methyl iodide (>98% pure) in tricaprylin for a total of 24 injections (total doses, 8.5, 21.3 and 44.0 mg/kg bw). A group of 30 untreated mice and a group of 160 tricaprylin-treated mice were used as controls. All survivors were killed 24 weeks after the first injection. Survival was 29/30 and 154/160 in the untreated and vehicle-treated control groups and 19/20 in the low-dose, 20/20 in the mid-dose and 11/20 in the high-dose groups. The proportions of mice with lung tumours were 6/29, 34/154, 4/19, 6/20 and 5/11 in the five groups, respectively [$p = 0.048$; one-sided Cochran-Armitage trend test using vehicle controls only]. The average numbers of lung tumours per mouse were 0.21, 0.22, 0.21, 0.30 and 0.55. In positive-control groups receiving a single intraperitoneal injection of 10 or 20 mg urethane, all animals developed lung tumours; the average number of lung tumours per animal was 8.1 in the low-dose and 17.8 in the high-dose group (Poirier et al., 1975).

3.2 Other relevant biological data

(a) *Experimental systems*

Toxic effects

The toxicity of methyl iodide has been reviewed (von Oettingen, 1964; Torkelson & Rowe, 1981).

The oral LD_{50} for methyl iodide in rats was 76 mg/kg bw (Johnson, 1966a). The subcutaneous LD_{50} was 110-220 mg/kg bw in rats (Buckell, 1950; Druckrey et al., 1970) and 0.78 mmol/kg (110 mg/kg) bw in mice (Kutob & Plaa, 1962). Exposure by inhalation to 85 000 mg/m³ methyl iodide was lethal to all exposed mice within 85 min (Buckell, 1950). Mice exposed continuously to $3 \times 10^{-6} - 3 \times 10^{-5}$ mol/l (426-4260 mg/m³) methyl iodide died within 24 h (Bachem, 1927). A 15-min exposure to 22 000 mg/l methyl iodide was reported to be lethal to rats (Torkelson & Rowe, 1981).

Toxic effects (which may be delayed) observed after exposure to methyl iodide include narcosis, congestion of the lungs, and liver and kidney damage (Buckell, 1950; Kutob & Plaa, 1962; von Oettingen, 1964). No death was observed after daily administration of oral doses of 30-50 mg/kg bw methyl iodide to rats on five days per week for a month (Johnson, 1966a).

Oral administration to rats of methyl iodide reduced nonprotein thiol concentrations in liver and kidney (Johnson, 1966a).

Effects on reproduction and prenatal toxicity
No data were available to the Working Group.

Absorption, distribution, excretion and metabolism
Approximately 1% of an oral dose of 76 mg/kg bw methyl iodide given to rats was expired unchanged within 30 min of administration (Johnson, 1966a). The urinary metabolites (2% of the total dose) detected after subcutaneous injection of rats with 50

mg/kg bw methyl iodide were *S*-methylcysteine, *N*-acetyl-*S*-methylcysteine, *S*-methylthioacetic acid and *N*-(methylthioacetyl)glycine. The authors concluded that these metabolites originated from *S*-methylglutathione (Barnsley & Young, 1965). *S*-Methylglutathione was isolated from livers of rats treated orally with 90 mg/kg bw methyl iodide; approximately 25% of an oral dose of 50 mg/kg bw was excreted into the bile as *S*-methylglutathione (Johnson, 1966a).

Liver and kidney homogenates catalysed the disappearance of methyl iodide in the presence of glutathione (Johnson, 1966b). *S*-Methylglutathione was converted to *S*-methylcysteine by kidney homogenate (Johnson, 1966a).

Methyl iodide reacted *in vitro* with 4-(*para*-nitrobenzyl)pyridine and *N*-7-deoxyguanosine (Hemminki *et al.*, 1980).

Mutagenicity and other short-term tests

Methyl iodide (10 μl/plate) gave a positive result in the *Escherichia coli polA* spot test in the absence of an exogenous metabolic system (S9) (Rosenkranz & Poirier, 1979).

Methyl iodide (tested at 10-50 μl/plate) was mutagenic to *Salmonella typhimurium* TA100 when assayed in dessicators in the absence of S9 (Simmon *et al.*, 1977). McCann *et al.* (1975) also reported that methyl iodide was weakly mutagenic to *S. typhimurium* TA100 [details not given] when assayed under similar test conditions. The compound was mutagenic to *S. typhimurium* TA1535 (2-10 μl/plate), but not to TA1538 (10 μl/plate), both in the presence and absence of uninduced rat-liver S9 (Rosenkranz & Poirier, 1979). However, Simmon (1979a) reported that methyl iodide (tested at up to 500 μg/plate) was not mutagenic to *S. typhimurium* TA1535, TA1536, TA1537, TA1538, TA98 or TA100 in a plate incorporation assay, but was positive in TA100 when tested in a desiccator. It was reported to be mutagenic to *E. coli* WP2 *uvrA* [details not given] (Hemminki *et al.*, 1980).

Methyl iodide (0.01-0.1 M) was reported to be nonmutagenic to *Aspergillus nidulans* (Moura Duarte, 1972). [The Working Group noted that the experimental design was not adequate in view of the volatility of the compound, because survival was 100%.]

Methyl iodide (0.1%, v/v) induced mitotic recombination in *Saccharomyces cerevisiae* D3 both in the presence and absence of Aroclor-induced rat-liver S9 (Simmon, 1979b).

It was reported in an abstract that methyl iodide induced DNA repair synthesis in cultured human lymphoblastoid cells, as measured by caesium chloride-gradient centrifugation [details not given] (Andrae *et al.*, 1979).

In the L5178Y/TK$^{+/-}$ mouse lymphoma assay, methyl iodide (7.5-15 μg/ml) induced a dose-dependent increase in trifluorothymidine-resistant colonies in the absence of a metabolic system (Moore *et al.*, 1985a). Analysis of resistant colonies suggested that they were the result of both mutations and chromosomal rearrangements (Moore *et al.*, 1985b). In earlier studies, the same authors had reported induction of bromodeoxyuridine-resistant mutants (Clive *et al.*, 1979), but no (Clive *et al.*, 1979) or low-level (Moore & Clive, 1982) induction of 6-thioguanine-resistant mutants at doses of 30-70 μg/ml or 10 μg/ml, respectively; Aroclor-induced rat-liver S9 was used in the study in which negative results were obtained. A dose-dependent increase in the frequency of ouabain-resistant mutants was reported in L5178Y cells treated with 1.9-3.6 µg/ml methyl iodide in the absence of a metabolic system (Amacher & Dunn, 1985). In the absence of a metabolic system, 6-thioguanine-resistant mutants were induced in a dose-dependent manner in Chinese hamster ovary cells treated with 0.1-1.5 µg/ml methyl iodide (Amacher & Zelljadt, 1984).

Treatment of Syrian hamster embryo cells with methyl iodide (tested at doses of 0.1-100 µg/ml) for eight days induced morphological transformation (Pienta et al., 1977). Treatment of C3H 10T1/2 mouse cells with 100-250 µg/ml methyl iodide for three days did not induce cell transformation (Oshiro et al., 1981).

(b) *Humans*

Toxic effects

Garland and Camps (1945) described a fatal case of poisoning in a worker exposed to methyl iodide fumes during its manufacture: severe neurological symptoms preceded death; all organs showed congestion at autopsy.

Non-fatal poisonings affecting workers in methyl iodide manufacture were characterized by the onset of neurological symptoms (such as vertigo, visual disturbances and weakness); these were followed by psychological disturbances and intellectual impairment, which had not been resolved entirely even after several months or years (Jaquet, 1901; Baselga-Monte et al., 1965; Appel et al., 1975).

Effects on reproduction and prenatal toxicity

No data were available to the Working Group.

Absorption, distribution, excretion and metabolism

After inhalation of trace amounts of ^{132}I-methyl iodide, the disappearance of radioactivity from the plasma, uptake in the thyroid and excretion in the urine resembled that of orally administered inorganic iodide (Morgan et al., 1967).

Mutagenicity and chromosomal effects

No data were available to the Working Group.

3.3 Case reports and epidemiological studies of carcinogenicity to humans

No data were available to the Working Group.

4. Summary of Data Reported and Evaluation

4.1 Exposure data

Methyl iodide is formed principally in the oceans. Smaller amounts are produced industrially for use as a chemical intermediate. Exposures occur from occupational use and from ubiquitous low-level exposure in ambient air and in water.

4.2 Experimental data

Methyl iodide was tested for carcinogenicity in one experiment in rats by subcutaneous administration and in a screening test for lung adenomas in strain A mice by intraperitoneal injection. It induced local sarcomas in rats after single or repeated subcutaneous injections; a marginally increased incidence of lung tumours was observed in mice.

No data were available to evaluate the reproductive effects or prenatal toxicity of methyl iodide to experimental animals.

Methyl iodide induces DNA damage and is mutagenic to bacteria in the presence or absence of an exogenous metabolic system. It induces mitotic recombination in yeast and mutations in cultured mammalian cells. It induces transformation in Syrian hamster embryo cells but not in C3H 10T1/2 cells.

Overall assessment of data from short-term tests: Methyl iodide[a]

	Genetic activity			Cell transformation
	DNA damage	Mutation	Chromosomal effects	
Prokaryotes	+	+		
Fungi/Green plants		+		
Insects				
Mammalian cells (*in vitro*)		+		+
Mammals (*in vivo*)				
Humans (*in vivo*)				
Degree of evidence in short-term tests for genetic activity: **Sufficient**				Cell transformation: Positive

[a]The groups into which the table is divided and the symbols '+' and '−' are defined on pp. 19-20 of the Preamble; the degrees of evidence are defined on pp. 20-21.

4.3 Human data

No data were available to evaluate the reproductive effects or prenatal toxicity of methyl iodide to humans.

No case report or epidemiological study of the carcinogenicity of methyl iodide to humans was available to the Working Group.

4.4 Evaluation[1,2]

There is *limited evidence* for the carcinogenicity of methyl iodide to experimental animals.

No evaluation could be made of the carcinogenicity of methyl iodide to humans.

5. References

ACF Chemiefarma NV (undated) *Product Data Sheet: Methyl Iodide*, Maarssen, The Netherlands

Amacher, D.E. & Dunn, E.M. (1985) Mutagenesis at the ouabain-resistance locus of 3.7.2C L5178Y cells by chromosomal mutagens. *Environ. Mutagenesis*, 7, 523-533

Amacher, D.E. & Zelljadt, I. (1984) Mutagenic activity of some clastogenic chemicals at the hypoxanthine guanine phosphoribosyl transferase locus of Chinese hamster ovary cells. *Mutat. Res.*, 136, 137-145

American Conference of Governmental Industrial Hygienists (1984) *Threshold Limit Values. Supplemental Documentation, 1984*, Cincinnati, OH, pp. 281-282

American Conference of Governmental Industrial Hygienists (1985) *TLVs Threshold Limit Values and Biological Exposure Indices for 1985-86*, 2nd ed., Cincinnati, OH, p. 24

Andrae, U., Jahnel, P. & Greim, H. (1979) Induction of DNA-repair synthesis in human lymphoblastoid cells by metabolically activated chemicals as short-term tests for DNA-damaging compounds (Abstract). *Mutat. Res.*, 64, 125

Appel, G.B., Galen, R., O'Brien, J. & Schoenfeldt, R. (1975) Methyl iodide intoxication. A case report. *Ann. intern. Med.*, 82, 534-536

Arbetarskyddsstyrelsens Fo¨rfattningssamling (National Swedish Board of Occupational Safety and Health) (1984) *Occupational Exposure Limit Values (AFS 1984:5)* (Swed.), Solna, p. 22

Bachem, C. (1927) Contribution to the toxicity of halogenoalkyls (Ger.). *Naunyn-Schmiedeberg's Arch. exp. Pathol. Pharmakol.*, 122, 69-76

Barnsley, E.A. & Young, L. (1965) Biochemical studies of toxic agents. *Biochem. J.*, 95, 77-81

Baselga-Monte, M., Estadella-Botha, S., Quer-Brossa, S. & Fornells-Martinez, E. (1965) Occupational intoxication by methyl iodide (Span.). *Med. Lav.*, 56, 592-595

---

[1]For definition of the italicized term, see Preamble, p. 18. See also 'General Remarks on the Substances Considered', pp. 37-38.
[2]The Working Group was aware that an ad-hoc group, convened in April 1978 to re-evaluate all chemicals considered in the first 17 volumes of *IARC Monographs* according to the present criteria, classified this chemical as one for which there was *sufficient evidence* of carcinogenicity to experimental animals (IARC, 1978).

Buckell, M. (1950) The toxicity of methyl iodide: I. Preliminary survey. *Br. J. ind. Med.*, 7, 122-124

The Chemical Daily Co. (1984) *JCW Chemicals Guide 84/85*, Tokyo, p. 270

Clive, D., Johnson, K.O., Spector, J.F.S., Batson, A.G. & Brown, M.M.M. (1979) Validation and characterization of the L5178Y/TK$^{+/-}$ mouse lymphoma mutagen assay system. *Mutat. Res.*, 59, 61-108

Davis, L.N., Strange, J.R., Hoecker, J.E., Howard, P.H. & Santodonato, J. (1977) *Investigation of Selected Potential Environmental Contaminants: Monohalomethanes (EPA-560/2-77-007)*, Washington DC, US Environmental Protection Agency, p. 16

Dickson, A.G. & Riley, J.P. (1976) The distribution of short-chain halogenated aliphatic hydrocarbons in some marine organisms. *Mar. Pollut. Bull.*, 7, 167-169

Direktoratet for Arbeidstilsynet (Directorate of Labour Inspection) (1981) *Administrative Norms for Pollution in Work Atmosphere, No. 361* (Norw.), Oslo, p. 16

Druckrey, H., Kruse, H., Preussmann, R., Ivankovic, S. & Lanschütz, C. (1970) Carcinogenic alkylating substances. III. Alkyl-halogenides, -sulphates, sulphonates and strained heterocyclic compounds (Ger.). *Z. Krebsforsch.*, 74, 241-273

Eller, P.M. (1985) *NIOSH Manual of Analytical Methods*, 3rd ed., 1st Suppl. (*DHHS (NIOSH) Pub. No. 84-100*), Washington DC, US Government Printing Office, pp. 1014-1 — 1014-3

Fairmount Chemical Co. (1984) *Specification Sheet: Methyl Iodide*, Newark, NJ

The Foxboro Co. (1985) *Ambient Air Instrumentation Sheet*, Foxboro, MA

Garland, A. & Camps, F.E. (1945) Methyl iodide poisoning. *Br. J. ind. Med.*, 2, 209-211

Grasselli, J.G. & Ritchey, W.M., eds (1975) *CRC Atlas of Spectral Data and Physical Constants for Organic Compounds*, Vol. 3, Cleveland, OH, CRC Press, p. 597

Hansch, C. & Leo, A. (1979) *Substituent Constants for Correlation Analysis in Chemistry and Biology*, New York, John Wiley & Sons, p. 173

Hawley, G.G., ed. (1981) *The Condensed Chemical Dictionary*, 10th ed., New York, Van Nostrand Reinhold, p. 681

Health and Safety Executive (1985) *Occupational Exposure Limits 1985* (*Guidance Note EH 40/85*), London, Her Majesty's Stationery Office, p. 14

Hemminki, K., Falck, K. & Vainio, H. (1980) Comparison of alkylation rates and mutagenicity of directly acting industrial and laboratory chemicals. Epoxides, glycidyl ethers, methylating and ethylating agents, halogenated hydrocarbons, hydrazine derivatives, aldehydes, thiuram, and dithiocarbamate derivatives. *Arch. Toxicol.*, 46, 277-285

IARC (1977) *IARC Monographs on the Evaluation of the Carcinogenic Risk of Chemicals to Man, Vol. 15, Some Fumigants, the Herbicides 2,4-D and 2,4,5-T, Chlorinated Dibenzodioxins and Miscellaneous Industrial Chemicals*, Lyon, pp. 245-254

IARC (1978) *Chemicals with Sufficient Evidence of Carcinogenicity in Experimental Animals. IARC Monographs Volumes 1-17* (*IARC int. tech. Rep. No. 78-003*), Lyon, p. 19

International Labour Office (1980) *Occupational Exposure Limits for Airborne Toxic Substances: A Tabular Compilation of Values from Selected Countries*, 2nd (rev.) ed. (*Occupational Safety and Health Series No. 37*), Geneva, pp. 148-149

International Technical Information Institute (1979) *Toxic and Hazardous Industrial Chemicals Safety Manual*, Tokyo, pp. 339-340

Jacquet, A. (1901) Bromomethyl poisoning. Medical clinic of Basel (Ger.). *Dtsch. Arch. klin. Med., 71*, 370-386

Johnson, M.K. (1966a) Metabolism of iodomethane in the rat. *Biochem. J., 98*, 38-43

Johnson, M.K. (1966b) Studies on glutathione-S-alkyltransferase of the rat. *Biochem. J., 98*, 44-56

Kutob, S.D. & Plaa, G.L. (1962) A procedure for estimating the hepatotoxic potential of certain industrial solvents. *Toxicol. appl. Pharmacol., 4*, 354-361

Lovelock, J.E. (1975) Natural halocarbons in the air and in the sea. *Nature, 256*, 193-194

Lovelock, J.E., Maggs, R.J. & Wade, R.J. (1973) Halogenated hydrocarbons in and over the Atlantic. *Nature, 241*, 194-196

Mazac, C.J. (1981) *Iodine and iodine compounds*. In: Grayson, M. & Eckroth, D., eds, *Kirk-Othmer Encyclopedia of Chemical Technology*, 3rd ed., Vol 13, New York, John Wiley & Sons, pp. 668-669

McCann, J., Choi, E., Yamasaki, E. & Ames, B.N. (1975) Detection of carcinogens as mutagens in the *Salmonella*/microsome test: assay of 300 chemicals. *Proc. natl Acad. Sci. USA, 72*, 5135-5139

Meister, R.T., ed. (1985) *Farm Chemical Handbook 85*, Willoughby, OH, Meister Publishing Co., p. C154

Moore, M.M. & Clive, D. (1982) The quantitation of $TK^{+/-}$ and $HGPRT^-$ mutants of $L5178Y/TK^{+/-}$ mouse lymphoma cells at varying times post-treatment. *Environ. Mutagenesis, 4*, 499-519

Moore, M.M., Clive, D., Howard, B.E., Batson, A.G. & Turner, N.T. (1985a) In situ analysis of trifluorothymidine-resistant (TFT^r) mutants of $L5178Y/TK^{+/-}$ mouse lymphoma cells. *Mutat. Res., 151*, 147-159

Moore, M.M., Clive, D., Hozier, J.C., Howard, B.E., Batson, A.G., Turner, N.T. & Sawyer, J. (1985b) Analysis of trifluorothymidine-resistant (TFT^r) mutants of $L5178Y/TK^{+/-}$ mouse lymphoma cells. *Mutat. Res., 151*, 161-174

Morgan, A., Morgan, D.J., Evans, J.C. & Lister, B.A. (1967) Studies on the retention and metabolism of inhaled methyl iodide — II. Metabolism of methyl iodide. *Health Phys., 13*, 1067-1074

Moura Duarte, F.A. (1972) Mutagenic effects of some esters of inorganic acids in *Aspergillus nidulans* (Eidam) winter (Span.). *Cienc. Cultura, 24*, 42-52

National Institute for Occupational Safety and Health (1984) *Monohalomethanes. Methyl Chloride CH_3Cl, Methyl Bromide CH_3Br, Methyl Iodide CH_3I* (*Current Intelligence Bull. 43; DHHS (NIOSH) Publ. No. 84-117*), Cincinnati, OH

National Institute for Occupational Safety and Health (1985) NIOSH recommendations for occupational safety and health standards. *Morbid. Mortal. wkly Rep. Suppl.*, *34*, 22S

von Oettingen, W.F. (1964) *The Halogenated Hydrocarbons of Industrial and Toxicological Importance*, Amsterdam, Elsevier, pp. 55-59

Oshiro, Y., Balwierz, P.S. & Molinary, S.V. (1981) Morphological transformation of C3H/10T1/2 CL8 cells by alkylating agents. *Toxicol. Lett.*, *9*, 301-306

Pienta, R.J., Poiley, J.A. & Lebherz, W.B. III (1977) Morphological transformation of early passage golden Syrian hamster embryo cells derived from cryopreserved primary cultures as a reliable *in vitro* bioassay for identifying diverse carcinogens. *Int. J. Cancer*, *19*, 642-655

Poirier, L.A., Stoner, G.D. & Shimkin, M.B. (1975) Bioassay of alkyl halides and nucleotide base analogs by pulmonary tumor response in strain A mice. *Cancer Res.*, *35*, 1411-1415

Prager, B. & Jacobson, P. (1918) *Beilstein's Handbook of Organic Chemistry* (Ger.)., 4th ed., Vol. 1, Syst. 5, Berlin (West), Springer-Verlag, pp. 69-70

Quimby, B.D., Delaney, M.F., Uden, P.C. & Barnes, R.M. (1979) Determination of trihalomethanes in drinking water by gas chromatography with a microwave plasma emission detector. *Anal. Chem.*, *51*, 875-880

Ragunathan, A.N., Muthu, M. & Majumder, S.K. (1974) Further studies on the control of internal fungi of sorghum by fumigation. *J. Food Sci. Technol. (India)*, *11*, 80-81

Rasmussen, R.A., Khalil, M.A.K., Gunawardena, R. & Hoyt, S.D. (1982) Atmospheric methyl iodide (CH_3I). *J. geophys. Res.*, *87*, 3086-3090

Rosenkranz, H.S. & Poirier, L.A. (1979) Evaluation of the mutagenicity and DNA-modifying activity of carcinogens and noncarcinogens in microbial systems. *J. natl Cancer Inst.*, *62*, 873-892

RSA Corporation (undated) *Data Sheet: Methyl Iodide*, Ardsley, NY

Sadtler Research Laboratories (1980) *The Sadtler Standard Spectra, Cumulative Index*, Philadelphia, PA

Scudamore, R.A. (1985) *Multi-residue gas chromatographic method for determination of fumigant residues in cereal grains and other foods*. In: Fishbein, L. & O'Neill, I.K., eds, *Environmental Carcinogens. Selected Methods of Analysis*, Vol. 7, *Some Volatile Halogenated Hydrocarbons* (*IARC Scientific Publications No. 68*), Lyon, International Agency for Research on Cancer, pp. 351-359

Shimkin, M.B. & Stoner, G.D. (1975) Lung tumors in mice: application to carcinogenesis bioassay. *Adv. Cancer Res.*, *21*, 1-58

Simmon, V.F. (1979a) In vitro mutagenicity assays of chemical carcinogens and related compounds with *Salmonella typhimurium*. *J. natl Cancer Inst.*, *62*, 893-899

Simmon, V.F. (1979b) In vitro assays for recombinogenic activity of chemical carcinogens and related compounds with *Saccharomyces cerevisiae* D3. *J. natl Cancer Inst.*, *62*, 901-909

Simmon, V.F., Kauhanen, K. & Tardiff, R.G. (1977) *Mutagenic activity of chemicals identified in drinking water.* In: Scott, D., Bridges, B.A. & Sobels, F.H., eds, *Progress in Genetic Toxicology*, Amsterdam, Elsevier/North-Holland Biomedical Press, pp. 249-258

Singh, H.B., Salas, L.J. & Cavanagh, L.A. (1977) Distribution, sources and sinks of atmospheric halogenated compounds. *J. Air Pollut. Control Assoc.*, *27*, 332-336

Singh, H.B., Salas, L.J. & Stiles, R.E. (1982) Distribution of selected gaseous organic mutagens and suspect carcinogens in ambient air. *Environ. Sci. Technol.*, *16*, 872-880

Singh, H.B., Salas, L.J. & Stiles, R.E. (1983) Methyl halides in and over the Eastern Pacific (40°N-32°S). *J. geophys. Res.*, *88*, 3684-3690

Torkelson, T.R. & Rowe, V.K. (1981) *Halogenated aliphatic hydrocarbons containing chlorine, bromine and iodine.* In: Clayton, G.D. & Clayton, F.E., eds, *Patty's Industrial Hygiene and Toxicology*, 3rd ed., Vol. 2B, New York, John Wiley & Sons, pp. 3446-3449

Työsuojeluhallitus (National Finnish Board of Occupational Safety and Health) (1981) *Airborne Contaminants in the Workplaces (Safety Bull. 3)* (Finn.), Tampere, p. 19

US International Trade Commission (1984) *Synthetic Organic Chemicals, US Production and Sales, 1983 (USITC Publ. 1588)*, Washington DC, US Government Printing Office, pp. 259, 294

US Occupational Safety and Health Administration (1985) Labor. *US Code fed. Regul.*, *Title 29*, Part 1910.1000

Verschueren, K. (1983) *Handbook of Environmental Data on Organic Chemicals*, 2nd ed., New York, Van Nostrand Reinhold, pp. 855-856

Voilleque, P.G. (1979) *Iodine Species in Reactor Effluents and in the Environment. Final Report (abstract)*, Rockville, MD, Science Applications, Nuclear Environmental Services Division

Weast, R.C., ed. (1985) *CRC Handbook of Chemistry and Physics*, 66th ed., Boca Raton, FL, CRC Press, pp. C-350, D-196

Windholz, M., ed. (1983) *The Merck Index*, 10th ed., Rahway, NJ, Merck & Co., p. 872

CHLOROFLUOROMETHANE

1. Chemical and Physical Data

1.1 Synonyms and trade names

Chem. Abstr. Services Reg. No.: 593-70-4
Chem. Abstr. Name: Chlorofluoromethane
IUPAC Systematic Name: Chlorofluoromethane
Synonyms: CFC 31; FC 31; R 31

1.2 Structural and molecular formulae and molecular weight

$$\begin{array}{c} H \\ | \\ Cl - C - F \\ | \\ H \end{array}$$

CH_2ClF Mol. wt: 68.48

1.3 Chemical and physical properties of the pure substance

(a) *Description*: Gas (Weast, 1985)

(b) *Boiling-point*: -9.1°C (Weast, 1985)

(c) *Melting-point*: -133°C (Smart, 1980)

(c) *Spectroscopy data*: Ultraviolet (Hubrich & Stuhl, 1980); infrared (Raman) (Grasselli & Ritchey, 1975); C-13 nuclear magnetic resonance (Somayajulu *et al.*, 1979) and mass spectral data (Grasselli & Ritchey, 1975) have been reported.

(d) *Solubility*: Soluble in chloroform (Weast, 1985)

(e) *Conversion factor*: mg/m³ = 2.80 × ppm[a]

1.4 Technical products and impurities

Chlorofluoromethane is not currently available as a commercial product.

2. Production, Use, Occurrence and Analysis

2.1 Production and use

(a) *Production*

Chlorofluoromethane is not produced in commercial or bulk quantities. The laboratory-scale synthesis of chlorofluoromethane can be accomplished by treatment of dichloromethane with potassium fluoride, sodium fluoride and ethylene glycol at 180-200°C for 2 h. This process yields 20% chlorofluoromethane as well as 17% difluoromethane (Hudlický, 1962).

(b) *Use*

There is no reported commercial use for chlorofluoromethane (Aviado & Micozzi, 1981).

(c) *Regulatory status and guidelines*

No specific regulation or guideline has been promulgated for the use of, or occupational exposure to, chlorofluoromethane.

2.2 Occurrence

(a) *Natural occurrence*

Chlorofluoromethane is not known to occur as a natural product.

(b) *Occupational exposure*

No data on exposure levels were available to the Working Group.

(c) *Air*

On the basis of its calculated rate of reaction with hydroxyl radical, the estimated tropospheric lifetime of chlorofluoromethane is reported to be 1.6 years (Makide & Rowland, 1981).

---

[a]Calculated from: mg/m³ = (molecular weight/24.45) × ppm, assuming standard temperature (25°C) and pressure (760 mm Hg)

(d) Other

Chlorofluoromethane has been identified as an impurity (3.8%) in commercial dichlorofluoromethane (FC-21) (Ratcliffe & Targett, 1969).

2.3 Analysis

No specific method for the analysis of chlorofluoromethane has been reported, but gas chromatographic methods developed for halocarbons may be applicable (Fishbein & O'Neill, 1985).

3. Biological Data Relevant to the Evaluation of Carcinogenic Risk to Humans

3.1 Carcinogenicity studies in animals

Oral administration

Rat: Groups of 36 male and 36 female SPF Alpk/Ap (Wistar-derived) rats, six weeks old, received 300 mg/kg bw chlorofluoromethane (FC 31; purity, 99.5%) as a 3% w/v solution in corn oil by gavage on five days per week for 52 weeks. Two control groups, one of which was housed separately, received corn oil alone, and a third control group received no treatment. All treated males had died by 100 weeks and females by 108 weeks; surviving controls were killed after 125 weeks. Mortality was greater in chlorofluoromethane-treated rats than in the control groups. The incidence of malignant forestomach and stomach neoplasms, probably all arising from the nonglandular region, was: males — combined controls, 1/104 (1 fibrosarcoma); chlorofluoromethane-treated, 33/36 [$p < 0.0001$] (13 squamous-cell carcinomas, five fibrosarcomas, 15 fibrosarcomas with squamous-cell carcinomas); females — combined controls, 1/104 (one papilloma); chlorofluoromethane-treated, 34/36 [$p < 0.0001$] (15 squamous-cell carcinomas, nine fibrosarcomas, eight fibrosarcomas with squamous-cell carcinomas, one poorly-differentiated sarcoma with squamous-cell carcinoma, one carcinoma not otherwise specified). The five chlorofluoromethane-treated rats that did not develop stomach neoplasms died between week 8 and week 79; all showed some degree of hyperplasia of the epithelium of the nonglandular stomach (Longstaff *et al.*, 1982, 1984).

3.2 Other relevant biological data

(a) *Experimental systems*

Toxic effects

It was reported in an abstract that rats exposed by inhalation to 1% (28 000 mg/m^3) chlorofluoromethane for 6 h per day on five days per week for two weeks showed moderate damage to the kidneys, adrenal glands, testes, epididymis and haematopoietic tissues (Trochimowicz et al., 1977).

In a study reported in an abstract, rats and cynomolgus monkeys were exposed by inhalation to 2% v/v (56 000 mg/m^3) and 1% v/v (28 000 mg/m^3) chlorofluoromethane, respectively, for 4 h. Central nervous system depression was observed in both species; one of the two monkeys died as a result of the exposure. In the same abstract it was reported that 5/8 monkeys exposed by inhalation to 0.5% (14 000 mg/m^3) chlorofluoromethane for 19-20 days died with severe epistaxis, and five had centrolobular to diffuse hepatocytic swelling (Coate et al., 1979).

Effects on reproduction and prenatal toxicity

In a study reported in an abstract, when rats were exposed to 0.1% (2800 mg/m^3) chlorofluoromethane for 6 h per day on days 6-15 of gestation, cervical ribs were found in 8/208 treated fetuses. [The Working Group noted that no information was provided on maternal toxicity or on the incidence of cervical ribs in the control pups.] In the same abstract, it was reported that rats but not monkeys exposed to 0.5% v/v (14 000 mg/m^3) chlorofluoromethane for 6 h per day for 20 days exhibited minimal to slight hypospermatogenesis. No such effect was observed in rats after exposure to 0.1% (2800 mg/m^3) chlorofluoromethane. After 65 exposures (13 weeks) to 0.1% (2800 mg/m^3) chlorofluoromethane, rats exhibited hypospermatogenesis, which was not reversed during a four-week recovery period (Coate et al., 1979).

Absorption, distribution, excretion and metabolism

Chlorofluoromethane is metabolized to carbon monoxide by Aroclor-induced rat hepatic microsomes and to formaldehyde by rat hepatic cytosolic fractions in the presence of glutathione (Green, 1983).

Mutagenicity and other short-term tests

Chlorofluoromethane (2.5-10%) was mutagenic to *Salmonella typhimurium* TA100 in the presence and absence of various exogenous metabolic systems (S9, cytosol and microsomes) from the liver of Aroclor-induced rats (Green, 1983). Longstaff et al. (1984) also reported a mutagenic effect in *S. typhimurium* TA1535 and TA100 in the presence and absence of a metabolic system [details not given].

Chlorofluoromethane (41% for 1, 2 or 3 h or 10-40% for 4 h) induced concentration- and time-dependent increases in 6-thioguanine-resistant mutants in Chinese hamster ovary cells in the presence and absence of S9 (Krahn et al., 1982).

It was reported in an abstract that no cytogenetic effect was observed in bone marrow harvested from rats exposed to 0.1% v/v chlorofluoromethane for 6 h per day for 13 weeks (65 exposures). In the same abstract, it was reported that 'no unequivocal dominant lethal effect' was observed in rats exposed to 0.1% chlorofluoromethane for 10 weeks (weekly matings of exposed males to unexposed females for 16 weeks). However, pregnancy rate was reduced throughout this period (Coate et al., 1979).

(b) *Humans*

No data were available to the Working Group.

3.3 Case reports and epidemiological studies of carcinogenicity to humans

No data were available to the Working Group.

4. Summary of Data Reported and Evaluation

4.1 Exposure data

Chlorofluoromethane has been reported as an impurity in dichlorofluoromethane, and thus limited exposures may occur.

4.2 Experimental data

Chlorofluoromethane was tested for carcinogenicity in one study in rats by oral administration by gavage at one dose level. High incidences of squamous-cell carcinomas and of fibrosarcomas of the forestomach and stomach were induced in animals of each sex.

No evaluation of the effects of chlorofluoromethane on reproduction or on prenatal toxicity in experimental animals could be made on the basis of the available data.

Chlorofluoromethane was mutagenic to *Salmonella typhimurium* and to cultured mammalian cells in the presence and absence of an exogenous metabolic system.

4.3 Human data

No data were available to evaluate the reproductive effects or prenatal toxicity of chlorofluoromethane to humans.

No case report or epidemiological study of the carcinogenicity of chlorofluoromethane to humans was available to the Working Group.

4.4 Evaluation[1]

There is *limited evidence* for the carcinogenicity of chlorofluoromethane to experimental animals.

No evaluation could be made of the carcinogenicity of chlorofluoromethane to humans.

---

[1]For definition of the italicized term, see Preamble, p. 18.

Overall assessment of data from short-term tests: Chlorofluoromethane[a]

	Genetic activity			Cell transformation
	DNA damage	Mutation	Chromosomal effects	
Prokaryotes		+		
Fungi/Green plants				
Insects				
Mammalian cells (*in vitro*)		+		
Mammals (*in vivo*)				
Humans (*in vivo*)				
Degree of evidence in short-term tests for genetic activity: **Limited**				Cell transformation: No data

[a]The groups into which the table is divided and the symbol '+' are defined on pp. 19-20 of the Preamble; the degrees of evidence are defined on pp. 20-21.

5. References

Aviado, D.M. & Micozzi, M.S. (1981) *Fluorine containing organic compounds*. In: Clayton, G.D. & Clayton, F.E., eds, *Patty's Industrial Hygiene and Toxicology*, 3rd ed., Vol. 2B, New York, John Wiley & Sons, pp. 3071-3115

Coate, W.B., Voelker, R., Kapp, R.W., Jr, Anderson, J. & Charm, J. (1979) Inhalation toxicity of monochloromonofluoromethane (Abstract No. 217). *Toxicol. appl. Pharmacol.*, *48*, A109

Fishbein, L. & O'Neill, I.K., eds (1985) *Environmental Carcinogens. Selected Methods of Analysis*, Vol. 7, *Some Volatile Halogenated Hydrocarbons* (*IARC Scientific Publications No. 68*), Lyon, International Agency for Research on Cancer

Grasselli, J.G. & Ritchey, W.M., eds (1975) *CRC Atlas of Spectral Data and Physical Constants for Organic Compounds*, Vol. 3, Cleveland, OH, CRC Press, p. 592

Green, T. (1983) The metabolic activation of dichloromethane and chlorofluoromethane in a bacterial mutation assay using *Salmonella typhimurium*. *Mutat. Res.*, *118*, 277-288

Hubrich, C. & Stuhl, F. (1980) The ultraviolet absorption of some halogenated methanes and ethanes of atmospheric interest. *J. Photochem.*, *12*, 93-107

Hudlický, M. (1962) *Chemistry of Organic Fluorine Compounds*, New York, The MacMillan Co., p. 108

Krahn, D.F., Barsky, F.C. & McCooey, K.T. (1982) CHO/HGPRT mutation assay: evaluation of gases and volatile liquids. *Environ. Sci. Res.*, 25, 91-103

Longstaff, E., Robinson, M., Bradbrook, C., Styles, J.A. & Purchase, I.F.H. (1982) Carcinogenicity of fluorocarbons: assessment by short-term in vitro tests and chronic exposure in rats (Abstract No. 374). *Toxicologist*, 2, 106

Longstaff, E., Robinson, M., Bradbrook, C., Styles, J.A. & Purchase, I.F.H. (1984) Genotoxicity and carcinogenicity of fluorocarbons: assessment by short-term in vitro tests and chronic exposure in rats. *Toxicol. appl. Pharmacol.*, 72, 15-31

Makide, Y. & Rowland, F.S. (1981) Tropospheric concentrations of methylchloroform, CH_3CCl_3, in January 1978 and estimates of the atmospheric residence times for hydrohalocarbons. *Proc. natl Acad. Sci. USA*, 78, 5933-5937

Ratcliffe, D.B. & Targett, B.H. (1969) A gas-chromatographic determination of organohalogen impurities in dichlorofluoromethane. *Analyst*, 94, 1028-1032

Smart, B.E. (1980) *Fluorine compounds, organic*. In: Grayson, M. & Eckroth, D., eds, *Kirk-Othmer Encyclopedia of Chemical Technology*, 3rd ed., Vol. 10, New York, John Wiley & Sons, pp. 856-870

Somayajulu, G.R., Kennedy, J.R., Vickrey, T.M. & Zwolinski, B.J. (1979) Carbon-13 chemical shifts for 70 halomethanes. *J. magn. Resonance*, 33, 559-568

Trochimowicz, H.J., Moore, B.E. & Chiu, T. (1977) Subacute inhalation toxicity studies on eight fluorocarbons (Abstract No. 161). *Toxicol. appl. Pharmacol.*, 41, 198-199

Weast, R.C., ed. (1985) *CRC Handbook of Chemistry and Physics*, 66th ed., Boca Raton, FL, CRC Press, p. C-349

CHLORODIFLUOROMETHANE

1. Chemical and Physical Data

1.1 Synonyms and trade names

Chem. Abstr. Services Reg. No.: 75-45-6

Chem. Abstr. Name: Chlorodifluoromethane

IUPAC Systematic Name: Chlorodifluoromethane

Synonyms: CFC 22; difluorochloromethane; difluoromonochloromethane; F 22; FC 22; monochlorodifluoromethane

Trade Names: Algeon 22; Algofrene 22; Algofrene Type 6; Arcton 4; Arcton 22; Daiflon 22; Dymel 22; Electro-CF 22; Flugene 22; Forane 22; Freon 22; Frigen 22; Genetron 22; Haltron 22; Isceon 22; Isotron 22; Khaladon 22; R 22; Ucon 22

1.2 Structural and molecular formulae and molecular weight

$$Cl-\underset{\underset{F}{|}}{\overset{\overset{H}{|}}{C}}-F$$

$CHClF_2$
Mol. wt: 86.47

1.3 Chemical and physical properties of the pure substance

(a) *Description*: Colourless, nearly odourless gas (Hawley, 1981; Sax, 1984; Weast, 1985)

(b) *Boiling-point*: -40.8°C (Weast, 1985)

(c) *Melting-point*: -146°C (Weast, 1985)

(d) *Density*: d^{-69} 1.4909 (Grasselli & Ritchey, 1975); 4.82 g/l (gas at boiling point) (Hawley, 1981)

(e) *Spectroscopy data*[a]: Ultraviolet (Hubrich & Stuhl, 1980), infrared (Sadtler Research Laboratories, 1980; prism [3701], grating [42670P]), C-13 nuclear magnetic resonance (Somayajulu *et al.*, 1979) and mass spectral data (Grasselli & Ritchey, 1975) have been reported.

(f) *Solubility*: Partly soluble in water (Hawley, 1981); soluble in acetone, chloroform and diethyl ether (Weast, 1985)

(g) *Volatility*: Vapour pressure, 10 atm (7600 mm Hg) at 24.0°C (Weast, 1985); vapour density, 3.87 (in air at 0°C) (Sax, 1984)

(h) *Stability*: Noninflammable (Hawley, 1981)

(i) *Reactivity*: Reacts with some metals (e.g., aluminium) (Sax, 1984)

(j) *Octanol/water partition coefficient (P)*: log P, 1.08 (Hansch & Leo, 1979)

(k) *Conversion factor*: mg/m^3 = 3.54 × ppm[b]

1.4 Technical products and impurities

Chlorodifluoromethane is typically available as a liquefied gas with a minimum purity of 99.9%. It is also available as a liquefied gas in an azeotropic mixture of 48.8% chlorodifluoromethane with 51.2% chloropentafluoroethane; this mixed gas is designated fluorocarbon-502 (Airco, 1982; Union Carbide Corp., 1982; American Chemical Society, 1985; Matheson Gas Products, 1985).

Dichlorofluoromethane (fluorocarbon 21) has been reported to be a contaminant of commercial chlorodifluoromethane (Aviado & Micozzi, 1981).

2. Production, Use, Occurrence and Analysis

2.1 Production and use

(a) *Production*

Chlorodifluoromethane is produced commercially by the reaction of chloroform with anhydrous hydrogen fluoride in the presence of an antimony chloride catalyst (Hawley, 1981). The traditional liquid-phase process for chlorodifluoromethane production is being replaced by a vapour-phase process that combines gaseous hydrogen fluoride with chloroform. Chromium oxide, ferric chloride and thorium tetrafluoride are used as

---

[a]In square brackets, spectrum number in compilation

[b]Calculated from: mg/m^3 = (molecular weight/24.45) × ppm, assuming standard temperature (25°C) and pressure (760 mm Hg)

catalysts for the vapour-phase process (Smart, 1980). The majority (85-90%) of the chloroform used in the USA is in the production of chlorodifluoromethane (Anon., 1981).

The cumulative worldwide production of chlorodifluoromethane from 1967 through 1977 was estimated to be 941 million kg (Panel on Stratospheric Chemistry and Transport, 1979). Worldwide production since 1950 is shown in Table 1.

Table 1. Worldwide production of chlorodifluoromethane (millions of kg)[a]

1950	1955	1960	1965	1970	1975	1977[b]
0.5	3.7	12.2	25.1	58.8	73.7	97.0

[a]From Barbour (1979), except when noted
[b]From Panel on Stratospheric Chemistry and Transport (1979)

In 1984, the US International Trade Commission (1985) reported the production of approximately 116 million kg of chlorodifluoromethane by five major manufacturers. Production in the USA increased by 8% annually between 1964 and 1974 and by 25% since 1974 (Smart, 1980), although it has remained relatively stable over the past five years (Table 2).

Table 2. US production of chlorodifluoromethane (millions of kg)[a]

1979	1980	1981	1982	1983	1984
96	103	114	79	107	116

[a]From US International Trade Commission (1980, 1981, 1982, 1983, 1984, 1985)

Chlorodifluoromethane is produced by two manufacturers each in France, the Federal Republic of Germany, Spain and the UK, and one in the Netherlands. In Japan, five producers of chlorofluorocarbons have the capacity to manufacture chlorodifluoromethane (The Chemical Daily Co., 1984); one producer has been reported in Australia (Benn Business Information Services, 1985).

(b) *Use*

Chlorodifluoromethane is used primarily as a refrigerant gas, and air-conditioning and refrigeration applications have been reported to account for 75% to over 90% of its use

(Smart, 1980; Anon., 1981). Such applications include portable air conditioners, central air conditioning, home freezers/refrigerators and automotive air conditioning.

Chlorodifluoromethane is also used to produce the fluorocarbon monomer tetrafluoroethylene (see IARC, 1979) by pyrolysis at 650-700°C (Smart, 1980; Anon., 1981). The tetrafluoroethylene monomer forms homopolymers or copolymers with other chlorofluorocarbons used in plastics and resins (Anon., 1981).

Unlike other chlorofluorocarbons, chlorodifluoromethane is not currently used to a large degree as a blowing agent for polyurethane resins or as an industrial solvent, but these applications have been reported previously (Hawley, 1981; US Environmental Protection Agency, 1981). The azeotropic mixture (F-502) of chlorodifluoromethane/chloropentafluoroethane (F-115) is also used as a refrigerant, primarily in food display cases, ice makers, home freezers and heat pumps (American Chemical Society, 1985).

(c) Regulatory status and guidelines

Occupational exposure limits for chlorodifluoromethane in 11 countries are presented in Table 3.

Table 3. Occupational exposure limits for chlorodifluoromethane[a]

Country	Year	Concentration (mg/m^3)	Interpretation[b]
Australia	1978	3500	TWA
Belgium	1978	3500	TWA
Finland	1981	3500	TWA
		4375	STEL
German Democratic Republic	1979	3000	TWA
		9000	STEL
The Netherlands	1978	3500	TWA
Norway	1981	3500	TWA
Sweden	1984	1800	TWA
		2500	STEL
Switzerland	1978	3500	TWA
UK	1985	3500	TWA
		4375	STEL
USA (ACGIH)	1985	3500	TWA
		4375	STEL
USSR	1977	3000	Ceiling

[a]From International Labour Office (1980); Direktoratet for Arbeidstilsylen (1981); Työsuojeluhallitus (1981); Arbetarskyddsstyrelsens Författningssamling (1984); American Conference of Governmental Industrial Hygienists (ACGIH) (1985); Health and Safety Executive (1985)

[b]TWA, time-weighted average; STEL, short-term exposure limit

2.2 Occurrence

(a) Natural occurrence

Chlorodifluoromethane is not known to occur as a natural product.

(b) Occupational exposure

In a fluorocarbon packaging and shipping plant, worker exposures to chlorodifluoromethane were found to be 4.7-13.5 ppm (17-48 mg/m^3) (Bales, 1978).

(c) Air

Of the estimated 941 million kg produced throughout the world between 1967 and 1977, approximately 44% (418 million kg) had been released to the atmosphere by 1978 (Panel on Stratospheric Chemistry and Transport, 1979). Emissions of chlorodifluoromethane to the troposphere in 1978 were calculated to be 50 million kg per year (21 million kg per year chlorine equivalents), with 40% reaching the stratosphere (Edwards *et al.*, 1982). In 1972, the atmospheric release rate in the northern hemisphere as a result of human activity was estimated to be 33 million kg per year (Derwent & Eggleton, 1978).

On the basis of its reaction with hydroxyl radicals, the estimated tropospheric lifetime for chlorodifluoromethane is reported to be 17 years (Makide & Rowland, 1981).

Analysis of the atmospheric burden of chlorodifluoromethane by gas chromatography/mass spectrometry gave an average global concentration of approximately 45 ppt (159 ng/m^3) in mid-1979; the northern hemisphere ambient concentration was 50 ppt (177 ng/m^3) and the southern hemisphere, 42 ppt (149 ng/m^3) (Rasmussen *et al.*, 1980). In May 1982, the average atmospheric concentration of chlorodifluoromethane in the lower troposphere measured over Point Barrow, Alaska was 73.2 ppt (259 ng/m^3) (Rasmussen & Khalil, 1983).

A study by Leifer *et al.* (1981) showed that levels of chlorodifluoromethane in air samples [31-55 ppt (110-195 ng/m^3)] collected over the State of Washington were not elevated two days after the eruption of the Mount St Helens volcano.

2.3 Analysis

Several methods have been reported for the analysis of chlorodifluoromethane in air, including gas chromatography/mass spectrometry (Bruner *et al.*, 1981), gas chromatography with electron capture detection (Shimohara *et al.*, 1979) and photothermal deflection spectrophotometry (Long & Bialkowski, 1985). Limits of detection are in the range of 0.1 ppb (0.4 µg/m^3) for the first two methods and of 1.3 ppb (4.6 µg/m^3) for the last.

3. Biological Data Relevant to the Evaluation of Carcinogenic Risk to Humans

3.1 Carcinogenicity studies in animals

(a) *Oral administration*

Rat: Groups of 36 male and 36 female SPF Alpk/Ap (Wistar-derived) rats, six weeks old, received 300 mg/kg bw chlorodifluoromethane (FC 22; purity, 99.5%) as a 3% solution in corn oil by gavage on five days per week for 52 weeks. The study was terminated after 125 weeks. Two control groups, one of which was housed separately, received corn oil alone and a third control group received no treatment (total numbers of control animals, 104 males and 104 females). No difference in body-weight gain or mortality was observed between chlorodifluoromethane-treated and control rats. No increase in the incidence of tumours was observed in any organ from treated rats when compared to either vehicle or untreated controls (Longstaff *et al.*, 1982, 1984).

(b) *Inhalation*

Mouse: Groups of 80 male and 80 female Alderley Park Swiss-derived mice [age unspecified] were exposed by inhalation to 0 (two groups), 1000, 10 000 or 50 000 ppm (0, 3540, 35 400 or 177 000 mg/m^3) chlorodifluoromethane (CFC 22; purity, >99.8%) for five hours per day on five days per week for up to 83 (males) or 94 (females) weeks, by which time approximately 80% of animals had died. There was no significant increase in the overall incidence of benign or malignant tumours in treated male or female animals compared to concurrent controls. A small but statistically significant dose-related increase [$p = 0.006$] in the incidence of hepatocellular carcinomas was observed in males (control, 0/74 and 1/77; low-dose, 0/78; mid-dose, 3/80; high-dose, 4/80). The tumour incidences in treated groups were all within the range of those in historical controls (2-10%) for this strain of mouse (Litchfield & Longstaff, 1984).

Rat: Groups of 60 male and 60 female Sprague-Dawley rats, eight weeks old, were exposed by inhalation to 0, 1000 or 5000 ppm (0, 3540 or 17 700 mg/m^3) chlorodifluoromethane (FC 22; purity unspecified) for four hours per day on five days per week for 104 weeks. No treatment-related tumour of the brain was observed (Maltoni *et al.*, 1982). [The Working Group noted that histopathology only of the brain was described and that other details were inadequately reported.]

Groups of 80 male and 80 female Alderley Park Wistar-derived rats [age unspecified] were exposed by inhalation to 0 (two groups), 1000, 10 000 or 50 000 ppm (0, 3540, 35 400 or 177 000 mg/m^3) chlorodifluoromethane (CFC 22; purity, >99.8%) for five hours per day, on five days per week for up to 118 (females) or 131 (males) weeks, by which time approximately 80% of animals had died. Body-weight gain was reduced in high-dose males up to week 80. Treatment did not affect the number of animals with benign tumours. Among males, the proportions of animals with malignant tumours were higher in treated groups (controls, 16/80 and 18/80; low-dose, 27/80; mid-dose, 22/80; high-dose, 33/80), due primarily to increases in the incidences of fibrosarcomas (controls, 5/80 and 7/80; low-dose, 8/80; mid-dose, 5/80; high-dose, 18/80). The numbers of animals in which such tumours involved the salivary glands were 1, 0, 1, 0 and 7, respectively. The increase in the overall incidences of fibrosarcomas occurred between weeks 105 and 130. In addition, four high-dose males had Zymbal-gland tumours, whereas no such tumour was found in males of the other groups. No increased incidence of malignant tumours was observed in treated females (Litchfield & Longstaff, 1984).

3.2 Other relevant biological data

(a) Experimental systems

A comprehensive review on the toxicology of chlorodifluoromethane is available (Litchfield & Longstaff, 1984).

Toxic effects

Exposure of rats, mice, guinea-pigs, dogs and rabbits by inhalation to concentrations of $>20\%$ v/v (708 g/m^3) chlorodifluoromethane for 0.5-4 h induced central-nervous-system depression and some deaths. Dogs were more resistant to these effects than rodents. Concentrations $\leqslant 20\%$ (708 g/m^3) were not lethal in any species tested (Poznak & Artusio, 1960; Karpov, 1963; Weigand, 1971; Sakata *et al.*, 1981; Clark & Tinston, 1982).

Cardiac toxicity (decreased heart rate and changes in electrocardiograms) was observed in rats exposed by inhalation to 15-60% (530-2120 g/m^3) chlorodifluoromethane for 2 min (Pantaleoni & Luzi, 1975). Myocardial depression was observed in monkeys exposed to 10 and 20% (354 and 708 g/m^3) chlorodifluoromethane (Belej *et al.*, 1974). Epinephrine-induced cardiac arrythmias were observed in dogs exposed by inhalation to 5% (177 g/m^3) chlorodifluoromethane and in mice exposed to 40% (1400 g/m^3) (Reinhardt *et al.*, 1971; Aviado & Belej, 1974). The EC$_{50}$ for cardiac sensitization in dogs was 14% (500 g/m^3) (Clark & Tinston, 1982).

No compound-related adverse effect was observed following inhalation exposure of rats, guinea-pigs, dogs or cats to 5% (177 g/m^3) chlorodifluoromethane for 3.5 h per day on five days per week for four weeks (Weigand, 1971) or of rats to 50 000 ppm (177 g/m^3) chlorodifluoromethane for 5 h per day for eight weeks (Lee & Suzuki, 1981).

As reported in an abstract, mild liver changes were observed in rabbits exposed to 6% (212 g/m^3) chlorodifluoromethane for 5 h per day on five days per week for 8-12 weeks; one of 14 rabbits developed supraventricular arrythmia (Van Stee & McConnell, 1977).

Exposure of dogs and rats by inhalation to 5000 and 10 000 ppm (17.7 and 35.4 g/m^3) chlorodifluoromethane for 6 h per day for 90 days had no effect on behaviour, body weight, haematology, biochemistry, organ weight or morphology; no effect was observed on the electrocardiogram or circulatory function of dogs (Leuschner *et al.*, 1983).

In chronic inhalation studies, no effect was observed on mortality, haematology or biochemistry in mice or rats exposed to 1000-50 000 ppm (3.54-177 g/m³) chlorodifluoromethane for 5 h per day on five days per week for up to 94 weeks (mice) or 131 weeks (rats). A decrease in body-weight gain up to week 80 was observed in rats, and hyperactivity in mice, exposed to 50 000 ppm (177 g/m³) chlorodifluoromethane (Litchfield & Longstaff, 1984).

Effects on reproduction and prenatal toxicity

Male Sprague Dawley rats were exposed to 0 or 50 000 ppm (0 or 177 g/m³) chlorodifluoromethane [purity unspecified] for 5 h per day for eight weeks and examined for testicular-weight changes and morphology, serum gonadotropins and sperm function. The weight of the prostate and coagulating glands were reduced in treated males. No other adverse effect was observed; the growth and development of embryos resulting from the breeding of the treated males were normal (Lee & Suzuki, 1981).

Studies to assess the teratogenic potential of chlorodifluoromethane in rats and rabbits have been reviewed (Litchfield & Longstaff, 1984).

CD rats were exposed to 0, 100, 1000 or 50 000 ppm (0, 0.354, 3.54 or 177 g/m³) chlorodifluoromethane for 6 h per day on days 6-15 of gestation. About one-half of the fetuses were selected for detailed histological observations of the eye and related structures. There were 646 rats (>6000 fetuses) in the control group and 418 (>4000 fetuses) in each exposure group. The highest dose level induced maternal (decreased body-weight gain) and fetal (slightly reduced body weight) toxicity and elevated the incidence of anophthalmia. The incidence of litters containing a fetus with anophthalmia increased from 1.65 per thousand litters in the control group to 15.67 per thousand at 50 000 ppm (177 g/m³) (Litchfield & Longstaff, 1984).

In a teratology study in rabbits, which was also described in the review article, groups of 14-16 New Zealand rabbits were exposed by inhalation to 0, 100, 1000 or 50 000 ppm (0, 0.354, 3.54 or 177 g/m³) chlorodifluoromethane for 6 h per day during days 6-18 of pregnancy. The high-dose level exerted maternal toxicity (lowered weight gain) during the first four days of treatment, but no other effect was noted in the does or fetuses (Litchfield & Longstaff, 1984).

Absorption, distribution, excretion and metabolism

Rats were exposed by inhalation to 10 000 and 50 000 ppm (35.4 and 177 g/m³) chlorodifluoromethane. The half-life for the clearance of chlorodifluoromethane from the blood, immediately after termination of exposure, was approximately 3 min (Litchfield & Longstaff, 1984). Within 1 min after rabbits were exposed to 10 and 20% (354 and 708 g/m³) chlorodifluoromethane, the blood concentration dropped by up to 50% (Sakata *et al.*, 1981).

Approximately 0.1 and 0.06% of an inhaled dose was recovered as $^{14}CO_2$ after exposure of rats to 500 and 10 000 ppm (1.77 and 35.4 g/m³) ^{14}C-chlorodifluoromethane, respectively, for 15-24 h by inhalation. The amount of radioactivity excreted in the urine was approximately 0.03 and 0.01% of the inhaled doses, respectively (Litchfield & Longstaff,

1984). Similar findings have been reported recently, indicating that there is little or no metabolism of chlorodifluoromethane in rats (Peter et al., 1986).

After incubation of ^{36}Cl-chlorodifluoromethane with Aroclor-induced rat-liver homogenates, no release of chloride was detected (Litchfield & Longstaff, 1984).

Mutagenicity and other short-term tests

Both 'pure' and 'crude' samples of chlorodifluoromethane were mutagenic to *Salmonella typhimurium* TA1535 when tested at a single dose level (50% in air) in the presence or absence of a metabolic system (S9) from Aroclor-induced rat liver (Longstaff & McGregor, 1978). [Results reported in this study for strain TA100 could not be evaluated.] Litchfield and Longstaff (1984) reported in a review that chlorodifluoromethane was mutagenic to strains TA1535 and TA100 in the presence and absence of S9 [details not given], and Bartsch *et al.* (1980) reported an increase in mutation frequency in *S. typhimurium* TA100 exposed to a concentration of 50% in air for 24 h [details not given]. However, Longstaff *et al.* (1982) reported that the compound was mutagenic to *S. typhimurium* TA1535, but not to TA1538, TA98 or TA100 [details not given].

Chlorodifluoromethane (20 mM solution generated at 500 ml/min; 1:1 air) was tested in assays for forward mutation in *Schizosaccharomyces pombe* and for mitotic gene conversion in *Saccharomyces cerevisiae*, in the presence and absence of mouse-liver S10; no activity was shown in either test system. Chlorodifluoromethane also gave negative results in a host-mediated assay in CD-1 mice administered a single dose of 816 mg/kg bw in oil by gavage, in which *S. pombe* or *S. cerevisiae* were injected into the venous orbital sinus and incubated for 5 or 15 h before harvesting and scoring (Loprieno & Abbondandolo, 1980). [The Working Group noted that the site of recovery of the cells was not indicated.]

Chlorodifluoromethane did not induce unscheduled DNA synthesis in human heteroploid EUE cells treated with a 20 mM solution generated at 500 ml/min (1:1 air) in the presence or absence of S10. At the same concentrations, the compound did not induce 6-thioguanine-resistant mutants in Chinese hamster V79 cells (Loprieno & Abbondandolo, 1980). It was reported in a review by Litchfield and Longstaff (1984) that chlorodifluoromethane (calculated concentrations of 33, 67 and 100%) did not induce reverse mutations at the hypoxanthine-guanine phosphoribosyl transferase (HGPRT) locus in Chinese hamster ovary cells in the presence or absence of metabolic activation.

No chromosomal damage was observed in the bone-marrow cells of CD1 mice receiving a single dose of 816 mg/kg bw chlorodifluoromethane in oil by gavage (Loprieno & Abbondandolo, 1980) nor in bone-marrow cells of rats following their exposure to concentrations of up to 150 000 ppm (531 g/m^3) by inhalation (data reported in a review by Litchfield & Longstaff, 1984). Several experimental regimes, employing different exposure and sampling times, were used; no consistent time- or treatment-related effect was observed.

There was no evidence for dominant lethality in a study in male Sprague-Dawley rats exposed to 50 000 ppm (177 g/m^3) chlorodifluoromethane for 5 h per day for eight weeks and then mated serially for 10 weeks (Lee & Suzuki, 1981).

Chlorodifluoromethane was tested in two studies of dominant lethal mutations in mice, in which male mice were exposed to various concentrations (10-100 000 ppm; 35.4-354 000 mg/m^3) for 6 h per day for five consecutive days. No consistent time- or treatment-related effect was observed (data reported in a review by Litchfield & Longstaff, 1984).

(b) Humans

Toxic effects

Heart palpitation was 3.6 times more common among hospital personnel using chlorodifluoromethane as a tissue-freezing agent than among other personnel in the same hospital (Speizer *et al.*, 1975). [The Working Group noted that there may have been over-reporting, since the study was prompted by a death believed to be associated with exposure to chlorodifluoromethane.]

An epidemiological study of 539 refrigeration workers exposed to a combination of chlorofluorocarbons, including chlorodifluoromethane, revealed five deaths due to heart or circulatory disorders compared with 9.62 expected (Szmidt *et al.*, 1981).

Effects on reproduction and prenatal toxicity

No data were available to the Working Group.

Absorption, distribution, excretion and metabolism

No data were available to the Working Group.

Mutagenicity and chromosomal effects

No data were available to the Working Group.

3.3 Case reports and epidemiological studies of carcinogenicity to humans

An epidemiological study was made of 539 refrigeration workers exposed to a combination of chlorofluorocarbons, including chlorodifluoromethane, for at least six months between 1950-1980 and followed up until 1980. There were six deaths due to cancer *versus* 5.7 expected, with two deaths from lung cancer *versus* 1.0 expected (Szmidt *et al.*, 1981). [The Working Group noted that the study was small and there were few deaths from cancer.]

4. Summary of Data Reported and Evaluation

4.1 Exposure data

Chlorodifluoromethane is produced extensively for use in refrigeration and air conditioning; significant quantities are subsequently released into the atmosphere, resulting in widespread, low-level human exposure. Occupational exposure to chlorodifluoromethane occurs during its production and use.

4.2 Experimental data

Chlorodifluoromethane was tested for carcinogenicity in one experiment in rats by oral administration by gavage and in experiments in rats and mice by inhalation exposure. No increase in tumour incidence was observed in rats after oral administration. The inhalation study in mice was inconclusive for males, and negative results were obtained for females. In the inhalation study in rats, males receiving the high dose had increased incidences of fibrosarcomas and Zymbal-gland tumours; negative results were obtained for female rats.

Chlorodifluoromethane causes malformations of the eyes of fetal rats, but has no reproductive effect in male rats and does not cause prenatal toxicity in rabbits following exposure by inhalation.

Chlorodifluoromethane is mutagenic to *Salmonella typhimurium* in the presence and absence of an exogenous metabolic system. It does not induce mutation or gene conversion in yeast, or DNA damage or mutation in cultured mammalian cells. It does not induce chromosomal damage in bone marrow or dominant lethal mutations in mice or rats treated *in vivo*.

Overall assessment of data from short-term tests: Chlorodifluoromethane[a]

	Genetic activity			Cell transformation
	DNA damage	Mutation	Chromosomal effects	
Prokaryotes		+		
Fungi/Green plants		–		
Insects				
Mammalian cells (*in vitro*)	–	–		
Mammals (*in vivo*)			–	
Humans (*in vivo*)				
Degree of evidence in short-term tests for genetic activity: **Inadequate**				Cell transformation: No data

[a]The groups into which the table is divided and the symbol '+' and – are defined on pp. 19-20 of the Preamble; the degrees of evidence are defined on pp. 20-21.

4.3 Human data

No data were available to evaluate the reproductive effects or prenatal toxicity of chlorodifluoromethane to humans.

A small study of workers exposed to a mixture of chlorofluorocarbons, including chlorodifluoromethane, was uninformative with regard to the carcinogenic hazard of this chemical.

4.4 Evaluation[1]

There is *limited evidence* for the carcinogenicity of chlorodifluoromethane to experimental animals.

There is *inadequate evidence* for the carcinogenicity of chlorodifluoromethane to humans.

5. References

Airco (1982) *Special Gases and Equipment*, Murray Hill, NJ

American Chemical Society (1985) *Chemcyclopedia 85*, Washington DC

American Conference of Governmental Industrial Hygienists (1985) *TLVs Threshold Limit Values and Biological Exposure Indices for 1985-86*, 2nd ed., Cincinnati, OH, p. 13

Anon. (1981) Chemical briefs. 3: Chloroform. *Chem. Purch.*, *17*, 69-70, 73

Arbetarskyddsstyrelsens Författningssamling (National Swedish Board of Occupational Safety and Health) (1984) *Occupational Exposure Limit Values (AFS 1984:5)* (Swed.), Solna, p. 10

Aviado, D.M. & Belej, M.A. (1974) Toxicity of aerosol propellants on the respiratory and circulatory systens. I. Cardiac arrythmia in the mouse. *Toxicology*, *2*, 31-42

Aviado, D.M. & Micozzi, M.S. (1981) *Fluorine containing organic compounds*. In: Clayton, G.D. & Clayton, F.E. eds, *Patty's Industrial Hygiene and Toxicology*, 3rd ed., Vol. 2B, New York, John Wiley & Sons, pp. 3071-3115

Bales, R.E. (1978) *Fluorocarbons — Worker Exposure in Four Facilities (PB 297772)*, Washington DC, National Technical Information Service, p. 22

Barbour, A.K. (1979) *Industrial aspects of fluorine chemistry*. In: Banks, R.E., ed., *Organofluorine Chemicals and Their Industrial Applications*, Chichester, Ellis Horwood, pp. 44-51

Bartsch, H., Malaveille, C., Camus, A.-M., Martel-Planche, G., Brun, G., Hautefeuille, A., Sabadie, N., Barbin, A., Kuroki, T., Drevon, C., Piccoli, C. & Montesano, R. (1980) Validation and comparative studies on 180 chemicals with *S. typhimirium* strains and V79 Chinese hamster cells in the presence of various metabolizing systems. *Mutat. Res.*, *76*, 1-50

---

[1]For definitions of the italicized terms, see Preamble, pp. 18 and 22.

Belej, M.A., Smith, D.G. & Aviado, D.M. (1974) Toxicity of aerosol propellants in the respiratory and circulatory systems. IV. Cardiotoxicity in the monkey. *Toxicology*, 2, 381-395

Benn Business Information Services (1985) *Chemical Industry Directory and Who's Who 1985*, Tunbridge Wells, Kent, p. A69

Bruner, F., Crescentini, G., Mangani, F., Brancaleoni, E., Cappiello, A. & Ciccioli, P. (1981) Determination of halocarbons in air by gas chromatography-high resolution mass spectrometry. *Anal. Chem.*, 53, 798-801

The Chemical Daily Co. (1984) *JCW Chemicals Guide 1984/1985*, Tokyo, p. 194

Clark, D.G. & Tinston, D.J. (1982) Acute inhalation toxicity of some halogenated and non-halogenated hydrocarbons. *Human Toxicol.*, 1, 239-247

Derwent, R.G. & Eggleton, A.E.J. (1978) Halocarbon lifetimes and concentration distributions calculated using a two-dimensional tropospheric model. *Atmos. Environ.*, 12, 1261-1269

Direktoratet for Arbeidstilsynet (Directorate of Labour Inspection) (1981) *Administrative Norms for Pollution in Work Atmosphere, No. 361* (Norw.), Oslo, p. 14

Edwards, P.R., Campbell, I. & Milne, G.S. (1982) The impact of chloromethanes in the environment. Part 1. The atmospheric chlorine cycle. *Chem. Ind.*, 16, 574-578

Grasselli, J.G. & Ritchey, W.M., eds (1975) *CRC Atlas of Spectral Data and Physical Constants for Organic Compounds*, Vol. 3, Cleveland, OH, CRC Press, p. 592

Hansch, C. & Leo, A. (1979) *Substituent Constants for Correlation Analysis in Chemistry and Biology*, New York, John Wiley & Sons, p. 172

Hawley, G.G., ed. (1981) *The Condensed Chemical Dictionary*, 10th ed., New York, Van Nostrand Reinhold, p. 236

Health and Safety Executive (1985) *Occupational Exposure Limits 1985 (Guidance Note EH 40/85)*, London, Her Majesty's Stationery Office, p. 10

Hubrich, C. & Stuhl, F. (1980) The ultraviolet absorption of some halogenated methanes and ethanes of atmospheric interest. *J. Photochem.*, 12, 93-107

IARC (1979) *IARC Monographs on the Evaluation of the Carcinogenic Risk of Chemicals to Humans*, Vol. 19, *Some Monomers, Plastics and Synthetic Elastomers, and Acrolein*, Lyon, pp. 285-301

International Labour Office (1980) *Occupational Exposure Limits for Airborne Toxic Substances: A Tabular Compilation of Values from Selected Countries*, 2nd (rev.) ed. (*Occupational Safety and Health Series No. 37*), Geneva, pp. 70-71

Karpov, B.C. (1963) The chronic toxicity of 'Freon' 22 (Russ). *Trudy Leningr. Sanit.-Gig. med. Inst.*, 75, 231-240

Lee, I.P. & Suzuki, K. (1981) Studies on the male reproductive toxicity of Freon 22. *Fundam. appl. Toxicol.*, 1, 266-270

Leifer, R., Sommers, K. & Guggenheim, S.F. (1981) Atmospheric trace gas measurements with a new clean air sampling system. *Geophys. Res. Lett.*, 8, 1079-1082

Leuschner, F., Neumann, B.-W. & Hübscher, F. (1983) Report on subacute toxicological studies with several fluorocarbons in rats and dogs by inhalation. *Arzneimittel-Forsch./Drug Res., 33*, 1475-1476

Litchfield, M.H. & Longstaff, E. (1984) The toxicological evaluation of chlorofluorocarbon 22 (CFC 22). *Food Chem. Toxicol., 22*, 465-475

Long, G.R. & Bialkowski, S.E. (1985) Saturation effects in gas-phase photothermal deflection spectrophotometry. *Anal. Chem., 57*, 1079-1083

Longstaff, E. & McGregor, D.B. (1978) Mutagenicity of a halocarbon refrigerant monochlorodifluoremethane (R-22) in *Salmonella typhimurium. Toxicol. Lett., 2*, 1-4

Longstaff, E., Robinson, M., Bradbrook, C., Styles, J.A. & Purchase, I.F.H. (1982) Carcinogenicity of fluorocarbons: assessment by short-term in vitro tests and chronic exposure in rats (Abstract No. 374). *Toxicologist, 2*, 106

Longstaff, E., Robinson, M., Bradbrook, C., Styles, J.A. & Purchase, I.F.H. (1984) Genotoxicity and carcinogenicity of fluorocarbons: assessment by short-term in vitro tests and chronic exposure in rats. *Toxicol. appl. Pharmacol., 72*, 15-31

Loprieno, N. & Abbondandolo, A. (1980) *Comparative mutagenic evaluation of some industrial compounds.* In: Norpoth, K.H. & Garner, R.C., eds, *Short-Term Mutagenicity Test Systems for Detecting Carcinogens*, Berlin (West), Springer-Verlag, pp. 333-356

Makide, Y. & Rowland, F.S. (1981) Tropospheric concentrations of methylchloroform, CH_3CCl_3, in January 1978 and estimates of the atmospheric residence times for hydrohalocarbons. *Proc. natl Acad. Sci. USA, 78*, 5933-5937

Maltoni, C., Ciliberti, A. & Carretti, D. (1982) Experimental contributions in identifying brain potential carcinogens in the petrochemical industry. *Ann. N.Y. Acad. Sci., 381*, 216-249

Matheson Gas Products (1985) *Matheson Gas Products Catalog 85*, Secaucus, NJ

Panel on Stratospheric Chemistry and Transport (1979) *Stratospheric Ozone Depletion by Halocarbons: Chemistry and Transport*, Washington DC, National Academy of Sciences

Pantaleoni, G.C. & Luzi, V. (1975) Cardiotoxicity of monochlorodifluoromethane (Ital.). *Russ. Med. sper., 22*, 265-269

Peter, H., Filser, J.G., Szentpály, L.V. & Wiegand, H.J. (1986) Different pharmacokinetics of dichlorofluoromethane (CFC 21) and chlorodifluoromethane (CFC 22). *Arch. Toxicol., 58*, 282-283

Poznak, A.V. & Artusio, J.F., Jr (1960) Anesthetic properties of a series of fluorinated compounds. I. Fluorinated hydrocarbons. *Toxicol. appl. Pharmacol., 2*, 363-373

Rasmussen, R.A. & Khalil, M.A.K. (1983) Natural and anthropogenic trace gases in the lower troposphere of the Arctic. *Chemosphere, 12*, 371-375

Rasmussen, R.A., Khalil, M.A.K., Penkett, S.A. & Prosser, N.J.D. (1980) $CHClF_2$ (F-22) in the earth's atmosphere. *Geophys. Res. Lett., 7*, 809-812

Reinhardt, C.F., Azar, A., Maxfield, M.E., Smith, P.E. & Mullin, L.S. (1971) Cardia arrythmias and aerosol 'sniffing'. *Arch. environ. Health, 22*, 265-279

Sadtler Research Laboratories (1980) *The Sadtler Standard Spectra, Cumulative Index*, Philadelphia, PA

Sakata, M., Kazama, H., Miki, A., Yoshida, A., Haga, M. & Morita, M. (1981) Acute toxicity of fluorocarbon-22: toxic symptoms, lethal concentration, and its fate in rabbit and mouse. *Toxicol. appl. Pharmacol., 59*, 64-70

Sax, N.I. (1984) *Dangerous Properties of Industrial Materials*, 6th ed., New York, Van Nostrand Reinhold, p. 701

Shimohara, K., Sueta, S., Tabata, T. & Shigemori, N. (1979) Determination of monochlorodifluoromethane in the ambient air by electron-capture gas chromatography with a back-flushing system (Jpn.). *Taiki Osen Gakkaishi, 14*, 31-37 [*Chem. Abstr., 91*, 128153y]

Smart, B.E. (1980) *Fluorine compounds, organic*. In: Grayson, M. & Eckroth, D., eds, *Kirk-Othmer Encyclopedia of Chemical Technology*, Vol. 10, New York, John Wiley & Sons, pp. 856-870

Somayajulu, G.R., Kennedy, J.R., Vickrey, T.M. & Zwolinski, B.J. (1979) Carbon-13 chemical shifts for 70 halomethanes. *J. magn. Reson., 33*, 559-568

Speizer, F.E., Wegman, D.H. & Ramirez, A. (1975) Palpitation rates associated with fluorocarbon exposure in a hospital setting. *New Engl. J. Med., 292*, 624-626

Szmidt, M., Axelson, O. & Edling, C. (1981) Cohort study of freon-exposed workers (Abstract No. 16) (Swed.). *Acta soc. med. suec. hyg., 90*, 77

Työsuojeluhallitus (National Finnish Board of Occupational Safety and Health) (1981) *Airborne Contaminants in the Workplaces (Safety Bull. No. 3)* (Finn.), Tampere, p. 16

Union Carbide Corp. (1982) *Specialty Gases & Related Products 1982*, Somerset, NJ

US Environmental Protection Agency (1981) Petition to remove chlorodifluoromethane from the list of toxic pollutants under Section 307(a)(1) of the Clean Water Act. *Fed. Regist., 46*, 2276-2278

US International Trade Commission (1980) *Synthetic Organic Chemicals, US Production and Sales, 1979 (USITC Publ. No. 1099)*, Washington DC, US Government Printing Office, p. 269

US International Trade Commission (1981) *Synthetic Organic Chemicals, US Production and Sales, 1980 (USITC Publ. No. 1292)*, Washington DC, US Government Printing Office, p. 265

US International Trade Commission (1982) *Synthetic Organic Chemicals, US Production and Sales, 1981 (USITC Publ. No. 1422)*, Washington DC, US Government Printing Office, p. 245

US International Trade Commission (1983) *Synthetic Organic Chemicals, US Production and Sales, 1982 (USITC Publ. No. 1422)*, Washington DC, US Government Printing Office, p. 261

US International Trade Commission (1984) *Synthetic Organic Chemicals, US Production and Sales, 1983* (*USITC Publ. No. 1588*), Washington DC, US Government Printing Office, p. 259

US International Trade Commission (1985) *Synthetic Organic Chemicals, US Production and Sales, 1984* (*USITC Publ. No. 1745*), Washington DC, US Government Printing Office, p. 258

Van Stee, E.W. & McConnell, E.E. (1977) Studies of the effects of chronic inhalation exposure of rabbits to chlorodifluoromethane (Abstract). *Environ. Health Perspect.*, *20*, 246-247

Weast, R.C., ed. (1985) *CRC Handbook of Chemistry and Physics*, 66th ed., Boca Raton, FL, CRC Press, pp. C-349, D-211

Weigand, W. (1971) Studies by inhalation of the toxicity of fluorinated derivatives of methane, ethane and cyclobutane (Ger.). *Zbl. Arbmed. Arbschutz*, *21*, 149-156

2-CHLORO-1,1,1-TRIFLUOROETHANE

1. Chemical and Physical Data

1.1 Synonyms and trade names

Chem. Abstr. Services Reg. No.: 75-88-7
Chem. Abstr. Name: 2-Chloro-1,1,1-trifluoroethane
IUPAC Systematic Name: 2-Chloro-1,1,1-trifluoroethane
Synonyms: 1-Chloro-2,2,2-trifluoroethane; CFC 133a; FC 133a; R 133a; 2,2,2-trifluorochloroethane; 1,1,1-trifluoro-2-chloroethane; 1,1,1-trifluoroethyl chloride

1.2 Structural and molecular formulae and molecular weight

$$\begin{array}{c} FH \\ || \\ F-C-C-Cl \\ || \\ FH \end{array}$$

$C_2H_2ClF_3$ Mol. wt: 118.49

1.3 Chemical and physical properties of the pure substance

(a) *Description*: Colourless gas

(b) *Boiling-point*: 6.93°C (Weast, 1985)

(c) *Melting-point*: -105.5°C (Weast, 1985)

(d) *Density*: d_4^0 1.389 (Weast, 1985)

(e) *Spectroscopy data*[a]: Ultraviolet (Hubrich & Stuhl, 1980), infrared (Sadtler Research Laboratories, 1980; prism [7457], grating [44459P]) and mass spectral data (Grasselli & Ritchey, 1975) have been reported.

---

[a] In square brackets, spectrum number in compilation

1.4 Technical products and impurities

As 2-chloro-1,1,1-trifluoroethane is used primarily as an intermediate, no information was available on technical products and impurities.

2. Production, Use, Occurrence and Analysis

2.1 Production and use

(a) Production

The synthetic route for the preparation of 2-chloro-1,1,1-trifluoroethane involves the fluorination of trichloroethylene with hydrogen fluoride in the presence of an antimony fluoride catalyst (McNeill, 1979). Commercial production of 2-chloro-1,1,1-trifluoroethane as an end-product is limited to one primary producer in the UK.

(b) Use

2-Chloro-1,1,1-trifluoroethane is used as intermediate in the synthesis of halothane (see IARC, 1976). Halothane (2-bromo-2-chloro-1,1,1-trifluoroethane) is commonly used as a general anaesthetic; it was first prepared by Suckling in 1953 and introduced into clinical practice by Johnstone in 1956. The industrial-scale liquid-phase process for the production of halothane involves direct bromination of 2-chloro-1,1,1-trifluoroethane at 500°C. Other synthetic routes for the industrial preparation of halothane do not utilize 2-chloro-1,1,1-trifluoroethane as an intermediate (Hudlický, 1971; Daley, 1972; Banks, 1982).

The acetolysis of 2-chloro-1,1,1-trifluoroethane, followed by hydrolysis, yields 2,2,2-trifluoroethanol. This alcohol has been used as a solvent for ionic reactions and as a working fluid in Rankine-cycle engines, which recover energy from waste heat sources (Astrologes, 1980a).

Trifluoroacetic acid can be prepared from 2-chloro-1,1,1-trifluoroethane by photochemical oxidation and hydrolysis (Astrologes, 1980b).

(c) Regulatory status and guidelines

No specific regulation or guideline has been promulgated for the use of, or occupational exposure to, 2-chloro-1,1,1-trifluoroethane.

2.2 Occurrence

(a) Natural occurrence

2-Chloro-1,1,1-trifluoroethane is not known to occur as a natural product.

(b) Occupational exposure

No data on exposure levels were available to the Working Group.

(c) *Air*

On the basis of its calculated rate of reaction with hydroxyl radical, the estimated tropospheric lifetime of 2-chloro-1,1,1-trifluoroethane is reported to be five years (Makide & Rowland, 1981).

(d) *Other*

2-Chloro-1,1,1-trifluoroethane is both an impurity (at 1 ml/m^3) (Chapman *et al.*, 1967) and a major mammalian metabolite of the anaesthetic agent halothane (Mukai *et al.*, 1977; Trudell *et al.*, 1982).

2.3 Analysis

Gas chromatography (with flame ionization detection) has been used to determine 2-chloro-1,1,1-trifluoroethane in tissues (headspace analysis) and in commercial products (Chapman *et al.*, 1967; Fiserova-Bergerova & Kawiecki, 1984).

3. Biological Data Relevant to the Evaluation of Carcinogenic Risk to Humans

3.1 Carcinogenicity studies in animals

Oral administration

Rat: Groups of 36 male and 36 female SPF Alpk/Ap (Wistar-derived) rats, six weeks old, received 300 mg/kg bw 2-chloro-1,1,1-trifluoroethane (FC 133a; purity, 99.5%) as a 3% w/v solution in corn oil by gavage on five days per week for 52 weeks. Two vehicle-control groups (one of which was housed separately) and one untreated control group were available. The study was terminated after 125 weeks. Body-weight gain was significantly reduced in treated males; mortality rates were comparable between the groups. The incidence of uterine carcinomas was 1/104 in the combined controls and 15/35 in treated females [$p < 0.001$]. The incidence of benign (often bilateral) interstitial-cell neoplasms of the testes was 16/104 in the combined controls and 29/36 in treated males [$p < 0.001$] (Longstaff *et al.*, 1982, 1984).

3.2 Other relevant biological data

(a) *Experimental systems*

Toxic effects

Groups of 36 male and 36 female Alpk/Ap rats were exposed to 0 or 300 mg/kg bw 2-chloro-1,1,1-trifluoroethane by gavage on five days per week for one year. Reduced growth and increased aggressive behaviour, arrest of spermatogenesis and seminiferous tubular atrophy were observed in males (Longstaff *et al.*, 1984).

Effects on reproduction and prenatal toxicity

No data were available to the Working Group.

Absorption, distribution, excretion and metabolism

No data were available to the Working Group.

Mutagenicity and other short-term tests

2-Chloro-1,1,1-trifluoroethane was not mutagenic to *Salmonella typhimurium* TA98 or TA100, in the presence and absence of a rat-liver metabolic system (S9) when incubated at concentrations of 0.1-50% for 48 h (Waskell, 1979) or at 0.5-10% for 8 h (Edmunds *et al.*, 1979). Longstaff *et al.* (1984) reported no mutagenic effect in *S. typhimurium* TA1535 or TA100 in the presence or absence of S9, when incubated for 72 h at concentrations up to 50%.

(*b*)   *Humans*

No data were available to the Working Group.

3.3 Case reports and epidemiological studies of carcinogenicity to humans

No data were available to the Working Group.

4. Summary of Data Reported and Evaluation

4.1 Exposure data

2-Chloro-1,1,1-trifluoroethane is used as a chemical intermediate in the production of the anaesthetic halothane. Human exposure occurs due to its presence as a low-level impurity in, and as a metabolite of, halothane.

4.2 Experimental data

2-Chloro-1,1,1-trifluoroethane was tested for carcinogenicity in one experiment in rats by oral administration by gavage at one dose level. Increased incidences of uterine carcinomas and benign testicular tumours were observed.

No data were available to evaluate the reproductive effects or prenatal toxicity of 2-chloro-1,1,1-trifluoroethane to experimental animals.

2-Chloro-1,1,1-trifluoroethane is not mutagenic to *Salmonella typhimurium* in the presence or absence of an exogenous metabolic system.

Overall assessment of data from short-term tests: 2-Chloro-1,1,1-trifluoroethane[a]

	Genetic activity			Cell transformation
	DNA damage	Mutation	Chromosomal effects	
Prokaryotes		−		
Fungi/Green plants				
Insects				
Mammalian cells (*in vitro*)				
Mammals (*in vivo*)				
Humans (*in vivo*)				
Degree of evidence in short-term tests for genetic activity: **Inadequate**				Cell transformation: No data

[a]The groups into which the table is divided and the symbol − are defined on pp. 19-20 of the Preamble; the degrees of evidence are defined on pp. 20-21.

4.3 Human data

No data were available to evaluate the reproductive effects or prenatal toxicity of 2-chloro-1,1,1-trifluoroethane to humans.

No case report or epidemiological study of the carcinogenicity of 2-chloro-1,1,1-trifluoroethane to humans was available to the Working Group.

4.4 Evaluation[1]

There is *limited evidence* for the carcinogenicity of 2-chloro-1,1,1-trifluoroethane to experimental animals.

No evaluation could be made of the carcinogenicity of 2-chloro-1,1,1-trifluoroethane to humans.

---

[1]For definition of the italicized term, see Preamble, p. 18.

5. References

Astrologes, G. (1980a) *Fluorine compounds, organic (Fluoroethanols)*. In: Grayson, M. & Eckroth, D., eds, *Kirk-Othmer Encyclopedia of Chemical Technology*, 3rd ed., Vol. 10, New York, John Wiley & Sons, pp. 871-874

Astrologes, G. (1980b) *Fluorine compounds, organic (Fluorinated acetic acids)*. In: Grayson, M. & Eckroth, D., eds, *Kirk-Othmer Encyclopedia of Chemical Technology*, 3rd ed., Vol. 10, New York, John Wiley & Sons, pp. 891-896

Banks, R.E., ed. (1982) *Preparation, Properties, and Industrial Applications of Organofluorine Compounds*, Chichester, Ellis Horwook, p. 160

Chapman, J., Hill, R., Muir, J., Suckling, C.W. & Viney, D.J. (1967) Impurities in halothane: their identities, concentrations and determination. *J. Pharm. Pharmacol.*, *19*, 231-239

Daley, R.D. (1972) *Halothane*. In: Florey, K., ed., *Analytical Profiles of Drug Substances*, Vol. 1, New York, Academic Press, pp. 119-147

Edmunds, H.N., Baden, J.M. & Simmon, V.J. (1979) Mutagenicity studies with volatile metabolites of halothane. *Anesthesiology*, *51*, 424-429

Fiserova-Bergerova, V. & Kawiecki, R.W. (1984) Effects of exposure concentrations on distribution of halothane metabolites in the body. *Drug Metab. Disposition*, *12*, 98-105

Grasselli, J.G. & Ritchey, W.M., eds (1975) *CRC Atlas of Spectral Data and Physical Constants for Organic Compounds*, Vol. 3, Cleveland, OH, CRC Press, p. 245

Hubrich, C. & Stuhl, F. (1980) The ultraviolet absorption of some halogenated methanes and ethanes of atmospheric interest. *J. Photochem.*, *12*, 93-107

Hudlický, M. (1971) *Organic Fluorine Chemistry*, New York, Plenum Press, pp. 67-79

IARC (1976) *IARC Monographs on the Evaluation of Carcinogenic Risk of Chemicals to Man*, Vol. 11, *Cadmium, Nickel, Some Epoxides, Miscellaneous Industrial Chemicals and General Considerations on Volatile Anaesthetics*, Lyon, pp. 285-293

Longstaff, E., Robinson, M., Bradbrook, C., Styles, J.A. & Purchase, I.F.H. (1982) Carcinogenicity of fluorocarbons: assessment by short-term in vitro tests and chronic exposure in rats (Abstract No. 374). *Toxicologist*, *2*, 106

Longstaff, E., Robinson, M., Bradbrook, C., Styles, J.A. & Purchase, I.F.H. (1984) Genotoxicity and carcinogenicity of fluorocarbons: assessment by short-term in vitro tests and chronic exposure in rats. *Toxicol. appl. Pharmacol.*, *72*, 15-31

Makide, Y. & Rowland, F.S. (1981) Tropospheric concentrations of methylchloroform, CH_3CCl_3, in January 1978 and estimates of the atmospheric residence times for hydrohalocarbons. *Proc. natl Acad. Sci. USA*, *78*, 5933-5937

McNeil, W.C., Jr (1979) *Trichloroethylene*. In: Grayson, M. & Eckroth, D., eds, *Kirk-Othmer Encyclopedia of Chemical Technology*, 3rd ed., Vol. 5, New York, John Wiley & Sons, pp. 745-753

Mukai, S., Morio, M., Fujii, K. & Hanaki, C. (1977) Volatile metabolites of halothane in the rabbit. *Anesthesiology*, *47*, 248-251

Sadtler Research Laboratories (1980) *The Sadtler Standard Spectra, Cumulative Index*, Philadelphia, PA

Trudell, J.R., Bösterling, B. & Trevor, A.J. (1982) Reductive metabolism of halothane by human and rabbit cytochrome P-450. Binding of 1-chloro-2,2,2-trifluoroethyl radical to phospholipids. *Mol. Pharmacol.*, *21*, 710-717

Waskell, L. (1979) Lack of mutagenicity of two possible metabolites of halothane. *Anesthesiology*, *50*, 9-12

Weast, R.C., ed. (1985) *CRC Handbook of Chemistry and Physics*, 66th ed., Boca Raton, FL, CRC Press, p. C-264

POLYBROMINATED BIPHENYLS

These substances were considered by a previous Working Group, in June 1978 (IARC, 1978). Since that time, new data have become available, and these have been incorporated into the monograph and taken into consideration in the present evaluation.

1. Chemical and Physical Data

1.1 Identity and general characteristics

Polybrominated biphenyls (PBBs) are synthetic chemicals that have found use as flame retardants. The technical-grade products are mixtures of brominated biphenyls of the general formula shown in Table 1. Also listed in Table 1 with their Chemical Abstracts Registry Numbers are some of the most important technical products and their major isomeric component. The composition of these products is discussed in more detail in section 1.2. There are 209 possible structural congeners of the brominated biphenyl structure containing two or more bromines; only a few of these have been synthesized individually and characterized (Di Carlo *et al.*, 1978).

PBBs with three or more bromines are solids (Norris *et al.*, 1973; De Kok *et al.*, 1977; Neufeld *et al.*, 1977; Di Carlo *et al.*, 1978), with low volatility decreasing with increasing bromine number. PBBs are virtually insoluble in water, soluble in fat (Kay, 1977) and slightly to highly soluble in various organic solvents, solubility also decreasing with increasing bromine number. These compounds are relatively stable and chemically unreactive, although several of the common isomers reportedly photodegrade with reductive debromination upon exposure to ultraviolet light (Norris *et al.*, 1973; Sundström *et al.*, 1976; Kay, 1977; Neufeld *et al.*, 1977).

The thermal degradation of hexabromobiphenyl (FireMaster FF-1) has been shown to produce polybrominated dibenzofurans, including the 2,3,7,8-tetra-substituted isomer (O'Keefe, 1978).

Ultraviolet, nuclear magnetic resonance and mass spectral data for various PBBs have been reported (De Kok *et al.*, 1977; Erickson *et al.*, 1977; Robertson *et al.*, 1984).

Table 1. Identification of polybrominated biphenyls

Structural and molecular formulae and molecular weights:

$C_{12}H_{(10-m-n)}Br_{(m+n)}$ Mol. wt: 154.21 + 78.90 (m+n)

Chemical Abstracts Names and Registry Numbers:

Hexabromobiphenyl

    Technical grades: FireMaster BP-6 (CAS No.: 59536-65-1)
                     FireMaster FF-1 (CAS No.: 67774-32-7)
    Major isomer: 2,2′,4,4′,5,5′-Hexabromo-1,1′-biphenyl (CAS No.: 59080-40-9)

Octabromobiphenyl

    Technical grade: Bromkal 80 (CAS No.: 61288-13-9)
    Major isomer: Unknown

Decabromobiphenyl

    Pure substance: 2,2′,3,3′,4,4′,5,5′,6,6′-Decabromo-1,1′-biphenyl (CAS No.: 13654-09-6)
    Technical grades: Adine 0102; Berkflam B 10; Flammex B 10 (CAS No.: 13654-09-6)

1.2 Technical products and impurities

PBBs are produced by the direct bromination of biphenyl; as would be expected, a number of different isomers are formed by this reaction. FireMaster BP-6 was marketed as technical-grade hexabromobiphenyl, although up to 22 different biphenyl isomers have been identified in this product (Robertson *et al.*, 1984). The quantity of hexabromobiphenyl in commercial FireMaster BP-6 has been reported to range from more than 60% to 90%, but the principal hexabromobiphenyl is the 2,2′,4,4′,5,5′-isomer (Sundström *et al.*, 1976; Di Carlo *et al.*, 1978). Three published analyses of FireMaster BP-6 are given in Table 2.

FireMaster FF-1 (white powder) is FireMaster BP-6 (brown flakes) to which 2% calcium silicate has been added as an anticaking agent (Di Carlo *et al.*, 1978).

Table 2. Composition (%) of FireMaster BP-6

Bromobiphenyl	Reference			
	Norström et al. (1976)[a]	Neufeld et al. (1977)[b]	Hass et al. (1978)[c]	Robertson et al. (1984)[a,c,d]
Tribromobiphenyl	–	–	–	0.1
Tetrabromobiphenyl	–	2	–	0.3
Pentabromobiphenyl	10	10	4.0	5.7
Hexabromobiphenyl	90	63	62.6	78.2
Heptabromobiphenyl	–	14	33.4	7.6

[a]Gas chromatography
[b]11% other bromobiphenyl
[c]Gas chromatography/mass spectrometry
[d]Total, 91.9%

Technical-grade octabromobiphenyl has also been shown to contain a number of brominated isomers. A sample of octabromobiphenyl obtained from a US producer (who synthesized PBBs only in experimental quantities (Neufeld et al., 1977)) contained primarily nonabromobiphenyl (60%), with 33%, 6% and 1% of the octabromobiphenyl, decabromobiphenyl and heptabromobiphenyl isomers, respectively (Waritz et al., 1977). A firm in the Federal Republic of Germany developed an octabromobiphenyl product, Bromkal 80, with 72% octabromobiphenyl (mixture of three isomers at 14, 16 and 46%; structures not specified), 27% heptabromobiphenyl, 1% hexabromobiphenyl and traces of nonabromobiphenyl (Norström et al., 1976). Another sample of commercial octabromobiphenyl was found to contain 45.2% octabromobiphenyl, 47.4% nonabromobiphenyl, 5.7% decabromobiphenyl and 1.8% heptabromobiphenyl (Norris et al., 1973).

Decabromobiphenyl products have been assayed at 96.8% decabromobiphenyl, 2.9% nonabromobiphenyl and 0.3% octabromobiphenyl (Di Carlo et al., 1978). The primary US producer of decabromobiphenyl reported a purity of more than 98%, with nonabromobiphenyl making up the remainder (Neufeld et al., 1977).

A number of brominated naphthalenes have been found as impurities in PBB mixtures. O'Keefe (1979) reported the presence of approximately 25 mg/kg hexabromonaphthalene and 1 mg/kg pentabromonaphthalene in FireMaster FF-1. FireMaster BP-6 has been found to contain 150 mg/kg pentabromonaphthalene and 70 mg/kg hexabromonaphthalene (Hass et al., 1978).

Neither bromodibenzofurans nor bromodibenzo-*para*-dioxins have been identified as contaminants of FireMaster FF-1 (at levels above 0.5 mg/kg) (O'Keefe, 1979) or FireMaster BP-6 (Hass et al., 1978) by gas chromatography/mass spectrometry.

Contaminants of the initial biphenyl feedstock may ultimately appear in commercial PBB mixtures. The biphenyl grade used for bromination by one producer contained less than 0.5% impurities, including toluene, naphthalene, methylene biphenyl (fluorene) and various methyl biphenyls (Neufeld et al., 1977).

2. Production, Use, Occurrence and Analysis

2.1 Production and use

(a) Production

The synthesis of PBBs generally involves the bromination of biphenyl. Patents were obtained by a firm in the Federal Republic of Germany in 1964 (Windholz, 1983) and in 1966 (Neufeld et al., 1977) for the bromination of biphenyl. A US firm patented a process for PBB production in 1973 which involves the reaction of an aromatic compound with bromine in the presence of a Lewis acid catalyst, such as aluminium chloride or aluminium bromide, and of solvent quantities of dibromomethane. Another process, patented by a US firm in 1974, consists of the bromination of biphenyl with bromine chloride in the presence of iron or a Friedel-Crafts catalyst in a closed vessel (Mumma & Wallace, 1975; Neufeld et al., 1977). The synthesis of research quantities (Neufeld et al., 1977) of PBBs has been described by Robertson et al. (1984) using the diazo coupling of a brominated aniline with an excess of a brominated benzene.

Hexabromobiphenyl

In 1970, a US firm began commercial production of a flame-retardant additive that contained various isomers of hexabromobiphenyl as the principal constituents. Approximately 9.5 thousand kg hexabromobiphenyl were manufactured in the first year of production (Neufeld et al., 1977). US production figures are presented in Table 3.

Table 3. Estimated US production of polybrominated biphenyls (thousands of kg)[a]

Polybrominated biphenyl	1970	1971	1972	1973	1974	1975	1976
Hexabromobiphenyl	9.5	84	1008	1766	2216	0	0
Octabromobiphenyl and decabromobiphenyl[b]	14.1	14.1	14.5	163	48.1	77.2	366

[a]From Neufeld et al. (1977)
[b]Mainly decabromobiphenyl

After producing a total of 5.1 million kg of hexabromobiphenyl from 1970 to 1974 (Neufeld et al., 1977), the sole US manufacturer of hexabromobiphenyl ceased production in November 1974 (Archer et al., 1979). This action was precipitated by an incident in May 1973 in which about 295 kg of hexabromobiphenyl, as FireMaster FF-1, were added to animal feed in Michigan (Fries, 1985). The resulting contamination of farm animals resulted in the destruction of thousands of cattle, hogs and sheep, and millions of chickens (Di Carlo et al., 1978). All of the remaining inventory of the manufacturer of FireMaster BP-6 and FF1 had been depleted by April 1975 (Neufeld et al., 1977).

Octabromobiphenyl and decabromobiphenyl

Commercial quantities of these compounds were produced in the US until 1979 (National Toxicology Program, 1982). One US firm initiated production of both octabromobiphenyl and decabromobiphenyl in 1970 (Archer et al., 1979). Another US manufacturer produced decabromobiphenyl from 1973 to prior to 1977. In the period from 1970 to 1976, an estimated combined total of 697 thousand kg decabromobiphenyl and octabromobiphenyl were produced in the USA (see Table 3). Of this total, 366 thousand kg were produced in 1976 (352 thousand kg decabromobiphenyl and 14 thousand kg octabromobiphenyl), all of which was exported to Europe (Neufeld et al., 1977).

One firm in the Federal Republic of Germany produced commercial quantities of octabromobiphenyl (Norström et al., 1976; Neufeld et al., 1977). In the UK, one firm made available octabromobiphenyl, and another produced 'polybromodiphenyl' as of 1980 (Baker et al., 1980). A third producer in the UK discontinued production of decabromobiphenyl prior to 1977 (Neufeld et al., 1977). One company in France currently produces commercial quantities of decabromobiphenyl.

(b) Use

PBBs were introduced as flame retardants for synthetic fibres and moulded thermoplastic parts in the early 1970s (Archer et al., 1979). Prior to November 1974, hexabromobiphenyl, the most commercially significant PBB in the USA, was incorporated into ABS (acrylonitrile-butadiene-styrene) plastics, coatings and lacquers, and polyurethane foam (Neufeld et al., 1977).

In 1974, approximately 55% of the total hexabromobiphenyl produced (as FireMaster BP-6) was used in flame retardant ABS resins (Mumma & Wallace, 1975). These plastics, which have a PBB content of about 10%, were used primarily for small appliance and automotive applications (Neufeld et al., 1977). Typical end-product uses of hexabromobiphenyl are summarized in Table 4 (Mumma & Wallace, 1975). Thermoplastics that contained hexabromobiphenyl as a flame retardant were not used in products intended for contact with food or feed (Neufeld et al., 1977).

One plant in the USA used an average of 34 thousand kg hexabromobiphenyl (FireMaster BP-6) per year as a flame retardant for cable coatings, until 1974, when they stopped using it; they had depleted their remaining stock by 1977 (Neufeld et al., 1977).

Table 4. Use of hexabromobiphenyl in the USA[a]

Industry	Approximate allocation of total FireMaster BP-6 produced (%)	Examples
Business machines and industrial equipment	48	Typewriters, calculators, microfilm readers and business machine housings
Electrical	35	Radio and television parts, thermostats, shavers and hand-tool housings
Fabricated products	12	Projector housings and film equipment cases
Transportation	1	Miscellaneous small automotive parts, e.g., electrical wire connectors, switch connectors, speaker grills
Miscellaneous	4	Small parts for electrical applications, motor housings and components for industrial equipment

[a]From Mumma and Wallace (1975)

Hexabromobiphenyl was used as a flame retardant in polyurethane foam for automobile upholstery. One US firm stopped using PBBs in polyurethane foams in 1972 due to the concern by automobile manufacturers over the environmental persistence of these compounds. Temperatures used during recycling of scrap material from automobiles were not sufficient to decompose the PBBs (Neufeld et al., 1977).

During the late 1960s and early 1970s, a number of US, German (Federal Republic), British and French patent applications were received for various PBB isomers. The majority of these pertained to the use of brominated biphenyls as flame retardants in polymeric materials. Other proposed applications included use in light-sensitive compositions to act as colour activators, as molecular-weight control agents for polybutadiene, as wood preservatives and in electrical insulation (Neufeld et al., 1977).

PBBs have been added to fibre polymers during spinning and finishing operations (Archer et al., 1979) but were not used in fabrics for clothing (Neufeld et al., 1977).

Octabromobiphenyl was present as the principal component of a flame retardant mixture manufactured by a firm in the Federal Republic of Germany (Norström et al., 1976) and was used as a flame retardant in ABS and polyolefin resins in the USA (Archer et al., 1979). Decabromobiphenyl was presumably used in similar flame retardant products (Neufeld et al., 1977).

(c) Regulatory status and guidelines

The US Food and Drug Administration (FDA) set guidelines of 1.0 mg/kg PBBs in the fat of milk, meat and poultry, 0.1 mg/kg in whole eggs and 0.3 mg/kg in animal feeds, in May 1974. In November 1974, the FDA reduced these guidelines to 0.3 mg/kg in the fat of milk, meat and poultry, and 0.05 mg/kg in whole eggs and animal feed. The FDA regulates PBBs as inadvertant environmental contaminants under the Food, Drug and Cosmetic Act (Di Carlo et al., 1978; National Toxicology Program, 1982).

The US Environmental Protection Agency (1984) requires notification of the manufacture or importation of PBBs under the Toxic Substances Control Act. Canada prohibited the manufacture and use of PBBs under the Environmental Contaminants Act (Anon., 1979).

2.2 Occurrence

(a) Natural occurrence

PBBs are not known to occur as natural products.

(b) Occupational exposure

No data on exposure levels were available to the Working Group.

(c) Water and sediments

PBBs were measured in effluents from the plant of the major US manufacturer and in the water and sediments of the river receiving the effluent. Concentrations in various plant effluents were as high as 104 μg/l, resulting in a discharge of an estimated 122 g PBBs/day. In the river-water, PBB concentrations ranged from 3.2 μg/l at 69 m downstream of the plant to 0.01 μg/l about 12.8 km downstream. Concentrations in samples taken 19.3 and 32.2 km downstream were below the sensitivity level of 0.01 μg/l. Sediment concentrations ranged from 77 000 μg/kg in the area of the plant outfalls to 100 μg/kg 38.4 km downstream (Hesse, 1975).

(d) Soil and plants

Only slight biodegradation in soil of any of six isomers studied (two pentabromobiphenyl, three hexabromobiphenyls and one heptabromobiphenyl) was found after 24 weeks (Jacobs et al., 1976).

Soil from a Michigan farm on which a dairy herd had 'moderate' PBB contamination (0.2-0.3 mg/l in the milk) was analysed to determine the concentration and persistence of PBBs under actual field conditions. Manure from the contaminated herd (PBB levels unknown) was spread in December 1974 or January 1975, and soil samples were taken the following October. The major PBB isomer, 2,2',4,4',5,5'-hexabromobiphenyl, was found in concentrations of 4.1, 5.7 and 4.4 μg/kg in samples from three different locations in the field, and peaks corresponding in retention time to six PBB components were found, indicating their persistence in soil (Jacobs et al., 1976).

In a greenhouse experiment, carrots and orchard grass were grown in soil with five concentrations of PBB (0, 0.1, 1.0, 10 and 100 mg/kg). No PBB was detected in the orchard grass cuttings or carrot tops, but detectable quantities were found in carrot roots grown in soil containing 10 and 100 mg/kg PBB, although the concentrations found (20-40 µg/kg) were too low to be quantified with precision (Jacobs *et al.*, 1976).

(e) Food, beverages and animal feeds

Following the accidental contamination of at least 865 000 kg of animal feed (Di Carlo *et al.*, 1978) with FireMaster FF-1 in Michigan, USA, in 1973, PBB levels ranging from 4000 to 13 500 mg/kg were detected in cattle feed supplement samples (Kay, 1977).

The average concentrations of PBB in milk fat of four Holstein cows that consumed 10 mg/day PBB (FireMaster BP-6) for 60 days reached a maximum steady state in 20 days, and levels of 2.78-3.39 mg/kg were maintained from day 30 to 60; these levels declined by 65-77% 15 days after feeding of PBB was stopped (Fries & Marrow, 1975).

Milk containing 0.1-0.3 µg/l (fat basis) PBBs was pasteurized and then analysed for PBB levels; the processing did not significantly affect the level in lipids (Murata *et al.*, 1977).

(f) Animals

Elevated PBB levels (up to a maximum of 1.33 mg/kg) were found in fish in a river in the vicinity of the major US manufacturing plant. Carp captured 4.8 km downstream showed a concentration of 1.26 mg/kg, and still measurable concentrations (0.09 mg/kg) were found in carp caught as far as 12.9 km downstream. Analysis of fathead minnows held in cages for one- and two-week intervals in the river showed that PBB uptake occurred rapidly: caged fish near the manufacturing plant outfalls and 1.6 km downstream accumulated 1.0 mg/kg PBB after only two weeks' exposure. Concentrations of PBBs in the flesh of three duck species (mallards, teal and wood ducks) collected within 3 km of the plant ranged from 5.3 to 29 mg/kg (fat basis) (Hesse, 1975).

(g) Human tissues

Human exposure to PBB has occurred through occupational contact and as a result of the accidental contamination of livestock feed in Michigan, USA, in 1973 (Williams *et al.*, 1984).

Gas chromatographic analyses of blood samples from 110 Michigan subjects who were exposed to PBB-contaminated farm products for more than six months since the accident of 1973 showed PBB levels ranging from 0.002 to 2.26 mg/l, with median values of 0.014 mg/l for adults and 0.035 mg/l for children. In comparison, unexposed adults and children had median values of 0.003 mg/l and 0.006 mg/l, respectively. Five exposed women had breast milk concentrations of PBB ranging from 0.21 to 92.66 mg/l; their corresponding blood levels of PBB ranged from 0.003 to 1.068 mg/l. In 13 exposed persons, adipose tissue levels of PBB ranged from 0.1 to 174 mg/kg and blood levels ranged from <0.002 to 1.07 mg/l (Humphrey & Hayner, 1975).

In 1976, 524 Michigan dairy farmers had a median concentration of 2.6 µg/l PBBs in serum, which was lower than that found in workers manufacturing PBBs (9.3 µg/l). In 56 Wisconsin farm residents examined as a control population, only two had detectable levels of PBBs (0.5 and 1.1 µg/l) (Wolff et al., 1978). An independent study conducted in 1976-1977 in Michigan on 3639 subjects (mainly farm residents and chemical workers) showed a median value of 3.0 µg/l PBBs in serum; the highest values were found among chemical workers and their families (4.5 µg/l) (Landrigan et al., 1979).

In 46 persons exposed to PBBs in the Michigan accident, serum PBB levels three years later ranged from 1 to 180 µg/l with an average of 14 µg/l, while fat levels ranged from 30 to 34 400 µg/kg with an average of 3260 µg/kg (Stross et al., 1979).

In 1977, PBB levels in serum were measured in 3683 persons, consisting of residents of quarantined farms with high-level PBB contamination after the 1973 Michigan accident, recipients of food products from quarantined farms, workers at the PBB manufacturing plant and their family members, and residents of farms with low-level PBB contamination. Levels ranged from <1 to 3150 µg/l, with a geometric mean of 4.1 µg/l and an arithmetic mean of 23.2 µg/l. In each age group, males had higher PBB levels than females (geometric mean values 5.8 µg/l and 2.8 µg/l, respectively). In 1978 and 1979, serum samples collected from 1631 of the 3683 individuals ranged from <1 to 57 µg/l, with a geometric mean of 6.4 µg/l and an arithmetic mean of 7.7 µg/l (Kreiss et al., 1982).

2.3 Analysis

Typical methods of analysis for PBBs in environmental samples are summarized in Table 5.

The analysis of PBBs by high-performance liquid chromatography, gas chromatography/mass spectrometry, ultraviolet spectrometry and nuclear magnetic resonance spectrometry has been reviewed (De Kok et al., 1977; Erickson et al., 1977). An AOAC (Association of Official Analytical Chemists) extraction technique for pesticides, followed by gel-permeation chromatography and gas chromatography/electron-capture detector has been used to determine PBBs in milk and dairy products (Fehringer, 1975). Ultraviolet irradiation of extracts followed by gas chromatography or mass spectrometry has been proposed to confirm the presence of PBBs (Erney, 1975; Murata et al., 1977; Trotter, 1977). Gas chromatography with detection/analysis by microwave-induced plasma spectrometry has been reported to show selectivity for PBBs at an absolute detection limit in the range of 3-19 ng (Mulligan et al., 1980; Eckhoff et al., 1983).

Gas chromatography with electron capture detection has been the primary analytical technique used for biological monitoring of body burden of PBBs in body fluids and fatty tissues (Willett et al., 1978; Burse et al., 1980; Roboz et al., 1985). Individual PBBs in serum can be quantified by selective-ion monitoring of bromide anions by mass spectrometry (Roboz et al., 1982, 1985).

Table 5. Methods for the analysis of PBBs

Sample matrix	Sample preparation	Assay procedure[a]	Limit of detection	Reference
Milk	Add ethanol; extract (petroleum ether and diethyl ether); centrifuge; evaporate ether phase to dryness; column chromatography	GC/EC	1.4 µg/kg	Willett et al. (1978)
Dairy products	Add water, potassium oxalate and ethanol; extract (diethyl ether and petroleum ether); evaporate to dryness; gel permeation chromatography; evaporate PBB fraction to dryness; dilute with isooctane; inject aliquot onto GC; Florisil column chromatography; confirm identity by TLC	GC/EC	7 µg/kg	Fehringer (1975)
Plant extracts	Extract (benzene, 2-propanol); wash with water to remove 2-propanol; dry; column chromatography; confirm identity by MS	GC/EC	10 µg/kg	Jacobs et al. (1976)
Blood	Add methanol; extract (hexane: diethyl ether); concentrate; wash (hexane); Florisil column chromatography	GC/EC	1 µg/l	Burse et al. (1980)
Bile, faeces and plasma	Add ethanol (except for faeces); extract (petroleum ether:diethyl ether); centrifuge; recover ether phase; evaporate to dryness; column chromatography	GC/EC	1.4 µg/kg (faeces) 1 µg/kg (plasma, bile)	Willett et al. (1978)
Soil	Extract (benzene:2-propanol); wash with water to remove 2-propanol; dry; column chromatography; confirm identity by MS	GC/EC	0.1 µg/kg	Jacobs et al. (1976)

[a]Abbreviations: GC/EC, gas chromatography/electron capture detector; TLC, thin-layer chromatography; MS, mass spectrometry

3. Biological Data Relevant to the Evaluation of Carcinogenic Risk to Humans

3.1 Carcinogenicity studies in animals

(a) Oral administration

Mouse: Groups of 22-27 male and 8-19 female B6C3F$_1$ mice, seven to eight weeks old, were administered 0 (control), 0.1, 0.3, 1.0, 3.0 or 10.0 mg/kg bw FireMaster FF-1 (FireMaster BP-6 plus 2% calcium trisilicate, lot No. 1312 FT, Batch 03) in 0.1 ml corn oil by gavage daily on five days per week for 25 consecutive weeks (for a total of 125 doses over a six-month period). Mice were held for an additional 24 months after treatment, and then the remaining 10% were killed. Male mice treated with the highest dose had a statistically significant ($p < 0.05$) reduction in survival (average survival, 688 days, compared with 784 days in controls). The incidence of hepatocellular carcinomas was significantly increased in mice receiving 10 mg/kg bw (males — control, 12/25; 10 mg/kg bw, 21/22; $p < 0.01$; females — 0/13, 7/8, $p < 0.01$) (Gupta *et al.*, 1983a; National Toxicology Program, 1983).

Rat: Groups of 20 male and 20 female Sherman (derived from Osborne-Mendel; Kimbrough *et al.*, 1981) rats, two or 2.5 months of age, were fasted for 15 h then administered a single dose by gavage of 0 (controls) or 1000 mg/kg bw FireMaster FF-1 (lot No. FH 7042) as a 5% solution in peanut oil. Groups of five animals were killed two, six, ten and 14 months after treatment. An increase in the incidence of neoplastic liver nodules was observed in treated animals (3/5 in females and 2/5 in males at 14 months) (Kimbrough *et al.*, 1977, 1978). [The Working Group noted that this was a preliminary study.]

Two groups of 65 female Sherman rats, two months of age, were fasted for 15 h and then received a single gastric intubation of 0 (controls) or 1000 mg/kg bw FireMaster FF-1 (lot No. 7042) as a 5% solution in corn oil. At terminal sacrifice at 25 months of age, 36 controls and 48 treated animals were still alive. The incidences of liver tumours were: trabecular hepatocellular carcinomas — controls, 0/53; treated, 24/58 [$p < 0.001$]; neoplastic nodules — controls, 0/53; treated, 42/58 [$p < 0.001$]; foci or altered areas — controls, 1/53; treated, 57/58 [$p < 0.001$] (Kimbrough *et al.*, 1981).

Groups of 19 and 16 female Sherman rats, four months old, received a single gastric intubation of 0 (controls) or 200 mg/kg bw FireMaster FF-1 (lot No. 7042) as a 5% solution in corn oil, respectively. Surviving rats were killed 22 months after treatment. The incidences of neoplastic nodules of the liver were: control, 0/19; treated, 5/16 [$p = 0.013$]; and those of altered areas: control, 1/19; treated 8/16 [$p = 0.004$] (Kimbrough *et al.*, 1981).

Two groups of 30 female Sherman rats, two months of age, received twice-weekly gastric intubations of 0 (controls) or 100 mg/kg bw FireMaster FF-1 (lot No. 7042) as a 5% solution in corn oil every three weeks for a total of 12 doses. At terminal sacrifice at 26 months of age, 13 controls and 20 treated animals were still alive. The incidences of liver tumours were: trabecular hepatocellular carcinomas — controls, 0/25; treated, 17/28 [$p < 0.001$]; neoplastic nodules — controls, 1/25, treated, 24/28 [$p < 0.001$]; foci or altered areas —controls, 1/25; treated, 23/28 [$p < 0.001$] (Kimbrough *et al.*, 1981).

Groups of 31-40 male and 11-21 female Fischer 344 rats, seven to eight weeks old, were administered 0 (control), 0.1, 0.3, 1.0, 3.0 or 10.0 mg/kg bw FireMaster FF-1 (FireMaster BP-6 plus 2.0% calcium trisilicate, lot No. 1312 FT, Batch 03) in 0.2 ml corn oil by gavage daily on five days per week for 25 consecutive weeks (total of 125 doses over a six-month period). Rats were maintained for an additional 23 months after treatment, and then the remaining 10% were killed. A dose-dependent reduction in survival time was observed in treated male rats (average survival time in days, 762, 744, 733, 677, 680, 615 in the six groups, respectively). No hepatocellular carcinoma was observed in male (0/33) or female (0/20) control rats. In contrast, 2/39, 0/40, 1/33, 7/33 ($p < 0.01$) and 7/31 ($p < 0.01$) of male rats treated with 0.1, 0.3, 1.0, 3.0 and 10 mg/kg bw FireMaster FF-1, respectively, and 3/19 and 7/20 ($p < 0.01$) of female rats treated with 3.0 and 10.0 mg/kg bw, respectively, had hepatocellular carcinomas; none was seen with lower doses in females. Cholangiocarcinomas were seen only in the highest-dose groups of males and females: 2/31 and 7/20 ($p < 0.01$). Statistically significant increases ($p < 0.01$) in the incidence of neoplastic nodules in the liver were found in females in the two highest-dose groups (5/19 and 8/20) and in males receiving 1.0 mg/kg bw (4/31), compared to 0/20 and 0/33 in female and male control rats, respectively. In addition, statistically significant increases ($p < 0.01$) in the incidence of atypical foci were observed in the livers of females in the highest dose group (8/20) and in males in the four highest dose groups (12/40, 11/31, 13/33 and 12/31, respectively) (Gupta et al., 1983a; National Toxicology Program, 1983).

(b) *Perinatal exposure*

Groups of 15 and 16 pregnant female Sherman rats were administered oral doses of 0 and 200 mg/kg bw FireMaster FF-1 (lot No. 7042) in corn oil on days 7 and 14 of gestation. All females were allowed to litter, and approximately 50 pups of each sex from each group were kept with the dams until weaning and thereafter maintained for a two-year follow-up period. The survival rate to weaning was significantly lower in the PBB-exposed group (due to loss of two and six pups in two litters); almost all pups survived to weaning in the other litters. Growth of treated male and female offspring was significantly lower than that of control offspring throughout the experiment. At two years of age, the incidences of trabecular hepatocellular carcinomas were 3/51 and 4/41 in female and male offspring exposed transplacentally and through milk; no such lesion was observed in control offspring (0/48 females, 0/42 males) (Groce & Kimbrough, 1984). [The Working Group noted that neither of these differences was statistically significant.]

(c) *Combined treatment*

PBBs (or constituent isomers) have been tested in combination with known carcinogens: with dietary 2-acetylaminofluorene in rats (Schwartz et al., 1980), by partial hepatectomy plus intraperitoneal administration of N-nitrosodiethylamine in rats (Jensen et al., 1982, 1983; Jensen, 1984; Sleight, 1985) and with 7,12-dimethylbenz[a]anthracene by skin application in mice (Berry et al., 1978). The PBBs had varied effects on the tumour-inducing effect of the carcinogen administered.

3.2 Other relevant biological data

(a) *Experimental systems*

Toxic effects

The toxicity of PBBs has been reviewed (Safe, 1984; Fries, 1985).

The LD_{50} value for a single oral dose of PBBs to rats was reported to be 21.5 g/kg bw (Michigan Chemical Corp., 1971). When rats were administered daily oral doses of 0.03, 0.1 0.3 or 1 g/kg bw FireMaster FF-1 on five days per week (for a total of 22 doses) and observed for 90 days, the LD_{50} values were 149 mg/kg bw per day (total, 3.28 g/kg bw) for males and 65 mg/kg bw per day (total, 1.43 g/kg bw) for females (Gupta & Moore, 1979). The dermal LD_{50} for hexabromobiphenyl was reported to be 5 g/kg bw in rabbits (Waritz *et al.*, 1977) and in the range of 2.15-10 g/kg bw in rats (Michigan Chemical Corp., 1971). A diet containing 6.25 mg/kg FireMaster FF-1 was lethal to 9/10 mink within an average of 210 days (Aulerich & Ringer, 1979). Pregnant Holstein heifers fed 25 g per day FireMaster BP-6 became moribund within 33-66 days (Durst *et al.*, 1977). Guinea-pigs fed a diet containing 500 mg/kg FireMaster BP-6 died within 15 days; at 100 mg/kg, only 2/6 survived for 30 days (Sleight & Sanger, 1976).

The acute oral LD_{50} for a technical preparation of octabromobiphenyl (containing 33% octabromobiphenyl and 60% nonabromobiphenyl) was reported to be more than 17 g/kg bw in male rats. Dermal administration of 10 g/kg bw of the same preparation to male rabbits produced no sign of toxicity (Waritz *et al.*, 1977). A single oral dose of 2 g/kg bw octabromobiphenyl (highest dose tested, containing 45.2% octa- and 47.4% nonabromo-biphenyls) caused no toxic effect during a 14-day follow-up in rats (Norris *et al.*, 1975). No sign of toxicity was observed in rats after inhalation exposure to 0.96 mg/l octabromo-biphenyl dust for 4 h (Waritz *et al.*, 1977).

No mortality was observed after oral or cutaneous administration of 5 g/kg bw decabromobiphenyl (96.8% pure) to rats during an observation period of 14 days, or in rats after inhalation exposure to 5 mg/l decabromobiphenyl dust for 6 h per day on five days per week for four weeks (Millischer *et al.*, 1979).

The main signs of chronic PBB poisoning in various animal species are weight loss, decrease in body-weight gain and loss of appetite (Jackson & Halbert, 1974; Durst *et al.*, 1977; Allen *et al.*, 1978; Aulerich & Ringer, 1979; Gupta *et al.*, 1981, 1983b).

In rats, morphological changes induced by FireMaster FF-1 (30-1000 mg/kg bw per day, up to 22 doses) included enlargement of the liver and atrophy of the thymus and spleen. Microscopic degenerative changes were seen in the liver, kidney, prostate and thyroid of male rats, and in the liver of females (Gupta & Moore, 1979; Gupta *et al.*, 1981). Increased tissue accumulation and urinary excretion of porphyrins have been reported in mice and rats (Kimbrough *et al.*, 1978; Goldstein *et al.*, 1979; Gupta & Moore, 1979; Gupta *et al.*, 1981; McCormack *et al.*, 1982; Gupta *et al.*, 1983b). Chronic progressive nephropathy was reported in male rats administered FireMaster FF-1 (1-10 mg/kg bw) on five days per week for six months and followed for life; gastric ulcers and hyperplastic gastropathy were also observed at the higher doses (Gupta *et al.*, 1983a).

Toxic effects in dairy cattle fed PBBs (either accidentally in the Michigan incident, or experimentally) included decreased milk production, excessive lachrymation and salivation, increased frequency in urination, abnormal hoof growth, thickening and wrinkling of the skin, matting and loss of hair, protracted delivery and underdevelopment of udders (Jackson & Halbert, 1974; Durst et al., 1977). Morphological changes included enlargement of the thymus and lymphocyte infiltration in several organs. Hepatocellular damage and degenerative changes in the kidney and testis, with decreased sperm counts, were also reported (Cook et al., 1978; Robl et al., 1978).

Weight loss and prolongation of menstrual cycles occurred within seven months in adult rhesus monkeys fed 0.3 mg/kg FireMaster FF1 in the diet. Diets containing 25 and 300 mg/kg caused alopecia, subcutaneous oedema, leukopenia, erythropenia, hypoproteinaemia, hyperplastic gastroenteritis with ulcerations, and an increase in serum alanine aminotransferase activity (Allen et al., 1978; Lambrecht et al., 1978).

In rats, 3,3',4,4',5,5'-hexabromobiphenyl was more toxic than 3,3',4,4'-tetrabromobiphenyl, which was more toxic than 2,2',5,5'-tetrabromobiphenyl (Robertson et al., 1983; Millis et al., 1985).

In rodents, increases in liver size have been demonstrated after administration of PBBs with a concomitant increase in the smooth endoplasmic reticulum of hepatic cells and increased activities of cytochrome P450-dependent monooxygenases, epoxide hydrolase and UDP-glucuronosyltransferases (Dent et al., 1976; Ahotupa & Aitio, 1978; Moore et al., 1978; Dannan et al., 1982; Parkinson et al., 1983). Microsomal enzyme induction has also been observed in rats in the F_1 and F_2 generations after perinatal exposure of the F_1 generation to PBBs (Dent et al., 1978; McCormack et al., 1981).

Congeners of PBBs differ qualitatively and quantitatively in their capacity to induce monooxygenases (Dent, 1978; Dannan et al., 1982, 1983; Parkinson et al., 1983).

FireMaster BP-6 enhanced the sensitivity of mice to hepatic and renal damage caused by chloroform and carbon tetrachloride (Kluwe et al., 1978; Ahmadizadeh et al., 1984).

Administration of FireMaster BP-6 to rats (5-500 mg/kg in the diet) for five weeks induced structural changes in the thyroid gland (Kasza et al., 1978). Decreased concentrations of serum thyroid hormones were observed in rats administered oral doses of 0.3-10 mg/kg bw per day on 125 days over a six-month period (Gupta et al., 1983a; National Toxicology Program, 1983) and in pigs administered 200 mg/kg in the diet (Werner & Sleight, 1981).

In a bioassay (see section 3.1), chronic administration to rats of FireMaster FF-1 for six months decreased haemoglobin content and mean corpuscular volume (National Toxicology Program, 1983). Similar findings and decreased erythrocyte counts were also observed following oral dosing over a 30-day period; however, comparable administration of 2,2',4,4',5,5'-hexabromobiphenyl (99% pure) had no haematological effect (Gupta et al., 1981).

FireMaster FF-1 (3 and 30 mg/kg bw per day by gavage; 22 doses) suppressed the mitogen response of splenic lymphocytes in rats and mice (Luster et al., 1978). Mice were more resistant than rats to the alterations in immune response induced by PBBs (Luster *et al.*, 1980).

Chronic administration of FireMaster FF-1 (3-30 mg/kg bw per day) caused a reduction in the performance of several neurophysiological and neurobehavioural tests by rats and mice (Tilson *et al.*, 1978; Tilson & Cabe, 1979).

Effects on reproduction and prenatal toxicity

White Leghorn chickens were exposed to FireMaster FF-1 in the diet for five weeks (0.2-3125 mg/kg in one study and 30-120 mg/kg in a second) and were observed for a further eight weeks. Feed with concentrations >125 mg/kg was rejected. Egg production, hatchability and offspring viability were reduced with levels of 45 mg/kg and above. Oedema of the abdominal and cervical regions was the predominant morphological finding in embryos and newly-hatched chicks (Polin & Ringer, 1978a,b).

In a classical teratology study, groups of 3-15 Wistar rats were administered a single oral dose of 40-800 mg/kg bw FireMaster BP-6 on one of days 6-14 of gestation. All treated females lost weight after treatment, and exposure to 800 mg/kg bw on day 7 of gestation resulted in maternal mortality. Embryonic death occurred after exposure to 800 mg/kg bw on days 7-12. The higher doses were reported to reduce fetal weight. Malformed fetuses were seen in all groups exposed to 800 mg/kg bw, regardless of the day of exposure. The most prevalent defects were cleft palate and diaphragmatic hernia after treatment on days 11-13 (Beaudoin, 1977). [The Working Group noted the lack of statistical analyses.]

In a classical teratology study, groups of Wistar rats [numbers unspecified] were administered FireMaster BP-6 orally on alternate days between day 0 and 14 of pregnancy (total doses: 0, 8, 40, 80 and 160 mg/animal). The highest dose was lethal to half of the dams, and the two higher doses appeared to interfere with implantation and cause embryonic death. It was stated that the two lowest doses markedly increased embryonic death rate, but no malformed embryo was produced (Beaudoin, 1979). [The Working Group noted the lack of statistical analyses.]

In a classical teratology study, Sprague-Dawley rats were given 0, 0.25, 0.50, 1.0, 5.0 or 10.0 mg/kg bw FireMaster BP-6 by oral intubation on days 7-15 of gestation. Group sizes varied from 30 in the controls to 6-8 in the treated groups. Doses above 0.50 mg/kg bw resulted in enlarged maternal livers; no other dose-related effect was observed in treated females or their fetuses. In a second experiment, eight pregnant rats were given 0 or 10 mg/kg bw PBBs per day on days 7-15 of gestation and offspring were examined postnatally. No effect on litter size at birth or on birth weight was observed. Pup mortality between birth and 21 days of age was 14.3% in the treated group and 1.5% in the control group. [The statistical significance of this finding was not indicated.] Growth rates of offspring from treated dams were inhibited. In a fostering study, both prenatal and postnatal components of growth retardation were observed; the greatest effect on growth was observed in pups exposed to PBBs *in utero* and nursed by treated dams (Harris *et al.*, 1978).

In a classical teratology study, groups of 9-16 Swiss ICR mice and 6-12 Sprague-Dawley rats were exposed to 0-1000 mg/kg FireMaster BP-6 in the diet on days 7-18 and 7-20 of gestation, respectively. No indication of maternal effects was reported. Mean fetal weight decreased with increasing dose. [No pairwise statistical comparison was performed.] A higher incidence of exencephaly and cleft palate was observed in treated offspring of mice when compared to a historical control group from another laboratory, but the effect was not significant when compared to concurrent controls (Corbett *et al.*, 1975).

In a classical teratology study, four groups of CD rats (23-27 per group) were fed 0, 100, 1000 or 10 000 mg/kg octabromobiphenyl in the diet on days 6-15 of gestation. Maternal body-weight gain during the exposure period was reduced in the highest-dose group. No dose-related trend in embryonic viability, growth or malformations was reported (Waritz *et al.*, 1977).

Perinatal exposure (day 8 of gestation to day 28 of lactation) of Sprague-Dawley rats to 0, 10 or 100 mg/kg FireMaster BP-6 in the diet did not affect litter size or weight at birth. However, postnatal growth and viability of the offspring were reduced with the highest dose (McCormack *et al.*, 1981).

Groups of 15-16 Sherman rats received FireMaster FF-1 (0 or 200 mg/kg bw per day) by stomach tube on days 7-14 of pregnancy. The survival rate to weaning was lower in the PBB-exposed group and, over a two-year observation period, growth of both male and female exposed offspring was reduced (Groce & Kimbrough, 1984).

Transient effects on development were observed in 10-day-old cultured embryos taken from Wistar rats administered 800 mg/kg bw FireMaster BP-6 on day 9 of gestation, but not in those exposed 4 h prior to dissection on day 10 (Fisher, 1980). Exposure *in vivo* on day 9 of gestation, 24 h prior to dissection, interfered with axial rotation, neural tube closure, limb bud development and somite differentiation in embryos cultured *in vitro*. Fewer effects on in-vitro development were observed in embryos exposed *in vivo* 4 h prior to dissection on day 10. The observed pattern of developmental alterations did not correlate with the effects of PBBs that had been observed previously *in vivo* at termination of pregnancy (Beaudoin, 1977; Beaudoin & Fisher, 1981).

Groups of six pregnant Holstein heifers were given daily oral doses of 0, 0.25, 250 or 25 000 mg FireMaster BP-6 for 60 days or until they became moribund; another pregnant heifer was given daily oral doses of 250 mg for 180 days. [The days of pregnancy on which dosing commenced were not specified.] Four non-pregnant heifers were given 250 mg FireMaster BP-6 for 202 days and bred immediately following dosing. All heifers fed 25 000 mg became moribund. Some treated heifers were rebred over an approximately five-year period. Fifteen of 75 calves born to treated cows or their offspring died or were stillborn; ten deaths occurred among calves born to heifers treated with 250 mg or to their offspring. Dystocia was the major cause of calf mortality; no other effect attributable to exposure to PBBs was observed in the offspring (Durst *et al.*, 1977; Willett *et al.*, 1982).

Seven rhesus monkeys fed a diet containing 0.3 mg/kg FireMaster FF-1 were mated after seven months of treatment. One animal aborted and one delivered a stillborn infant on day 154; the remainder delivered offspring at term (Allen *et al.*, 1978).

Absorption, distribution, excretion and metabolism

More than 90% of ^{14}C-2,2',4,4',5,5'-hexabromobiphenyl (99% pure; single 1 mg/kg bw dose or multiple doses up to 30 mg/kg bw) was absorbed from the gastrointestinal tract in rats. Tissue distribution was similar following oral or intravenous administration. After a single intravenous dose (1 mg/kg bw), 2,2',4,4',5,5'-hexabromobiphenyl was initially distributed into highly perfused tissues, but later redistributed to adipose tissue; within four days, adipose tissue contained >60% of the total body burden. Excretion was slow: after an intravenous dose, cumulative excretion in the faeces was 6.6% by day 42; excretion in the urine was <0.1% of the dose (Matthews *et al.*, 1977; Tuey & Matthews, 1978, 1980). Qualitatively similar results were obtained in rats given FireMaster FF-1 (Domino *et al.*, 1980, 1982). A half-time for the disappearance of 2,2',4,4',5,5'-hexabromobiphenyl from the blood of rats given a single oral dose of 10 mg/kg bw FireMaster FF-1 was 145 days (Domino *et al.*, 1982). Apart from 2,2',4,5,5'-pentabromobiphenyl, for which disappearance was more rapid, all the other major components of FireMaster FF-1 closely resembled 2,2',4,4',5,5'-hexabromobiphenyl in their kinetics (Domino *et al.*, 1980). In dairy cows exposed to FireMaster BP-6 in the Michigan incident, concentrations of PBBs in adipose tissue differed in different parts of the body (Fries *et al.*, 1978a).

After oral administration, 3,3',4,4'-tetrabromobiphenyl disappeared from the liver and fat of rats much faster than did 3,3',4,4',5,5'-hexabromobiphenyl (Millis *et al.*, 1985).

When pregnant guinea-pigs were given an intragastric dose of FireMaster FF-1 on approximately day 65 of gestation, the concentrations of 2,2',4,4',5,5'-hexabromobiphenyl in adipose tissue at term (67-68 days) were similar in fetuses and dams. When the dams were given a dose of FireMaster FF-1 on day 1 after delivery, the concentrations of 2,2',4,4',5,5'-hexabromobiphenyl in the lungs, liver, kidneys and fat of the pups were similar to those in the dams (Ecobichon *et al.*, 1983). When pregnant rats were given 50 mg/kg FireMaster BP-6 in the diet from day 8-21 of gestation, hepatic concentrations of total hexabromobiphenyls in pups at day 21 of gestation were on average approximately 1/20 of those in the dams. However, in pups of nontreated dams nursed by dams fed 50 mg/kg FireMaster BP-6 on days 1-14 *post partum*, hepatic concentrations of total hexabromobiphenyls were on average approximately eight times higher than those in the dams at day 14 *post partum* (Rickert *et al.*, 1978).

Tissue bromine concentrations were studied in rats fed 1, 10, 100 or 1000 mg/kg technical octabromobiphenyl in the diet for four weeks. The bromine content in the liver and muscles reached a constant level within two weeks, whereas the content in fat continued to rise for four weeks. During six weeks after cessation of treatment, bromine concentrations in the liver and muscle gradually decreased, whereas the bromine content in the fat of the highest-dose group continued to increase during 18 weeks of recovery (Lee *et al.*, 1975). When rats were fed 0.1 mg/kg bw per day technical octabromobiphenyl for 180 days, the bromine concentration increased throughout the treatment period in both the liver and fat. After treatment of rats with 1 mg/kg bw per day for 90 days, the bromine content in the liver decreased rapidly, whereas practically no decrease was seen in fat, during 90 days' recovery. After administration of a single oral dose (1 mg/kg bw) of ^{14}C-octabromobiphenyl to rats, 61.9% of the dose was found in the faeces during the first 24 h, and a further 7% during the next 24 h. After 16 days, 26% of the dose was retained in the body. Less than 1% of the radioactivity was detected in urine or expired air (Norris *et al.*, 1975).

When dairy cows were fed 10 mg FireMaster BP-6 per day, the concentrations of hexa-, hepta- and octabromobiphenyls in the milk reached a steady state within 30 days; the level of hexabromobiphenyl was approximately 25 times higher than that of hepta- plus octabromobiphenyls, whereas the ratio of hexabromobiphenyl:hepta- plus octabromobiphenyl in the feed was approximately 5. After cessation of treatment, a fast initial disappearance phase occurred, and the concentration of hepta- and octabromobiphenyls in the milk fell to below the limit of detection. Following its initial rapid elimination, hexabromobiphenyl disappeared more slowly, with a half-time of approximately 60 days (Fries & Marrow, 1975; Fries et al., 1978b).

Little metabolism of several tetra-, penta- or hexabromobiphenyls was demonstrated in rat-liver microsomes in vitro (Mills et al., 1985). When microsomes from animals treated with phenobarbital or methylcholanthrene were used, metabolism of some di-, tri- and tetrabromobiphenyls was demonstrated (Millis et al., 1985; Mills et al., 1985).

Following incubation of ^{14}C-PBBs with rat-liver microsomes, no binding to exogenous DNA was detected, and the extent of irreversible binding to protein was very small (Dannan et al., 1978).

Mutagenicity and other short-term tests

Hexabromobiphenyl (tested at up to 10 000 μg/plate) was not mutagenic to *Salmonella typhimurium* TA1535, TA1537, TA98 or TA100 (preincubation procedure) in the presence or absence of an Aroclor-induced rat- or hamster-liver exogenous metabolic system (S9) (Haworth et al., 1983).

A commercial PBB flame retardant, 'Decabromobiphenyle' (96.8% decabromobiphenyl, 2.9% nonabromobiphenyl and 0.3% octabromobiphenyl), was not mutagenic to *S. typhimurium* TA1535, TA1537 or TA1538 when assayed in spot tests at up to 1000 μg/plate, both in the presence and absence of Aroclor-induced rat-liver S9. The compound also gave negative results in the host-mediated assay with strain TA1538, in mice receiving oral doses of 5, 10 and 20 g/kg bw (Millischer et al., 1979).

When tested at concentrations of 10^{-7} to 10^{-3}M, FireMaster FF-1 did not induce DNA repair synthesis (as measured by autoradiography) in primary cultures of hepatocytes from rats, mice or hamsters. 6-Thioguanine-resistant mutants were not induced by concentrations of 5×10^{-6} to 10^{-3} M in a rat-liver epithelial-cell line, nor by concentrations of 10^{-5} to 10^{-4} M in human fibroblasts cocultivated with rat hepatocytes as the metabolic system (Williams et al., 1984). [No data on survival were given.] Kavanagh et al. (1985) tested several PBBs for mutation of Chinese hamster V79 cells to 6-thioguanine (HGPRT locus)- and oubain (Na-K ATPase locus)-resistance and for mutation of WB rat-liver cells at the HGPRT locus. FireMaster BP-6 (1-40 μg/ml) was tested for mutation at both loci in V79 cells with and without a postmitochondrial supernatant fraction (S15) from Aroclor-induced rat liver; 3,3',4,4'-tetrabromobiphenyl (1-10 μg/ml) was tested for mutation at the HGPRT locus in V79 cells with and without S15; 2,2',4,4',5,5'-hexabromobiphenyl was tested for mutation at the HGPRT locus in V79 cells (20-50 μg/ml) and in WB cells (5-20 μg/ml); and 3,3',4,4',5,5'-hexabromobiphenyl was tested for mutation at the HGPRT locus in V79 cells (7-12 μg/ml) and in WB cells (1-5 μg/ml). All the chemicals were tested to levels that reduced cell survival, except FireMaster BP-6, with which survival was unaffected. No mutagenicity was seen.

It was reported in an abstract (Mirsalis et al., 1985) that PBBs [composition not given] did not induce unscheduled DNA synthesis (as measured by autoradiography) in an in-vivo/in-vitro study in $B6C3F_1$ mice [details not given].

No chromosomal abnormality was seen in bone-marrow or spermatogonial cells of rats fed diets containing 5, 50 or 500 mg/kg FireMaster BP-6 for 34-37 days (Garthoff et al., 1977). [The Working Group noted that an insufficient number of cells were examined in the spermatogonial test.] The commercial flame retardant 'Decabromobiphenyle' did not increase the number of micronuclei in polychromatic erythrocytes in mouse bone marrow after total oral doses of 5, 10 or 20 g/kg bw given as two doses 24 h apart (Millischer et al., 1979).

Wertz and Ficsor (1978) studied chromosomal aberrations in bone marrow of mice administered by gavage 0, 50, or 500 mg/kg bw FireMaster [FF-1] from a bag found in the grain mill where the contamination occurred in the Michigan accident. No increase in the incidence of chromosomal or chromatid breaks was observed, but after 24 h and 48 h, 50 mg/kg bw and 500 mg/kg bw, respectively, significantly increased the number of gaps.

Intercellular communication between Chinese hamster V79 cells, measured by metabolic cooperation between $HGPRT^+$ and $HGPRT^-$ cells, was inhibited by FireMaster BP-6 (Trosko et al., 1981). Tsushimoto et al. (1982) studied seven PBB congeners (2,2',4,5,5'-penta, 3,3',4,5,5'-penta, 2,3',4,4',5,5'-hexa-, 2,2',4,4',5,5'-hexa-, 3,3',4,4',5,5'-hexa-, 2,2',3,4,4',5,5'-hepta-, and 2,2',3,3',4,4',5,5'-octabromobiphenyl) purified from commercial products for their ability to inhibit metabolic cooperation in Chinese hamster V79 cells. Those congeners with two bromine substitutions on carbons in positions *ortho* to the biphenyl bridge carbons (leading to a co-planar configuration of the phenyl groups) caused a dose-dependent inhibition of metabolic cooperation at non-toxic concentrations. Congeners in which only one *ortho* carbon was brominated were cytotoxic at high levels but appeared to inhibit metabolic cooperation before cytotoxicity occurred. The 3,3',4,5,5'-pentabromobiphenyl, with no *ortho* substitution, showed only a cytotoxic effect. FireMaster FF-1 (7.5×10^{-5} to 7.5×10^{-4} M) inhibited gap-junctional communication between primary cultures of rat hepatocytes and a 6-thioguanine-resistant, adult rat-liver epithelial-cell line (Williams et al., 1984). [Assays measuring inhibition of metabolic cooperation are considered as possible tests to detect promoting activity of chemicals (Griesemer et al., 1986)].

(b) Humans

Toxic effects

Most of the data on the toxicity of PBBs are derived from studies carried out after the contamination of human food through their unintentional addition (as FireMaster FF-1, lot no. 7042), instead of magnesium oxide, to farm feed in Michigan in 1973, resulting in exposure of large numbers of the rural population Michigan.

Higher rates of dermatological, neurological and musculoskeletal disorders were reported in a group of 933 Michigan farmers and residents than in 229 unexposed Wisconsin farmers used as controls (Anderson et al., 1978a), and a high prevalence of abnormal liver function tests (serum glutamic oxaloacetic transaminase and serum glutamic pyruvic transaminase) was observed among 614 Michigan adults compared with 141 Wisconsin adults (Anderson et al., 1978b).

A variety of tests have been used to assess the immune status of PBB-exposed Michigan farmers. The prevalence of carcinoembryonic antigen titres (>2.5 ng/ml) was slightly increased in the Michigan populations compared to the Wisconsin population used as controls and was associated with higher serum levels of PBBs (Anderson et al., 1978c). In addition to reduced numbers of T- and B-lymphocytes in peripheral blood, decreases in the response to T- and B-lymphocyte mitogens and in a T-lymphocyte proliferation test were found in up to 40% of the exposed subjects in comparison with control populations of non-exposed Wisconsin farmers and New York City residents (Bekesi et al., 1978). The immune status of Michigan farmers was re-examined in 1981: 17/40 subjects still suffered from an altered immune state, as indicated by lowered number of T-lymphocytes and reduced response to T-lymphocyte mitogens. Increases in immunoglobulin A and G levels which were seen in 1976 were still present in 1981. The persistent change in immunocompetence suggests either permanent damage to the immune system or continuing action of PBBs persisting in the adipose tissues (Roboz et al., 1985).

The toxicity of PBBs in children was studied in 1976-1977 by comparing the symptoms reported in 343 Michigan children to those in 72 Wisconsin control children. During the years prior to 1972, no appreciable difference was observed, whereas an increase in neurobehavioural, gastrointestinal and immunological disorders was evident in Michigan children from 1974-1976; they declined somewhat from 1975 to 1976 (Barr, 1978).

A group of 55 workers who had been actively employed in the Michigan plant producing FireMaster BP-6 from 1970-1974 volunteered for medical examination; all had serum levels of PBBs above 1 mg/l. An increased prevalence of respiratory symptoms and skin disorders was reported in this group compared to the available data on PBB-exposed farmers in Michigan (Anderson et al., 1978d).

Among 35 workers in a plant manufacturing decabromobiphenyl and its oxide from 1973-1978 and who volunteered for a comprehensive medical examination, there were four cases of subclinical primary hypothyroidism; no such disorder was detected among 89 non-exposed referent subjects (Bahn et al., 1980).

Effects on reproduction and prenatal toxicity

A study of sperm counts, motility and morphology was conducted on two groups of subjects exposed to PBBs (11 workers involved in manufacture and 41 Michigan farmers and users) and a reference group of 52 graduate students. No major difference was found among the three groups (Rosenman et al., 1979). [The Working Group noted the low power of the study to detect an effect of PBBs on spermatogenesis.]

Absorption, distribution, excretion and metabolism

In necropsy analyses of Michigan residents, the highest concentrations of PBBs were detected in perinephric fat. Ratios of the concentrations in various organs to that in fat were: 0.45-0.56 for adrenal glands, atheromatous aorta and thymus; 0.1-0.28 for pancreas, liver and left ventricle of the heart; and 0.02-0.09 for kidneys, lung, brain, skeletal muscle, thyroid gland and nonatheromatous aorta (Miceli *et al.*, 1985). In studies on Michigan residents, the ranges of ratios of PBB levels in adipose tissue:blood were 140-260 in women and in male workers in a chemical plant manufacturing PBBs and 320-329 in male farm workers and other men; the range of the ratios of cord blood:maternal serum was 0.1-0.14, and that of milk:maternal serum was 107-119 (Eyster *et al.*, 1983). In samples collected from Michigan residents between 1974 and 1977, the range of the ratios of adipose tissue:serum was 320-358 (Wolff *et al.*, 1979; Tuey & Matthews, 1980); in lactating women, the ratio was 100 (Brilliant *et al.*, 1978). In a population study in Michigan, the serum PBB concentrations were lower in women than in men (Kreiss *et al.*, 1982). Faecal and bile PBB concentrations were lower than serum concentrations in Michigan residents (Eyster *et al.*, 1983).

No decrease in serum PBB concentrations was detected over a period of 12 months in 92 Michigan farmers exposed during the 1973 accident (Wolff *et al.*, 1979). In a study on 816 Michigan residents followed for one year (1977-1978) and on 832 followed for two years (1977-1979), the median serum PBB concentration decreased from 4 to 3 $\mu g/l$ in both groups (Kreiss *et al.*, 1982). The half-time for PBBs in human fat has been estimated to be at least 7.8 years (Miceli *et al.*, 1985).

For levels of PBBs in human tissues, see also section 2.2(*g*).

Mutagenicity and chromosomal effects

No data were available to the Working Group.

3.3 Case reports and epidemiological studies of carcinogenicity to humans

The cohort studied by Wong *et al.* (1984) (described in the monograph on methyl bromide, p. 187) also had exposure to PBBs. Due to the lack of quantitative data, potential exposures of workers to PBBs were categorized as 'routine' and 'non-routine'. Of the 91 workers potentially exposed on a 'routine' basis, none died during the study period; among the 237 'non-routinely' exposed male workers, two deaths were observed *versus* 6.4 expected, one of which was due to cancer of the large intestine (1 observed *versus* 0.1 expected).

4. Summary of Data Reported and Evaluation

4.1 Exposure data

Polybrominated biphenyls were produced primarily in the 1970s as flame retardants. Occupational exposure occurred during their production and use. The accidental addition of a hexabromobiphenyl product (FireMaster FF-1) to a large quantity of farm animal feed in Michigan (USA) in 1973 resulted in widespread population exposure.

4.2 Experimental data

The carcinogenicity of a commercial preparation of polybrominated biphenyls (FireMaster FF-1, various lots), composed primarily of hexabromobiphenyl with smaller amounts of penta- and heptabrominated isomers, was tested by oral administration by gavage in one strain of mice and two strains of rats and by perinatal exposure in one strain of rats. It produced malignant hepatic tumours in mice of each sex. In rats of each sex, it produced benign and malignant hepatic tumours, including cholangiocarcinomas, after single or multiple administrations.

Short-term exposures of rats to high doses of commercial preparations of polybrominated biphenyls induce embryonic death, growth retardation and malformation; malformations have not been observed following exposure to lower doses throughout gestation. Pre- and perinatal exposure of rats to polybrominated biphenyls reduces postnatal growth and viability.

Various commercial preparations of polybrominated biphenyls are not mutagenic to bacteria in the presence or absence of exogenous metabolic systems or in a host-mediated assay. Polybrominated biphenyls do not induce DNA damage, or mutation, in cultured mammalian cells or chromosomal aberrations in rat or mouse bone marrow, nor micronuclei in mouse bone marrow, but they do inhibit junctional intercellular communication in cultured mammalian cells.

Overall assessment of data from short-term tests: Polybrominated biphenyls[a]

	Genetic activity			Cell transformation
	DNA damage	Mutation	Chromosomal effects	
Prokaryotes		–		
Fungi/Green plants				
Insects				
Mammalian cells (*in vitro*)	–	–		
Mammals (*in vivo*)			–	
Humans (*in vivo*)				
Degree of evidence in short-term tests for genetic activity: **No evidence**				Cell transformation: No data

[a]The groups into which the table is divided and the symbol '–' are defined on pp. 19-20 of the Preamble; the degrees of evidence are defined on pp. 20-21.

4.3 Human data

No relevant data were available to evaluate the reproductive effects or prenatal toxicity of polybrominated biphenyls to humans.

A small cohort study of chemical workers potentially exposed to polybrominated biphenyls together with other chemicals was uninformative with regard to cancer.

4.4 Evaluation[1]

There is *sufficient evidence*[2] for the carcinogenicity of commercial mixtures of polybrominated biphenyls to experimental animals.

There is *inadequate evidence* for the carcinogenicity of polybrominated biphenyls to humans.

5. References

Ahmadizadeh, M., Kuo, C.-H., Echt, R. & Hook, J.B. (1984) Effect of polybrominated biphenyls, *beta*-naphthoflavone and phenobarbital on arylhydrocarbon hydroxylase activities and chloroform-induced nephrotoxicity and hepatotoxicity in male C47bl/6J and DBA/2J mice. *Toxicology*, *31*, 343-352

Ahotupa, M. & Aitio, A. (1978) Effect of polybrominated biphenyls on drug metabolizing enzymes in different tissues of C57 mice. *Toxicology*, *11*, 309-314

Allen, J.R., Lambrecht, L.K. & Barsotti, D.A. (1978) Effects of polybrominated biphenyls in nonhuman primates. *J. Am. vet. Med. Assoc.*, *173*, 1485-1489

Anderson, H.A., Lilis, R., Selikoff, I.J., Rosenman, K.D., Valciukas, J.A. & Freedman, S. (1978a) Unanticipated prevalence of symptoms among dairy farmers in Michigan and Wisconsin. *Environ. Health Perspect.*, *23*, 217-226

Anderson, H.A., Holstein, E.C., Daum, S.M., Sarkozi, L. & Selikoff, I.J. (1978b) Liver function tests among Michigan and Wisconsin dairy farmers. *Environ. Health Perspect.*, *23*, 333-339

Anderson, H.A., Rosenman, K.D. & Snyder, J. (1978c) Carcinoembryonic antigen (CEA) plasma levels in Michigan and Wisconsin dairy farmers. *Environ. Health Perspect.*, *23*, 193-197

Anderson, H.A., Wolff, M.S., Fischbein, A. & Selikoff, I.J. (1978d) Investigation of the health status of Michigan Chemical Corporation employees. *Environ. Health Perspect.*, *23*, 187-191

---

[1]For definition of the italicized terms, see Preamble, pp. 18 and 22.
[2]In the absence of adequate data on humans, it is reasonable, for practical purposes, to regard chemicals or exposures for which there is *sufficient evidence* of carcinogenicity in animals as if they presented a carcinogenic risk to humans.

Anon. (1979) Registration SOR/79-351, April 1977 (P.C. 1979-1206, April 11, 1979). *Canada Gaz.* (Part II), *113*, 1563

Archer, S.R., Blackwood, T.R. & Collins, C.S. (1979) *Status Assessment of Toxic Chemicals: Polybrominated Biphenyls (EPA-600/2-79-210k)*, Cincinnati, OH, US Environmental Protection Agency

Aulerich, R.J. & Ringer, R.K. (1979) Toxic effects of dietary polybrominated biphenyls on mink. *Arch. environ. Contam. Toxicol.*, *8*, 487-498

Bahn, A.K., Mills, J.L., Snyder, P.J., Gann, P.H., Houten, L., Bialik, O., Hollmann, L. & Utiger, R.D. (1980) Hypothyroidism in workers exposed to polybrominated biphenyls. *New Engl. J. Med.*, *302*, 31-33

Baker, M.J., Gandenberger, C.L., Gandenberger, R. & Merz, J.B., eds (1980) *Chemical Sources Europe*, Mountain Lakes, NJ, Chemical Sources Europe, p. 377

Barr, M., Jr (1978) Pediatric health aspects of PBBs. *Environ. Health Perspect.*, *23*, 291-294

Beaudoin, A.R. (1977) Teratogenicity of polybrominated biphenyls in rats. *Environ. Res.*, *14*, 81-86

Beaudoin, A.R. (1979) Embryotoxicity of polybrominated biphenyls. *Teratol. Testing*, *2*, 211-222

Beaudoin, A.R. & Fisher, D.L. (1981) An in vivo/in vitro evaluation of teratogenic action. *Teratology*, *23*, 57-61

Bekesi, J.G., Holland, J.F., Anderson, H.A., Fischbein, A.S., Rom, W., Wolff, M.S. & Selikoff, I.J. (1978) Lymphocyte function of Michigan dairy farmers exposed to polybrominated biphenyls. *Science*, *199*, 1207-1209

Berry, D.L., DiGiovanni, J., Juchau, M.R., Bracken, W.M., Gleason, G.L. & Slaga, T.J. (1978) Lack of tumor-promoting ability of certain environmental chemicals in a two-stage mouse skin tumorigenesis assay. *Res. Commun. chem. Pathol. Pharmacol.*, *20*, 101-108

Brilliant, L.B., Wilcox, K., van Amburg, G., Eyster, J., Isbister, J., Bloomer, A.W., Humphrey, H. & Price, H. (1978) Breast-milk monitoring to measure Michigan's contamination with polybrominated biphenyls. *Lancet*, *ii*, 643-646

Burse, V.W., Needham, L.L., Liddle, J.A., Bayse, D.D. & Price, H.A. (1980) Interlaboratory comparison for results of analysis for polybrominated biphenyls in human serum. *J. anal. Toxicol.*, *4*, 22-26

Cook, H., Helland, D.R., Van der Weele, B.H. & DeJong, R.J. (1978) Histotoxic effects of polybrominated biphenyls in Michigan dairy cattle. *Environ. Res.*, *15*, 82-89

Corbett, T.H., Beaudoin, A.R., Cornell, R.G., Anver, M.R., Schumacher, R., Endres, J. & Szwabowska, M. (1975) Toxicity of polybrominated biphenyls (Firemaster BP-6) in rodents. *Environ. Res.*, *10*, 390-396

Dannan, G.A., Moore, R.W. & Aust, S.D. (1978) Studies on the microsomal metabolism and binding of polybrominated biphenyls (PBBs). *Environ. Health Perspect.*, *23*, 51-61

Dannan, G.A., Sleight, S.D. & Aust, S.D. (1982) Toxicity and microsomal enzyme induction effects of several polybrominated biphenyls of Firemaster. *Fundam. appl. Toxicol.*, *2*, 313-321

Dannan, G.A., Guengerich, F.P., Kaminsky, L.S. & Aust, S.D. (1983) Regulation of cytochrome P-450. Immunochemical quantitation of eight isoenzymes in liver microsomes of rats treated with polybrominated biphenyls congeners. *J. biol. Chem.*, *258*, 1282-1288

De Kok, J.J., De Kok, A., Brinkman, U.A.T. & Kok, R.M. (1977) Analysis of polybrominated biphenyls. *J. Chromatogr.*, *142*, 367-383

Dent, J.G. (1978) Characteristics of P-450 and mixed function oxidase enzymes following treatment with PBBs. *Environ. Health Perspect.*, *23*, 301-307

Dent, J.G., Netter, K.J. & Gibson, J.E. (1976) The induction of hepatic microsomal metabolism in rats following acute administration of a mixture of polybrominated biphenyls. *Toxicol. appl. Pharmacol.*, *38*, 237-249

Dent, J.G., McCormack, K.M., Rickert, D.E., Cagen, S.Z., Melrose, P. & Gibson, J.E. (1978) Mixed function oxidase activities in lactating rats and their offspring following dietary exposure to polybrominated biphenyls. *Toxicol. appl. Pharmacol.*, *46*, 727-735

Di Carlo, F.J., Seifter, J. & DeCarlo, V.J. (1978) *Assessment of the Hazards of Polybrominated Biphenyls* (*EPA-560/6-77-037; PB 285 532*), Washington DC, US Environmental Protection Agency

Domino, E.F., Fivenson, D.P. & Domino, S.E. (1980) Differential tissue distribution of various polybrominated biphenyls of Firemaster FF-1 in male rats. *Drug Metab. Disposition*, *8*, 332-336

Domino, L.E., Domino, S.E. & Domino, E.F. (1982) Toxicokinetics of 2,2',4,4',5,5'-hexabromobiphenyl in the rat. *J. Toxicol. environ. Health*, *9*, 815-833

Durst, H.I., Willett, L.B., Brumm, C.J. & Mercer, H.D. (1977) Effects of polybrominated biphenyls on health and performance of pregnant Holstein heifers. *J. Dairy Sci.*, *60*, 1294-1300

Eckhoff, M.A., Ridgeway, T.H. & Caruso, J.A. (1983) Polychromator system for multielement determination by gas chromatography with helium plasma atomic emission spectrometric detection. *Anal. Chem.*, *55*, 1004-1009

Ecobichon, D.J., Hidvegi, S., Comeau, A.M. & Cameson, P.H. (1983) Transplacental and milk transfer of polybrominated biphenyls to perinatal guinea pigs from treated dams. *Toxicology*, *28*, 51-63

Erickson, M.D., Zweidinger, R.A. & Pellizzari, E.D. (1977) *Analysis of a Series of Samples for Polybrominated Biphenyls (PBBs)* (*EPA-560/6-77-020; US NTIS PB-273196*), Washington DC, US Environmental Protection Agency

Erney, D.R. (1975) Confirmation of polybrominated biphenyl residues in feeds and dairy products, using an ultraviolet irradiation-gas-liquid chromatographic technique. *J. Assoc. off. anal. Chem.*, *58*, 1202-1205

Eyster, J.T., Humphrey, H.E.B. & Kimbrough, R.D. (1983) Partitioning of polybrominated biphenyls (PBBs) in serum, adipose tissue, breast milk, placenta, cord blood, biliary fluid, and feces. *Arch. environ. Health*, *38*, 47-53

Fehringer, N.V. (1975) Determination of polybrominated biphenyl residues in dairy products. *J. Assoc. off. anal. Chem.*, *58*, 978-982

Fisher, D.L. (1980) Effect of polybrominated biphenyls on the accumulation of DNA, RNA, and protein in cultured rat embryos following maternal administration. *Environ. Res.*, *23*, 334-340

Fries, G.F. (1985) The PBB episode in Michigan: an overall appraisal. *Crit. Rev. Toxicol.*, *16*, 105-156

Fries, G.F. & Marrow, G.S. (1975) Excretion of polybrominated biphenyls into the milk of cows. *J. Dairy Sci.*, *58*, 947-951

Fries, G.F., Cook, R.M. & Prewitt, L.R. (1978a) Distribution of polybrominated biphenyls residues in the tissues of environmentally contaminated dairy cows. *J. Dairy Sci.*, *61*, 420-425

Fries, G.F., Marrow, G.S. & Cook, R.M. (1978b) Distribution and kinetics of PBB residues in cattle. *Environ. Health Perspect.*, *23*, 43-50

Garthoff, L.H., Friedman, L., Farber, T.M., Locke, K.K., Sobotka, T.J., Green, S., Hurley, N.E., Peters, E.L., Story, G.E., Moreland, F.M., Graham, C.H., Keys, J.E., Taylor, M.J., Scalera, J.V., Rothlein, J.E., Marks, E.M., Cerra, F.E., Rodi, S.B. & Sporn, E.M. (1977) Biochemical and cytogenetic effects in rats caused by short-term ingestion of Aroclor 1254 or FireMaster BP6. *J. Toxicol. environ. Health*, *3*, 769-796

Goldstein, J.A., Linko, P.C., Levy, L.A., McKinney, J.D., Gupta, B.N. & Moore, J.A. (1979) A comparison of a commercial polybrominated biphenyl mixture, 2,4,5,2',4',5'-hexabromobiphenyl and 2,3,6,7-tetrabromonaphthalene as inducers of liver microsomal drug-metabolizing enzymes. *Biochem. Pharmacol.*, *28*, 2947-2956

Groce, D.F. & Kimbrough, R.D. (1984) Stunted growth, increased mortality, and liver tumors in offspring of polybrominated biphenyls (PBB) dosed Sherman rats. *J. Toxicol. environ. Health*, *14*, 695-706

Gupta, B.N. & Moore, J.A. (1979) Toxicologic assessments of a commercial polybrominated biphenyl mixture in the rat. *Am. J. vet. Res.*, *40*, 1458-1468

Gupta, B.N., McConnell, E.E., Harris, M.W. & Moore, J.A. (1981) Polybrominated biphenyl toxicosis in the rat and mouse. *Toxicol. appl. Pharmacol.*, *57*, 99-118

Gupta, B.N., McConnell, E.E., Moore, J.A. & Haseman, J.K. (1983a) Effects of a polybrominated biphenyl mixture in the rat and mouse. II. Life-time study. *Toxicol. appl. Pharmacol.*, *68*, 19-35

Gupta, B.N., McConnell, E.E., Goldstein, J.A., Harris, M.W. & Moore, J.A. (1983b) Effects of polybrominated biphenyl mixture in the rat and mouse. I. Six-month exposure. *Toxicol. appl. Pharmacol.*, *68*, 1-18

Harris, S.J., Cecil, H.C. & Bitman, J. (1978) Embryotoxic effects of polybrominated biphenyls (PBB) in rats. *Environ. Health Perspect.*, *23*, 295-300

Hass, J.R., McConnell, E.E. & Harvan, D.J. (1978) Chemical and toxicological evaluation of FireMaster BP-6. *J. agric. Food Chem.*, *26*, 94-99

Haworth, S., Lawlor, T., Mortelmans,K., Speck, W. & Zeiger, E. (1983) *Salmonella* mutagenicity test results for 250 chemicals. *Environ. Mutagenesis, Suppl. 1*

Hesse, J.L. (1975) Polybrominated biphenyls: an agricultural incident and its consequences. *Trace Subst. environ. Health*, *9*, 63

Humphrey, H.E.B. & Hayner, N.S. (1975) Polybrominated biphenyls: an agricultural incident and its consequences. II. An epidemiological investigation of human exposure. *Trace Subst. environ. Health*, *9*, 57-62

IARC (1978) *IARC Monographs on the Evaluation of the Carcinogenic Risk of Chemicals to Humans*, Vol. 18, *Polychlorinated Biphenyls and Polybrominated Biphenyls*, Lyon, pp. 107-124

Jackson, T.F. & Halbert, F.L. (1974) A toxic syndrome associated with the feeding of polybrominated biphenyl-contaminated protein concentrate to dairy cattle. *J. Am. vet. Med. Assoc.*, *165*, 437-439

Jacobs, L.W., Chou, S.-F. & Tiedje, J.M. (1976) Fate of polybrominated biphenyls (PBBs) in soils. Persistence and plant uptake. *J. agric. Food Chem.*, *24*, 1198-1201

Jensen, R.K. (1984) *Pathologic Effects and Hepatic Tumor Promoting Ability of Fire-Master BP-6, 3,3',4,4',5,5'-Hexabromobiphenyl and 2,2',4,4',5,5'-Hexabromobiphenyl in the Rat*, PhD Thesis, East Lansing, MI, Michigan State University [*Diss. Abstr. int.*, *44*, 2712-B]

Jensen, R.K., Sleight, S.D., Goodman, J.I., Aust, S.D. & Trosko, J.E. (1982) Polybrominated biphenyls as promoters in experimental hepatocarcinogenesis in rats. *Carcinogenesis*, *3*, 1183-1186

Jensen, R.K., Sleight, S.D., Aust, S.D., Goodman, J.I. & Trosko, J.E. (1983) Hepatic tumor-promoting activity of 3,3',4,4',5,5'-hexabromobiphenyl: The interrelationship between toxicity, induction of hepatic microsomal drug metabolizing enzymes, and tumor-promoting ability. *Toxicol. appl. Pharmacol.*, *71*, 163-176

Kasza, L., Collins, W.T., Capen, C.C., Garthoff, L.H. & Friedman, L. (1978) Comparative toxicity of polychlorinated biphenyl and polybrominated biphenyl in the rat thyroid gland: Light and electron microscopic alterations after subacute dietary exposure. *J. environ. Pathol. Toxicol.*, *1*, 587-599

Kavanagh, T.J., Rubinstein, C., Liu, P.L., Chang, C.-C., Trosko, J.E. & Sleight, S.D. (1985) Failure to induce mutations in Chinese hamster V79 cells and WB rat liver cells by the polybrominated biphenyls, FireMaster BP-6, 2,2',4,4'5,5'-hexabromobiphenyl, 3,3',4,4',5,5'-hexabromobiphenyl and 3,3',4,4'-tetrabromobiphenyl. *Toxicol. appl. Pharmacol.*, *79*, 91-98

Kay, K. (1977) Polybrominated biphenyls (PBB) environmental contamination in Michigan, 1973-1976. *Environ. Res.*, *13*, 74-93

Kimbrough, R.D., Burse, V.W., Liddle, J.A. & Fries, G.F. (1977) Toxicity of polybrominated biphenyl. *Lancet*, *ii*, 602-603

Kimbrough, R.D., Burse, V.W. & Liddle, J.A. (1978) Persistent liver lesions in rats after a single oral dose of polybrominated biphenyls (FireMaster FF-1) and concomitant PBB tissue levels. *Environ. Health Perspect.*, *23*, 265-273

Kimbrough, R.D., Groce, D.F., Korver, M.P. & Burse, V.W. (1981) Induction of liver tumors in female Sherman strain rats by polybrominated biphenyls. *J. natl Cancer Inst.*, *66*, 535-542

Kluwe, W.M., McCormack, K.M. & Hook, J.B. (1978) Potentiation of hepatic and renal toxicity of various compounds by prior exosure to polybrominated biphenyls. *Environ. Health Perspect.*, *23*, 241-246

Kreiss, K., Roberts, C. & Humphrey, H.E.B. (1982) Serial PBB levels, PCB levels, and clinical chemistries in Michigan's PBB cohort. *Arch. environ. Health*, *37*, 141-147

Lambrecht, L.K., Barsotti, D.A. & Allen, J.R. (1978) Responses of nonhuman primates to a polybrominated biphenyl mixture. *Environ. Health Perspect.*, *23*, 139-145

Landrigan, P.J., Wilcox, K.R., Jr, Silva, J., Jr, Humphrey, H.E.B., Kauffman, C. & Heath, C.W., Jr (1979) Cohort study of Michigan residents exposed to polybrominated biphenyls: epidemiologic and immunologic findings. *Ann. N.Y. Acad. Sci.*, *320*, 284-294

Lee, K.P., Herbert, R.R., Sherman, H., Aftosmis, J.G. & Waritz, R.S. (1975) Bromine tissue residue and hepatotoxic effects of octabromobiphenyl in rats. *Toxicol. appl. Pharmacol.*, *34*, 115-127

Luster, M.I., Faith, R.E. & Moore, J.A. (1978) Effects of polybrominated biphenyls (PBB) on immune response in rodents. *Environ. Health Perspect.*, *23*, 227-232

Luster, M.I., Boorman, G.A., Harris, M.W. & Moore, J.A. (1980) Laboratory studies on polybrominated biphenyl-induced immune alterations following low-level chronic or pre/postnatal exposure. *Int. J. Immunopharmacol.*, *2*, 69-80

Matthews, H.B., Kato, S., Morales, N.M. & Tuey, D.B. (1977) Distribution and excretion of 2,4,5,2',4',5'-hexabromobiphenyl, the major component of Firemaster BP-6. *J. Toxicol. environ. Health*, *3*, 599-605

McCormack, K.M., Lepper, L.F., Wilson, D.M. & Hook, J.B. (1981) Biochemical and physiological sequelae to perinatal exposure to polybrominated biphenyls: a multi-generation study in rats. *Toxicol. appl. Pharmacol.*, *59*, 300-313

McCormack, K.M., Stickney, J.L., Bonhaus, D.W. & Hook, J.B. (1982) Cardiac and hepatic effects of pre- and postnatal exposure to polybrominated biphenyls in rats. *J. Toxicol. environ. Health*, *9*, 13-26

Miceli, J.N., Nolan, D.C., Marks, B. & Hariharan, M. (1985) Persistence of polybrominated biphenyls (PBB) in human post-mortem tissue. *Environ. Health Perspect.*, *60*, 399-403

Michigan Chemical Corp. (1971) *FireMaster BP-6. A New Fire Retardant Additive*, 2nd ed., St Louis, MI

Millis, C.D., Mills, R.A., Sleight, S.D. & Aust, S.D. (1985) Toxicity of 3,4,5,3',4',5'-hexabrominated biphenyl and 3,4,3',4'-tetrabrominated biphenyl. *Toxicol. appl. Pharmacol.*, *78*, 88-95

Millischer, R., Girault, F., Heywood, R., Clarke, G., Hossack, D. & Clair, M. (1979) Decabromobiphenyl: toxicological study (Fr.). *Toxicol. Eur. Res.*, *2*, 155-161

Mills, R.A., Millis, C.D., Dannan, G.A., Guengerich, F.P. & Aust, S.D. (1985) Studies on the structure-activity relationships for the metabolism of polybrominated biphenyls by rat liver microsomes. *Toxicol. appl. Pharmacol.*, *78*, 96-104

Mirsalis, J., Tyson, K., Loh, E., Bakke, J., Hamilton, C., Spak, D., Steinmetz, K. & Spalding, J. (1985) Induction of unscheduled DNA synthesis (UDS) and cell proliferation in mouse and rat hepatocytes following in vivo treatment (Abstract). *Environ. Mutagenesis*, *7* (Suppl. 3), 73

Montesano, R., Bartsch, H., Vainio, H., Wahrendorf, J., Wilbourn, J. & Yamasaki, H., eds (1986) *Long-term and Short-term Screening Assays for Carcinogens — A Critical Appraisal (IARC Scientific Publications No. 83)*, Lyon, International Agency for Research on Cancer (in press)

Moore, R.W., Dannan, G.A. & Aust, S.D. (1978) Induction of drug metabolizing enzymes in polybrominated biphenyl-fed lactating rats and their pups. *Environ. Health Perspect.*, *23*, 159-165

Mulligan, K.J., Caruso, J.A. & Fricke, F.L. (1980) Determination of polybrominated biphenyl and related compounds by gas-liquid chromatography with a plasma emission detector. *Analyst*, *105*, 1060-1067

Mumma, C.E. & Wallace, D.D. (1975) *Survey of Industrial Processing Data, Task II. Pollution Potential of Polybrominated Biphenyls (EPA-560/3-75-004)*, Washington DC, US Environmental Protection Agency

Murata, T., Zabik, M.E. & Zabik, M. (1977) Polybrominated biphenyls in raw milk and processed dairy products. *J. Dairy Sci.*, *60*, 516-520

National Toxicology Program (1982) *Third Annual Report on Carcinogens*, Washington DC, US Government Printing Office, pp. 247-248

National Toxicology Program (1983) *Carcinogenesis Studies of Polybrominated Biphenyl Mixture (FireMaster FF-1) (CAS No. 67774-32-7) in F344/N Rats and B6C3F1 Mice (Gavage Studies) (Tech. Rep. Ser. No. 244)*, Research Triangle Park, NC, US Department of Health and Human Services

Neufeld, M.L., Sittenfield, M. & Wolk, K.F. (1977) *Market Input/Output Studies, Task IV, Polybrominated Biphenyls (EPA-560/6-77-017; PB 271 915)*, Washington DC, US Environmental Protection Agency

Norris, J.M., Ehrmantraut, J.W., Gibbons, C.L., Kociba, R.J., Schwetz, B.A., Rose, J.Q., Humiston, C.G., Jewett, G.L., Crummett, W.B., Gehring, P.J., Tirsell, J.B. & Brosier, J.S. (1973) Toxicological and environmental factors involved in the selection of decabromodiphenyl oxide as a fire retardant chemical. In: Golub, M.A. & Parkin, J.A., eds, *Applied Polymer Symposium No. 22*, New York, John Wiley & Sons, pp. 195-219

Norris, J.M., Kociba, R.J., Schwetz, B.A., Rose, J.Q., Humiston, C.G., Jewett, G.L., Gehring, P.J. & Mailhes, J.B. (1975) Toxicology of octabromobiphenyl and decabromobiphenyl oxide. *Environ. Health Perspect.*, *11*, 153-161

Norström, Å., Andersson, K. & Rappe, C. (1976) Major components of some brominated aromatics used as flame retardants. *Chemosphere*, *4*, 255-261

O'Keefe, P.W. (1978) Formation of brominated dibenzofurans from pyrolysis of the polybrominated biphenyl fire retardant, FireMaster FF-1. *Environ. Health Perspect.*, *23*, 347-350

O'Keefe, P.W. (1979) Trace contaminants in a polybrominated biphenyl fire retardant and a search for these compounds in environmental samples. *Bull. environ. Contam. Toxicol.*, *22*, 420-425

Parkinson, A., Safe, S.H., Robertson, L.W., Thomas, P.E., Ryan, D.E., Reik, L.M. & Levin, W. (1983) Immunochemical quantitation of cytochrome P-450 isozymes and epoxide hydrolase in liver microsomes from polychlorinated or polybrominated biphenyl-treated rats. A study of structure-activity relationships. *J. biol. Chem.*, *258*, 5967-5976

Polin, D. & Ringer, R.K. (1978a) PPB fed to adult female chickens: its effect on egg production, reproduciton, viability of offspring, and residues in tissues and eggs. *Environ. Health Perspect.*, *23*, 283-290

Polin, D. & Ringer, R.K. (1978b) Polybrominated biphenyls in chicken eggs vs. hatchability. *Proc. Soc. exp. Biol. Med.*, *159*, 131-135

Rickert, D.E., Dent, J.G., Cagen, S.Z., McCormack, K.M., Melrose, P. & Gibson, J.E. (1978) Distribution of polybrominated biphenyls after dietary exposure in pregnant and lactating rats and their offspring. *Environ. Health Perspect.*, *23*, 63-66

Robertson, L.W., Andres, J.L., Safe, S.H. & Lovering, S.L. (1983) Toxicity of 3,3',4,4'- and 2,2',5,5'-tetrabromobiphenyl: correlation of activity with aryl hydrocarbon hydroxylase induction and lack of protection by antioxidants. *J. Toxicol. environ. Health*, *11*, 81-91

Robertson, L.W., Safe, S.H., Parkinson, A., Pellizzari, E., Pochini, C. & Mullin, M.D. (1984) Synthesis and identification of highly toxic polybrominated biphenyls in the fire retardant FireMaster BP-6. *J. agric. Food Chem.*, *32*, 1107-1111

Robl, M.G., Jenkins, D.H., Wingender, R.J. & Gordon, D.E. (1978) Toxicity and residue studies in dairy animals with FireMaster FF-1 (polybrominated biphenyls). *Environ. Health Perspect.*, *23*, 91-97

Roboz, J., Greaves, J., Holland, J.F. & Bekesi, J.G. (1982) Determination of polybrominated biphenyls in serum by negative chemical ionization mass spectrometry. *Anal. Chem.*, *54*, 1104-1108

Roboz, J., Greaves, J. & Bekesi, J.G. (1985) Polybrominated biphenyls in model and environmentally contaminated human blood: protein binding and immunotoxicological studies. *Environ. Health Perspect.*, *60*, 107-113

Rosenman, K.D., Anderson, H.A., Selikoff, I.J., Wolff, M.S. & Holstein, E. (1979) Spermatogenesis in man exposed to polybrominated biphenyl (PBB). *Fert. Steril.*, *32*, 209-213

Safe, S. (1984) Polychlorinated biphenyls (PCBs) and polybrominated biphenyls (PBBs): biochemistry, toxicology, and mechanism of action. *Crit. Rev. Toxicol.*, *13*, 319-395

Schwartz, E.L., Kluwe, W.M., Sleight, S.D., Hook, J.B. & Goodman, J.I. (1980) Inhibition of *N*-2-fluorenylacetamide-induced mammary tumorigenesis in rats by dietary polybrominated biphenyls. *J. natl Cancer Inst.*, *64*, 63-67

Sleight, S. (1985) Effects of PCBs and related compounds on hepatocarcinogenesis in rats and mice. *Environ. Health Perspect.*, *60*, 35-39

Sleight, S.D. & Sanger, V.L. (1976) Pathologic features of polybrominated biphenyl toxicosis in the rat and guinea pig. *J. Am. vet. Med. Assoc.*, *169*, 1231-1235

Stross, J.K., Nixon, R.K. & Anderson, M.D. (1979) Neuropsychiatric findings in patients exposed to polybrominated biphenyls. *Ann. N.Y. Acad. Sci.*, *320*, 368-372

Sundström, G., Hutzinger, O. & Safe, S. (1976) Identification of 2,2′,4,4′,5,5′-hexabromobiphenyl as the major component of flame retardant FireMaster®. *Chemosphere*, *1*, 11-14

Tilson, H.A. & Cabe, P.A. (1979) Studies on the neurobehavioural effects of polybrominated biphenyls in rats. *Ann. N.Y. Acad. Sci.*, *320*, 325-336

Tilson, H.A., Cabe, P.A. & Mitchell, C.L. (1978) Behavioral and neurological toxicity of polybrominated biphenyls in rats and mice. *Environ. Health Perspect.*, *23*, 257-263

Trosko, J.E., Dawson, B. & Chang, C.-C. (1981) PBB inhibits metabolic cooperation in Chinese hamster cells *in vitro*: its potential as a tumor promoter. *Environ. Health Persp.*, *37*, 179-182

Trotter, W.J. (1977) Confirming low levels of hexabromobiphenyl by gas-liquid chromatography of photolysis products. *Bull. environ. Contam. Toxicol.*, *18*, 726-733

Tsushimoto, G., Trosko, J.E., Chang, C.-C. & Aust, S.D. (1982) Inhibition of metabolic cooperation in Chinese hamster V79 cells in culture by various polybrominated biphenyl (PBB) congeners. *Carcinogenesis*, *3*, 181-185

Tuey, D.B. & Matthews, H.B. (1978) Pharmacokinetics of hexabromobiphenyl disposition in the rat (Abstract No. 275). *Toxicol. appl. Pharmacol.*, *45*, 337

Tuey, D.B. & Matthews, H.B. (1980) Distribution and excretion of 2,2′,4,4′,5,5′-hexabromobiphenyl in rats and man: pharmacokinetic model predictions. *Toxicol. appl. Pharmacol.*, *53*, 420-431

US Environmental Protection Agency (1984) Polybrominated biphenyls (PBBs). *US Code fed. Regul., Title 40*, Part 704.195, pp. 579-581

Waritz, R.S., Aftosmis, J.G., Culik, R., Dashiell, O.L., Faunce, M.M., Griffith, F.D., Hornberger, C.S., Lee, K.P., Sherman, H. & Tayfun, F.O. (1977) Toxicological evaluations of some brominated biphenyls. *Am. ind. Hyg. Assoc. J.*, *38*, 307-320

Werner, P.R. & Sleight, S.D. (1981) Toxicosis in sows and their pigs caused by feeding rations containing polybrominated biphenyls to sows during pregnancy and lactation. *Am. J. vet. Res.*, *42*, 183-188

Wertz, G.F. & Ficsor, G. (1978) Cytogenetic and teratogenic test of polybrominated biphenyls in rodents. *Environ. Health Perspect.*, *23*, 129-132

Willett, L.B., Brumm, C.J. & Williams, C.L. (1978) Method for extraction, isolation, and detection of free polybrominated biphenyls (PBBs) from plasma, faeces, milk, and bile using disposable glassware. *J. agric. Food Chem.*, *26*, 122-125

Willett, L.B., Durst, H.I., Lieu, T.-T.Y., Shanbacher, F.L. & Moorhead, P.D. (1982) Performance and health of offspring of cows experimentally exposed to polybrominated biphenyls. *J. Dairy Sci.*, *65*, 81-91

Williams, G.M., Tong, C. & Telang, S. (1984) Polybrominated biphenyls are nongenotoxic and produce an epigenetic membrane effect in cultured liver cells. *Environ. Res.*, *34*, 310-320

Windholz, M., ed. (1983) *The Merck Index*, 10th ed., Rahway, NJ, Merck & Co., p. 1091

Wolff, M.S., Aubrey, B., Camper, F. & Haymes, N. (1978) Relation of DDE and PBB serum levels in farm residents, consumers, and Michigan Chemical Corporation employees. *Environ. Health Perspect.*, *23*, 177-181

Wolff, M.S., Anderson, H.A., Rosenman, K.D. & Selikoff, I.J. (1979) Equilibrium of polybrominated biphenyl (PBB) residues in serum and fat of Michigan residents. *Bull. environ. Contam. Toxicol.*, *21*, 775-781

Wong, O., Brocker, W., Davis, H.V. & Nagle, G.S. (1984) Mortality of workers potentially exposed to organic and inorganic brominated chemicals, DBCP, TRIS, PBB, and DDT. *Br. J. ind. Med.*, *41*, 15-24

AMITROLE

This substance was considered by previous Working Groups, in February 1974 (IARC, 1974) and February 1982 (IARC, 1982). Since that time, new data have become available, and these have been incorporated into the monograph and taken into consideration in the present evaluation.

1. Chemical and Physical Data

1.1 Synonyms and trade names

Chem. Abstr. Services Reg. No.: 61-82-5
Chem. Abstr. Name: 1*H*-1,2,4-Triazol-3-amine
IUPAC Systematic Name: 3-Amino-*s*-triazole
Synonyms: Aminotriazole; 2-aminotriazole; 3-aminotriazole; 3-amino-1*H*-1,2,4-triazole; 5-amino-1*H*-1,2,4-triazole; 2-amino-1,3,4-triazole; 3-amino-1,2,4-triazole; 5-amino-1,2,4-triazole; 1,2,4-triazole-3-amine
Trade Names: Amitrol; Amitrol 90; Amizol; AT; 3,A-T; ATA; Azaplant; Cytrol; Cytrole; ENT 25445; Herbidal total; Weedazol

1.2 Structural and molecular formulae and molecular weight

$C_2H_4N_4$ Mol. wt: 84.08

1.3 Chemical and physical properties of the pure substance

(a) *Description*: White crystalline solid from water, ethanol or acetoxyethanol (Hawley, 1981; Weast, 1985)

(b) *Melting-point*: 159°C (Weast, 1985)

(c) *Spectroscopy data[a]*: Infrared (Sadtler Research Laboratories, 1980; prism [8667, 50957], grating [21258, 50957]), nuclear magnetic resonance (Sadtler Research Laboratories, 1980; proton [9499, 22958], C-13 [6254]) and mass spectral data (Grasselli & Ritchey, 1975) have been reported.

(d) *Solubility*: Soluble in water (28 g/100 ml at 25°C), ethanol (26 g/100 g at 23°C), chloroform and methanol; slightly soluble in ethyl acetate; insoluble in acetone and diethyl ether (Anon., 1981; Hawley, 1981; US Environmental Protection Agency, 1982; Windholz, 1983; Weast, 1985)

(e) *Reactivity*: Weak base; forms chelates with metal ions such as iron and copper (Anon., 1981)

(f) *Octanol/water partition coefficient (P)*: log P, 0.52 (ICIS Chemical Information System, 1985 [ISHOW])

1.4 Technical products and impurities

In the USA, formulations are available containing amitrole as an active ingredient at concentrations of 4.9 (with 9.9% simazine), 15.0, 21.6 and 90% (Union Carbide Agricultural Products Co., 1984a,b,c, 1985a,b). Technical-grade amitrole of 95 or 97% purity can be obtained (Steinhoff *et al.*, 1983; Aldrich Chemical Co., 1984; Japan Chemical Week, 1984; Eastman Kodak Co., 1985). Formulations containing amitrole include combinations with ammonium thiocyanate and an activator or one or more of the following additional herbicides: simazine, diuron, Fenac® (2,3,6-trichlorophenylacetic acid), atrazine, bromacil, 2,4-D (see IARC, 1977, 1982), linuron and MCPA (see IARC, 1983) (Worthing, 1983; Anon., 1985). Amitrole products are marketed as soluble powders and liquids, and are also available in pressurized containers (Union Carbide Agricultural Products Co., 1984a,b,c; Anon., 1985; Eastman Kodak Co., 1985; Union Carbide Agricultural Products Co., 1985a,b).

---

[a]In square brackets, spectrum number in compilation

2. Production, Use, Occurrence and Analysis

2.1 Production and use

(a) Production

The synthesis of amitrole was first reported by Thiele and Manchot in 1898, involving the reaction of aminoguanidine with formic acid (Carter, 1976). The current industrial production process, described by Allen and Bell (1946) and patented in 1954 by Allen, involves the same reaction, in which an aminoguanidine salt is heated to 100-120°C with formic acid in an inert solvent (Carter, 1976; Sittig, 1980).

Amitrole is currently manufactured or formulated by four European firms (one in the Federal Republic of Germany, two in the UK and one in France) and by one company in Israel (Anon., 1985). Two companies in Japan produce approximately 18 000 kg annually (Japan Chemical Week, 1984).

Technical-grade amitrole is not produced in the USA, and current annual imports range from 227 000 to 363 000 kg. Use of amitrole in the USA has changed little in the last ten years (US Environmental Protection Agency, 1984a).

(b) Use

Amitrole was introduced as a herbicide and plant-growth regulator in the USA in 1954, and was originally registered in the USA for use in food crop production (Carter, 1976). Public use of the compound began in 1955 (Hazardous Materials Advisory Committee, 1974), but the US Environmental Protection Agency cancelled all food crop registrations for amitrole in 1971 (Carter, 1976; US Environmental Protection Agency, 1984a). Nonetheless, amitrole remains an important specialty herbicide in the USA.

Among the foliage-active herbicides, amitrole is unique in its mode of action. Application of the compound to foliage results in the destruction of chlorophyll, and it is thus not selective for any one class of weeds (Carter, 1976). Amitrole is therefore used primarily as a brush killer (US Environmental Protection Agency, 1984a). It is used around established apple and pear trees between harvest and the following summer, as a nonselective herbicide before planting kale, maize, oilseed rape, potatoes and wheat, on fallow land and in other non-crop situations (Worthing, 1983).

Application to roadsides accounts for approximately 50% of US use of amitrole, although amitrole accounts for only 5% of all herbicides used on roadsides. It is used on approximately 41 000 ha of US land, typically at a rate of 2.8 to 4.4 kg/ha. It is used to lesser extents in industrial, utility and commercial locations (US Environmental Protection Agency, 1984a,b). Amitrole has also been reported to have been used as a photographic reagent (Hawley, 1981).

(c) Regulatory status and guidelines

A Joint Meeting of the Food and Agriculture Organization Panel of Experts on Pesticide Residues and the Environment, and the WHO Expert Committee on Pesticide

Residues (Food and Agriculture Organization, 1978) confirmed in 1977 a conditional admissible dietary intake (ADI) of 0-0.00003 mg/kg bw for human exposure to amitrole residues in foods, established by the same body in 1974. A conditional tolerance of 0.02 mg/kg for residues of amitrole in raw agricultural commodities of plant origin was recommended by the same body in 1974 (WHO, 1975).

Since amitrole may not be used in food production systems, no US tolerance for residues of amitrole or its metabolites has been established by the US Environmental Protection Agency. The American Conference of Governmental Industrial Hygienists (1985), however, has recommended a threshold limit value (TLV) of 0.2 mg/m^3 for occupational exposure averaged over an 8-h workday. The Federal Republic of Germany has adopted the same standard for occupational exposure to amitrole (Deutsche Forschungsgemeinschaft, 1985).

2.2 Occurrence

(a) Natural occurrence

Amitrole is not known to occur as a natural product.

(b) Occupational exposure

No data on exposure levels were available to the Working Group.

(c) Air

Amitrole was released from the stack of a production plant during dry crushing and bagging operations, and atmospheric levels of up to 100 μg/m^3 were measured in the vicinity of the plant (Alary *et al.*, 1985).

(d) Water

Amitrole levels determined in water samples collected in a river downstream from the discharge of an aeration pond in the vicinity of a production plant were 0.6-1.2 mg/l. The concentrations in the wastewaters from the aeration pond were 50-200 mg/l (Alary *et al.*, 1985).

(e) Foods

Residues of amitrole proportional to the rate of application were found in mature apples at harvest when amitrole had been applied directly to the foliage and/or fruit three months earlier (Table 1). No residue was found on fruit after ground cover applications of amitrole (Schubert, 1971).

2.3 Analysis

Selected methods for the analysis of amitrole in wastewater, in pesticide formulations and in food crops are identified in Table 2. The spectrophotometric procedure of Alary *et al.* (1985) has also been used for the analysis of air samples collected on glass-fibre filters; detection limits in air were not reported. The analysis of amitrole in urine has been proposed for biological monitoring (Archer, 1984).

Table 1. Amitrole residues in apples at harvest (October) after direct application (July) of amitrole[a]

Rate of amitrole application (g/l)	Amitrole residues (mg/kg) in apples when applied to:		
	Foliage only	Fruit only	Foliage and fruit
0.0	<0.02	<0.02	<0.02
0.6	0.50	0.38	0.92
1.2	1.05	1.08	1.85
2.4	2.00	2.25	4.95
4.8	3.88	4.65	7.00

[a]From Schubert (1971)

Table 2. Methods for the analysis of amitrole

Sample matrix	Sample preparation	Assay procedure[a]	Limit of detection	Reference
Water	Adjust pH to 12.5; chromatograph (Amberlite IRC 50), eluting amitrole with 0.2 M hydrochloric acid; diazotize (NaNO$_2$/HCl); convert to azo dye [sulphamic acid, N-(1-naphthyl)ethylenediamine]	Spectrophotometry (520 nm)	50 μg/l	Alary et al. (1985)
Pesticide formulations	Dissolve sample in water; add dilute sodium hydroxide and potassium ferrocyanide/sodium nitroferricyanide/hydrogen peroxide colour reagent	Spectrophotometry (634nm)	not given	US Environmental Protection Agency (1982)
	Dissolve sample in water or aqueous dimethylformamide; acidify (dilute HCl); titrate to second end-point with dilute NaOH	Titrimetry	not given	Gentry et al. (1984)
Air	Trap on glass-fibre filter; add 0.2 M hydrochloric acid; extract in an ultrasonic bath; diazotize (NaNO$_2$/HCl); convert to azo dye [sulphamic acid, N-(1-naphthyl)-ethylenediamine]	Spectrophotometry (520 nm)	not given	Alary et al. (1985)

Table 2 (contd)

Sample matrix	Sample preparation	Assay procedure[a]	Limit of detection	Reference
Crops	Blend crop with Celite 545 and ethanol; filter; wash with ethanol; adsorb on Amberlite 120 (H); wash; add NH_4OH; clean (acetonitrile); add $NaNO_2$, sulphamic acid, N-(1-naphthyl)ethylenediamine	Spectro-photometry (455 nm)	0.025 mg/kg	Storherr & Burke (1961)
Potatoes and beetroots	Extract (ethanol); adsorb on resin; digest with acetonitrile and acid; diazotize (sodium nitrite); react with 8-amino-1-naphthol-3,6-disulphonic acid; chromatograph on polycaprolactam column	HPLC/VIS (546 nm)	0.005 mg/kg	Løkke (1980)
Soil and vineleaves	*Soil sample*: extract (acetonitrile with aqueous ammonia); centrifuge; boil to remove ammonia; adjust to pH 6; filter; concentrate; nitrosate (rinse with water then glacial acetic acid; add sodium nitrite; dilute) *Leaf sample*: cut and crush to a paste; extract (acetonitrile); mix and decant; centrifuge; evaporate to dryness; elute (polycar A.T. GC); nitrosate as above	HPLC/UV	0.17 μg/ml	Soulier *et al.* (1980)
Urine	Add HCl and potassium permanganate; diazotize (sodium nitrite); react with 1-naphthol-3,6-disulphonic acid disodium salt; add sodium acetate	HPLC/VIS (440 nm)	200 μg/l	Archer (1984)

[a]Abbreviations: HPLC/VIS, high-performance liquid chromatography/visible spectrophotometric detection; HPLC/UV, high-performance liquid chromatography/ultraviolet detection

3. Biological Data Relevant to the Evaluation of Carcinogenic Risk to Humans

3.1 Carcinogenicity studies in animals

(a) Oral administration

Mouse: In a preliminary note, it was reported that groups of 18 male and 18 female (C57BL/6×C3H/Anf)F_1 and (C57BL/6×AKR)F_1 mice, seven days of age, were administered 1000 mg/kg bw (maximum tolerated dose) amitrole [purity unspecified] in distilled water daily by stomach tube until four weeks of age. Subsequently, animals were fed diets containing 2192 mg/kg amitrole (maximum tolerated dose) until the end of the observation period (53-60 weeks). Thyroid tumours were reported to have occurred in nearly all of the treated mice (64/72). Liver tumours were observed in 34/36 (C57BL/6×C3H/Anf)F_1 mice and in 33/36 (C57BL/6×AKR)F_1 treated mice. In pooled control groups, 8/166 (C57BL/6×CeH/Anf)F_1 mice and 6/172 (C57BL/6×AKR)F_1 mice had liver tumours (Innes et al., 1969). [The Working Group noted that thyroid tumours are uncommon in these strains of mice.]

Groups of acatalasemic and normal catalase C3H mice were treated at weaning with neutron irradiation, or fed a diet containing amitrole [purity unspecified], or both. Dietary amitrole was administered at a concentration of 1% (10 000 mg/kg) during the first four weeks of a five-week cycle; this cycle was repeated continuously. The groups that received amitrole alone or amitrole following neutron irradiation had a 100% incidence of liver tumours (29/29, 33/33), whereas among the group treated with neutrons but not fed amitrole in the diet, 2/37 mice had liver tumours. In another study, among female acatalasemic C3H mice carrying the murine mammary tumour virus (MMTV), for 28 mice fed 1% amitrole for 12 weeks beginning at weaning there was a statistically significant delay in the initial appearance of mammary tumours, when compared to 29 mice fed the standard diet. This result was stated to have been confirmed in four of five repetitions of this experiment (Feinstein et al., 1978). [The Working Group noted that, in the first experiment, there was no untreated control group.]

Groups of 75 male and 75 female NMRI mice, six weeks of age, were fed diets containing 0, 1, 10 or 100 mg/kg amitrole (technical grade, 97% pure) for lifetime. There was no difference in body weights or survival (average, >600 days) between amitrole-treated and control mice and no indication of a carcinogenic effect (Steinhoff et al., 1983).

Groups of B6C3F_1 mice were nursed by dams fed diets containing 500 mg/kg amitrole [purity unspecified] from delivery until weaning and then maintained on a standard diet for 90 weeks. In 45 males, six hepatocellular adenomas and four hepatocellular carcinomas were observed; no liver tumour occurred in 55 females. Among untreated controls observed for 90 weeks, one hepatocellular adenoma and no carcinoma was observed in a group of 98 males; neither was observed among 96 females (Vesselinovitch, 1983). [The Working Group noted that the increases in the incidences of carcinomas ($p = 0.009$) and adenomas ($p = 0.004$) were statistically significant.]

Groups of B6C3F_1 mice were fed diets containing 500 mg/kg amitrole [purity unspecified] continously from weaning until 90 weeks. Among the 55 males, nine hepatocellular adenomas and 11 hepatocellular carcinomas were observed; among the 49 females, there were five hepatocellular adenomas and four hepatocellular carcinomas. Among untreated controls held for 90 weeks, one hepatocellular adenoma and no carcinoma was observed in a group of 98 males; neither was observed among 96 females (Vesselinovitch, 1983). [The Working Group noted that the increases in the incidences of carcinomas in males ($p < 0.001$) and females ($p = 0.012$) and of adenomas in males ($p < 0.001$) and females ($p = 0.004$) were statistically significant.]

Rat: Groups of rats [numbers, sex, strain and age unspecified] were fed diets containing 0, 10, 50 or 100 mg/kg amitrole [purity unspecified] for 104 weeks. Thyroid adenomas were observed in 1/10, 2/15 (one 'adenocarcinomatous') and 17/26 (four 'adenocarcinomatous') rats in the low-, mid- and high-dose groups, respectively. No thyroid tumour was found in the five controls examined (Jukes & Shaffer, 1960).

Male and female white rats (weighing 100-120 g) were each administered 20-25 mg amitrole per day in the drinking-water or were fed diets containing 250 or 500 mg amitrole per day for life (5-23 months). Of the group receiving amitrole in the drinking-water, eight were alive at the time of appearance of the first thyroid tumour, and three thyroid and six liver tumours were seen. Of the groups receiving amitrole in the diet, 10 and 11 rats, respectively, were alive at the time of the first tumour in the two groups; two thyroid and eight liver tumours were seen in the group fed 250 mg amitrole; five thyroid and ten liver tumours were seen in the group receiving 500 mg amitrole (Napalkov, 1962). [The Working Group noted that no data on controls were reported.]

Groups of rats [initial numbers, sex, strain and age unspecified] were fed diets containing 0, 10, 50 or 100 mg/kg amitrole [purity unspecified] for 104 weeks. Thyroid adenomas were observed in 15/27 rats in the high-dose group and in 1-3/27 in the other two groups. Mammary and other tumours were distributed in a random fashion (Hodge *et al.*, 1966). [The Working Group noted that the experiment was inadequately reported and that control data were not included.]

Six groups of female Wistar rats, weighing approximately 200 g, received the following treatments: Group 1 (40 rats) received 2500 mg/l amitrole [purity unspecified] in the drinking-water for 70 weeks; Group 2 (30 rats) received a partial thyroidectomy plus, two weeks later, 2500 mg/l amitrole in the drinking-water; Group 3 (30 rats) received a partial thyroidectomy followed by autoimplantation of the resected thyroid tissue and, two weeks later, administration of amitrole in the drinking-water; Group 4 (10 rats) served as untreated controls; Group 5 (10 rats) received a partial thyroidectomy with no further treatment; and Group 6 (10 rats) received a partial thyroidectomy with autoimplantation but no other treatment. Rats surviving more than 30 weeks comprised the effective animals. Premature deaths were largely a result of infection; survival at 30 weeks ranged from 70-80% of rats not receiving amitrole and 47-65% of amitrole-treated rats. No gross or microscopic pathology was observed among rats not treated with amitrole. Goitres developed in all amitrole-treated rats in Groups 1, 2 and 3, including the ten autografts which 'took' in Group 3. Invasive tumours of the thyroid were found in 19/26, 14/14 and 10/10 rats in Groups 1, 2 and 3, respectively. These results were significantly different ($p < 0.001$) from those in controls (0/7, 0/7 and 0/8). Papillary adenoma nodules were observed in the thyroid in 3/26, 1/14 and 1/10 rats in Groups 1, 2 and 3, respectively (Tsuda *et al.*, 1976).

Groups of female Wistar rats, weighing approximately 200 g, received the following treatments: Group 1 (20 rats) received 2500 mg/l amitrole [purity unspecified] in the drinking-water and a standard diet; Group 2 (20 rats) was a fed a low-iodine (0.25 mg/kg) diet; and a further group (20 rats) was fed a standard diet (containing 5 mg/kg iodine) and served as untreated controls. The experiment was terminated at 60 weeks; rats surviving 30 weeks or more were considered to be the effective animals. Body-weight gain was reduced in

Groups 1 and 2. Follicular carcinomas of the thyroid were observed in 9/13 rats treated with amitrole only, in 4/9 rats fed the low-iodine diet and in 0/16 untreated controls (Tsuda *et al.*, 1978).

Two groups of 12 male Wistar rats, ten and 14 weeks of age, respectively, were fed a diet containing 2000 mg/kg amitrole [purity unspecified] for 12 weeks; a further group of 12 rats was fed a standard diet and served as untreated controls. All animals were sacrificed after 26 weeks of age. No thyroid tumour was found (Iwata, 1981; Hiasa *et al.*, 1982). [The Working Group noted the small numbers of animals used, the short duration of the study and that these groups were controls to a combined treatment study; see section 3.1(*e*).]

Groups of 75 male and 75 female Wistar rats, six weeks of age, were fed diets containing 0, 1, 10 or 100 mg/kg amitrole (technical grade, 97% pure) for lifetime. No difference was observed in body-weight gain, and average survival exceeded 900 days in all groups. Marked increases in the incidences of pituitary and thyroid tumours were observed in the high-dose group: benign pituitary tumours in females (36/75 high-dose; 14/74 controls), benign thyroid tumours in males (45/75; 5/75) and females (44/75; 7/74), and malignant thyroid tumours in males (18/75; 3/75) and females (28/75, 0/74) (Steinhoff *et al.*, 1983).

Hamster: Groups of 76 male and 76 female golden hamsters, six weeks old, were fed diets containing 0, 1, 10 or 100 mg/kg amitrole (technical grade, 97% pure) for lifetime. No difference was observed in the body-weight gains or survival of control and low- or mid-dose animals. Reduction in body-weight gain and a statistically significant ($p \leqslant 0.0002$) reduction in survival was observed in the high-dose groups. There was no indication of a carcinogenic effect (Steinhoff *et al.*, 1983).

(*b*) *Skin application*

Mouse: Groups of 50 male and 50 female C3H/Anf mice, two to four months old, received weekly skin applications of 0.1 or 10 mg analytical-grade amitrole in 0.2 ml acetone:methanol mixture (65:35) for life. The median survival times ranged from 44-57 weeks. No skin tumour was observed (Hodge *et al.*, 1966).

(*c*) *Subcutaneous administration*

Rat: A group of 12 male and female random-bred white rats, weighing 100-120 g, received twice-weekly subcutaneous injections of 125 mg amitrole [purity unspecified] in water for 11 months and were observed up to 23 months. In the seven rats alive at the appearance of the first tumour, five liver and five thyroid tumours were observed (Napalkov, 1962). [The Working Group noted that no data on controls were reported.]

(*d*) *Transplacental exposure*

Mouse: Female B6C3F$_1$ mice were fed diets containing 500 mg/kg amitrole [purity unspecified] from the 12th day of gestation until delivery, and the offspring were maintained on a standard diet for 90 weeks. Four hepatocellular adenomas and two hepatocellular carcinomas were observed in 74 males, but no such tumour occurred in 83 females. These results were considered to be not significantly different from those seen with untreated B6C3F$_1$ mice held for 90 weeks, in which one hepatocellular adenoma and no carcinoma was observed in 98 males and neither was observed in 96 females (Vesselinovitch, 1983).

(e) Combined treatment

Rat: Two groups of 30 male albino rats, two to three months of age, were fed 0.06% 4-dimethylaminoazobenzene (DAB) in the diet, and one group also received intraperitoneal injections of 1000 mg/kg bw amitrole [purity unspecified] every second day as a 10% solution in water. The surviving 16 DAB-treated and 19 DAB-plus-amitrole-treated rats were killed at 21 weeks. The incidences of liver tumours were 12/16 in the group receiving DAB alone and 4/19 in the group that received DAB plus amitrole ($p < 0.01$). Liver carcinomas produced by DAB alone were mostly hepatocellular carcinomas, whereas those in the group treated with DAB plus amitrole were hepatocellular carcinomas and cholangiocarcinomas (Hoshino, 1960).

Groups of 12 six-week-old male Wistar rats received the following treatments: Group 1 received four weekly subcutaneous injections of 700 mg/kg bw *N*-bis(2-hydroxypropyl)nitrosamine (DHPN) followed by a diet containing 2000 mg/kg amitrole [purity unspecified] for a total of 12 weeks; Group 2 received four weekly injections of DHPN only; Group 3 was fed a diet containing 2000 mg/kg amitrole beginning at week 4 for 12 weeks; Group 4 received eight weekly injections of DHPN followed by a diet containing 2000 mg/kg amitrole for 12 weeks; Group 5 received eight weekly injections of DHPN only; Group 6 was fed a diet containing 2000 mg/kg amitrole beginning at week 8; and Group 7 was fed a standard diet and served as untreated controls. All animals were killed after 20 weeks. No thyroid tumour was found in rats in Groups 2, 3, 6 or 7, but a significantly increased incidence ($p < 0.05$) of thyroid tumours compared with that in controls was observed in rats in Group 1 (9/11) and in rats in Group 4 (12/12) and Group 5 (7/12) (Iwata, 1981; Hiasa *et al.*, 1982).

A group of 75 male Wistar-Furth rats was castrated at 40 days of age and divided into six groups: Group 1 (five rats) received no further treatment and served as controls; Group 2 (ten rats) was given 1500 mg/l amitrole [purity unspecified] in the drinking-water starting at 47 days after castration; Group 3 (ten rats) received subcutaneous implantation of a pellet on the back containing 5 mg diethylstilboestrol (DES) plus 45 mg cholesterol, which was replaced every two months; Group 4 (11 rats) received implantation of the DES pellet plus amitrole in the drinking-water; Group 5 (20 rats) received implantation of the DES pellet followed by administration of 5 mg *N*-nitrosobutylurea (NBU) per day in the drinking-water for 30 days starting at 50-55 days of age; and Group 6 (19 rats) received implantation of the DES pellet followed by administration of NBU and, subsequently (seven days after NBU treatment), amitrole in the drinking-water. Rats surviving beyond 230 days of age were considered to be effective animals; all survivors were killed at 14 months of age. Rats in Groups 3 and 5 developed numerous hepatocellular carcinomas and neoplastic nodules (4/9 and 15/17 animals) and pituitary tumours (7/9 and 12/17). Addition of amitrole to these regimens (Groups 4 and 6, respectively) had no effect on the incidence of pituitary tumours

(8/11 and 10/14) but slightly (Group 4; 2/11) and significantly (Group 6; 3/14) reduced the incidences of hepatocellular carcinomas and neoplastic nodules. No such tumour was found in controls (Group 1) or in rats receiving amitrole only (Group 2) (Sumi *et al.*, 1985).

3.2 Other relevant biological data

(a) Experimental systems

Toxic effects

Amitrole generally has little acute toxicity to experimental animals. Acute oral LD_{50} values have been reported to be 14.7 g/kg bw in mice and 25 g/kg bw in male rats (Kröller, 1966). An oral dose of 2 g/kg bw administered to sheep was fatal [details not given] (Hapke *et al.*, 1965). No pronounced toxicity was seen in mice, cats or dogs after intravenous injections of 1.6, 1.7 or 1.2 g/kg bw, respectively; mice tolerated an intraperitoneal injection of 4 g/kg bw amitrole (Kröller, 1966). No sign of acute toxicity was seen in specific pathogen-free adult male or female rats given amitrole (99% pure) in aqueous solution by stomach tube (4.0 g/kg bw) or by dermal application (2.5 g/kg bw) (Gaines *et al.*, 1973). Intraperitoneal administration of 4.0 g/kg bw amitrole every 4 h for 24 h to male rats did not induce toxic effects (Kato, 1967).

Adult specific pathogen-free rats given 500, 1000 or 5000 mg/kg amitrole (99% pure) in the diet for 107-110 days gained 14-26% less body weight than did controls. No reduction in weight gain was observed in rats fed 25 or 100 mg/kg amitrole in the diet for 240-247 days (Gaines *et al.*, 1973). The weight gain of specific pathogen-free rats, mice and golden hamsters was not affected by life-long administration of 10 mg/kg technical-grade amitrole (97% pure) in the diet. However, administration of 100 mg/kg amitrole in the diet to golden hamsters resulted in a slight reduction in body weight (Steinhoff *et al.*, 1983).

Amitrole is goitrogenic, apparently by inhibition of peroxidase-dependent iodide oxidation in the thyroid (Alexander, 1959; Strum & Karnovsky, 1971). Male rats administered 0.04% (400 mg/l) amitrole in the drinking-water developed goitre within seven days (Strum & Karnovsky, 1971). In female rats administered 2.5 mg/ml amitrole in the drinking-water, a small increase in the size of the thyroid was visible after three days, and its size had doubled by ten days (Tsuda *et al.*, 1973). Continuous feeding of 100 mg/kg amitrole in the diet caused goitre in rats of each sex within three months, 25 mg/kg caused goitre in 4/10 females sacrificed at 240 days, while 10 mg/kg were without goitrogenic effect within 24 months (Gaines *et al.*, 1973; Steinhoff *et al.*, 1983). Continuous feeding (up to 18 months) of 100 mg/kg amitrole in the diet caused goitre in mice but not in golden hamsters; 10 mg/kg were without effect in either species (Steinhoff *et al.*, 1983).

In rats fed 0, 0.25, 0.5, 2, 10 or 50 mg/kg amitrole in the diet for 11-13 weeks, several measures of thyroid function were affected: reduced serum protein-bound iodine concentrations and thyroid iodine uptake were observed in groups receiving $\geqslant 2$ mg/kg in the diet. Morphological changes in the thyroid were observed at the two higher doses (Fregly, 1968).

Amitrole inhibits catalase in rat liver and kidney (Heim et al., 1955). Large doses of amitrole (1 g/kg bw intraperitoneally) reduced rat hepatic triglyceride concentrations (Ishii et al., 1976).

Effects on reproduction and prenatal toxicity

Mallard eggs were immersed on day 3 of incubation in one of a series of graded concentrations of amitrole in aqueous emulsion. Amitrole reduced embryonic growth at concentrations below the LC_{50} level, but was teratogenic only at concentrations above the LC_{50} level (Hoffman & Albers, 1984).

In a breeding study, groups of ten male and female Sherman rats were exposed to 0, 500 or 1000 mg/kg amitrole (99% pure) in the diet for 55 days prior to breeding. Litter size and weight as well as postnatal viability of the offspring were reduced in both treated groups. Reductions in the relative weights of the thymus and spleen were observed in weanling pups from the group exposed to 500 mg/kg (Gaines et al., 1973).

No effect on offspring growth or viability was observed when groups of male and female Sherman rats were exposed to 0, 25 or 100 mg/kg amitrole in the diet for two generations (Gaines et al., 1973).

Groups of eight Sherman rats received 0, 20 or 100 mg/kg bw amitrole orally by gavage on days 7-15 of gestation. No effect on postnatal growth or viability was observed in the offspring (Gaines et al., 1973).

Absorption, distribution, excretion and metabolism

Amitrole is rapidly and almost completely absorbed from the gut and lungs (Fang et al., 1964, 1966; Burton et al., 1974; Tjälve, 1975; Brown & Schanker, 1983). After intravenous and intragastric administration of radiolabelled amitrole to mice, radioactivity accumulated in the bone marrow, spleen, liver and gut mucosa (Tjälve, 1975; Fujii et al., 1984). Following administration of an intravenous dose of radiolabelled compound to mice, the radioactivity disappeared from the blood, heart, lung, muscle, salivary glands and brain with a half-time of approximately 1.2 h (Fujii et al., 1984). After oral administration to rats, disappearance of amitrole from different organs followed first-order kinetics with a half-time of 2-4 h. It was excreted mainly in the urine; a small, variable amount was found in the faeces (Fang et al., 1964). Amitrole was also found in the faeces after both oral and parenteral administration to mice (Tjälve, 1975). This compound is excreted mostly unchanged (Fang et al., 1966); 6% of a 50 mg/kg bw oral dose was excreted in the urine of rats as 3-amino-5-mercapto-1,2,4-triazole and 3-amino-1,2,4-triazoyl-(5)-mercapturic acid (Grunow et al., 1975).

When mice were given an intravenous dose (3.4 mg/kg bw) of ^{14}C-amitrole, approximately 10% [estimated from data presented by Fujii et al. (1984)] of the radioactivity appeared to be irreversibly bound to liver tissue. The bound radioactivity was apparently localized mainly centrilobularly and the amount decreased very little during 24 h (Fujii et al., 1984).

Mutagenicity and other short-term tests

The amitrole used in the studies described in this section was of reagent grade, unless otherwise stated.

Amitrole gave negative results in the *Escherichia coli* K12 prophage induction test at doses of up to 2000 µg/plate in the presence of a metabolic system (S9) from Aroclor-induced rat liver (Mamber *et al.*, 1984). It induced prophage λ cIts857 (at 15 000 µg/ml) but not wild-type prophage λ (at up to 40 000 µg/ml) in *E. coli* in the presence and absence of phenobarbital-induced rat-liver S9 (Ho & Ho, 1979). In an International Collaborative Program study[1], amitrole was tested in two laboratories using two different λ phage induction assays; in one assay, a questionable result was obtained in a spot test with 250 µg, in the presence of Aroclor-induced rat S9; a negative result [dose not given] was obtained with and without S9 in the second assay (de Serres & Ashby, 1981).

Amitrole (25 and 50 µg/plate) did not induce rII mutants in T4 bacteriophage, nor, at 400 µg/plate, reversion of bacteriophages AP72 and N17 to T4 phenotype (Andersen *et al.*, 1972).

Amitrole gave negative results in the *Bacillus subtilis rec* assay [assumed dose, 20 µg/disc] (Shirasu *et al.*, 1976). It also gave negative results in a spot test (1 mg) in a *B. subtilis rec* assay in the presence of polychlorinated biphenyl-induced rat-liver S9. In a parallel study, four laboratories tested amitrole in repair tests in *E. coli* using a number of different isogenic repair-deficient strains both with and without Aroclor-induced rat-liver S9. No positive effect was seen (de Serres & Ashby, 1981). Negative results were obtained in an *E. coli* WP2/WP100 *rec* assay when amitrole was tested in a quantitative spot test at up to 4000 µg/ml in the presence of Aroclor-induced rat-liver S9 (Mamber *et al.*, 1983). Amitrole gave negative results in the *E. coli polA* assay at 5 mg/disc (Bamford *et al.*, 1976) and at 250 µg/plate (Rosenkranz & Poirier, 1979).

Mutagenicity tests were carried out with a variety of *E. coli* WP2 strains (wild-type, *uvr*A, *uvr*A pKM101) in different laboratories using plate, suspension and fluctuation assays, with and without S9. Positive results were obtained in one laboratory using the *uvr*A pKM101 strain and 30% Aroclor-induced rat S9, but not in another which used 10% Kanechlor-induced rat S9. The other strains gave negative results. It was noted in a summary evaluation of the tests that all other investigators evaluated the responses in *E. coli* as negative (de Serres & Ashby, 1981).

In a four-laboratory collaborative study, amitrole (98% pure) was tested in *E. coli* WP2 *uvr*A at doses of up to 333 µg/plate with and without S9 from Aroclor-induced and uninduced B6C3F$_1$ mice, Fischer 344 rats and Syrian hamsters. No mutagenic response was obtained (Dunkel *et al.*, 1984).

---

[1]Amitrole was studied in the International Collaborative Program for the evaluation of short-term tests for carcinogenicity (de Serres & Ashby, 1981). Thirty assay systems were included, and more than 50 laboratories contributed data. Data from this programme is derived from the reports of the working groups, supplemented, in most cases, with data from the reports of individual investigators.

Amitrole (20 µg/disc) gave negative results in a reversion assay with two tryptophan-requiring strains of *E. coli* (B/r *try* WP2 and WP2 *try hcr*) (Shirasu *et al.*, 1976).

This compound has been tested extensively for mutagenicity in *Salmonella typhimurium*, consistently yielding negative results. Plate incorporation assays were performed with strains TA1535, TA1536, TA1537, TA1538, TA98 and TA100 in the presence and absence of various metabolic systems (Aroclor-induced and uninduced rat, mouse and hamster S9) at concentrations of up to 5000 µg/plate. It also gave negative results in spot tests (Andersen *et al.*, 1972; McCann *et al.*, 1975; Shirasu *et al.*, 1976; Carere *et al.*, 1978; Rosenkranz & Poirier, 1979; Simmon, 1979a; Dunkel *et al.*, 1984; Mamber *et al.*, 1984). Amitrole was also tested in 15 different laboratories, using various combinations of *S. typhimurium* TA1535, TA1537, TA1538, TA98 and TA100 with and without induced rat-liver S9 or a primary culture of rat hepatocytes in a number of protocols (plate test, preincubation test, fluctuation test). No mutagenic activity was seen in any of the tests. A suspension assay using *S. typhimurium* TM677 (mutation to 8-azaguanine resistance) also gave negative results when amitrole was tested at doses of up to 100 µg/ml with Aroclor- and phenobarbital-induced S9 (de Serres & Ashby, 1981).

This chemical was assayed in a host-mediated assay in mice, using *S. typhimurium* and *Saccharomyces cerevisiae* D3 as indicator organisms. Doses of 12, 125, and 250 mg/kg bw were administered by intramuscular injection in experiments with *S. typhimurium* TA1530 and TA1538; an oral dose of 1585 mg/kg bw was given in experiments with *S. typhimurium* TA1535; and 5000 mg/kg bw were given orally with the yeast. The authors reported that, at low doses, amitrole 'appeared' to be mutagenic to strains TA1530 and TA1538 and showed borderline activity toward the yeast (Simmon *et al.*, 1979). Amitrole (2900 µmol/kg bw) alone was not mutagenic in a host-mediated assay with *S. typhimurium* TA1950 when administered by gavage to male NMRI mice (Braun *et al.*, 1977).

[The Working Group noted that since amitrole inhibits histidine synthesis in *S. typhimurium* (Hilton *et al.*, 1965), mutation assays measuring histidine reversion in these bacteria are not suitable for testing this compound.]

Amitrole (0.4%) failed to induce reverse mutation in three tryptophan-requiring strains of *S. typhimurium* LT2 *trp* (A8, B4, C3) (Bamford *et al.*, 1976).

Amitrole was tested in a number of yeast assays. In one laboratory, it gave positive results in a DNA-repair assay using *rad* mutants of *S. cerevisiae* without S9 at concentrations of 100 and 300 µg/ml. No effect was seen with Aroclor-induced S9. Four laboratories using different strains of *S. cerevisiae* (XV185-14C, D4, D7, PG-154, PG-155) obtained no mutation, mitotic recombination or mitotic gene conversion, with or without S9, at doses of up to 12.5 mg/ml. In one laboratory, a positive response was obtained for mitotic gene conversion in *S. cerevisiae* JD1 at 300 µg/ml, with and without S9. Amitrole induced mitotic aneuploidy in *S. cerevisiae* at 50 µg/ml, with and without S9 (de Serres & Ashby, 1981). Concentrations of up to 5% w/v failed to induce mitotic recombination in *S. cerevisiae* D3 (Simmon, 1979b).

Amitrole (1000 µg) was weakly mutagenic in a spot test in *Streptomyces coelicolor* (forward mutation to streptomycin resistance) (Carere *et al.*, 1978). In a spot test in

Aspergillus nidulans, this compound (2000 μg) failed to induce forward mutation to 8-azaguanine resistance in strain 35; weak induction of mitotic recombination (spot test with 2000 μg) and mitotic nondisjunction (with 0.4 mg/ml) was found with strain P (Bignami *et al.*, 1977).

Amitrole gave negative results in assays with *Drosophila melanogaster* for sex chromosome nondisjunction (females reared on medium containing 10 μg/ml amitrole) and sex-linked recessive lethal mutations (males reared on medium containing 10 μg/ml amitrole). Amitrole was highly toxic at early larval stages (LD_{50}, 40 μg/ml in medium), and prolongation of development time was observed even with 10 μg/ml (Laamanen *et al.*, 1976). Amitrole did not induce sex-linked recessive lethal mutations in *D. melanogaster* fed 2000 μg/ml (de Serres & Ashby, 1981). Treatment of adult male Canton-S *D. melanogaster* with amitrole by feeding (10 000, 20 000 and 35 000 μg/ml) or injection (0.2-0.3 μl of a 10 000-μg/ml solution) did not induce sex-linked recessive lethal mutations (Woodruff *et al.*, 1985).

Amitrole induced unscheduled DNA synthesis, measured by scintillation counting of incorporated ^{3}H-thymidine, in HeLa cells in the presence of phenobarbital-induced rat-liver S9. A dose-response relationship was reported, but dose levels and response values were not given (de Serres & Ashby, 1981). It was reported in an abstract that amitrole induced unscheduled DNA synthesis in EUE cells [details not given] (Benigni & Dogliotti, 1980).

Concentrations of 0.3-10 μg/ml amitrole induced a dose-dependent increase in the incidence of ouabain- and 6-thioguanine-resistant mutants in cultured Syrian hamster embryo cells (Tsutsui *et al.*, 1984).

Sister chromatid exchanges were reported in Chinese hamster ovary cells treated with up to 10 μg/ml amitrole in the presence of S9 (de Serres & Ashby, 1981). [The Working Group was unable to evaluate this study due to inadequate presentation of the data.]

Amitrole was tested in cultured human lymphocytes at concentrations of 0.2% w/v for 72 h and 1.0% for 6 and 24 h; no significant difference between test cultures and controls was observed with respect to aneuploidy or structural chromosomal aberrations (Meretoja *et al.*, 1976).

The induction of micronuclei in $B6C3F_1$ and CD1 mouse bone-marrow cells was investigated in two laboratories. Two intraperitoneal injections of 2800 μg/mouse and 125, 250 or 500 mg/kg bw were given to the two strains, respectively. Micronuclei were not induced in any of the 500 ($B6C3F_1$) or 1500 (CD1) cells analysed (de Serres & Ashby, 1981).

Amitrole (tested at 0.1, 1.0, 10.0 and 100.0 μg/ml) transformed Syrian hamster embryo cells in culture, only with the two intermediate doses, although sufficient survivors were observed at the highest dose (Pienta *et al.*, 1977). Positive results were also obtained by Inoue *et al.* (1981) and Tsutsui *et al.* (1984). In Rauscher murine leukaemia virus-infected Fischer 344 rat embryo cells, a dose-response relationship for cell transformation was observed when amitrole was tested at 10, 80, 100 and 1200 μg/ml (Dunkel *et al.*, 1981).

Amitrole, administered as a single gavage dose of 11.9 mmol/kg bw to male Sprague-Dawley rats, did not induce unscheduled DNA synthesis (assayed *in vitro*), as measured by autoradiography, in an in-vivo/in-vitro test of rat hepatocytes (Kornbrust *et al.*, 1984).

The commercial herbicide, Cytrol (which contains amitrole as its active ingredient), induced a statistically significant increase in the incidence of chromosomal abnormalities (mainly fragments and anaphase bridges) in root tips of barley after treatment of seeds for 2, 4 or 6 h with solutions containing 100, 200 or 300 μg/ml Cytrol (Wuu & Grant, 1966). Chromosomal aberrations (mainly fragments and anaphase bridges) were induced in root tips of *Vicia faba* exposed to solutions of 10, 50 and 100 μg/ml Cytrol for 3, 6 and 12 h. The intermediate dose produced the highest percentage of chromosome abnormalities (Wuu & Grant, 1967).

(b) Humans

Toxic effects

No sign of intoxication was observed in a 39-year-old woman who ingested a commercial preparation containing 30% amitrole (about 20 mg/kg bw) and 56% 3,4-dichlorophenyl-N,N'-dimethylurea (diuron) (Geldmacher-von Mallinckrodt & Schmidt, 1970).

Effects on reproduction and prenatal toxicity

No data were available to the Working Group.

Absorption, distribution, excretion and metabolism

Unchanged amitrole (1 g/l) was found in the urine of a 39-year-old woman who had ingested a commercial preparation containing 30% amitrole (estimated dose, 20 mg/kg bw) and 56% 3,4-dichlorophenyl-N,N'-dimethylurea (diuron). Fifty percent of the estimated dose was excreted in the urine within a few hours after ingestion, but no metabolite was detected (Geldmacher-von Mallinckrodt & Schmidt, 1970).

Mutagenicity and chromosomal effects

An analysis of lymphocyte chromosomes of agricultural workers in Idaho, USA, with extensive occupational exposure to pesticides was reported by Yoder *et al.* (1973). A group of 26 herbicide-exposed workers was compared with 16 controls. The list of the most commonly used herbicides comprised 14 formulations, but the predominant exposures were to amitrole, 2,4-D and atrazine. Blood samples were drawn both off-season and mid-season and cultured for 48 h. Only 25 metaphases were examined from each. From off-season to mid-season the mean number of chromatid gaps in the herbicide-exposed group increased four fold (from 0.38 ± 0.10 to 1.38 ± 0.22 per person per 25 cells). In the same group chromatid breaks increased 25 fold (from 0.07 ± 0.05 to 1.81 ± 0.35). The off-season aberration frequencies were, however, very low as compared with the control group (off-season, 0.63 ± 0.22 for gaps and 0.31 ± 0.12 for breaks), although it is possible to use subjects as their own controls. The authors concluded that the increased incidence of chromosomal aberrations was probably due to exposure to herbicides, although it was not possible to distinguish which herbicide formulations were responsible. [The Working Group noted the small number of cells examined and that possible confounding factors were not taken into consideration.]

Sister chromatid exchange (SCE) frequency was studied in 57 herbicide and pesticide sprayers in New Zealand (Crossen *et al.*, 1978). Amitrole was mentioned as one of the 30 formulations most commonly encountered in the study. Overall, there was no difference in SCE frequency between the control group and the sprayers (mean rates, 7.65 *versus* 8.48). The sprayers were divided into three groups: those with no protection, those with some protection (either clothing, gloves or respirator) and those with full protection. Those with no protection had a significantly higher mean SCE rate than the control group (mean rate, 9.03 *versus* 7.65), but the authors noted that the length and level of exposure could also have influenced the findings. There was no difference in SCE rate between those who used herbicides exclusively and those using both herbicides and other pesticides. [The Working Group noted that confounding factors such as smoking were not taken into consideration.]

3.3 Case reports and epidemiological studies of carcinogenicity to humans

A cohort of 348 Swedish railroad workers exposed for 45 days or more to amitrole, 2,4-D or 2,4,5-T were investigated in a follow-up study from 1957-1972 and again in 1978 (Axelson & Sundell, 1974; Axelson *et al.*, 1980). There was a deficit of deaths from all causes (45 observed, 49 expected) but an excess from malignant neoplasms (17 observed, 11.9 expected). In a subcohort exposed to amitrole but not 2,4-D or 2,4,5-T, there were five deaths from cancer (two lung cancers, one pancreatic cancer, one reticulum-cell sarcoma and one maxillary sinus cancer), with 3.3 expected; three of the deaths (with 2.0 expected) occurred in those first exposed ten years or more before death. In a subcohort exposed to both amitrole and chlorophenoxy herbicides (2,4-D or 2,4,5-T), there were six deaths from cancer with 2.9 expected, of which all six (with 1.8 expected; $p < 0.005$) occurred in those first exposed ten years or more before death. The men were also exposed to other organic (e.g., monuron and diuron) and inorganic chemicals (e.g., potassium chlorate). The results obtained during the follow-up period 1972-1978, in which the exposure assessment could not be related to knowledge of the disease, are consistent with the results from the whole period 1957-1978, reported above, i.e., a statistically significant excess of all tumours was observed in the subcohort exposed to both amitrole and chlorophenoxy herbicides but not in the subcohort exposed to amitrole but not to 2,4-D or 2,4,5-T.

4. Summary of Data Reported and Evaluation

4.1 Exposure data

Amitrole has been widely produced since the 1950s for use as a herbicide. Occupational exposures occur during production, formulation and application of this herbicide, and nonoccupational exposures may occur from residues in food.

4.2 Experimental data

Amitrole was tested in mice by oral administration, skin application and transplacental exposure, in rats by oral and subcutaneous administration, and in hamsters by oral administration. In mice, thyroid and liver tumours were produced after oral administration; no skin tumour was observed after topical application. The study by transplacental exposure yielded inconclusive results. In rats, amitrole induced thyroid and pituitary tumours after oral administration. No carcinogenic effect was observed in hamsters.

Dietary exposure of breeding pairs of rats to amitrole reduces growth and viability of offspring.

Amitrole does not induce DNA damage in bacteria but may have an effect in yeast. It is not mutagenic to *Salmonella typhimurium* or *Escherichia coli*. Amitrole induces aneuploidy in yeast, but not mutation in yeast or *Aspergillus nidulans*. However, conflicting results were obtained in assays for mitotic gene conversion and recombination. It is weakly mutagenic to *Streptomyces coelicolor*. It does not induce sex-linked recessive lethal mutations or nondisjunction in *Drosophila melanogaster*. Amitrole induced mutations at two loci in one mammalian cell line. Amitrole does not induce unscheduled DNA synthesis in hepatocytes of rats exposed *in vivo*. No aneuploidy or chromosomal aberration is found in cultured human lymphocytes. Micronuclei are not induced in mouse bone marrow. Cell transformation is induced in mammalian cells. A commercial preparation of amitrole induces chromosomal abnormalities in plants.

Overall assessment of data from short-term tests: Amitrole[a]

	Genetic activity			Cell transformation
	DNA damage	Mutation	Chromosomal effects	
Prokaryotes	−	−		
Fungi/Green plants	?	?	+	
Insects		−	−	
Mammalian cells (*in vitro*)		+	−	+
Mammals (*in vivo*)	−		−	
Humans (*in vivo*)			?	
Degree of evidence in short-term tests for genetic activity: **Limited**				Cell transformation: Positive

[a]The groups into which the table is divided and the symbols '+', '−', and ? are defined on pp. 19-20 of the Preamble; the degrees of evidence are defined on pp. 20-21.

4.3 Human data

Two cytogenetic studies of occupational exposure to a number of herbicides, including amitrole, were available. A possible increase in the frequency of chromosomal aberrations was seen in one study, and an increased frequency of sister chromatid exchanges occurred in one group of workers in the other. The role of amitrole itself cannot be evaluated from these studies.

No data were available to evaluate the reproductive effects or prenatal toxicity of amitrole to humans.

In a small cohort study of Swedish railroad workers who had sprayed herbicides, there was a statistically significant excess of all cancers among those exposed to both amitrole and chlorophenoxy herbicides, but not among those exposed mainly to amitrole.

4.4 Evaluation[1]

There is *sufficient evidence*[2] for the carcinogenicity of amitrole to experimental animals. There is *inadequate evidence* for the carcinogenicity of amitrole to humans.

5. References

Alary, J., Bourbon, P., Escrieut, C. & Vandaele, J. (1985) Spectrophotometric determination of guanazole and aminotriazole in waters from an aminotriazole production plant. *Environ. Technol. Lett.*, 6, 93-100

Aldrich Chemical Co. (1984) *1984-1985 Aldrich Catalog/Handbook of Fine Chemicals*, Milwaukee, WI, p. 82

Alexander, N.M. (1959) Antithyroid action of 3-amino-1,2,4-triazole. *J. biol. Chem.*, 234, 148-150

Allen, C.F.H. & Bell, A. (1946) 3-Amino-1,2,4-triazole (1,2,4-triazole, 3-amino). *Org. Synth.*, 26, 11-12

American Conference of Governmental Industrial Hygienists (1985) *TLVs Threshold Limit Values and Biological Exposure Indices for 1985-86*, 2nd ed., Cincinnati, OH, p. 36

Andersen, K.J., Leighty, E.G. & Takahashi, M.T. (1972) Evaluation of herbicides for possible mutagenic properties. *J. agric. Food Chem.*, 20, 649-656

Anon. (1981) *Datensammlung zur Toxicologie der Herbizide* (Data compilation on toxicology of herbicides) (Ger.), 3rd ed., Weinheim, Verlag Chemie

---

[1]For definitions of the italicized terms, see Preamble, pp. 18 and 22.
[2]In the absence of adequate data on humans, it is reasonable, for practical purposes, to regard chemicals or exposures for which there is *sufficient evidence* of carcinogenicity in animals as if they presented a carcinogenic risk to humans.

Anon. (1985) *Farm Chemicals Handbook*, Willoughby, OH, Meister Publ. Co., p. C14

Archer, A.W. (1984) Determination of 3-amino-1,2,4-triazole (amitrole) in urine by ion-pair high-performance liquid chromatography. *J. Chromatogr.*, *303*, 267-271

Axelson, O. & Sundell, L. (1974) Herbicide exposure, mortality and tumor incidence. An epidemiological investigation on Swedish railroad workers. *Work Environ. Health*, *11*, 21-28

Axelson, O., Sundell, L., Andersson, K., Edling, C. & Hogstedt, C. (1980) Herbicide exposure and tumor mortality. An updated epidemiologic investigation on Swedish railroad workers. *Scand. J. Work Environ. Health*, *6*, 73-79

Bamford, D., Sorsa, M., Gripenberg, U., Laamanen, I. & Meretoja, T. (1976) Mutagenicity and toxicity of amitrole. III. Microbial tests. *Mutat. Res.*, *40*, 197-202

Benigni, R. & Dogliotti, E. (1980) UDS studies on selected environmental chemicals (Abstract No. 97). *Mutat. Res.*, *74*, 217

Bignami, M., Aulicino, F., Velcich, A., Carere, A. & Morpurgo, G. (1977) Mutagenic and recombinogenic action of pesticides in *Aspergillus nidulans*. *Mutat. Res.*, *46*, 395-402

Braun, R., Schöneich, J. & Ziebarth, D. (1977) In vivo formation of *N*-nitroso compounds and detection of their mutagenic activity in the host-mediated assay. *Cancer Res.*, *37*, 4572-4579

Brown, R.A., Jr & Schanker, L.S. (1983) Absorption of aerosolized drugs from the rat lung. *Drug Metab. Disposition*, *11*, 355-360

Burton, J.A., Gardiner, T.H. & Schanker, L.S. (1974) Absorption of herbicides from the rat lung. *Arch. environ. Health*, *29*, 31-33

Carere, A., Ortali, V.A., Cardamone, G., Torracca, A.M. & Raschetti, R. (1978) Microbiological mutagenicity studies of pesticides in vitro. *Mutat. Res.*, *57*, 277-286

Carter, M.C. (1976) *Amitrole*. In: Kearney, P.C. & Kaufman, D.D., eds, *Herbicides: Chemistry, Degradation and Mode of Action*, New York, Marcel Dekker, pp. 377-398

Crossen, P.E., Morgan, W.F., Horan, J.J. & Stewart, J. (1978) Cytogenetic studies of pesticide and herbicide sprayers. *N. Z. med. J.*, *88*, 192-195

Deutsche Forschungsgemeinschaft (German Research Community) (1985) *Maximal Concentrations in Workplaces and Biological Tolerance Values for Substances in the Work Environment 1984* (Ger.), Vol. 21, Weinheim, Verlagsgesellschaft mbH, p. 18

Dunkel, V.C., Pienta, R.J., Sivak, A. & Traul, K.A. (1981) Comparative neoplastic transformation responses of Balb/3T3 cells, Syrian hamster embryo cells, and Rauscher murine leukemia virus-infected Fischer 344 rat embryo cells to chemical carcinogens. *J. natl Cancer Inst.*, *67*, 1303-1315

Dunkel, V.C., Zeiger, E., Brusick, D., McCoy, E., McGregor, D., Mortelmans, K., Rosenkranz, H.S. & Simmon, V.F. (1984) Reproducibility of microbial mutagenicity assays. I. Tests with *Salmonella typhimurium* and *Escherichia coli* using a standardized protocol. *Environ. Mutagenesis*, *6* (Suppl. 2)

Eastman Kodak Co. (1985) *Kodak Laboratory Chemicals*, Rochester, NY, p. 29

Fang, S.C., George, M. & Yu, T.C. (1964) Metabolism of 3-amino-1,2,4-triazole-5-C14 by rats. *J. agric. Food Chem.*, *12*, 219-223

Fang, S.C., Khanna, S. & Rao, A.V. (1966) Further study on the metabolism of labeled 3-amino-1,2,4-triazole (ATA) and its plant metabolites in rats. *J. agric. Food Chem.*, *14*, 262-265

Feinstein, R.N., Fry, R.J.M. & Staffeldt, E.F. (1978) Carcinogenic and antitumor effects of aminotriazole on acatalasemic and normal catalase mice. *J. natl Cancer Inst.*, *60*, 1113-1116

Food and Agriculture Organization (1978) *FAO Plant Production and Protection Paper. Pesticides Residues in Food: 1977 Evaluations. The Monographs, 10 Sup.*, Rome, pp. 11-14

Fregly, M.J. (1968) Effect of aminotriazole on thyroid function in the rat. *Toxicol. appl. Pharmacol.*, *13*, 271-286

Fujii, T., Miyazaki, H. & Hashimoto, M. (1984) Autoradiographic and biochemical studies of drug distribution in the liver. III. [^{14}C]Aminotriazole. *Eur. J. Drug Metab. Pharmacokinet.*, *9*, 257-265

Gaines, T.B., Kimbrough, R.D. & Linder, R.E. (1973) The toxicity of amitrole in the rat. *Toxicol. appl. Pharmacol.*, *26*, 118-129

Geldmacher-von Mallinckrodt, M. & Schmidt, H.P. (1970) Toxicity and metabolism of aminotriazole in man (Ger.). *Arch. Toxikol.*, *27*, 13-18

Gentry, G.M., Jackson, E.R., Jensen, T.L., Jung, P.D., Launer, J.E. & Torma, L., eds (1984) Pesticide Formulations. Miscellaneous Pesticides. In: Williams, W., ed., *Official Methods of Analysis of the Association of Official Analytical Chemists*, Arlington, VA, Association of Official Analytical Chemists, p. 146

Grasselli, J.G. & Ritchey, W.M., eds (1975) *CRC Atlas of Spectral Data and Physical Constants for Organic Compounds*, Vol. 4, Cleveland, OH, CRC Press, p. 705

Grunow, W., Altman, H.-J. & Böhme, C. (1975) Metabolism of 3-amino-1,2,4-triazole in rats (Ger.). *Arch. Toxicol.*, *34*, 315-324

Hapke, H.-J., Rüssel, H. & Ueberschär, S. (1965) Hazards in the use of aminotriazole (Ger.). *Dtsch. Tierärzt. Wochenschr.*, *72*, 204-206

Hawley, G.G., ed. (1981) *The Condensed Chemical Dictionary*, 10th ed., New York, Van Nostrand Reinhold, p. 55

Hazardous Materials Advisory Committee (1974) *Herbicide Report; Chemistry and Analysis, Environmental Effects, Agricultural and Other Applied Uses*, Washington DC, US Environmental Protection Agency, p. 13

Heim, W.G., Appelman, D. & Pyfrom, H.T. (1955) Production of catalase changes in animals with 3-amino-1,2,4-triazole. *Science*, *122*, 693-694

Hiasa, Y., Ohshima, M., Kitahori, Y., Yuasa, Y., Fujita, T. & Iwata, C. (1982) Promoting effects of 3-amino-1,2,4-triazole on the development of thyroid tumors in rats treated with *N*-bis(2-hydroxypropyl)nitrosamine. *Carcinogenesis*, *3*, 381-384

Hilton, J.L., Kearney, P.C. & Ames, B.N. (1965) Mode of action of the herbicide, 3-amino-1,2,4-triazole (amitrole): inhibition of an enzyme of histidine biosynthesis. *Arch. Biochem. Biophys.*, *112*, 544-547

Ho, Y.L. & Ho, S.K. (1979) The induction of a mutant prophage λ in *Escherichia coli*: a rapid screening test for carcinogens. *Virology*, *99*, 257-264

Hodge, H.C., Maynard, E.A., Downs, W.L., Ashton, J.K. & Salerno, L.L. (1966) Tests on mice for evaluating carcinogenicity. *Toxicol. appl. Pharmacol.*, *9*, 583-596

Hoffman, D.J. & Albers, P.H. (1984) Evaluation of potential embryotoxicity and teratogenicity of 42 herbicides, insecticides, and petroleum contaminants to mallard eggs. *Arch. environ. Contam. Toxicol.*, *13*, 15-27

Hoshino, M. (1960) Effect of 3-amino-1,2,4-triazole on the experimental production of liver cancer. *Nature*, *186*, 174-175

IARC (1974) *IARC Monographs on the Evaluation of Carcinogenic Risk of Chemicals to Man*, Vol. 7, *Some Anti-thyroid and Related Substances, Nitrofurans and Industrial Chemicals*, Lyon, pp. 31-43

IARC (1977) *IARC Monographs on the Evaluation of the Carcinogenic Risk of Chemicals to Man*, Vol. 15, *Some Fumigants, the Herbicides 2,4-D and 2,4,5-T, Chlorinated Dibenzodioxins and Miscellaneous Industrial Chemicals*, Lyon, pp. 111-138

IARC (1982) *IARC Monographs on the Evaluation of the Carcinogenic Risk of Chemicals to Humans*, Suppl. 4, *Chemicals, Industrial Processes and Industries Associated with Cancer in Humans, IARC Monographs, Volumes 1 to 29*, Lyon, pp. 38-40, 101-103

IARC (1983) *IARC Monographs on the Evaluation of the Carcinogenic Risk of Chemicals to Humans*, Vol. 30, *Miscellaneous Pesticides*, Lyon, pp. 255-269

ICIS Chemical Information System (1985) *Information System for Hazardous Organics in Water* (ISHOW) and *Environmental Fate* (ENVIROFATE), Washington DC, Information Consultants

Innes, J.R.M., Ulland, B.M., Valerio, M.G., Petrucelli, L., Fishbein, L., Hart, E.R., Pallotta, A.J., Bates, R.R., Falk, H.L., Gart, J.J., Klein, M., Mitchell, I. & Peters, J. (1969) Bioassay of pesticides and industrial chemicals for tumorigenicity in mice: a preliminary note. *J. natl Cancer Inst.*, *42*, 1101-1114

Inoue, K., Katoh, Y. & Takayama, S. (1981) In vitro transformation of hamster embryo cells by 3-(*N*-salicyloyl)amino-1,2,4-triazole. *Toxicol. Lett.*, *7*, 211-215

Ishii, H., Suga, T. & Niinobe, S. (1976) Effect of 3-amino-1,2,4-triazole on lipid metabolism in the rat. *Biochem. Pharmacol.*, *25*, 1438-1440

Iwata, C. (1981) Effect of hemithyroidectomy and 3-amino-1,2,4-triazole on thyroid tumor induced by *N*-bis(2-hydroxypropyl)nitrosamine in rats (Jpn.). *J. Nara med. Assoc.*, *32*, 670-684

Japan Chemical Week, ed. (1984) *Specialty Chemicals Handbook 1984/1985*, Tokyo, The Chemical Daily Co., p. 17

Jukes, T.H. & Shaffer, C.B. (1960) Antithyroid effects of aminotriazole. *Science*, *132*, 296-297

Kato, R. (1967) Effect of administration of 3-aminotriazole on the activity of microsomal drug-metabolizing enzyme systems of rat liver. *Jpn. J. Pharmacol.*, *17*, 56-63

Kornbrust, D.J., Barfnecht, T.R., Ingram, P. & Shelburne, J.D. (1984) Effect of di(2-ethylhexyl)phthalate on DNA repair and lipid peroxidation in rat hepatocytes and on metabolic cooperation in Chinese hamster V-79 cells. *J. Toxicol. environ. Health*, *13*, 99-116

Kröller, E. (1966) Use and properties of 3-amino-1,2,4-triazole in relation to its residues in foodstuffs (Ger.). *Residue Rev.*, *12*, 162-192

Laamanen, I., Sorsa, M., Bamford, D., Gripenberg, U. & Meretoja, T. (1976) Mutagenicity and toxicity of amitrol. I. Drosophila tests. *Mutat. Res.*, *40*, 185-190

Løkke, H. (1980) Determination of amitrole by ion-pair high-performance liquid chromatography. *J. Chromatogr.*, *200*, 234-237

Mamber, S.W., Bryson, V. & Katz, S.E. (1983) The *Escherichia coli* WP2/WP100 rec assay for detection of potential chemical carcinogens. *Mutat. Res.*, *119*, 135-144

Mamber, S.W., Bryson, V. & Katz, S.E. (1984) Evaluation of the *Escherichia coli* K12 inductest for detection of potential chemical carcinogens. *Mutat. Res.*, *130*, 141-151

McCann, J., Choi, E., Yamasaki, E. & Ames, B.N. (1975) Detection of carcinogens as mutagens in the *Salmonella*/microsome test: assay of 300 chemicals. *Proc. natl Acad. Sci. USA*, *72*, 5135-5139

Meretoja, T., Gripenberg, U., Bamford, D., Laamanen, I. & Sorsa, M. (1976) Mutagenicity and toxicity of amitrole. II. Human lymphocyte culture tests. *Mutat. Res.*, *40*, 191-196

Napalkov, N.P. (1962) Blastomogenic action of 3-amino-1,2,4-triazole (Russ.). *Gig. Tr. prof. Zabol.*, *6*, 48-51

Pienta, R.J., Poiley, J.A. & Lebherz, W.B., III (1977) Morphological transformation of early passage golden Syrian hamster embryo cells derived from cryopreserved primary cultures as a reliable *in vitro* bioassay for identifying diverse carcinogens. *Int. J. Cancer.*, *19*, 642-655

Rosenkranz, H.S. & Poirier, L.A. (1979) Evaluation of the mutagenicity and DNA-modifying activity of carcinogens and noncarcinogens in microbial systems. *J. natl Cancer Inst.*, *62*, 873-892

Sadtler Research Laboratories (1980) *The Sadtler Standard Spectra, Cumulative Index*, Philadelphia, PA

Schubert, O.E. (1971) Residues of amitrole in apple fruit following ground cover, fruit and leaf applications. *Proc. West Virginia Acad. Sci.*, *43*, 29-35

de Serres, F.J. & Ashby, J., eds (1981) *Evaluation of Short-Term Tests for Carcinogens. Report of the International Collaborative Program*, Vol. 1, *Progress in Mutation Research*, Amsterdam, Elsevier/North-Holland Publishing Co.

Shirasu, Y., Moriya, M., Kato, K., Furuhashi, A. & Kada, T. (1976) Mutagenicity screening of pesticides in the microbial system. *Mutat. Res.*, *40*, 19-30

Simmon, V.F. (1979a) In vitro mutagenicity assays of chemical carcinogens and related compounds with *Salmonella typhimurium*. *J. natl Cancer Inst.*, *62*, 893-899

Simmon, V.F. (1979b) In vitro assays for recombinogenic activity of chemical carcinogens and related compounds with *Saccharomyces cerevisiae* D3. *J. natl Cancer Inst.*, *62*, 901-909

Simmon, V.F., Rosenkranz, H.S., Zeiger, E. & Poirier, L.A. (1979) Mutagenic activity of chemical carcinogens and related compounds in the intraperitoneal host-mediated assay. *J. natl Cancer Inst.*, *62*, 911-918

Sittig, M., ed. (1980) *Pesticide Manufacturing and Toxic Materials Control Encyclopedia*, Park Ridge, NJ, Noyes Data Corp., pp. 54-55

Soulier, J., Farines, M. & Vicens, G. (1980) *Microanalysis of aminotriazole by high performance liquid chromatography*. In: *Pergamon Series in Environmental Sciences*, Vol. 3, Oxford, Pergamon Press, pp. 203-209

Steinhoff, D., Weber, H., Mohr, U. & Boehme, K. (1983) Evaluation of amitrole (aminotriazole) for potential carcinogenicity in orally dosed rats, mice, and golden hamsters. *Toxicol. appl. Pharmacol.*, *69*, 161-169

Storherr, R.W. & Burke, J. (1961) Determination of 3-amino-1,2,4-triazole in crops. *J. Assoc. off. anal. Chem.*, *44*, 196-199

Strum, J.M. & Karnovsky, M.J. (1971) Aminotriazole goiter. Fine structure and localization of thyroid peroxidase activity. *Lab. Invest.*, *24*, 1-12

Sumi, C., Yokoro, K. & Matsushima, R. (1985) Inhibition by 3-amino-1H-1,2,4-triazole of hepatic tumorigenesis induced by diethylstilbestrol alone or combined with *N*-nitrosobutylurea in WF rats. *J. natl Cancer Inst.*, *74*, 1329-1334

Tjälve, H. (1975) The distribution of labelled aminotriazole in mice. *Toxicology*, *3*, 49-67

Tsuda, H., Takahashi, M., Fukushima, S., Endo, Y. & Hikosaka, Y. (1973) Fine structure and localization of peroxidase activity in aminotriazole goiter. *Nagoya med. J.*, *18*, 183-190

Tsuda, H., Hananouchi, M., Tatematsu, M., Hirose, M., Hirao, K., Takahashi, M. & Ito, N. (1976) Tumorigenic effect of 3-amino-1H-1,2,4-triazole on rat thyroid. *J. natl Cancer Inst.*, *57*, 861-864

Tsuda, H., Takahashi, M., Murasaki, G., Ogiso, T. & Tatematsu, M. (1978) Effect of 3-amino-1H-1,2,4-triazole or low iodine diet on rat thyroid carcinogenesis induced by ethylenethiourea. *Nagoya med. J.*, *23*, 83-92

Tsutsui, T., Maizumi, H. & Barrett, J.C. (1984) Amitrole-induced cell transformation and gene mutations in Syrian hamster embryo cells in culture. *Mutat. Res.*, *140*, 205-207

Union Carbide Agricultural Products Co. (1984a) *Chemical Guide. Amitrole-T. Liquid Herbicide*, Research Triangle Park, NC, pp. 41-43

Union Carbide Agricultural Products Co. (1984b) *Chemical Guide. Liquid Amizine. Herbicide*, Research Triangle Park, NC, pp. 46-47

Union Carbide Agricultural Products Co. (1984c) *Chemical Guide. Amizol. Industrial Herbicide*, Research Triangle Park, NC, pp. 121-123

Union Carbide Agricultural Products Co. (1985a) *Chemical Guide. Amizine. Herbicide*, Research Triangle Park, NC, pp. 44-45

Union Carbide Agricultural Products Co. (1985b) *Material Safety Data Sheet, Amitrol-T, Liquid Herbicide*, Research Triangle Park, NC

US Environmental Protection Agency (1982) *Determination of amitrole by visible (colorimetric) spectroscopy*. In: Bontoyan, W.R. & Looker, J.B., eds, *Manual of Chemical Methods for Pesticides and Devices*, Arlington, VA, Association of Official Analytical Chemists

US Environmental Protection Agency (1984a) *Amitrole: Pesticide Registration Standard and Guidance Document*, Washington DC, Office of Pesticides and Toxic Substances

US Environmental Protection Agency (1984b) Amitrole; special review of pesticide products. *Fed. Regist.*, *49*, 20546-20549

Vesselinovitch, S.D. (1983) Perinatal hepatocarcinogenesis. *Biol. Res. Pregnancy Perinatology*, *4*, 22-25

Weast, R.C., ed. (1985) *CRC Handbook of Chemistry and Physics*, 66th ed., Boca Raton, FL, CRC Press, p. C-529

WHO (1975) *1974 Evaluations of Some Pesticide Residues in Food. The Monographs* (*WHO Pestic. Res. Ser., No. 4*), Geneva, pp. 3-49

Windholz, M., ed. (1983) *The Merck Index*, 10th ed., Rahway, NJ, Merck & Co., p. 74

Woodruff, R.C., Mason, J.M., Valencia, R. & Zimmering, S. (1985) Chemical mutagenesis testing in *Drosophila*. V. Results of 53 coded compounds tested for the National Toxicology Program. *Environ. Mutagenesis*, *7*, 677-702

Worthing, C.R., ed. (1983) *The Pesticides Manual*, 7th ed., Croydon, British Crop Protection Council, p. 300

Wuu, K.D. & Grant, W.F. (1966) Morphological and somatic chromosomal aberrations induced by pesticides in barley (*Hordeum vulgare*). *Can. J. genet. Cytol.*, *8*, 481-501

Wuu, K.D. & Grant, W.F. (1967) Chromosomal aberrations induced in somatic cells of *Vicia faba* by pesticides. *Nucleus*, *10*, 37-46

Yoder, J., Watson, M. & Benson, W.W. (1973) Lymphocyte chromosome analysis of agricultural workers during extensive occupational exposure to pesticides. *Mutat. Res.*, *21*, 335-340

OCCUPATIONAL EXPOSURES TO CHLOROPHENOLS

These exposures were considered by a previous Working Group, in February 1982 (IARC, 1982a). Individual compounds — pentachlorophenol and 2,4,5- and 2,4,6-trichlorophenol — were also evaluated earlier (IARC, 1979a, 1982b). Since that time, new data have become available and these have been incorporated into the monograph and taken into consideration in the present evaluation.

1. Historical Perspectives

Although all of the 19 possible chlorinated phenol structural isomers are available commercially (Freiter, 1979), only five have been of major significance, primarily as intermediates in the production of herbicidal chlorophenoxy acids. These are 2,4-dichloro-, 2,4,5-trichloro-, 2,4,6-trichloro-, 2,3,4,6-tetrachloro- and pentachlorophenol; the last has been used since the late 1930s as a wood preservative, with greatest use during the 1970s.

Earlier products contained high levels of higher chlorinated dibenzodioxins and dibenzofurans. Due to concern over the health effects of these contaminants, efforts were made in the late 1970s and 1980s to control and minimize these levels.

2. Production, Use, Occurrence and Analysis

2.1 Production

The chlorophenols to which there are important occupational exposures are given in Table 1.

(a) Processes

Although pentachlorophenol, 2,3,4,6-tetrachlorophenol and 2,4,6-trichlorophenol have been produced by the direct chlorination of phenol since the 1800s, variations in the catalysts, solvents and process conditions used that have been introduced since then may affect the type and levels of impurities (see section 2.2). In one process, molten phenol is chlorinated at 1.3 atm pressure and at temperatures of 65-130°C (optimum, 105°C) for approximately 3 h, until the melting-point of the product reaches 95°C. This product is a

Table 1. Identification of chlorophenols

Chemical name [Chem. Abstr. Services Reg. No.]	Chem. Abstr. and IUPAC Systematic Names [Synonym]	Structural and molecular formulae and molecular weight
2,4-Dichlorophenol [120-83-2]	2,4-Dichlorophenol [DCP]	$C_6H_4Cl_2O$ Mol. wt: 163.00
2,4,5-Trichlorophenol [95-95-4]	2,4,5-Trichlorophenol [TCP]	$C_6H_3Cl_3O$ Mol. wt: 197.46
2,4,6-Trichlorophenol [88-06-2]	2,4,6-Trichlorophenol	$C_6H_3Cl_3O$ Mol. wt: 197.46
2,3,4,6-Tetrachlorophenol [58-90-2]	2,3,4,6-Tetrachlorophenol [TECP]	$C_6H_2Cl_4O$ Mol. wt: 231.89

Table 1 (contd)

Chemical name [Chem. Abstr. Services Reg. No.]	Chem. Abstr. and IUPAC Systematic Names [Synonym]	Structural and molecular formulae and molecular weight
Pentachlorophenol [87-86-5]	Pentachlorophenol [PCP]	C_6HCl_5O Mol. wt: 266.35

mixture of tri- and tetrachlorophenols. Further chlorination is effected by adding an aluminium chloride catalyst and gradually increasing the temperature until the reactor product reaches a melting point of 174°C. The total reaction time is generally 8-10 h. The reaction product is technical-grade pentachlorophenol. If the desired products are tri- or tetrachlorophenols, the reactor contents are batch distilled when the product melting-point is 95°C. Vacuum distillation is carried out at approximately 60 mm pressure (0.08 atm); 2,4,6-trichlorophenol is removed at about 160°C, and 2,3,4,6-tetrachlorophenol distills at about 190°C (Herrick et al., 1979). The tri- and tetrachlorophenols may be used without further purification. In other processes for the chlorination of phenol, solvents (e.g., carbon tetrachloride; Rappe et al., 1978a; see IARC, 1979b, 1982c) and catalysts (such as iron, potassium and tellurium salts) have been used (Freiter, 1979).

Because the 5-position in phenol is not activated for electrophilic chlorination, 2,4,5-trichlorophenol is produced by hydrolysis of tetrachlorobenzene. In one procedure, 1,2,4,5-tetrachlorobenzene is treated with methanolic sodium hydroxide in an autoclave at 160°C for several hours. This synthetic process was patented in Germany in 1925 and is the basis for current production processes for 2,4,5-trichlorophenol (Windholz, 1983), although higher temperatures (180-200°C) and other solvents (ethylene glycol) have also been used (Milnes, 1971; Holmstedt, 1980). Hydrolysis has been used to transform hexachlorobenzene to pentachlorophenol (Sittig, 1980).

2,4-Dichlorophenol is produced in the same basic way as the penta-, tetra- and 2,4,6-trichlorophenols. Off-gases from the primary reactor, which contain chlorine and hydrogen chloride, are bubbled into a scrubber-reactor containing phenol at a temperature of approximately 120°C. Within the scrubber-reactor, 2,4-dichlorophenol and 2,6-dichlorophenol are formed, which are separated by vacuum distillation (Herrick et al., 1979).

(b) *Production volumes and producers*

Pentachlorophenol has been manufactured in many countries, including Canada, Czechoslovakia, the Federal Republic of Germany, France, Japan, the UK and the USA. In the USA, demand has been declining due to health concerns, and all but one producer have closed their plants. Production of pentachlorophenol in the USA peaked in 1974 and 1979 at 21.4 and 22.2 million kg, respectively. The current annual demand, at 15-20 million kg, exceeds the current capacity of the sole manufacturer in the USA and requires some importation (Mannsville Chemical Products Corp., 1983; Lee, 1985). There have been three producers in the Federal Republic of Germany; however, at least one plant is expected to cease pentachlorophenol production in 1986 (Anon., 1985a). France currently produces pentachlorophenol at two production sites (Herrick *et al.*, 1979). A plant in Wales (UK) with a capacity of 3 million kg ceased production in 1978 (Mannsville Chemical Products Corp., 1983). Commercial production of pentachlorophenol was started in Japan in 1960. In 1966, production was reported to be 14.5 million kg. In 1984, there were five manufacturers of pentachlorophenol (The Chemical Daily Co., 1984), but production volumes were not available.

Compared to pentachlorophenol, significantly less 2,3,4,6-tetrachlorophenol is produced worldwide. The chemical was produced in the USA and in Finland, but its production was discontinued in the late 1970s or early 1980s. One company in France also produced 2,3,4,6-tetrachlorophenol (Herrick *et al.*, 1979).

2,4,6-Trichlorophenol production was first reported in the USA in 1950 (US Tariff Commission, 1951). In 1979, the TSCA inventory listed two US companies, which manufactured 25 thousand kg of the compound (National Toxicology Program, 1984). The USA imported 250 kg of 2,4,6-trichlorophenol in 1980 (US International Trade Commission, 1981). There is currently no producer of the chemical in the USA. 2,4,6-Trichlorophenol is produced by one manufacturer in the Federal Republic of Germany and one in the UK. It was produced in Denmark and Sweden, and there were two producers in Japan (The Chemical Daily Co., 1984).

Commercial 2,4,5-trichlorophenol production in the USA was first reported in 1950 (US Tariff Commission, 1951), and in the late 1970s there were two producers. Due to the health hazards associated with 2,3,7,8-tetrachlorodibenzo-*para*-dioxin (TCDD) present in 2,4,5-trichlorophenol, however, the chemical and pesticide products derived from it have been banned in many countries. The one manufacturing plant in the UK and the only one in Italy ceased production in 1976 (Anon., 1977; Holmstedt, 1980). There has been no manufacture of 2,4,5-trichlorophenol in the USA since 1983 (Lee, 1985). There was one producer each in Austria and the Federal Republic of Germany, but both governments have ordered cessation of 2,4,5-trichlorophenol production (Lee, 1985).

2,4-Dichlorophenol was produced by three companies in the USA in 1982 (Anon., 1985b). The US International Trade Commission (1984) reported the existence of two manufacturers during 1983; based on the Commission's reporting criterion, this implies that at least 2300 kg of 2,4-dichlorophenol were produced during that year. US production in 1974 was estimated at 24 000 tonnes (Freiter, 1979). The Federal Republic of Germany,

France, Spain and the UK each have one producer of 2,4-dichlorophenol, and it is produced by two companies in Japan (The Chemical Daily Co., 1984).

2.2 Technical products and impurities

(a) Major chemical components

The structures of the chlorophenols considered are given in Table 1. These chemicals are generally marketed as unmodified phenols or as their sodium salts.

Pentachlorophenol is generally sold as a technical grade of 85% minimum purity. Specifications for pentachlorophenol and sodium pentachlorophenate products from one manufacturer in 1984 are given in Table 2. [Chlorinated dibenzodioxin and dibenzofuran impurities are discussed in detail in section 2.2(*b*).] A higher purity product, containing 98% pentachlorophenol, was produced in the USA, but production has been discontinued (Lee, 1985). Pentachlorophenol is sold as a dry product, as a concentrated (40%) solution in petroleum hydrocarbons, and as a 5% ready-to-use solution in petroleum hydrocarbons. The sodium salt is sold as a dry powder and is readily soluble in water.

The tri- and tetrachlorophenols produced by direct chlorination of phenol have essentially similar chlorophenol contaminants, varying in concentrations with source and batch. 2,4-Dichlorophenol produced by the chlorination of phenol would be expected to contain the *ortho*- and *para*-monochlorophenols as well as 2,6-dichlorophenol. The compound is available in the USA in a 99% pure form; the levels of impurities in the typical product were not available (Aldrich Chemical Co., 1984).

The exact composition of commercial grade 2,4,5-trichlorophenol was not available. However, the composition of the chlorophenoxy herbicide derivative, 2,4,5-trichlorophenoxyacetic acid (2,4,5-T) made from 2,4,5-trichlorophenol, indicates that low levels of dichlorophenol isomers are present (Sundström *et al.*, 1979). The nature of the impurities may depend upon the solvent and reaction conditions used in the production process.

(b) Chlorinated dibenzodioxin and dibenzofuran impurities

Commercial formulations of chlorophenols contain a series of nonpolar impurities including polychlorinated dibenzodioxins (PCDDs) (see IARC, 1977, 1982d) and dibenzofurans (PCDFs). There are 75 possible structural congeners of PCDDs (Esposito *et al.*, 1980) and 135 of PCDFs (Rappe, 1984). The individual isomers that have been tested vary greatly in biological action and toxicity. Special attention has been given to Cl_4-Cl_6 congeners in which the 2, 3, 7 and 8 positions are substituted with chlorine. The levels of PCDDs and PCDFs found in technical products depends on the production process, production conditions and purification processes used and may vary from batch to batch.

In the 1970s, a series of studies was performed to investigate the levels of PCDDs and PCDFs in various chlorophenol products (Firestone *et al.*, 1972; Firestone, 1977). In these early studies, a non-validated analytical method was used, and, due to lack of reference compounds and standards, isomer-specific analyses could not be performed, so that levels

Table 2. Specifications for technical pentachlorophenol and sodium pentachlorophenate products[a]

	Pentachlorophenol	Sodium pentachlorophenate
Pentachlorophenol (or Na salt)	85% min	82 ± 3%
Tetrachlorophenol (or Na salt)	10% max	4.5 ± 2%
Trichlorophenol (or Na salt)	0.1% max	0.1% max
Chlorophenoxyphenols	7% max	2 ± 1%
Hexachlorobenzene	50 mg/kg max	40 mg/kg max

[a]From Dynamit Nobel Chemicals (1984)

were reported as, e.g., total hexa-CDDs and total hepta-CDDs. The values reported are summarized in Tables 3, 4 and 5.

Buser and Bosshardt (1976) surveyed the PCDD and PCDF contents of pentachlorophenol and its sodium salt from commercial sources in Switzerland (Table 6). The samples were grouped into two series: those containing <1 μg/g and those containing >3 μg/g hexa-CDD. Samples with high levels of PCDDs also had high levels of PCDFs. For most samples, these contaminants occurred in the order tetra-<penta-<hexa-<hepta-<octa-CDD and tetra-≃penta-<hexa-<hepta-≃octa-CDF. The combined levels of PCDDs and PCDFs were in the ranges 2-16 and 1-26 μg/g, respectively, for the first series of samples, and 120-500 and 85-570 μg/g, respectively, for the second series of samples. The maximum levels of octa-CDD and octa-CDF were 370 and 300 μg/g, respectively.

Analysis of some of the pentachlorophenol sodium salt samples showed the unexpected presence of 0.06-0.25 μg/g of a tetra-CDD (Buser & Bosshardt, 1976); this impurity was later identified by Buser and Rappe (1978) as the unusual 1,2,3,4-tetra-CDD isomer.

Pentachlorophenol sodium salt samples with a high PCDD content (hexa-CDD ≥ 1 μg/g) were reanalysed on a high-resolution gas chromatographic column for the presence of individual PCDD isomers (Buser, 1978). As reported earlier, all samples showed an almost identical pattern of hexa- and hepta-CDD isomers. The major hexa-CDD isomers were identified as the toxic 1,2,3,6,7,8-hexa-CDD, and, in addition, 1,2,4,6,8,9- and 1,2,3,6,7,9-hexa-CDD. These three isomers were always present in an almost constant isomeric ratio of 50:40:10. Both of the hepta-CDD isomers were present in these samples in a ratio of 15:85, with 1,2,3,4,6,7,8-hepta-CDD as the major constitutent. All hexa-CDD isomers found in these samples were dimerization products of 2,3,4,6-tetrachlorophenol, the assumed precursor of pentachlorophenol in the chlorination process starting from phenol. Although the actual methods by which these samples were produced were not known, this result indicates that phenol chlorination processes were probably used.

Table 3. Chlorinated dibenzo-*para*-dioxins in various mono-, di-, tri-, tetra- and pentachlorophenol products by electron capture gas chromatography[a,b]

Sample[c]	Dioxin	Level found (μg/g)	Date recorded
2-CP	none	—	April 1967
2,4-DCP	none	—	April 1970
2,6-DCP	none	—	—
2,4,5-TCP-Na	none	—	Sept. 1967
2,4,5-TCP-Na	2,7-dichloro	0.7	June 1969
	2,3,7,8-tetrachloro[d]	1.4	
2,4,5-TCP-Na	1,3,6,8-tetrachloro	0.3	June 1969
	2,3,7,8-tetrachloro[d]	6.2	
2,4,5-TCP[d]	pentachloro	1.5	July 1970
2,4,5-TCP[d]	none	—	July 1970
2,4,5-TCP[d]	2,3,7,8-tetrachloro	0.07	July 1970
2,4,6-TCP	2,3,7-trichloro	93	—
	1,3,6,8-tetrachloro	49	
2,3,4,6-TeCP	hexachloro[d]	15	—
	hexachloro[d]	14	
	heptachloro[d]	5.1	
	octachloro	0.2	
2,3,4,6-TeCP	hexachloro[d]	4.1	March 1967
2,3,4,6-TeCP	none	—	—
PCP-Na	hexachloro[d]	14	Sept. 1967
	heptachloro	5.4	
	heptachloro[d]	9.1	
	octachloro	3.8	
PCP-Na	hexachloro[d]	20	June 1969
	heptachloro	1.3	
	heptachloro[d]	10	
	octachloro	3.3	
PCP	hexachloro	0.96	May 1970
	hexachloro[d]	38	
	heptachloro	10	
	heptachloro[d]	39	
	octachloro	15	
PCP	hexachloro[d]	35	July 1970
	heptachloro[d]	23	
PCP	hexachloro	0.03	March 1967
	hexachloro	0.14	

Table 3 (contd)

Sample[c]	Dioxin	Level found (μg/g)	Date recorded
PCP	hexachloro[d]	13	June 1969
	heptachloro	12	
	heptachloro[d]	35	
PCP	hexachloro[d]	0.9	May 1970
	heptachloro[d]	0.5	
	heptachloro[d]	1.6	
	octachloro[d]	5.3	
PCP	hexachloro[d]	15	July 1970
	heptachloro[d]	23	
	octachloro	15	

[a]From Firestone et al. (1972)

[b]The Working Group was aware that some 2,4,5-trichlorophenol products might have been purified, due to their subsequent use in the production of hexachlorophene.

[c]2-CP, 2-chlorophenol; 2,4-DCP, 2,4-dichlorophenol; 2,6-DCP, 2,6-dichlorophenol; 2,4,5-TCP-Na, 2,4,5-trichlorophenol sodium salt; 2,4,5-TCP, 2,4,5-trichlorophenol; 2,4,6-TCP, 2,4,6-trichlorophenol; 2,3,4,6-TeCP, 2,3,4,6-tetrachlorophenol; PCP-Na, pentachlorophenol sodium salt; PCP, pentachlorophenol

[d]Confirmed by combined gas chromatography-mass spectrometry

Rappe et al. (1978a) analysed the PCDF content of 2,4,6-trichlorophenol and 2,3,4,6-tetrachlorophenol, two of the most commonly used chlorophenol formulations on the Scandinavian market, and a pentachlorophenol from the USA. Both the tri- and tetrachlorophenol were prepared by the chlorination of phenol; the production method for the pentachlorophenol was not known. Although the chlorophenols differed as to origin and method of synthesis, the same penta-CDF, hexa-CDF and hepta-CDF isomers were found as the main PCDF components, although in somewhat different proportions. The main isomers were 1,2,4,6,8-penta-CDF, 1,2,3,4,6,8-, 1,2,4,6,7,8- and 1,2,4,6,8,9-hexa-CDF, and 1,2,3,4,6,7,8- and 1,2,3,4,6,8,9-hepta-CDF.

2.3 Use

The chlorophenols are used primarily as pesticides and as intermediates in the production of pesticides.

Pentachlorophenol is a fungicide used primarily for wood preservation (see IARC, 1981a). In the USA, 90% of pentachlorophenol is used for direct application to wood; the remaining 10% is converted to sodium pentachlorophenate, which is also used primarily to

Table 4. Chlorinated dibenzofurans detected in chlorophenols by combined gas chromatography-mass spectrometry[a]

Sample	Chlorinated dibenzofurans					
	3 Cl	4 Cl	5 Cl	6 Cl	7 Cl	8 Cl
2-CP		+				
2,4-DCP						
2,6-DCP						
2,4,5-TCP-Na						
2,4,5-TCP-Na	+[b]	+[c]	+[d]			
2,4,5-TCP		+[c]				
2,4,5-TCP						
2,4,5-TCP	+					
2,4,5-TCP						
2,4,6-TCP		+	+	+		
2,3,4,6-TeCP				+	+	+
2,3,4,6-TeCP			+	+		
2,3,4,6-TeCP		+		+		
PCP-Na				+	+	
PCP-Na			+	+	+	
PCP				+	+	
PCP				+		
PCP						
PCP		+	+	+	+	
PCP				+	+	+
PCP				+		

[a]From Firestone et al. (1972); abbreviations as in legend to Table 3
[b]Data suggest component is a trichlorodimethoxy-dibenzofuran
[c]Data suggest component is a tetrachlorodimethoxy-dibenzofuran
[d]Data suggest component is a pentachlorodimethoxy-dibenzofuran

Table 5. Polychlorinated dibenzodioxins (CDD) and polychlorinated dibenzofurans (CDF) found in pentachlorophenol products manufactured in the USA[a]

Sample	Level found (μg/g)
Hexa-CDDs	0.03-23
Hepta-CDDs	0.6-180
Octa-CDD	ND[b]-3600
Tetra-CDFs	<0.02-0.45
Penta-CDFs	<0.03-0.25
Hexa-CDFs	<0.03-36
Hepta-CDFs	<0.1-320
Octa-CDFs	<0.1-210

[a]From Firestone (1977)
[b]ND, not detected; detection limit, 1 μg/g

Table 6. PCDDs and PCDFs in commercial pentachlorophenol (PCP) samples in Switzerland[a]

Sample	Manufacturer	Appearance	PCDD (µg/g)					PCDF (µg/g)				
			Tetra-	Penta-	Hexa-	Hepta-	Octa-	Tetra-	Penta-	Hexa-	Hepta-	Octa-
PCP	b	powder, white	<0.01	<0.03	<0.03	1.0	3.2	<0.02	<0.03	0.15	1.7	1.3
			<0.01	<0.03	<0.03	1.1	2.8	<0.02	<0.03	0.10	1.4	1.2
PCP-Na	A	powder, cream-coloured	0.16	0.03	<0.03	0.3	1.2	<0.02	<0.03	0.20	1.2	3.0
			0.23	0.03	<0.03	0.3	1.2	<0.02	<0.03	0.10	1.0	2.5
PCP-Na	A	powder, cream-coloured	0.12	0.03	<0.03	0.3	1.5	<0.02	0.03	0.70	1.3	2.1
PCP-Na	A	pellets, cream-coloured	0.08	0.03	0.25	2.8	5.1	0.02	0.13	4.1	13	8.6
PCP	B	granules, yellow	<0.01	<0.03	0.15	1.1	5.5	0.45	0.03	0.30	0.5	0.2
PCP	B	powder, off-white	<0.01	<0.03	0.03	0.6	8.0	<0.02	<0.03	<0.03	<0.1	<0.1
PCP-Na	C	granules, light brown	0.06	0.03	0.40	4.2	11	0.02	0.08	1.2	3.6	3.9
PCP-Na	c	pellets, light brown	0.25	0.08	0.03	0.4	1.5	<0.02	0.03	0.75	2.3	4.1
PCP	d	flakes, greyish	<0.02	<0.03	9.5	125	160	<0.02	0.05	15	95	105
			<0.02	<0.03	7.2	150	200	<0.02	0.05	19	110	120
PCP	d	flakes, grey	<0.02	<0.03	9.1	180	280	0.05	0.25	36	320	210
			<0.02	<0.03	9.0	240	250	0.05	0.20	39	280	230
PCP-Na	d	granules, light brown	0.05	<0.03	3.4	40	115	<0.02	0.05	11	50	24
			0.05	<0.03	3.4	36	105	<0.02	0.03	11	44	29

Table 6 (contd)

Sample	Manufacturer	Appearance	PCDD (µg/g)					PCDF (µg/g)				
			Tetra-	Penta-	Hexa-	Hepta-	Octa-	Tetra-	Penta-	Hexa-	Hepta-	Octa-
PCP	B	flakes, greyish brown	<0.02	<0.03	10.0	130	210	0.20	0.20	13	70	55
PCP	B	flakes, greyish brown	<0.02	<0.03	5.4	130	370	0.07	0.20	9.1	60	65
PCP	C	flakes, light brown	<0.02	<0.03	5.2	95	280	0.02	0.40	28	200	230
PCP	C	flakes, light brown	<0.02	<0.03	3.3	27	90	<0.02	0.25	12	65	75
PCP	D	flakes, light brown	<0.02	<0.03	3.1	50	135	0.04	0.65	23	140	150
PCP	D	flakes, light brown	<0.02	<0.03	4.2	54	210	<0.02	0.10	23	160	140
PCP	c	flakes, light brown	<0.02	<0.03	3.1	54	170	<0.02	0.05	23	180	250
PCP	c	flakes, light brown	<0.02	<0.03	3.8	90	290	0.02	0.35	30	200	300

[a]From Buser and Bosshardt (1976)
[b]Sample from laboratory chemical supplier, analytical quality
[c]Manufacturer unknown
[d]Sample from laboratory chemical supplier, technical quality

preserve green lumber from fungal and mould spore infestation. The majority of US use of pentachlorophenol has been to preserve poles for power transmission lines and other utilities and cross arms (Mannsville Chemical Products Corp., 1983). In 1978, about 20% of all treated timber and lumber in the USA was treated with pentachlorophenol (US Environmental Protection Agency, 1984a). In the USA and Canada, the major treatment process is high-pressure impregnation in closed systems, whereas in Scandinavia dipping of wood in open vats is more common.

Pentachlorophenol has also been used as a preservative for leather (see IARC, 1981b), burlap, cordage, starches, dextrins and glues. It is also used as an insecticide on masonry for termite control, and has been used in rice fields (Vineis *et al.*, 1986), as a preharvest weed defoliant on seed crops and as a preservative for beans used for replanting. The sodium salt is used as a slimicide in pulp and paper mills, in cooling-tower water, in evaporation condensers, and, in the petroleum industry, as a slimicide in secondary oil recovery injection water (Freiter, 1979; Esposito *et al.*, 1980).

The alkali salts of 2,3,4,6-tetrachlorophenol have been used in the USA for the control of sapstain fungus in wood, which cause the wood to absorb water, leading to accelerated decay. Sodium tetrachlorophenate has been used for this purpose in the USA for over 40 years but has been replaced recently by sodium pentachlorophenate (US Environmental Protection Agency, 1984a).

The major use of 2,4,5-trichlorophenol has been as an intermediate in the manufacture of industrial and agricultural chemicals, including 2,4,5-T (see monograph on occupational exposures to chlorophenoxy herbicides, p. 357), 2-(2,4,5-trichlorophenoxy)propanoic acid (silvex), *O,O*-dimethyl *O*-(2,4,5-trichlorophenyl)phosphorothioic acid ester (ronnel) and hexachlorophene. 2,4,5-Trichlorophenol and its salts have been used as preservatives in the textile industry, in the adhesives industry for polyvinyl acetate emulsions, in the leather industry, and in the automotive industry for preservation of rubber gaskets. The water-soluble sodium salt has been used to preserve casein-derived adhesives, to preserve and stabilize metal cutting fluids and foundry core washes, and as an antimicrobial agent in cooling tower water and in pulp and paper mills (Freiter, 1979).

2,4,6-Trichlorophenol is used primarily as an unisolated intermediate in the production of 2,3,4,6-tetrachlorophenol and pentachlorophenol. It can also be used directly as a germicide, bactericide, glue and wood preservative, and antimildew treatment (Sittig, 1985). No data were available on the extent of this use.

2,4-Dichlorophenol is used principally as an intermediate in the production of herbicides such as 2,4-dichlorophenoxyacetic acid (2,4-D; see monograph on occupational exposures to chlorophenoxy herbicides, p. 357) and its derivatives (Sittig, 1985).

2.4 Regulatory status and guidelines

In 1978, the US Environmental Protection Agency began an action on pesticide uses of pentachlorophenol. In December 1984, that Agency prepared to cancel registrations of most non-wood uses and to modify the terms and conditions of registration for the remaining uses. All herbicide uses were cancelled, and its uses in pulp and paper mills and in the petroleum industry have been allowed under the conditions that protective clothing be worn by workers and that the hexachlorodibenzo-*para*-dioxin content be reduced to 1 µg/g in formulated products (US Environmental Protection Agency, 1984b, 1985).

In October 1983, the US Environmental Protection Agency (1983) cancelled registration of all pesticide products containing, or derived from 2,4,5-trichlorophenol. Existing stocks of these pesticides could still be used until supplies were depleted.

Its major derivative, 2,4,5-T, has been banned in many countries (see monograph on occupational exposures to chlorophenoxy herbicides, p. 369).

Occupational exposure limits for pentachlorophenol in 17 countries are presented in Table 7.

Regulations pertaining to the pesticide use of chlorophenols are not reviewed or reported in this monograph.

2.5 Occupational exposure

Table 8 summarizes chlorophenol measurements made in the urine of exposed workers in various industries and occupations. Generally, no indication is given of when the samples were taken, but levels would be expected to be highest towards the end of a working week. Pentachlorophenol levels seen in urine after unspecified nonoccupational exposures are reported to vary considerably (<0.01-1.8 mg/l), but, in general, are <0.05 mg/l (Detrick, 1977). It should be noted that both analytical methods and timing of sample collection may differ in the listed studies; for example, the hydrolysis phase of the analytical procedure is decisive for comparability of results (Edgerton & Moseman, 1979).

(a) Production plants

In pentachlorophenol manufacture, mean urinary concentrations ranging from 0.11 to 2.38 mg/l have been reported (Table 8). The average blood pentachlorophenol level of workers in a pentachlorophenol plant was 4.73 µg/ml, as compared to 2.23 µg/ml in a sodium chlorophenate plant; short-term levels of airborne pentachlorophenol exceeded 500 µg/m^3 during some work-phases (Bauchinger *et al.*, 1982). In another pentachlorophenol plant, the highest airborne concentration was 50 µg/m^3 (Zober *et al.*, 1981). In a plant manufacturing pentachlorophenol and its sodium salt, average airborne concentrations in the range of 0.5 to 16.5 mg/m^3 were found, with short-term exposures up to 100 mg/m^3 (Baxter, 1984).

Rappe and Nygren (1984) measured PCDD and PCDF levels in blood plasma of a group of workers manufacturing a chlorophenol formulation. Neither PCDDs nor octa-CDFs were found; levels of hepta-CDFs are shown in Table 9. The highest levels were found for those workers with the longest exposure.

(b) Wood preservation and other industrial uses of chlorophenols

Urinary levels of pentachlorophenol and tetrachlorophenol in wood treatment workers have been studied in the USA, Canada, Sweden and Finland (see Table 8). In pressure and non-pressure treatment plants, the mean pentachlorophenol levels in urine ranged from 0.08

Table 7. Occupational exposure limits for pentachlorophenol[a]

Country	Year	Concentration (mg/m^3)	Interpretation[b]
Australia	1978	0.5	TWA
Belgium	1978	0.5	TWA
Finland	1981	0.5	TWA
		1.5	STEL
German Democratic Republic	1979	0.5	TWA
		1.0	STEL
Germany, Federal Republic of	1985	0.5	TWA
Hungary	1974	0.2	TWA
Italy	1978	0.5	TWA
Japan	1978	0.5	TWA
The Netherlands	1978	0.5	TWA
Norway	1981	0.5	TWA
Poland	1976	0.5	Ceiling
Romania	1975	0.5	TWA
		1.0	Ceiling
Sweden	1984	0.5	TWA
		1.5	STEL
Switzerland	1978	0.5	TWA
UK	1985	0.5	TWA
		1.5	STEL
USA	1985		
ACGIH		0.5	TWA
		1.5	STEL
OSHA		0.5	TWA
USSR	1977	0.1	TWA
Yugoslavia	1971	0.5	Ceiling

[a]From International Labour Office (1980); Directoratet for Arbeidstilsynet (1981); Työsuojeluhallitus (1981); Arbetarskyydsstyrelsens Förgattningssamling (1984); Deutsche Forschungsgemeinschaft (1985); American Conference of Governmental Industrial Hygienists (ACGIH) (1985); Health and Safety Executive (1985); US Occupational Safety and Health Administration (OSHA) (1985)

[b]TWA, time-weighted average; STEL, short-term exposure limit

to 1.6 mg/l and from 0.02 to 2.8 mg/l, respectively. In sawmills, the corresponding concentrations varied from 0.20 to 3.2 mg/l among exposed workers. Values exceeding 1 mg/l have been reported during dipping and spraying and during loading of newly treated timber (Kauppinen & Lindroos, 1985). Embree *et al.* (1984) demonstrated a gradient of exposure from 0.230 mg/l in urine and 0.919 µg/l in serum in those directly handling wood, to 0.139 mg/l in urine and 0.354 µg/l in serum for those working in a sawmill but not in manual contact.

Table 8. Concentrations of chlorophenols in the urine of exposed workers by industry and activity

Industry and activity (country)	Substances measured[a]	Concentration in urine (mg/l) Mean (range)	No. of samples (of workers)	Reference
Production of pentachlorophenol (PCP) or its sodium salt (PCP-Na)				
PCP production plant (FRG)	PCP	0.11[b] (0.01-1.22)[c]	(18)	Triebig et al. (1981)
PCP and PCP-Na production plant (Brazil)	PCP	1.20 (0.35-3.40)[c]	9	Siqueira & Fernicola (1981)
PCP production plant (FRG)	PCP	2.38 (NS)[c]	(8)	Bauchinger et al. (1982)
PCP-Na production plant (FRG)	PCP	0.84 (NS)[c]	(14)	Bauchinger et al. (1982)
Pressure treatment of wood				
Pressure treatment plant (Hawaii, USA)	PCP	1.6 (NS)[c]	99 (11)	Casarett et al. (1969)
Pressure treatment plant (USA)	PCP	0.14 (0.04-0.76)[c]	30 (5)	Wyllie et al. (1975)
Pressure treatment plant (USA)	PCP	1.24 (0.17-5.57)[c]	NS	Arsenault (1976)
Pressure treatment plant (Hawaii, USA)	PCP	0.27 (<0.01-2.40)[c]	(23)	Klemmer et al. (1980)
	PCP	0.08[c] (NS)		
Pressure treatment plant (Canada)	TeCP	0.11[c] (NS)	8	Embree et al. (1984)
sawmill	PCP	0.08[c] (NS)	8	Embree et al. (1984)
Pressure treatment plant (Canada)	PCP	0.95 (<0.01-5.2)[c]	11	Markel et al. (1977)
Non-pressure treatment of wood				
Wood treatment plant (Hawaii, USA) vat dipping	PCP	2.6 (NS)[c]	136 (11)	Casarett et al. (1969)
Wood industry (USA)	PCP			Cranmer & Freal (1970)
carpentry		0.02[c] (NS)	(1)	
boat building		0.06 (NS)	(1)	
spraying		NS (0.13-0.27)[c]	(2)	
Spraying of timber (USA)	PCP	0.98 (0.13-2.58)[c]	NS	Arsenault (1976)
Dipping of timber (USA)	PCP	2.83 (0.12-9.68)[c]	NS	Arsenault (1976)

Table 8 (contd)

Industry and activity (country)	Substances measured[a]	Concentration in urine (mg/l) Mean (range)	No. of samples (of workers)	Reference
Non-pressure treatment of wood (contd)				
Wood treatment plant (Hawaii, USA) vat dipping, spraying, brushing	PCP	1.31 (0.09-3.3)[c]	(18)	Begley et al. (1977)
Wood treatment plant (Hawaii, USA) vat dipping	PCP	0.95 (<0.01-7.80)[c]	(18)	Klemmer et al. (1980)
Wood preservation and chlorophenol manufacture (Finland)	TeCP	≃1 (<0.2-10)	1376	Pekari & Aitio (1982)
	PCP	≃0.3 (<0.3-1)	1376	
Sawmills (Sweden)	CPs			Rappe et al. (1982)
loading of newly treated timber		3.2 (0.12-10.3)[c]	12	
handling of imported timber treated with chlorophenols		0.20 (0.03-0.50)[c]	22	
Sawmills (Finland)	CPs	0.3 (<0.01-3.9)	86	Kauppinen & Lindroos (1985)
Other industries				
Pest control (Hawaii, USA)	PCP	1.8 (0.003-35.7)[c]	210 (130)	Bevenue et al. (1967)
Pest control and farming (Hawaii, USA)	PCP	0.01 (<0.01-0.40)[c]	210	Klemmer et al. (1980)
Application of herbicides (USA)	PCP	0.01 (NS)	NS	Draper (1982)
Textile industry (Sweden)	CPs			Rappe et al. (1982)
impregnation of fabrics		0.30 (0.01-0.80)[c]	15	
sewing of impregnated fabrics		0.20 (0.01-0.35)[c]	20	
Tannery (Sweden)	CPs	2.7 (0.10-10.5)[c]	20	Rappe et al. (1982)

[a]PCP, pentachlorophenol; PCP-Na, pentachlorophenate sodium; CPs, chlorophenols; TeCP, 2,3,4,6-tetrachlorophenol; NS, not specified

[b]Median

[c]The Working Group noted that, as no acid hydrolysis of conjugates was included in these analyses, results reported probably underestimate total chlorophenol levels.

Table 9. Levels of heptachlorinated dibenzofurans (hepta-CDF) measured in blood plasma of individual workers exposed during the manufacture of chlorophenol[a]

Length of exposure (years)	Level of hepta-CDF (pg/g)
< 0.5	3
0.5	5
1	191
2	19
3	46-47
4	42
5	20
6+3	184
13	132
18	197

[a]From Rappe and Nygren (1984)

During routine tasks, the mean concentration of airborne pentachlorophenol in wood treatment plants was <300 µg/m³ (Wyllie et al., 1975; Arsenault, 1976; Todd & Timbie, 1983). Inhalation exposure to higher concentrations is possible during certain short workphases: opening of a pressurized cylinder (up to 1700 µg/m³) (Todd & Timbie, 1983), inside heated kilns in sawmills (up to 17 000 µg/m³ for unspecified chlorophenols) (Kauppinen & Lindroos, 1985) and emptying of bags of pentachlorophenols (up to 1500 µg/m³) (Markel & Lucas, 1975). In one saw mill, the air concentration of 2,4,6-trichlorophenol ranged from <2 to 120 µg/m³ (Kauppinen & Lindroos, 1985).

In addition to those industries listed in Table 8, chlorophenols have been measured in the air during brush application (15-40 µg/m³), during spraying of wood rafters (9-19 µg/m³) and in plywood plants (<10 µg/m³) (Kauppinen, 1984, 1986). The levels of chlorophenols to which workers are exposed through inhalation of wood dust containing chlorophenols have been reported to be <10 µg/m³ (Levin et al., 1976; Kauppinen & Lindroos, 1985).

Rappe et al. (1982) investigated the levels of PCDDs and PCDFs in samples of blood plasma taken from workers exposed to chlorophenols in a sawmill and in textile and leather industries (Table 10). The highest levels of chlorophenols in urine and of PCDDs and PCDFs in blood were found for workers exposed directly to liquid chlorophenol formulations (textile industry and tannery) and aqueous solutions of chlorophenates (sawmills). In the textile industry, a 100-fold difference was observed for workers in different job categories.

Table 10. Levels of PCDDs and PCDFs in blood samples from workers in a sawmill after exposure to 2,3,4,6-tetrachlorophenate and in textile and leather industries after exposure to pentachlorophenol or its derivatives[a]

Job	Chlorophenols in urine (µg/ml)	PCDDs in blood (pg/g)				PCDFs in blood (pg/g)			
		penta	hexa	hepta[b]	octa	penta	hexa	hepta[c]	octa
Sawmill									
Loader[d]	0.04	–	<3	<2	5	–	<3	40	<3
Loader[d]	0.03	–	3	10	18	–	<3	18	<3
Cleaner[d]	<0.02	–	<3	2	5	–	<3	30	<3
Packager[d]	<0.05	–	<3	<2	<3	–	<3	7	<3
Textile	3.12	1	1	59	304	1	1	33	10
	<0.01	<1	<1	<1	3	<1	<1	<1	<2
	<0.01	<1	<1	1	10	10	<1	<1	<2
	0.42	<1	<1	15	105	<1	<1	<1	<2
	0.16	<1	<1	6	30	1	<1	<1	<2
Tannery[e]	0.55[f]	–	<3	7	20	–	<3	7	<3
	0.04[f]	–	3	30	80	–	3	18	7
	0.03[f]	–	<3	4	12	–	<3	3	<3
	–	–	<2	2	7	–	<3	3	<3

[a]From Rappe et al. (1982)
[b]Major isomer, 1,2,3,4,6,7,8-hepta-CDD
[c]Major isomer, 1,2,3,4,6,7,8-hepta-CDF
[d]Blood sampling six months after last exposure
[e]Blood sampling eight months after last exposure
[f]Urine sampling six months after last exposure

(c) *Incineration*

Olie et al. (1977) reported on the occurrence of PCDDs and PCDFs in fly ash from three municipal incinerators in the Netherlands. They found several peaks by gas chromorography-mass spectrometry, but no quantification or isomer identification was possible due to lack of synthetic standards. Buser and Bosshardt (1978) studied fly ash from a municipal incinerator and an industrial heating facility, both in Switzerland. In the former, the level of PCDDs was 0.2 µg/g and that of PCDFs 0.1 µg/g. In the industrial incinerator, the levels were 0.6 µg/g and 0.3 µg/g, respectively.

The findings of Olie et al. (1977) and Buser and Bosshardt (1978) have been confirmed in subsequent reports and reviews (Buser et al., 1978a,b; Lustenhouwer et al., 1980). Fewer data have been reported on the levels of PCDDs and PCDFs in other incineration products, such as particulates and flue-gas condensate.

Rappe et al. (1978b) studied the burning extracts of materials impregnated with purified 2,4,6-tri- and pentachlorophenates and two commercial formulations of 2,3,4,6-tetrachlorophenate (containing 2,4,6-and pentachlorophenates). In addition to the expected dimerization products, present at total levels of mg/g chlorophenate burned, the highly toxic TCDD and 1,2,3,7,8-penta-CDD were also present in burning extracts from penta- and 2,3,4,6-tetrachlorophenate, at levels exceeding 10 μg/g chlorophenate. Levels of most PCDFs were generally much decreased (in contrast to PCDDs), although the levels of a few individual PCDFs increased; e.g., that of the major tetra-CDF (an unknown isomer) increased during burning by more than 100-fold, and two isomers were found that had not been identified in the starting materials. 2,3,7,8-Tetra-CDF, which is considered to be the most toxic of all PCDFs, was only a minor component in all the samples.

Rappe and Marklund (1983) analysed samples of baghouse ash and bottom ash from an industrial boiler burning pentachlorophenol-containing waste and sludge. Many PCDDs and PCDFs were identified; the total level of PCDDs was about 5 μg/g ash and that of PCDFs about 2.5 μg/g. TCDD was a very minor component. The level of PCDFs in baghouse ash was about half that of PCDDs, although known precursors of PCDFs are reported as only minor constituents of commercial chlorophenols (chlorinated diphenyl ethers).

2.6 Analytical methods

Typical methods for the analysis of chlorophenols in occupational environments are listed in Table 11.

Since chlorophenols are effectively absorbed through the skin, in most workplace situations air measurements alone may not give an accurate estimate of the total exposure of a worker. Therefore, biological monitoring of chlorophenol or dioxin levels in body fluids may be preferable. Hydrolysis is required to break down glucuronic acid conjugates of chlorophenols in urine. Most of the older analytical procedures, however, did not include the hydrolysis step, and the results thus mainly represent the concentration of 'free' chlorophenols in urine. The 'total' concentration may be considerably higher than the 'free' concentration (Edgerton & Moseman, 1979; Pekari & Aitio, 1982).

Occupational exposure to chlorophenols has also been estimated on the basis of blood samples (see, e.g., Bevenue et al., 1968; Rivers, 1972; Gossler & Schaller, 1978).

Methods for the analysis of chlorinated dibenzodioxins and dibenzofurans have been reviewed (Rappe, 1984). In the 1970s, when most of the analyses of commercial products were performed, the methods used did not allow isomer-specific determination of PCDD and PCDF isomers, and results were reported as total levels. At the end of the 1970s and during the 1980s, standard compounds became available, and isomer-specific methods were possible. Use of ^{13}C-labelled surrogates facilitates determinations of isomers at ppt (ng/kg or ng/l) or sub-ppt levels in biological matrices such as adipose tissue and blood serum.

Table 11. Methods for the analysis of chlorophenols

Sample matrix	Substances measured[a]	Sample preparation	Assay procedure[b]	Limit of detection[c]	Reference
Air	PCP	Collect in bubbler (ethylene glycol); acidify; extract (benzene); methylate (diazomethane)	GC/EC	<1 µg/m³	Wyllie et al. (1975)
	PCP	Collect on filter/bubbler (ethylene glycol); add methanol	HPLC/UV	ND	National Institute for Occupational Safety and Health (1978)
	PCP	Collect in silica gel tube; desorb (methanol in diethylether); methylate (diazomethane)	GC/EC	ND	Todd & Timbie (1983)
	CPs	Collect in bubbler (toluene); extract (borax); acetylate; extract (hexane)	GC/EC	1 µg/m³	Kauppinen & Lindroos (1985)
Airborne particulates	CPs	Collect on filter; Soxhlet extract (diethylether); methylate (diazomethane); dissolve (acetone); extract (hexane)	GC/EC	ND	Levin & Nilsson (1977)
Urine	PCP	Acidify; extract (petroleum ether); wash (water); methylate (diazomethane)	GC/EC	ND	Bevenue et al. (1966)
	PCP (organic tissues)	Acidify; extract (hexane/borax); acetylate (acetic anhydride/pyridine)	GC/EC	ND	Rudling (1970)
	PCP	Wash (hexane in alkaline solution); acidify; extract (hexane); alkylate (diazoalkane)	GC/EC	2 µg/l	Cranmer & Freal (1970)
	PCP	Acidify; extract (benzene); methylate (diazomethane)	GC	10 µg/l	Rivers (1972)
	PCP	Acidify; extract (benzene); methylate (diazomethane)	GC	1 µg/l	Gossler & Schaller (1978)
	PCP	Acid hydrolysis; extract (benzene); methylate (diazomethane); clean up column	GC/EC	1 µg/l	Edgerton & Moseman (1979)
	TeCP, PCP	Acid hydrolysis; extract (hexane/isopropanol); evaporate; dissolve (methanol/water)	HPLC/UV	27 µg/l	Pekari & Aitio (1982)

[a]PCP, pentachlorophenol; CPs, chlorophenols; TeCP, tetrachlorophenols

[b]GC, gas chromatography; EC, electron capture detection; HPLC, high-performance liquid chromatography; UV, ultraviolet detection

[c]ND, not determined

3. Biological Data Relevant to the Evaluation of Carcinogenic Risk to Humans

3.1 Biological effects in animals

Evaluations of evidence for carcinogenicity in animals and evidence for genetic activity in short-term tests of chlorophenols previously considered in the *IARC Monographs* are listed in Table 12. No attempt was made to update these data.

Table 12. Chlorophenols and their main impurity considered in this monograph which have previously been evaluated in the *IARC Monographs*[a]

Chemical	Evidence for carcinogenicity in animals	Evidence for genetic activity in short-term tests
Pentachlorophenol	inadequate	inadequate
2,4,5-Trichlorophenol	inadequate	no data
2,4,6-Trichlorophenol	sufficient	no data
TCDD	sufficient	inadequate

[a]From IARC (1982b)

3.2 Biological effects in humans other than cancer

(a) *Toxic effects*

The toxicological effects of occupational exposure to chlorophenols have been reviewed (IARC, 1979a; Ahlborg & Thunberg, 1980; Sterling *et al.*, 1982). The acute toxicity of chlorophenols, especially pentachlorophenol, is manifested by hyperthermia, convulsions and rapid death. Most data on the toxicity of chlorophenols come from studies in occupational environments, and the effects described include skin irritation, rashes and sometimes chloracne, indications of liver damage and neurological effects (Fielder *et al.*, 1982).

Zober *et al.* (1981) have reported immunological changes in woodworkers exposed to pentachlorophenol, and McGovern (1982) observed marked T-cell suppression in several persons exposed to chlorophenols. Aplastic anaemia has been associated with exposure to pentachlorophenol (Roberts, 1981). Baader and Bauer (1951) mentioned haematological changes in workers in a plant manufacturing pentachlorophenol.

Chloracne — similar to acne vulgaris in appearance — is seen in some heavily exposed workers (see Kimmig & Schulz, 1957; Arsenault, 1976; Suskind, 1985). It is one of the most prominent features of exposure to TCDD, which occurs as an impurity in commercial chlorophenol preparations.

(b) *Effects on reproduction and prenatal toxicity*

An interviewer-administered questionnaire study was carried out in a chemical company in Michigan, USA, in which 737 birth outcomes of wives of employees with potential exposure to dioxins and chlorophenols (pentachlorophenol and 2,4,5-trichlorophenol) were compared with those of wives of employees with no such exposure (Townsend *et al.*, 1982). No significant association was observed with regard to reproductive events in the study group as a whole. The statistically significant findings of increased incidences of spontaneous abortions in certain subgroups were considered to be attributable to chance.

A study of male workers manufacturing 2,4,5-T and with potential exposure to 2,4,5-trichlorophenol and TCDD in Nitro, WV, USA, included information concerning birth outcomes (Suskind & Hertzberg, 1980). The study population comprised 189 exposed and 155 nonexposed workers. Perinatal mortality was more frequent among newborns of exposed fathers, but not significantly so. No difference was found for either miscarriages or birth defects.

(c) *Absorption, distribution, excretion and metabolism*

Several cases of acute pentachlorophenol intoxication have resulted from dermal exposure (Truhaut *et al.*, 1952; Menon, 1958; Robson *et al.*, 1969). Among sawmill workers, the highest urinary concentrations of chlorophenols were found after dermal exposure (Kauppinen & Lindroos, 1985). Urine and serum concentrations of penta- and tetrachlorophenols were higher in workers exposed both dermally and by inhalation than in those exposed by inhalation only. However, in the absence of dermal exposure, workers exposed by inhalation were found to have higher serum chlorophenol levels than nonexposed workers, indicating that pulmonary absorption may also be important (Embree *et al.*, 1984). Absorption of pentachlorophenol from the gastrointestinal tract is rapid (Braun *et al.*, 1979).

In necropsy material from the general population, highest concentrations of pentachlorophenol were detected in the liver, brain and kidney (medians: 67, 47 and 43 ng/g, respectively) (Grimm *et al.*, 1981). The concentrations generally reported in fat are lower (range of median values, 10-23 ng/g) (Shafik, 1973; Ohe, 1979; Morgade *et al.*, 1980; Grimm *et al.*, 1981).

After administration of a low oral dose of pentachlorophenol (0.1 mg/kg bw) to volunteers, the kinetics fitted an open one-compartment model, with extensive enterohepatic circulation. In the urine, 86% of the dose was excreted in seven days; the half-times for elimination of pentachlorophenol and pentachlorophenyl glucuronide were 33 and 13 h, respectively (Braun *et al.*, 1979). After repeated occupational exposures, slower urinary elimination of pentachlorophenol has been reported (Begley *et al.*, 1977). In an abstract, the half-lives for the urinary excretion of 2,4,6-tri-, 2,3,4,6-tetra- and pentachlorophenol were reported to average 18 h, 5.1 days and 16 days, respectively (Pekari *et al.*, 1985). Kalman and Horstman (1983) reported that the minimum estimated half-time for urinary elimination of 2,3,4,6-tetrachlorophenol in two occupationally exposed workers was approximately two days.

Of total urinary pentachlorophenol, 40% or more has been found to be in the form of conjugates (Edgerton & Moseman, 1979; Needham *et al.*, 1981). Human liver microsomes metabolize pentachlorophenol to the glucuronide conjugate (Lilienblum, 1985). In an abstract, it was reported that 2,3,4,6-tetrachlorophenol and 2,4,6-trichlorophenol occur in the urine almost exclusively as conjugates (Pekari *et al.*, 1985).

With the exception of conjugates, the only other metabolite of pentachlorophenol detected in humans is tetrachlorohydroquinone (Ahlborg *et al.*, 1974). Human liver microsomes metabolize pentachlorophenol to this compound (Juhl *et al.*, 1985).

2,4-Dichlorophenol, 2,4,6-trichlorophenol (Judis, 1982) and pentachlorophenol (Hoben *et al.*, 1976) have been shown to bind reversibly to human serum albumin. The binding affinity appears to be related to lipophilicity (Judis, 1982).

(d) *Mutagenicity and chromosomal effects*

The genetic effects of chlorophenols have been reviewed (Wassom *et al.*, 1977/1978; Exon, 1984).

Peripheral blood lymphocytes were taken each month from January to May from six workers exposed to pentachlorophenol in a wood treatment plant in Idaho, USA and from four control subjects. During this time, exposure varied considerably. Levels of pentachlorophenol residues in the serum and urine were one to two orders of magnitude higher in the exposed individuals than in the single control measured (serum: 348-3963 *versus* 38-68 μg/l; urine: 41-760 *versus* 3-4 μg/l). Lymphocytes were cultured for 48 hours and only 25 cells were examined from each person. The incidence of chromosomal aberrations (breaks and gaps) was not statistically different in the two groups (Wyllie *et al.*, 1975). [The Working Group noted the small number of workers and the small sample size of cells.]

Bauchinger *et al.* (1982) examined lymphocyte chromosomes from 22 male workers employed at two plants of a pentachlorophenol-producing factory in the Federal Republic of Germany. The mean concentrations of pentachlorophenol in the blood and urine of workers at one plant (producing pentachlorophenol) was 4.73 and 2.38 μg/ml, and in the second plant (producing the sodium salt), 2.23 and 0.84 μg/ml, respectively. A small, but significant increase in the frequency of dicentrics (0.0016 *versus* 0.0005 per cell) and acentrics (0.0057 *versus* 0.0022 per cell) was observed in exposed workers as compared with a group of 22 matched controls (300 cells were examined from each worker and 500 from each control). No difference was found for other categories of damage (gaps, chromatid breaks and chromatid exchanges). No significant increase in the incidence of sister chromatid exchanges (SCEs) was found in pentachlorophenol-exposed workers who smoked as compared with smoking controls (9.41 *versus* 8.89 per cell). Within the control group, smokers had a higher incidence of SCEs than nonsmokers (8.89 *versus* 7.60). [The Working Group noted the unusual pattern of chromosomal damage in the exposed groups, similar to that after exposure to radiation.]

Chromosome studies were made on a group of workers in England ten years after accidental exposure to crude 2,4,5-trichlorophenol in which TCDD was a trace contaminant (Blank *et al.*, 1983). Group A was the control, group B had 'possibly been exposed', and group C was 'known to be exposed' (with chloracne). Fifty cells were examined from each subject. Chromosomal aberrations (in preparations from 48-h cultures) were scored in 124 individuals (31 in group A, 55 in group B, and 38 in group C). SCEs were scored in 40 individuals (8 in group A, 20 in group B, and 12 in group C). No significant difference was found between the groups either as regards chromosomal damage or SCEs. The mean number of chromosomal aberrations per cell was 0.0755 in group A, 0.0807 in group B and 0.0858 in group C. The mean number of SCEs per cell was 8.0 in group A, 8.5 in group B and 8.1 in group C. [The Working Group noted that the number of aberrant cells was not given nor was the type of aberration described.]

3.3 Case reports and epidemiological studies of carcinogenicity to humans

(a) Cohort studies

Mortality was reported for a small cohort of 204 workers involved in the manufacture of 2,4,5-T between 1950 and 1971 (Ott et al., 1980) and followed up to 1976, where reported exposures included 2,4,5-trichlorophenol. There were five deaths (7.0 expected) among those with one or more years of exposure, including one from cancer (1.3 expected).

Zack and Gaffey (1983) reported the mortality status of 884 white men employed for at least one year between 1955-1977 by a chemical plant in Nitro, WV, USA, involved in the production of trichlorophenol and 2,4,5-T. 4-Aminobiphenyl, a human bladder carcinogen (see IARC, 1982e), was produced from 1941-1952 in this plant. There were nine cases of bladder cancer, with 0.91 expected; deaths from cancer other than of the bladder were not in excess. One case of liposarcoma was reported among workers assigned to 2,4,5-T operations. An accident during trichlorophenol production which took place in this plant was reported by Zack and Suskind (1980) (see below).

In a cohort study of workers in two Danish chemical plants (Lynge, 1985) (described in the monograph on occupational exposure to chlorophenoxy herbicides, p. 388), the only potential exposure to 2,4,5-trichlorophenol was between 1951 and 1959, when small amounts were produced or purchased to make 2,4,5-T. No overall increase in cancer incidence rate was observed, but there were statistically significantly increased risks of soft-tissue sarcoma and lung cancer in different subcohorts. [The Working Group noted that 2,4-dichlorophenol is an intermediate in the production of 2,4-D, which was produced by the larger of the two plants.]

Cook et al. (1986) examined mortality between 1940 and 1979 for 2189 men involved in the manufacture of 2,4,5-trichlorophenol and 2,4,5-T; work histories were classified according to exposure to TCDD. There were 298 deaths observed (standardized mortality ratio [SMR], 91) and 61 cases of cancer (SMR, 96). Five cases of non-Hodgkin's lymphoma were seen (SMR, 238; 95% confidence interval [CI], 77-556), but there was no evidence of a dose-response relationship for TCDD exposure. [The Working Group noted that no account was taken of latency in the analysis.]

Three studies have described cancer occurrence among workers following accidents in trichlorophenol-producing plants, with peak exposures to TCDD. A high proportion of persons developed chloracne or acne-like lesions. Cook et al. (1980) observed three cancer deaths (1.6 expected) among 61 male employees involved in a 1964 accident in Michigan and followed up to the end of 1978. One death was reported to be from a fibrosarcoma. In the Federal Republic of Germany (Thiess et al., 1982), 74 workers were involved in an accident in 1953 in a plant producing 2,4,5-trichlorophenol. Follow-up through 1980 revealed three deaths from stomach cancer, with relative risks of the order of 4-5 depending on the comparison group; there was no excess of cancers at other sites combined. Zack and Suskind (1980) reported cancer outcomes of a cohort of 121 males involved in a 1949 accident in a 2,4,5-trichlorophenol plant in West Virginia, USA. Between 1949-1978, follow-up revealed nine cancer deaths, with 9.04 expected. Three of these were lymphatic or haematopoietic in origin (0.88 expected [$p = 0.047$]), and one was a primary dermal fibrous histiocytoma (0.15 expected).

Exposure to chlorophenols and chlorophenoxy herbicides in humans and death from soft-tissue sarcoma

[The Working Group noted that it is difficult to evaluate mortality from soft-tissue sarcoma in relation to exposure to chlorophenols and chlorophenoxy herbicides from cohort studies and case reports, as reviewed by Fingerhut et al. (1984).

[A pathology review of histological specimens confirmed only five of seven cases as soft-tissue sarcomas, indicating possible overascertainment of soft-tissue sarcomas in mortality studies. Underascertainment is also a possibility, as shown by Fingerhut et al. (1984) and by Lynge (1985), for employees of a chemical plant producing chlorophenoxy herbicides. The limited validity of death certificates of soft-tissue sarcoma has been documented in the USA (Percy et al., 1981).

[The Working Group also noted that revision of diagnosis for observed cases in mortality studies without identical revision of reference rates increases the complexity of interpreting results of such investigations.]

(b) *Case-control studies*

(i) *Soft-tissue sarcoma*

A Swedish case-control study of soft-tissue sarcoma and exposure to chlorophenoxy herbicides and chlorophenols (Hardell & Sandström, 1979) is described in the monograph on occupational exposures to chlorophenoxy herbicides (p. 390). When patients and controls with exposure to chlorophenoxy herbicides were excluded, the relative risk estimate for chlorophenol exposure was 6.6 ($p < 0.001$), with seven cases and six controls exposed [95% CI, 2.1-20.6].

A study in southern Sweden (also described in the monograph on occupational exposures to chlorophenoxy herbicides, p. 390) found a relative risk estimate for exposure to chlorophenols of 3.3 (95% CI, 1.3-8.1) (with 11 cases and eight controls exposed), when patients and controls with exposure to chlorophenoxy herbicides were excluded (Eriksson et al., 1981).

A New Zealand study of soft-tissue sarcoma (referred to in the monograph on occupational exposures to chlorophenoxy herbicides, p. 391) found an odds ratio of 1.6 (90% CI, 0.5-5.2) for potential exposure to chlorophenols for five days or more, more than ten years prior to diagnosis (Smith et al., 1984). Work in pelt-treatment departments (where 2,4,6-trichlorophenol has been used) or in tanneries (where pentachlorophenol and 2,4,6-trichlorophenol are used) yielded an odds ratio of 7.2 (six exposed cases; $p = 0.04$). When meat works and tanneries were contacted, it was found that two of the cases could not have been exposed to chlorophenols and exposure of a third was unlikely, while two could have been exposed to 2,4,6-trichlorophenol and one to pentachlorophenol.

(ii) *Malignant lymphoma*

A case-control study of exposure to chlorophenols, chlorophenoxy herbicides and other chemicals among 169 cases of malignant lymphoma and 338 controls has been reported in Sweden (Hardell et al., 1981). The study design, including ascertainment of exposure, was similar to that of the Swedish soft-tissue sarcoma studies described in the monograph on occupational exposures to chlorophenoxy herbicides (p. 392). Relative risk estimates of 2.2 [95% CI, 1.1-4.4] for low-grade chlorophenol exposure and 7.6 [95% CI, 3.5-17.4] for high-grade exposure were found, after excluding those exposed to chlorophenoxy herbicides. The low-grade classification involved continuous exposure for not more than one week, or repeated brief exposures for not more than one month; longer exposures were classified as high-grade. No 'noticeable difference' in excess risk could be demonstrated between Hodgkin's disease and non-Hodgkin's lymphoma.

A New Zealand case-control study of non-Hodgkin's lymphoma involving 83 cases, 168 controls with other cancer and 228 general population controls, found an odds ratio of 1.2 (90% CI, 0.5-2.9) for potential exposure to chlorophenols when using other cancer patients as controls, and an odds ratio of 1.4 (90% CI, 0.5-3.7) when using general population controls (Pearce et al., 1986). The odds ratio for fencing work, which involves exposure to chemicals such as copper-chrome arsenate as well as pentachlorophenol, was 2.0 (90% CI, 1.3-3.0). The odds ratio for slaughterhouse employment, which involves potential exposure to 2,4,6-trichlorophenol, was 1.8 (90% CI, 1.1-3.1); however, only four of the 19 cases who had worked in the plant reported working in the pelt department, where 2,4,6-trichlorophenol is used.

(iii) *Nasal and nasopharyngeal cancer*

Hardell et al. studied 44 cases of nasal cancer and 27 cases of nasopharyngeal cancer in northern Sweden and compared the reported frequency of exposure to chlorophenol (and other chemicals) with that of the combined 541 referents from earlier studies (Hardell & Sandström, 1979; Eriksson et al., 1981) from the Umeå region. Exposure was assessed in the same way as in the previous studies (see above). A relative risk estimate of 6.7 (95% CI, 2.8-16.2) was found for exposure to chlorophenols of more than one week, continuously or intermittently for more than one month. The most frequent occupations in which exposure occurred were sawmilling and carpentery. After controlling for exposure to wood dust, an odds ratio of 6.7 (95% CI, 2.9-15.6) was obtained.

A study of 167 sinonasal cancer cases and 167 colorectal cancer controls carried out in Denmark, Finland and Sweden found an association with woodwork (Hernberg et al., 1983). Two cases and no referent had probably been exposed to chlorophenols in addition to wood dust. A study of 839 cases of sinonasal cancer and 2465 controls from the Danish Cancer Registry classified these according to wood dust and chlorophenol exposure. A relative risk of 0.6 (95% CI, 0.3-1.2) was seen after adjustment for exposure to wood dust (Olsen & Møller-Jensen, 1984). [The Working Group noted that classification of chlorophenol exposure was based on occupational title and might not have been accurate.]

(iv) *Colon and liver cancer*

A study on colon cancer (described in the monograph on occupational exposures to chlorophenoxy herbicides, p. 393; Hardell, 1981) found a relative risk estimate of 1.8 (95% CI, 0.6-5.3) for high-grade chlorophenol exposure, based on six exposed cases and 13 (out of 541) exposed referents.

In a case-control study on primary liver cancer and several chemical exposures (Hardell et al., 1984; described in the monograph on occupational exposures to chlorophenoxy herbicides, p. 393), the risk ratio for high-grade exposure to chlorophenols was 2.2 (95% CI, 0.7-7.3).

4. Summary of Data Reported and Evaluation

4.1 Exposure data

Several chlorophenols and their salts have been widely produced since the 1950s and used as wood preservatives, fungicides, slimicides, weed-killers and as precursors for chlorophenoxy herbicides. Widespread occupational exposure to chlorophenols and their chlorinated dibenzodioxin and dibenzofuran impurities is known to have occurred, especially in manufacturing plants and in wood-treatment applications. Increased urinary levels of chlorophenols and increased concentrations in adipose tissue of some chlorinated dibenzodioxins and dibenzofurans have been measured in workers exposed in sawmills and tanneries and in the textile industry. Skin absorption is believed to be a major route of exposure in these occupations. Burning of chlorophenol-containing materials in industrial or municipal incinerators may lead to the formation of various dibenzodioxin and dibenzofuran congeners.

4.2 Experimental data

Previous IARC evaluations of the carcinogenicity to experimental animals of several individual chlorophenols and of their impurity, 2,3,7,8-tetrachloro-*para*-dibenzodioxin (TCDD), are summarized in section 3.1.

4.3 Human data

Two studies among the wives of the workers at two chemical plants did not show an association between pregnancy outcomes and paternal exposure to 2,4,5-trichlorophenol, pentachlorophenol and TCDD and other dioxins.

Three studies have been published in which cytogenetic effects were investigated in workers exposed occupationally to chlorophenols. In two of the studies, no difference was seen between exposed and control subjects; but in one of these studies the persons were examined ten years after exposure. The other study showed increased incidences of dicentric and acentric chromosomal aberrations, but not of gaps, chromatid breaks or sister chromatid exchanges.

Several cohort studies have been conducted among chemical industry workers with potential exposure to 2,4,5-trichlorophenol, TCDD and other chemicals. Mortality rates for all cancers combined were not elevated. In a Danish cohort study, there may have been exposure to chlorophenols, present as intermediates in the production of chlorophenoxy herbicides. No increase in the incidence of cancers at all sites combined was observed, but there were statistically significantly increased risks of soft-tissue sarcoma and lung cancer in different subcohorts.

Two case-control studies conducted in different regions of Sweden showed a statistically significant association between exposure to chlorophenols and soft-tissue sarcoma; a study from New Zealand did not.

A statistically significant association between malignant lymphoma and exposure to chlorophenols was identified in a Swedish case-control study. A case-control study of non-Hodgkin's lymphoma in New Zealand suggested a possible association with fencing work, but not with other occupational exposures to chlorophenols.

A case-control study in Sweden detected a significant association between nasal and nasopharyngeal cancer and exposure to chlorophenols, independent of exposure to wood dust.

4.4 Evaluation[1]

There is *limited evidence* for the carcinogenicity of occupational exposure to chlorophenols to humans.

5. References

Ahlborg, U.G. & Thunberg, T.M. (1980) Chlorinated phenols: occurrence, toxicity, metabolism and environmental impact. *Crit. Rev. Toxicol.*, 7, 1-35

---

[1]For definition of the italicized term, see Preamble, p. 22.

Ahlborg, U.G., Lindgren, J.-E. & Mercier, M. (1974) Metabolism of pentachlorophenol. *Arch. Toxicol.*, *32*, 271-281

Aldrich Chemical Co. (1984) *1984-1985 Catalog Aldrich Handbook of Fine Chemicals*, Milwaukee, WI, p. 374

American Conference of Governmental Industrial Hygienists (1985) *TLVs Threshold Limit Values and Biological Exposure Indices for 1985-86*, 2nd ed., Cincinnati, OH, p. 26

Anon. (1977) Coalite develops dioxin disposal system. *Eur. Chem. News*, *31*, 19

Anon. (1985a) Dynamit Nobel wins PCP fight. *Manuf. Chem.*, *56*, 13

Anon. (1985b) *Thomas Register of American Manufacturers and Thomas Register Catalog File*, 72nd ed., New York, Thomas Publishing Co., p. 3969/DIE

Arbetarskyddsstyrelsens Författningssamling (National Swedish Board of Occupational Safety and Health) (1984) *Occupational Exposure Limit Values (AFS 1984:5)* (Swed.), Solna, pp. 24, 26

Arsenault, R.D. (1976) Pentachlorophenol and contained chlorinated dibenzodioxins in the environment. A study of environmental fate, stability, and significance when used in wood preservation. *Proc. Am. Wood-Preserv. Assoc.*, *72*, 122-148

Baader, E.W. & Bauer, H.J. (1951) Industrial intoxication due to pentachlorophenol. *Ind. Med. Surg.*, *20*, 286-290

Bauchinger, M., Dresp, J., Schmid, E. & Hauf, R. (1982) Chromosome changes in lymphocytes after occupational exposure to pentachlorophenol (PCP). *Mutat. Res.*, *102*, 83-88

Baxter, R.A. (1984) Biochemical study of pentachlorophenol workers. *Ann. occup. Hyg.*, *28*, 429-438

Begley, J., Reichert, E.L., Rashad, M.N., Klemmer, H.W. & Siemsen, A.W. (1977) Association between renal function tests and pentachlorophenol exposure. *Clin. Toxicol.*, *11*, 97-106

Bevenue, A., Wilson, J., Potter, E.F., Song, M.K., Beckman, H. & Mallett, G. (1966) A method for the determination of pentachlorophenol in human urine in picogram quantities. *Bull. environ. Contam. Toxicol.*, *1*, 257-266

Bevenue, A., Wilson, J., Casarett, L.J. & Klemmer, H.W. (1967) A survey of pentachlorophenol content in human urine. *Bull. environ. Contam. Toxicol.*, *2*, 319-332

Bevenue, A., Emerson, M.L., Casarett, L.J. & Yauger, W.L., Jr (1968) A sensitive gas chromatographic method for the determination of pentachlorophenol in human blood. *J. Chromatogr.*, *38*, 467-472

Blank, C.E., Cooke, P. & Potter, A.M. (1983) Investigations for genotoxic effects after exposure to crude 2,4,5-trichlorophenol. *Br. J. ind. Med.*, *40*, 87-91

Braun, W.H., Blau, G.E. & Chenoweth, M.B. (1979) *The metabolism/pharmacokinetics of pentachlorophenol in man, and a comparison with the rat and monkey*. In: Deichmann, W.B., ed., *Toxicology and Occupational Medicine*, Amsterdam, Elsevier/North-Holland, pp. 289-296

Buser, H.-R. (1978) *Polychlorinated Dibenzo-p-dioxins and Dibenzofurans: Formation, Occurrence and Analysis of Environmentally Hazardous Compounds*, Thesis, Umeå, University of Umeå

Buser, H.-R. & Bosshardt, H.P. (1976) Determination of polychlorinated dibenzo-*p*-dioxins and dibenzofurans in commercial pentachlorophenols by combined gas chromatography-mass spectrometry. *J. Assoc. off. anal. Chem., 59*, 562-569

Buser, H.R. & Bosshardt, H.P. (1978) Polychlorinated dibenzo-*p*-dioxins, dibenzofurans and benzene in ashes of a municipal and industrial incinerator (Ger.). *Mitt. Geb. Lebensmittelunters. Hyg., 69*, 191-199

Buser, H.R. & Rappe, C. (1978) Identification of substitution patterns in polychlorinated dibenzo-*p*-dioxins (PCDDs) by mass spectrometry. *Chemosphere, 7*, 199-211

Buser, H.R., Bosshardt, H.-P. & Rappe, C. (1978a) Identification of polychlorinated dibenzo-*p*-dioxin isomers found in fly ash. *Chemosphere, 7*, 165-172

Buser, H.R., Bosshardt, H.-P., Rappe, C. & Lindahl, R. (1978b) Identification of polychlorinated dibenzofuran isomers in fly ash and PCB pyrolyses. *Chemosphere, 7*, 419-429

Casarett, L.J., Bevenue, A., Yauger, W.L., Jr & Whalen, S.A. (1969) Observations on pentachlorophenol in human blood and urine. *Am. ind. Hyg. Assoc. J., 30*, 360-366

The Chemical Daily Co. (1984) *JCW Chemicals Guide '84-85*, Tokyo, p. 312

Cook, R.R., Townsend, J.C., Ott, M.G. & Silverstein, L.G. (1980) Mortality experience of employees exposed to 2,3,7,8-tetrachlorodibenzo-*p*-dioxin (TCDD). *J. occup. Med., 22*, 530-532

Cook, R.R., Bond, G.G., Olsen, R.A., Ott, M.G. & Gondek, M.R. (1986) Evaluation of the mortality experience of workers exposed to the chlorinated dioxins. *Chemosphere* (in press)

Cranmer, M. & Freal, J. (1970) Gas chromatographic analysis of pentachlorophenol in human urine by formation of alkyl ethers. *Life Sci., 9*, 121-128

Detrick, R.S. (1977) Pentachlorophenol. Possible sources of human exposure. *Forest Prod. J., 27*, 13-16

Deutsche Forschungsgemeinschaft (German Research Community) (1985) *Maximal Concentrations in the Workplace and Biological Tolerance Values for Substances in the Work Environment 1985* (Ger.), Vol. 20, Weinheim, Verlagsgesellschaft mbH, p. 47

Direktoratet for Arbeidstilsynet (Directorate of Labour Inspection) (1981) *Administration Norms for Pollution in Work Atmosphere, No. 361* (Norw.), Oslo, p. 18

Draper, W.M. (1982) A multiresidue procedure for the determination and confirmation of acidic herbicide residues in human urine. *J. agric. Food Chem., 30*, 227-231

Dynamit Nobel Chemicals (1984) *Technical Information: Witophen P (Pentachlorophenol), Witophen N (Sodium Pentachlorophenate)*, Troisdorf, Federal Republic of Germany

Edgerton, T.R. & Moseman, R.F. (1979) Determination of pentachlorophenol in urine: the importance of hydrolysis. *J. agric. Food Chem., 27*, 197-199

Embree, V., Enarson, D.A., Chan-Yeung, M., DyBuncio, A., Dennis, R. & Leach, J. (1984) Occupational exposure to chlorophenates: toxicology and respiratory effects. *Clin. Toxicol.*, *22*, 317-329

Eriksson, M., Hardell, L., Berg, N.O., Möller, T. & Axelson, O. (1981) Soft-tissue sarcomas and exposure to chemical substances: a case-referent study. *Br. J. ind. Med.*, *38*, 27-33

Esposito, M.P., Tiernan, T.O. & Dryden, F.E. (1980) *Dioxins (EPA-600/2-80-197)*, Cincinnati, OH, US Environmental Protection Agency

Exon, J.H. (1984) A review of chlorinated phenols. *Vet. hum. Toxicol.*, *26*, 508-520

Fielder, R.J., Sorrie, G.S., Bishop, C.M., Jones, R.B. & Van Den Heuvel, M.J. (1982) *Toxicity Review*, Vol. 5, *Pentachlorophenol*, London, Her Majesty's Stationery Office

Fingerhut, M.A., Halperin, W.E., Honchar, P.A., Smith, A.B., Groth, D.H. & Russell, W.O. (1984) An evaluation of reports of dioxin exposure and soft tissue sarcoma pathology among chemical workers in the United States. *Scand. J. Work Environ. Health*, *10*, 299-303

Firestone, D. (1977) *Chemistry and analysis of phentachlorophenol and its contaminants*. In: Reynolds, H.L. & Yess, N., eds, *FDA By-lines No. 2, September 1977*, Washington DC, US Food and Drug Administration, pp. 57-89

Firestone, D., Ress, J., Brown, N.L., Barron, R.P. & Damico, J.N. (1972) Determination of polychlorodibenzo-*p*-dioxins and related compounds in commercial chlorophenols. *J. Assoc. off. anal. Chem.*, *55*, 85-92

Freiter, E.R. (1979) *Chlorophenols*. In: Grayson, M. & Eckroth, D., eds, *Kirk-Othmer Encyclopedia of Chemical Technology*, 3rd ed., Vol. 5, New York, John Wiley & Sons, pp. 864-872

Gossler, K. & Schaller, K.H. (1978) Quantitative determination of pentachlorophenol in urine and plasma by gas chromatography (Ger.). *Fresenius' Z. anal. Chem.*, *290*, 111-112

Grimm, H.-G., Schellmann, B., Schaller, K.-H. & Gossler, K. (1981) Pentachlorophenol concentrations in tissues and body fluids of normal people (Ger.). *Zbl. Bakt. Hyg. B.*, *174*, 77-90

Hardell, L. (1981) Relation of soft-tissue sarcoma, malignant lymphoma and colon cancer to phenoxy acids, chlorophenols and other agents. *Scand. J. Work Environ. Health*, *7*, 119-130

Hardell, L. & Sandström, A. (1979) Case-control study: soft-tissue sarcomas and exposure to phenoxyacetic acids or chlorophenols. *Br. J. Cancer*, *39*, 711-717

Hardell, L., Eriksson, M., Lenner, P. & Lundgren, E. (1981) Malignant lymphoma and exposure to chemicals, especially organic solvents, chlorophenols and phenoxy acids: a case-control study. *Br. J. Cancer*, *43*, 169-176

Hardell, L., Johansson, B. & Axelson, O. (1982) Epidemiological study of nasal and nasopharyngeal cancer and their relation to phenoxy acid or chlorophenol exposure. *Am. J. ind. Med.*, *3*, 247-257

Hardell, L., Bengtsson, N.O., Jonsson, U., Eriksson, S. & Larsson, L.G. (1984) Aetiological aspects on primary liver cancer with special regard to alcohol, organic solvents and acute intermittent porphyria — an epidemiological investigation. *Br. J. Cancer*, *50*, 389-397

Health and Safety Executive (1985) *Occupational Exposure Limits 1985* (*Guidance Note EH 40/85*), London, Her Majesty's Stationery Office, p. 17

Hernberg, S., Westerholm, P., Schultz-Larsen, K., Degerth, R., Kuosma, E., Englund, A., Engzell, U., Hansen, H.S. & Mutanen, P. (1983) Nasal and sinonasal cancer. Connection with occupational exposures in Denmark, Finland and Sweden. *Scand. J. Work Environ. Health*, *9*, 315-326

Herrick, E.C., Goldfarb, A.S., Fong, C.V., Konz, J. & Walker, P. (1979) *Hazards Associated with Organic Chemical Manufacturing: Chlorophenols by Chlorination of Phenol* (*MITRE Technical Report No. MTR-78W00364-05*), McLean, VA, The MITRE Corp.

Hoben, H.J., Ching, S.A., Young, R.A. & Casarett, L.J. (1976) A study of the inhalation of pentachlorophenol by rats. V. A protein binding study of pentachlorophenol. *Bull. environ. Contam. Toxicol.*, *16*, 225-232

Holmstedt, B. (1980) Prolegomena to Seveso. Ecclesiastes I 18. *Arch. Toxicol.*, *44*, 211-230

IARC (1977) *IARC Monographs on the Evaluation of the Carcinogenic Risk of Chemicals to Man*, Vol. 15, *Some Fumigants, the Herbicides 2,4-D and 2,4,5-T, Chlorinated Dibenzodioxins and Miscellaneous Industrial Chemicals*, Lyon, pp. 41-102

IARC (1979a) *IARC Monographs on the Evaluation of the Carcinogenic Risk of Chemicals to Humans*, Vol. 20, *Some Halogenated Hydrocarbons*, Lyon, pp. 303-325, 349-367

IARC (1979b) *IARC Monographs on the Evaluation of the Carcinogenic Risk of Chemicals to Humans*, Vol. 20, *Some Halogenated Hydrocarbons*, Lyon, pp. 371-399

IARC (1981a) *IARC Monographs on the Evaluation of the Carcinogenic Risk of Chemicals to Humans*, Vol. 25, *Wood, Leather and Some Associated Industries*, Lyon, pp. 63-76

IARC (1981b) *IARC Monographs on the Evaluation of the Carcinogenic Risk of Chemicals to Humans*, Vol. 25, *Wood, Leather and Some Associated Industries*, Lyon, p. 228

IARC (1982a) *IARC Monographs on the Evaluation of the Carcinogenic Risk of Chemicals to Humans*, Suppl. 4, *Chemicals, Industrial Processes and Industries Associated with Cancer in Humans, IARC Monographs, Volumes 1 to 29*, Lyon, pp. 88-89

IARC (1982b) *IARC Monographs on the Evaluation of the Carcinogenic Risk of Chemicals to Humans*, Suppl. 4, *Chemicals, Industrial Processes and Industries Associated with Cancer in Humans, IARC Monographs, Volumes 1 to 29*, Lyon, pp. 205-206, 249-250

IARC (1982c) *IARC Monographs on the Evaluation of the Carcinogenic Risk of Chemicals to Humans*, Suppl. 4, *Chemicals, Industrial Processes and Industries Associated with Cancer in Humans, IARC Monographs, Volumes 1 to 29*, Lyon, pp. 74-75

IARC (1982d) *IARC Monographs on the Evaluation of the Carcinogenic Risk of Chemicals to Humans*, Suppl. 4, *Chemicals, Industrial Processes and Industries Associated with Cancer in Humans, IARC Monographs, Volumes 1 to 29*, Lyon, pp. 238-243

IARC (1982e) *IARC Monographs on the Evaluation of the Carcinogenic Risk of Chemicals to Humans*, Suppl. 4, *Chemicals, Industrial Processes and Industries Associated with Cancer in Humans, IARC Monographs, Volumes 1 to 29*, Lyon, pp. 37-38

International Labour Office (1980) *Occupational Exposure Limits for Airborne Toxic Substances: A Tabular Compilation of Values from Selected Countries*, 2nd (rev.) ed. (*Occupational Safety and Health Series No. 37*), Geneva, pp. 166-167

Judis, J. (1982) Binding of selected phenol derivatives to human serum proteins. *J. pharm. Sci., 71*, 1145-1147

Juhl, U., Witte, I. & Butte, W. (1985) Metabolism of pentachlorophenol to tetrahydroquinone by human liver homogenate. *Bull. environ. Contam. Toxicol., 35*, 596-601

Kalman, D.A. & Horstman, S.W. (1983) Persistence of tetrachlorophenol and pentachlorophenol in exposed woodworkers. *J. Toxicol. clin. Toxicol., 20*, 343-352

Kauppinen, T. (1984) *Nordic Expert Group for Documentation of Occupational Exposure Limits. 54. Chlorophenols (Arbete och Hälsa No. 46)* (Swed.), Solna, Arbetarskyddsstyrelsen

Kauppinen, T. (1986) Occupational exposure to chemical agents in the plywood industry. *Ann. occup. Hyg., 30*, 19-29

Kauppinen, T. & Lindroos, L. (1985) Chlorophenol exposure in sawmills. *Am. ind. Hyg. Assoc. J., 46*, 34-38

Kimmig, J. & Schulz, K.H. (1957) Occupational acne (so-called chloracne) due to chlorinated aromatic cyclic ether (Ger.). *Dermatologica, 115*, 540-546

Klemmer, H.W., Wong, L., Sato, M.M., Reichert, E.L., Korsak, R.J. & Rashad, M.N. (1980) Clinical findings in workers exposed to pentachlorophenol. *Arch. environ. Contam. Toxicol., 9*, 715-725

Lee, A. (1985) *Analysis of Technical Information to Support RCRA Rules for Dioxins-containing Waste Streams (Contract No. 5W-6242-NASX)*, Washington DC, US Environmental Protection Agency

Levin, J.-O. & Nilsson, C.-A. (1977) Chromatographic determination of polychlorinated phenols, phenoxyphenols, dibenzofurans and dibenzodioxins in wood-dust from worker environments. *Chemosphere, 7*, 443-448

Levin, J.-O., Rappe, C. & Nilsson, C.-A. (1976) Use of chlorophenols as fungicides in sawmills. *Scand. J. Work Environ. Health, 2*, 71-81

Lilienblum, W. (1985) Formation of pentachlorophenol glucuronide in rat and human liver microsomes. *Biochem. Pharmacol., 34*, 893-894

Lustenhouwer, J.W.A., Olie, K. & Hutzinger, O. (1980) Chlorinated dibenzo-p-dioxins and related compounds in incinerator effluents: A review of measurements and mechanisms of formation. *Chemosphere, 9*, 501-522

Lynge, E. (1985) A follow-up study of cancer incidence among workers in manufacture of phenoxy herbicides in Denmark. *Br. J. Cancer, 52*, 259-270

Mannsville Chemical Products Corp. (1983) *Chemical Products Synopsis: Pentachlorophenol*, Cortland, NY

Markel, H.L., Jr & Lucas, J.B. (1975) *Health Hazard Determination Report No. 74-117-251, Weyerhaeuser Treating Plant, De Queen, Arkansas*, Cincinnati, OH, National Institute for Occupational Safety and Health

Markel, H.L., Jr, Ligo, R.N. & Lucas, J.B. (1977) *Health Hazard Evaluation Determination Report No. 75-117-372, Koppers Company, Inc., North Little Rock, Arkansas*, Cincinnati, OH, National Institute for Occupational Safety and Health

McGovern, J.J., Jr (1982) Apparent immunotoxic response to phenolic compounds. *Food Chem. Toxicol., 20*, 496

Menon, J.A. (1958) Tropical hazards associated with the use of pentachlorophenol. *Br. med. J., i*, 1156-1158

Milnes, M.H. (1971) Formation of 2,3,7,8-tetrachlorodibenzodioxin by thermal decomposition of sodium 2,4,5-trichlorophenate. *Nature, 232*, 395-396

Morgade, C., Barquet, A. & Pfaffenberger, C.D. (1980) Determination of polyhalogenated phenolic compounds in drinking water, human blood serum and adipose tissue. *Bull. environ. Contam. Toxicol., 24*, 257-264

National Institute for Occupational Safety and Health (1978) *Pentachlorophenol S-297 (NIOSH Manual of Analytical Methods, Vol. 4, 2nd ed.; DHEW (NIOSH) Publ. No. 78-175)*, Cincinnati, OH, US Department of Health, Education, and Welfare, pp. S297-1 — S297-8

National Toxicology Program (1984) *Fourth Annual Report on Carcinogens*, Washington DC, US Department of Health and Human Services, pp. 439-441

Needham, L.L., Cline, R.E., Head, S.L. & Liddle, J.A. (1981) Determining pentachlorophenol in body fluids by gas chromatography after acetylation. *J. anal. Toxicol., 5*, 283-286

Ohe, T. (1979) Pentachlorophenol residues in human adipose tissue. *Bull. environ. Contam. Toxicol., 22*, 287-292

Olie, K., Vermeulen, P.L. & Hutzinger, O. (1977) Chlorodibenzo-*p*-dioxins and chlorodibenzofurans are trace components of fly ash and flue gas of some municipal incinerators in The Netherlands. *Chemosphere, 8*, 455-459

Olsen, J.H. & Møller-Jensen, O. (1984) Nasal cancer and chlorophenols. *Lancet, ii*, 47-48

Ott, M.G., Holder, B.B. & Olson, R.D. (1980) A mortality analysis of employees engaged in the manufacture of 2,4,5-trichlorophenoxyacetic acid. *J. occup. Med., 22*, 47-50

Pearce, N.E., Smith, A.H., Howard, J.K., Sheppard, R.A., Giles, H.J. & Teague, C.A. (1986) Non-Hodgkin's lymphoma and exposure to phenoxyherbicides, chlorophenols, fencing work and meat works employment. A case-control study. *Br. J. ind. Med., 43*, 75-83

Pekari, K. & Aitio, A. (1982) A simple liquid chromatographic method for the analysis of penta- and tetrachlorophenols in urine of exposed workers. *J. Chromatogr., 232*, 129-136

Pekari, K., Järvisalo, J. & Aitio, A. (1985) *Kinetics of urinary excretion of 2,4,6-tri-, 2,3,4,6-tetra- and pentachlorophenol in workers exposed in lumber treatment* (Abstract). In: *Proceedings of the 26th Congress of the European Society of Toxicology, Kuopio, Finland, June 16-19, 1985*, Kuopio, University of Kuopio, p. 193

Percy, C., Stanek, E. & Gloeckler, L. (1981) Accuracy of cancer death certificates and its effect on cancer mortality statistics. *Am. J. publ. Health, 71*, 242-250

Rappe, C. (1984) Analysis of polychlorinated dioxins and furans. *Environ. Sci. Technol., 18*, 78A-90A

Rappe, C. & Marklund, S. (1983) *Thermal degradation of pesticides and xenobiotics: formation of polychlorinated dioxins and dibenzofurans*. In: Miyamoto, J., ed., *IUPAC Pesticide Chemistry*, Oxford, Pergamon, pp. 317-322

Rappe, C. & Nygren, M. (1984) *Chemical analysis of human samples. Identification and quantification of polychlorinated dioxins and dibenzofurans*. In: de Serres, F.J. & Pero, R.W., eds, *Individual Susceptibility of Genotoxic Agents in the Human Population*, New York, Plenum, pp. 305-314

Rappe, C., Garå, A. & Buser, H.R. (1978a) Identification of polychlorinated dibenzofurans (PCDFs) in commercial chlorophenol formulations. *Chemosphere, 7*, 981-991

Rappe, C., Marklund, S., Buser, H.R. & Bosshardt, H.-P. (1978b) Formation of polychlorinated dibenzo-*p*-dioxins (PCDDs) and dibenzofurans (PCDFs) by burning or heating chlorophenates. *Chemosphere, 7*, 269-281

Rappe, C., Nygren, M., Buser, H.-R. & Kauppinen, T. (1982) *Occupational exposure to polychlorinated dioxins and dibenzofurans*. In: Hutzinger, O., Frei, R.W., Merian, E. & Pocchiari, F., eds, *Chlorinated Dioxins and Related Compounds — Impact on the Environment*, New York, Pergamon Press, pp. 495-513

Rivers, J.B. (1972) Gas chromatographic determination of pentachlorophenol in human blood and urine. *Bull. environ. Contam. Toxicol., 8*, 294-296

Roberts, H.J. (1981) Aplastic anaemia due to pentachlorophenol. *New Engl. J. Med., 305*, 1650-1651

Robson, A.M., Kissane, J.M., Elvick, N.H. & Pundavela, L. (1969) Pentachlorophenol poisoning in a nursery for newborn infants. I. Clinical features and treatment. *J. Pediatr., 75*, 309-316

Rudling, L. (1970) Determination of pentachlorophenol in organic tissues and water. *Water Res., 4*, 533-537

Shafik, T.M. (1973) The determination of pentachlorophenol and hexachlorophene in human adipose tissue. *Bull. environ. Contam. Toxicol., 10*, 57-63

Siqueira, M.E.P.B. & Fernicola, N.A.G.G. (1981) Determination of pentachlorophenol in urine. *Bull. environ. Contam. Toxicol., 27*, 380-385

Sittig, M., ed. (1980) *Pesticide Manufacturing and Toxic Materials Control Encyclopedia*, Park Ridge, NJ, Noyes Data Corp., pp. 598-602

Sittig, M., ed. (1985) *Handbook of Toxic and Hazardous Chemicals and Carcinogens*, 2nd ed., Park Ridge, NJ, Noyes Publications, pp. 327-328, 886-887

Smith, A.H., Pearce, N.E., Fisher, D.O., Giles, H.J., Teague, C.A. & Howard, J.K. (1984) Soft tissue sarcoma and exposure to phenoxyherbicides and chlorophenols in New Zealand. *J. natl Cancer Inst., 73*, 1111-1117

Sterling, T.D., Stoffman, L.D., Sterling, D.A. & Maté, G. (1982) Health effects of chlorophenol wood preservatives on sawmill workers. *Int. J. Health Serv., 12*, 559-571

Sundström, G., Jensen, S., Jansson, B. & Erne, K. (1979) Chlorinated phenoxyacetic acid derivatives and tetrachlorodibenzo-*p*-dioxin in foliage after application of 2,4,5-trichlorophenoxyacetic acid esters. *Arch. environ. Contam. Toxicol., 8*, 441-448

Suskind, R.R. (1985) Chloracne, the hallmark of dioxin intoxication. *Scand. J. Work Environ. Health, 11*, 165-171

Suskind, R.R. & Hertzberg, V.J. (1980) Human health effects of 2,4,5-T and its toxic contaminants. *J. Am. med. Assoc., 251*, 2372-2380

Thiess, A.M., Frentzel-Beyme, R. & Link, R. (1982) Mortality study of persons exposed to dioxin in a trichlorophenol-process accident that occurred in the BASF AG on November 17, 1953. *Am. J. ind. Med., 3*, 179-189

Todd, A.S. & Timbie, C.Y. (1983) *Industrial Hygiene Surveys of Occupational Exposure to Wood Preservative Chemicals (DHHS (NIOSH) Publ. No. 83-106)*, Cincinnati, OH, US Department of Health and Human Services

Townsend, J.C., Bodner, K.M., van Peenen, P.F.D., Olson, R.D. & Cook, R.R. (1982) Survey of reproductive events of wives of employees exposed to chlorinated dioxins. *Am. J. Epidemiol., 115*, 695-713

Triebig, G., Krekeler, H., Gossler, K. & Valentin, H. (1981) Investigations on neurotoxicity of chemical substances at the workplace. II. Determination of motor and sensory nerve conduction velocity in persons occupationally exposed to pentachlorophenol (Ger.). *Int. Arch. occup. environ. Health, 48*, 357-367

Truhaut, R., l'Epée, P. & Boussemart, E. (1952) Research on pentachlorophenol toxicology. II. Occupational poisonings in the wood industry. Observations of two deaths (Fr.). *Arch. Mal. prof., 13*, 567-569

Työsuojeluhallitus (National Finnish Board of Occupational Safety and Health) (1981) *Airborne Contaminants in the Workplaces* (*Safety Bull. 3*) (Finn.), Tampere, p. 21

US Environmental Protection Agency (1983) 2,4,5-T and silvex products: intent to cancel registrations of pesticide products containing 2,4,5-T and silvex; revocation of notices of intent to hold a hearing to determine whether certain uses of 2,4,5-T or silvex should be cancelled. Enforcement policy on transfer, distribution, sale and importation of unregistered products. *Fed. Regist.*, *48*, 48434-48437

US Environmental Protection Agency (1984a) *Wood Preservative Pesticides: Creosote, Pentachlorophenol, and the Inorganic Arsenicals* (*Position Document 4*), Washington DC, p. 32

US Environmental Protection Agency (1984b) Pentachlorophenol; preliminary notice of determination concluding the rebuttable presumption against registration of pesticide products containing pentachlorophenol for non-wood preservative uses; proposed notice of intent to cancel such registrations; notice of availability of Position Document 2/3. *Fed. Regist.*, *49*, 48367-48372

US Environmental Protection Agency (1985) Creosote, pentachlorophenol, and inorganic arsenicals; decision to postpone effective dates. *Fed. Regist.*, *50*, 4269-4270

US International Trade Commission (1981) *Imports of Benzenoid Chemicals and Products, 1980* (*USITC Publ. 1163*), Washington DC, US Government Printing Office, p. 108

US International Trade Commission (1984) *Synthetic Organic Chemicals, United States Production and Sales, 1983* (*USITC Publ. 1588*), Washington DC, US Government Printing Office, p. 38

US Occupational Safety and Health Administration (1985) Labor. *US Code fed. Regul.*, Title 29, Part 1910.1000

US Tariff Commission (1951) *Synthetic Organic Chemicals, US Production and Sales, 1950* (*Report No. 173, Second Series*), Washington DC, US Government Printing Office, p. 130

Vineis, P., Terracini, B., Ciccone, G., Cignetti, A., Colombo, E., Donna, A., Maffi, L., Pisa, R., Ricci, P., Zanini, E. & Comba, P. (1986) Phenoxy herbicides and soft-tissue sarcomas in female-rice weeders: a population-based case-control study. *Scand. J. Work Environ. Health* (in press)

Wassom, J.S., Huff, J.E. & Loprieno, N. (1977/1978) A review of the genetic toxicology of chlorinated dibenzo-*p*-dioxins. *Mutat. Res.*, *47*, 141-160

Windholz, M., ed. (1983) *The Merck Index*, 10th ed., Rahway, NJ, Merck & Co., pp. 1021, 1378

Wyllie, J.A., Gabica, J., Benson, W.W. & Yoder, J. (1975) Exposure and contamination of the air and employees of a pentachlorophenol plant, Idaho — 1972. *Pestic. Monit. J.*, *9*, 150-153

Zack, J.A. & Gaffey, W.R. (1983) A mortality study of workers employed at the Monsanto company plant in Nitro, West Virginia. *Environ. Sci. Res.*, *26*, 575-591

Zack, J.A. & Suskind, R.R. (1980) The mortality experience of workers exposed to tetrachlorodibenzodioxin in a trichlorophenol process accident. *J. occup. Med.*, *22*, 11-14

Zober, A., Schaller, K.H., Gossler, K. & Krekeler, H.J. (1981) Pentachlorophenol and liver function: a pilot-study on occupationally exposed groups (Ger.). *Int. Arch. occup. environ. Health*, *48*, 347-356

OCCUPATIONAL EXPOSURES TO CHLOROPHENOXY HERBICIDES

These exposures were considered by a previous Working Group, in February 1982 (IARC, 1982a). Individual compounds — 2,4-dichlorophenoxyacetic acid (2,4-D), 2,4,5-trichlorophenoxyacetic acid (2,4,5-T) and (4-chloro-2-methylphenoxy)acetic acid (MCPA) — were also considered earlier (IARC, 1977a, 1982b, 1983). Since that time, new data have become available, and these have been incorporated into the monograph and taken into consideration in the present evaluation.

1. Historical Perspectives

The first chlorophenoxy acetic herbicides, 2,4-D, 2,4,5-T and MCPA, were introduced for agricultural use in the mid-1940s; the chlorophenoxy propionic acid derivatives, 2-(2,4,5-trichlorophenoxy)propanoic acid (silvex), 2-(4-chloro-2-methylphenoxy)-propanoic acid (mecoprop) and 2-(2,4-dichlorophenoxy)propanoic acid (dichlorprop), have been used since the mid 1950s and early 1960s. By the mid 1960s, chlorophenoxy herbicides, generally formulated as esters or amine salts, were the most widely used class of commercial herbicides. Agent Orange, a mixture of butyl esters of 2,4-D and 2,4,5-T, was used extensively during the US intervention in Viet Nam in 1961-1971, for defoliation.

The identification and characterization of the chlorinated dibenzo-*para*-dioxins (IARC, 1977b, 1982c), and especially the toxic 2,3,7,8-tetra isomer, in chlorophenoxy herbicides such as 2,4,5-T in the late 1960s resulted in greatly diminished use of this class of herbicides.

2. Production, Use, Occurrence and Analysis

2.1 Production

The synthesis of 2,4-D and 2,4,5-T was first reported in 1941 (Pokorny, 1941). 2,4-D is currently produced by the reaction of 2,4-dichlorophenol with the sodium salt of monochloroacetic acid, typically followed by an acid treatment to convert the 2,4-D salt to an acid (Sittig, 1980). 2,4-D has also been prepared commercially by the chlorination of phenoxyacetic acid (International Programme on Chemical Safety, 1984). 2,4-D esters and amine salts are produced by reaction with suitable alcohols or amines. 2,4,5-T and its derivatives are manufactured by the same process, using 2,4,5-trichlorophenol (Sittig, 1980). Chlorophenoxy ester herbicides can also be produced by direct reaction of chlorophenol with appropriate chloroacetic esters (International Programme on Chemical Safety, 1984). Commercial production and marketing of these compounds in the USA began in 1944 (Hamner & Tukey, 1944; US Tariff Commission, 1946).

MCPA was first produced commercially in 1945 (Hayes, 1982; Windholz, 1983). It is made by chlorination of 2-methylphenoxyacetic acid in 1,2-dichloropropane at 60-100°C, sometimes with iodine and ferric chloride catalysts. The product precipitates from the solvent after cooling. An alternate synthetic pathway involves reaction of 4-chloro-*ortho*-cresol with chloroacetic acid (Sittig, 1980).

Commercial production of mecoprop was started in 1957 in the UK (Worthing, 1977). It is produced by condensation of 2-chloropropanoic acid with 4-chloro-*ortho*-cresol (Sittig, 1980). Silvex was first marketed in 1953 in the USA (Worthing, 1977) and is manufactured by reaction of 2,4,5-trichlorophenol with sodium 2-chloropropanoate (Sittig, 1980). A related propanoic acid derivative, dichlorprop, was first described in 1944 and was first marketed by a UK company in 1961 (Worthing, 1983). It is produced commercially by condensation of 2-chloropropanoic acid with 2,4-dichlorophenol, or by chlorination of 2-phenoxypropanoic acid (Sittig, 1980).

Nine US companies manufacture technical or formulated 2,4-D products. In Europe, the Federal Republic of Germany, France, the UK and Spain each have two manufacturers of 2,4-D products, and the Netherlands, the German Democratic Republic and Austria each have one commercial producer. One manufacturer has been identified in Australia, Brazil, India, Japan, and the Philippines. Two manufacturers of 2,4-D products have been reported in Argentina (Anon., 1985).

Estimates of US production, exports and imports of all forms of 2,4-D for the past several years are presented in Table 1. Estimated use of 2,4-D in other countries is presented in Table 2.

Table 1. US production, exports and imports of 2,4-D (all forms) (millions of kg)[a]

	1977	1978	1979	1980	1981	1982	1983	1984
Production	12.6	9.0	17.7	17.5	14.9	8.8	7.7	NR[b]
Exports	NR	3.0	5.4	3.4	2.9	5.8	4.7	7.1
Imports	1.2	1.6	0.9	0.2	0.1	3.2	4.9	NR

[a]From US International Trade Commission (1978a,b, 1979a,b, 1980a,b, 1981a,b, 1982a,b, 1983a,b, 1984a,b, 1985; US Department of Commerce, 1979, 1980, 1981, 1982, 1983, 1984, 1985)
[b]NR, not reported

Table 2. Consumption pattern of 2,4-D, MCPA and 2,4,5-T herbicides in some countries (tonnes), 1974-1982[a,b]

Country	2,4-D				MCPA				2,4,5-T			
	1974-1976	1980	1981	1982	1974-1976	1980	1981	1982	1974-1976	1980	1981	1982
Canada	3454				1317				46			
Mexico	33	1500	1350	1350						40	30	50
Argentina	2267	1880	1520	1520	103	154	104	104			20	
Suriname		37	53									
Uruguay	140	190	138	69	4	7	9	2				
Cyprus	5	4	18		6	17	1		0.1	0.1		
India	447	338	400									
Japan		130	198	159		133	108	83				
Jordan		200	725	100								
Korea, Republic of	21	2	9	10								
Kuwait	0.1											
Pakistan		29	0.2									
Turkey	1478	1297	848	890								
Austria	169	274	236	200	134	193	212	127	51	41	29	39
Czechoslovakia	294	102	81		1920	3647	2183		11	9	11	
Denmark	210	308		324	595	615		760	7	3		
Finland	37				1003				54			
Greece	371				125							
Hungary	2579	2179	1996	1485	2238	2987	3292	3342		31	20	29
Iceland		0.1	0.1	0.2	333							
Italy	541					1	1					
Malta												
Norway	109	17	16	19	280	215	203	249				

Table 2 (contd)

Country	2,4-D				MCPA				2,4,5-T			
	1974-1976	1980	1981	1982	1974-1976	1980	1981	1982	1974-1976	1980	1981	1982
Poland	1263	1016	1160	1916	1704	1438	2275	1399	10			
Portugal	5	15	0.4		52	39	22					
Sweden	111	33	38		1443	1343	1524					
Samoa		1										
Mauritius	98				11				0.1	0.1		
Sierra Leone	1								19			
Zimbabwe		3	4			27	30					

[a]From Food and Agriculture Organization (1984)
[b]Missing values do not necessarily indicate non-consumption.

Over the last 20 years, 13 US companies at one time produced 2,4,5-T (Esposito et al., 1980). As a result of the gradual cancellation of its various pesticide registrations since 1970, 2,4,5-T is no longer produced in the USA. Three manufacturers have been identified in the Federal Republic of Germany, one in the UK, one in Australia (Anon., 1985), and one in New Zealand (Smith & Pearce, 1986). In 1981, 708 tonnes of 2,4,5-T were used in New Zealand (Smith et al., 1983).

As late as 1975, US companies produced large quantities of 2,4,5-T (5.7 million kg) for export (US Department of Commerce, 1975). US imports of 2,4,5-T increased from 20 000 kg in 1979 to 278 000 kg in 1983 (US International Trade Commission, 1980b, 1984b). Consumption patterns in some other countries are summarized in Table 2.

MCPA is not produced in the USA. Five manufacturers have been identified in the UK, where the chemical was first produced; France and the Federal Republic of Germany each have two producers, and there is one manufacturer in each of the following countries: Austria, the Netherlands, Japan, Argentina, Australia (Anon., 1985) and Denmark (Lynge, 1985). Imports of MCPA into the USA have declined steadily from 0.95 million kg in 1976 to 0.24 million kg in 1983 (US International Trade Commission, 1977, 1984b). The US Environmental Protection Agency estimated MCPA consumption for domestic usage in 1980 at 2.1-2.9 million kg (Holtorf, 1982). MCPA consumption patterns in other countries are presented in Table 2.

Mecoprop is produced by three companies in the USA, three companies in the UK, two companies in the Federal Republic of Germany, and one firm each in the Netherlands, France (Anon., 1985) and Denmark (Lynge, 1985). Silvex, although once widely produced, is now manufactured by only two companies — one in Austria and one in the UK. In 1978, silvex and related salts were produced by three US companies (US International Trade Commission, 1979a). Dichlorprop is produced by four manufacturers in the UK, two in the Federal Republic of Germany, one in the Netherlands (Anon., 1985) and two in Denmark (Lynge, 1985).

2.2 Technical products and impurities

(a) Major chemical components

The structures of the acid forms of commercially important chlorophenoxy herbicides and pertinent identifying information are given in Table 3. These herbicides are typically formulated as esters or amine salt derivatives. Chlorophenoxy herbicide formulations used in agriculture and forestry may also contain organic solvents, emulsifiers, inert ingredients and other additives (Plimmer, 1980; Anon., 1985; Leng, 1986).

Common derivatives of 2,4-D include amine and alkali metal salts and esters. Of the amines, the dimethylamine salt is produced in highest quantities. The other amine salts that are produced include diethanolamine, triethanolamine, trimethylamine, oleylpropylenediamine, dodecyl-/tetradodecylamine and heptylamine derivatives (Anon., 1985). Ester derivatives include ethyl, isooctyl, butoxyethyl, ethyl hexyl and mixed butyl (Que Hee & Sutherland, 1981; Anon., 1985). High-purity sodium and lithium salts are also marketed

Table 3. Identification of chlorophenoxy herbicides

Common name [Chem. Abstr. Serv. Reg. No.]	Chem. Abstr. Name IUPAC Systematic Name [Synonym]	Structural and molecular formulae and molecular weight
2,4-D [94-75-7]	(2,4-Dichlorophenoxy)-acetic acid (2,4-Dichlorophenoxy)-acetic acid [2,4-D acid]	$C_8H_6Cl_2O_3$ Mol. wt: 221.04
2,4-DP [120-36-5]	2-(2,4-Dichlorophenoxy)-propanoic acid 2-(2,4-Dichlorophenoxy)-propionic acid [Dichlorprop]	$C_9H_8Cl_2O_3$ Mol. wt: 235.05
2,4,5-T [93-76-5]	(2,4,5-Trichlorophenoxy)-acetic acid (2,4,5-Trichlorophenoxy)-acetic acid	$C_8H_5Cl_3O_3$ Mol. wt: 255.49
Silvex [93-72-1]	2-(2,4,5-Trichloro-phenoxy)propanoic acid 2-(2,4,5-Trichloro-phenoxy)propionic acid [2,4,5-TP; Fenoprop]	$C_9H_7Cl_3O_3$ Mol. wt: 269.53

Table 3. (contd)

Common name [Chem. Abstr. Serv. Reg. No.]	Chem. Abstr. Name IUPAC Systematic Name [Synonym]	Structural and molecular formulae and molecular weight
MCPA [94-74-6]	(4-Chloro-2-methyl-phenoxy)acetic acid (4-Chloro-*ortho*-tolyloxy)acetic acid [Agroxone, Metaxon]	$C_9H_9ClO_3$ Mol. wt: 200.63
MCPP [93-65-2]	2-(4-Chloro-2-methyl-phenoxy)propanoic acid 2-(4-Chloro-*ortho*-tolyloxy)propionic acid [Mecoprop]	$C_{10}H_{11}ClO_3$ Mol. wt: 214.6

(Anon., 1985). Formulated 2,4-D products normally contain one derivative, but may also contain other herbicides, especially other chlorophenoxyacetic or chlorophenoxypropanoic derivatives. Appendix A in Volume 15 of the *IARC Monographs* (IARC, 1977c) should be consulted for a more comprehensive listing of formulations.

The composition of technical-grade 2,4-D depends on the process by which it is produced and, when 2,4-dichlorophenol is used, on the purity of that compound. Ranges of impurities, other than dioxins and dibenzofurans, that are present in typical technical grades of 2,4-D are listed in Table 4.

N-Nitrosamines were reported to occur in earlier amine formulations of 2,4-D. *N*-Nitrosodimethylamine (see IARC, 1978) has been detected at levels of up to 0.3 mg/l in dimethylamine salts (Ross *et al.*, 1977).

2,4,5-T is formulated as products similar to those with 2,4-D. The most commonly marketed formulations contain mixed butyl esters; others include the isooctyl ester, the ethyl ester and the dimethylamine salt (Anon., 1985). A comprehensive list of 2,4,5-T formulations appears in Appendix A of Volume 15 of the *IARC Monographs* (IARC, 1977c); however, it should be noted that production of many formulations has been discontinued due to increasing regulatory constraints. Impurities that occur in technical products of 2,4,5-T are similar to those encountered in 2,4-D; Table 5 gives the major components of two formulations of technical-grade 2,4,5-T.

Table 4. Typical levels of 2,4-D and major impurities in technical-grade 2,4-D[a]

Component	Range (%)
2,4-D	94-99
2,6-Dichlorophenoxyacetic acid	0.5-1.5
2-Chlorophenoxyacetic acid	0.1-0.5
4-Chlorophenoxyacetic acid	0.2-0.8
Bis(2,4-dichlorophenoxy)acetic acid	0.1-2.0
Phenoxyacetic acid	trace-0.2
2,4-Dichlorophenol	0.1-0.6
2,6-Dichlorophenol	0.001-0.048
2,4,6-Trichlorophenol	0.001-0.14
2-Chlorophenol	0.0004-0.04
4-Chlorophenol	0.0004-0.005
Water	0.1-0.8

[a]From International Programme on Chemical Safety (1984)

Table 5. Composition of two typical 2,4,5-T formulations[a]

Constituent[b]	Sample[c]	
	A	B
2,4,5-T acid (mg/l)	72	20
2,4,5-T isobutyl ester (g/l)	ND	150
2,4,5-T butyl ester (g/l)	ND	365
2,4,5-T 2-butoxyethyl ester (g/l)	500	ND
2,4,5-Trichlorophenol (μg/g)	250	460
2,5-D 2-butoxyethyl ester (g/l)	2.5	_[d]
2,4-D 2-butoxyethyl ester (g/l)	1.2	_[d]
3,4-D 2-butoxyethyl ester (g/l)	<0.4	_[d]
Tetrachlorodibenzo-*para*-dioxin (μg/g)	0.06	0.12

[a]From Sundström *et al.* (1979)

[b]Amounts calculated on the basis of weight or volume of formulation

[c]A, tractor-sprayed sample; B, aeroplane-sprayed sample; ND, not detected

[d]Not investigated

MCPA is marketed alone or in combination with other phenoxy herbicides. Potassium, sodium and dimethylamine salts, and ethyl, butyl, isooctyl and butoxyethyl esters are used in formulated products (Anon., 1985). An international listing of MCPA formulated products is given by Que Hee and Sutherland (1981).

The sodium salt of MCPA is available as a 75-80% soluble powder. MCPA is also marketed as 24-60% aqueous concentrates (salts) or emulsifiable liquids (Hayes, 1982). In the USA, the amine salt formulations have been most widely used (Soderquist & Crosby, 1975). Crude MCPA is 85-95% pure (Hayes, 1982); technical products of 94-96% purity (National Research Council, 1977) and 85-99% purity (Worthing, 1983) have been reported. The following ranges of impurities have been found (National Research Council, 1977): 2-methyl-6-chlorophenoxyacetic acid (1.5-3%); a mixture of 2-methyl-4,6-dichlorophenoxyacetic acid, 2-methylphenoxyacetic acid, 2-chlorophenoxyacetic acid, 2,6-dimethyl-4-chlorophenoxyacetic acid and 4-chlorophenoxyacetic acid (0.5-1.5%); chloro-*ortho*-cresol (0.5%); and water (1.0%). One commercial sample of MCPA was reported to contain approximately 4% 4-chloro-*ortho*-cresol (Hattula *et al.*, 1979).

Mecoprop is marketed alone or in combination with other herbicides. Formulations employ dimethylamine and diethanolamine salts, butyl and isooctyl esters and potassium salts (Anon., 1985). Commercial formulations containing mecoprop have been listed (Que Hee & Sutherland, 1981; Anon., 1985). Mecoprop technical products are mixtures of the (+) and (−) stereoisomers, of which only the (+) form is biologically active (Worthing, 1983).

Silvex, also known as fenoprop and 2,4,5-TP, is currently marketed to a very limited extent; only two US products in current use are known to contain silvex (Anon., 1985). Formerly marketed products contained mixtures of esters (Que Hee & Sutherland, 1981).

Dichlorprop, or 2,4-DP, is also used to a more limited extent than 2,4-D or MCPA. It is available as isooctyl and butyl esters, and as potassium or amine salts, alone or in combination with other similar herbicides. Esters are marketed as emulsions, and salts as aqueous solutions. Commercial formulations containing dichlorprop have been listed (Anon., 1985). As for mecoprop, technical-grade dichlorprop contains a mixture of (+) and (−) stereoisomers, of which only the (+) form is biologically active (Worthing, 1983).

(b) Chlorinated dibenzodioxin and dibenzofuran impurities

Commercial formulations of chlorophenoxy herbicides contain a series of nonpolar impurities including polychlorinated dibenzodioxins (PCDDs) (see IARC, 1977b) and polychlorinated dibenzofurans (PCDFs). (See also the monograph on occupational exposures to chlorophenols, section 2.2(*b*), p. 323.)

In 1973, Edmunds *et al.* reported the results of analyses of 80 samples of 2,4,5-T formulations in the range of 100% ester and 32 samples in the range of 50% ester formulations, obtained from stocks delivered to forest areas in the UK between January 1967 and April 1970 (mainly 1969 and 1970). The results are given in Table 6. The maximum value found was 28.3 μg/g of 2,3,7,8-tetrachlorodibenzo-*para*-dioxin (TCDD).

Young *et al.* (1978) reported levels of TCDD found in more than 450 samples of the herbicide Agent Orange that were placed in storage in the USA and in the Pacific before 1978 (Table 7). Since Agent Orange was formulated as a 1:1 mixture of the butyl esters of 2,4,5-T and 2,4-D, the levels of TCDD in individual 2,4,5-T batches manufactured and used in the 1960s could have been as high as about 100 μg/g; the waste streams from a purification process could be even more highly contaminated. The weighted mean concentrations of TCDD in Agent Orange equal 1.98 μg/g. The level of TCDD in the single

Table 6. Concentrations (μg/g) of TCDD in 2,4,5-T alkyl ester herbicide formulations[a]

TCDD concentration (range)	Number of samples in range	
	100% ester	50% ester
<0.005	25	8
0.05-0.09	14	0
0.10-0.19	19	5
0.20-0.29	9	3
0.30-0.39	6	1
0.40-0.49	2	1
>0.50	5 (1.74, 1.70, 1.51, 1.48, 1.30)[b]	14 (28.3, 27.2, 1.30, 1.26, 1.17, 0.95, 0.91, 0.80, 0.75, 0.65, 0.60, 0.58, 0.58, 0.55)[b]

[a]From Edmunds et al. (1973)
[b]Concentrations in individual samples with >0.5 μg/g TCDD

Table 7. Concentrations (μg/g) of TCDD in samples of Agents Orange and Purple[a]

Source of samples	Number of samples		Concentration of TCDD	
	Orange	Purple	Range	Mean
Johnston Atoll inventory, 1972[b]	200	(4)[c]	0.05-47	1.91
Johnston Atoll inventory, 1974	10		0.07-5.3	1.68
NCBC, Gulfport inventory, 1972[d]	42		0.05-13.3	1.77
NCBC, Gulfport inventory, 1975	238		0.02-15	2.11
Eglin AFB archived sample		1[e]	–	45
Eglin AFB inventory, 1972	2		–	0.04

[a]From Young et al. (1978)
[b]Surplus Agent Orange was shipped from South Viet Nam to Johnston Atoll (near Hawaii) for storage in April 1972.
[c]Four of 200 samples may have been Agent Purple.
[d]The Naval Construction Battalion Center (NCBC), Gulfport, Mississipi, USA, served as a storage site for surplus Agent Orange from 1969 to 1977.
[e]Agent Purple was used extensively in the evaluation of aerial spray equipment on Test Area C-52, Eglin Air Force Base (AFB) Reservation, Florida, USA, 1962-1964.

sample of Agent Purple (Table 7) was also quite high (45 μg/g). Agent Purple is a mixture of n-butyl-2,4 D (50%), n-butyl-2,4,5-T (30%) and isobutyl-2,4,5-T (20%).

In analysis using high-resolution gas chromatography/mass spectrometry and mass spectrometry, Rappe et al. (1978), Norström et al. (1979) and Rappe and Buser (1981) reported that in other samples of Agent Orange, as well as in European and US 2,4,5-T formulations from the 1950s and 1960s, TCDD was the dominant compound of this group (Table 8). Only minor amounts of other PCDDs and PCDFs were found, and particularly lower chlorinated PCDDs in samples of Agent Orange. The analytical methods used in these studies of phenoxy herbicides are not isomer-specific; however, studies using isomer-specific methods have confirmed that the 2,3,7,8-isomer is the major tetra-CDD isomer in 2,4,5-T formulations (Buser & Rappe, 1978).

Table 8. Levels of TCDD (μg/g) in 2,4,5-T acid and 2,4,5-T ester formulations[a]

Sample	Location	TCDD
2,4,5-T acid	1952, Sweden	1.10
2,4,5-T ester	unknown, Sweden	0.50
2,4,5-T ester	unknown, Sweden	<0.05
2,4,5-T ester	1960, Sweden	0.40
2,4,5-T ester	1962, Finland	0.95
2,4,5-T ester	1966, Finland	0.10
2,4,5-T ester	1967, Finland	<0.05
2,4,5-T ester	1967, Finland	0.22
2,4,5-T ester	1967, Finland	0.18
2,4,5-T acid	1964, USA	4.8
2,4,5-T acid	1969, USA	6.0
Agent Orange	unknown, USA	0.12
Agent Orange	unknown, USA	1.1
Agent Orange	unknown, USA	5.1

[a]From Rappe *et al.* (1978); Norström *et al.* (1979); Rappe and Buser (1981)

Table 9 shows levels of TCDD in 2,4,5-T manufactured by the sole New Zealand producer. The average levels have decreased steadily since 1971, the first year for which such data were available (Smith & Pearce, 1986).

As a result of government regulations and general awareness of the toxicity of dioxins, efforts were made during the 1970s to control and minimize the formation of TCDD during 2,4,5-T production. In 16 samples of 2,4-D esters and amine salts from Canada analysed for the presence of PCDDs, eight out of nine esters and four out of seven amine salts were found to be contaminated, the esters having significantly higher levels than the amine salts. The tetra-CDD observed was the 1,3,6,8-isomer, as verified by gas chromatography with a synthetically prepared authentic standard (210-1752 ng/g in the esters, and 20-278 ng/g in the amine salts) (Cochrane *et al.*, 1982).

2.3 Use

The chlorophenoxy herbicides and their derivatives and analogues function by mimicking the action of a natural plant hormone, indoleacetic acid. Absorption and

Table 9. Average levels of TCDD (ng/g) in 2,4,5-T produced in New Zealand[a]

Year	TCDD
1971	950
1972	470
1973	47
1974	33
1975	24
1976	27
1977	31
1978	22
1979	13
1980	14
1981	7.3
1982	8.5
1983	5.3
1984	5.9
1985	4.7

[a]From Smith and Pearce (1986)

translocation of these compounds are necessary for herbicidal activity, and all herbicides in this class must be applied to the foliage of actively growing plants. Chlorophenoxy herbicides are used primarily for selective control of broadleaf weeds in cereal grains, pastures and turf and for removing unwanted brushy species in rangeland, forests and noncropland. Rates of application range from as low as 0.25 kg/ha in grain crops to as high as 16 kg/ha for spot treatment of individual trees in rights-of-way. Very dilute solutions of 2,4-D and silvex derivatives have also been used as growth regulators in fruit orchards. Chlorophenoxy herbicides are applied alone or as mixtures with other herbicides, in solutions, dispersions, or emulsions in water and/or oil, using equipment that produces large droplets to avoid spray drift (Hayes, 1982; Leng, 1986).

Registrations in the USA for 2,4-D and 2,4,5-T in the 1940s included many food crops, and use of these pesticides in the USA was up to nearly 17 million kg annually by 1960. By the mid-1960s, chlorophenoxy herbicides were the most important single class of herbicides. In 1966-1969, they were used for weed control on over 62 million acres (25 million ha) of US agricultural land, and annual US usage of all chlorophenoxy herbicides was nearly 20 million kg (Hazardous Materials Advisory Committee, 1974). It was also during the 1960s that 2,4-D and 2,4,5-T, principally as Agent Orange, were heavily used in South Viet Nam and Cambodia for defoliation of forests by the US Armed Forces. From 1961 to 1971, mixtures of 2,4-D and 2,4,5-T n-butyl esters and 2,4-D and picloram tri-isopropanolamine salts were applied at rates of up to 28.6 kg/ha to an estimated 2 million hectares of Vietnamese forests (almost 20% of the forested land area of South Viet Nam). It has been estimated that a total of 25 million kg of 2,4-D and 21 million kg of 2,4,5-T were applied during this time (Westing, 1971).

US production and use of 2,4,5-T and 2,4-D decreased markedly in 1969 and the early 1970s due to governmental restrictions on their use (Hazardous Materials Advisory Committee, 1974; Grant, 1979). In 1974, an estimated 452 000 kg of 2,4,5-T were used in the USA, the majority of which was applied to rangeland and pasture (US Environmental Protection Agency, 1979a).

Although MCPA has never attained the level of consumption of 2,4-D or 2,4,5-T, it has found specialized use for weed control in cereal grain production. US consumption of MCPA was 2.1-2.9 million kg in 1980 (Holtorf, 1982); most (70-71%) was used on wheat and rice.

Mecoprop is used similarly as a post-emergence herbicide for control of cleavers and chickweed in cereal grains (Worthing, 1983).

Silvex has been recommended for control of aquatic weeds, weeds in pasture, sugar cane and rice, and especially for brush. The triethanolamine salt has been used to reduce preharvest dropping of apples (Hayes, 1982; Meister, 1983). In California, for example, only approximately 360 kg were used in 1983 on pasture, rangeland and landscaped areas (California Department of Agriculture, 1984). All US registrations for silvex were cancelled in 1983 (US Environmental Protection Agency, 1983).

Dichlorprop has been used and is still recommended for removal of brush on rangeland and rights-of-way, and for control of aquatic weeds (Anon., 1985). Approximately 3150 kg of dichlorprop were used in California for these applications in 1983 (California Department of Agriculture, 1984).

2.4 Regulatory status and guidelines

By 1974, the US Environmental Protection Agency had cancelled all registrations for chlorophenoxy herbicides, except those pertaining to uses other than on foods and on rice paddies, pastures and rangelands (Anon., 1974). In 1983, all registrations for 2,4,5-T were cancelled, and this chemical can therefore no longer be used legally for any purpose in the USA (US Environmental Protection Agency, 1983). Registrations have also been cancelled in, e.g., Sweden, the Netherlands, the USSR and Australia (Anon., 1983). 2,4,5-T was banned in Italy in 1970 (Vineis *et al.*, 1986) and in the Federal Republic of Germany in 1985.

Occupational exposure limits for 2,4-D in 14 countries and for 2,4,5-T in nine countries have been reported and are presented in Table 10. Regulations pertaining to the pesticide use of chlorophenoxy herbicides are not reviewed or reported comprehensively in this monograph.

2.5 Occupational exposure

Exposure to chlorophenoxy herbicides may occur through inhalation, skin contact or ingestion. In most cases, the predominant route of occupational exposure has been by the absorption of spills or aerosol droplets through the skin (Leng *et al.*, 1982; International Programme on Chemical Safety, 1984). Measurements are usually reported in terms of

Table 10. Occupational exposure limits for 2,4-D and 2,4,5-T[a]

Country	Year	Concentration (mg/m^3)	Interpretation[b]
Australia	1978	10	TWA
Belgium	1978	10	TWA
Finland[c]	1981	10	TWA
		20	STEL
Germany, Federal Republic of	1985	10	TWA
Hungary[c]	1974	10	TWA
Japan[c]	1978	10	TWA
The Netherlands	1978	10	TWA
Norway	1981	5	TWA
Romania[c]	1975	5	TWA
		10	Ceiling
Switzerland	1978	10	TWA
UK	1985	10	TWA
		20	STEL
USA	1985		
ACGIH		10	TWA
		20	STEL
OSHA		10	TWA
USSR[c]	1977	1	Ceiling
Yugoslavia	1971	10	Ceiling

[a]From International Labour Office (1980); Direktoratet for Arbeidstilsynet (1981); Työsuojeluhallitus (1981); American Conference of Governmental Industrial Hygienists (ACGIH) (1985); Deutsche Forschungsagemeinschaft (1985); Health and Safety Executive (1985); US Occupational Safety and Health Administration (OSHA) (1985)

[b]TWA, time-weighted average; STEL, short-term exposure limit

[c]2,4-D only

herbicide concentrations in the breathing zone air or in the urine of exposed workers. Table 11 gives measurements of 2,4-D, 2,4,5-T and MCPA in the urine of workers in various industries and occupations. Monitoring of air, water and food outside areas of herbicide use has shown that the 2,4-D intake of the general population is below the present detection limits (International Programme on Chemical Safety, 1984). No quantitative data on background levels of chlorophenoxy compounds in urine and human tissues were available to the Working Group. However, the background levels of several 2,3,7,8-substituted PCDDs and PCDFs have been reported from Canada by Ryan et al. (1985) and from Sweden by Nygren et al. (1986) (Tables 12 and 13). The isomers and levels were very similar in the two studies.

Table 11. Concentrations (mg/l) of chlorophenoxy herbicides in the urine of exposed workers, by industry and activity

Industry and activity (country)	Substance measured	Concentration in urine[a] Mean (range)	No. of reported measurements (No. of workers)	Reference
Production of herbicides				
Formulation of 2,4-D derivatives (Turkey)	2,4-D	1.37 (0.06-9.51)	(15)	Vural & Burgaz (1984)
Forestry				
Ground application (Australia)	2,4,5-T			Simpson et al. (1978)
injector gun		0.26 (0.23-0.31)	(3)	
knapsack mister and power spray		0.99 (0.16-1.74)	(5)	
Tractor spraying (Sweden)	2,4-D	8 (3-14)	(4)	Kolmodin-Hedman & Erne (1980)
	2,4,5-T	3.5 (1-11)	(4)	
Ground spraying (USA)	2,4,5-T			Leng et al. (1982)
backpack crew		6.29 (0.85-17.0)	(4)	
foremen		1.16 (0.03-3.80)	(4)	
Ground spraying (New Zealand)	2,4,5-T	0.61 (0.02-22.2)	124 (5)	Ferry et al. (1982)
Aerial application (USA) helicopter crew	2,4-D	<0.04* (NS)	524 (18)	Lavy et al. (1982)
Ground application in forest (Finland)	2,4-D + MCPA			Kangas et al. (1984)
knapsack spraying		3.63 (0.06-10.5)	20 (1)	
brush saw spraying		2.12 (0.15-4.13)	7 (1)	
tractor spraying, driver		1.49 (0.09-3.33)	9 (1)	
tractor spraying, assistant		1.53 (0.11-4.66)	7 (1)	
ultra-low-volume spraying		2.99 (0.27-6.33)	8 (1)	

Table 11 (contd)

Industry and activity (country)	Substance measured	Concentration in urine[a] Mean (range)	No. of reported measurements (No. of workers)	Reference
Forestry (contd)				
Ground application along electric transmission line (Canada)	2,4-D			Libich et al. (1984)
gun, roadside		1.42 (0.04-8.15)	53 (12)	
gun, right-of-way		1.72 (0.15-5.45)	37 (8)	
mist blowers, right-of-way		2.55 (0.44-5.07)	9 (3)	
all-terrain vehicle with four spray guns				
site A		6.17 (0.27-32.7)	25 (7)	
site B		3.16 (0.63-12.4)	20 (5)	
Aerial application (Canada)	2,4-D			Frank et al. (1985)
mixer/loaders		0.33 (0.02-0.84)	18 (3)	
supervisor		0.01	18 (1)	
balloon men		(0.24-0.26)	18 (2)	
Agriculture				
Ground-boom spraying of grass pasture (USA)	2,4-D			Draper & Street (1982)
tractor drivers		4.8 (0.12-20)	8 (2)	
tractor sprayers		4.2 (<0.06-12)	8 (2)	
Application (USA)	2,4,5-T	0.02 (NS)		Draper (1982)
	2,4-D	0.006 (NS)		

Table 11 (contd)

Industry and activity (country)	Substance measured	Concentration in urine[a] Mean (range)	No. of reported measurements (No. of workers)	Reference
Agriculture (contd)				
Tractor spraying (Sweden)				Kolmodin-Hedman et al. (1983a)
spraymen	2,4-D	(ND-0.21)	(2)	
	MCPA	1.7* (0.48-12.2)	(9)	
farmers	MCPA	0.31* (ND-3.7)	(24)	
Airboat application (USA)	2,4-D	0.41 (0.12-0.67)	(4)	Nigg & Stamper (1983)
airboat crew				
Aerial application (Turkey)	2,4-D			Vural & Burgaz (1984)
helicopter pilots		(ND-1.09)	(4)	
mixer		0.58	(1)	
flagmen		(1.01-1.92)	(2)	
supervisors		(ND-1.16)	(6)	

[a] Abbreviations: ND, not detected; NS, not specified; *, median

Table 12. Concentrations of major dioxins and furans (ng/kg) in A, ten human adipose tissue samples collected in 1980 from deceased hospital patients in eastern Ontario (Canada) and in B, 46 human adipose tissue samples collected in 1976 from accident victims across Canada[a]

Analyte[b]	A				B		
	Average ± SD	Range	No. positive No. analysed		Average ± SD	Range[c]	No. positive No. analysed
TCDD	10.0 ± 4.9	3.0-17.8	10/10		6.2 ± 2.6	ND, 2.0-12.7	21/46
2,3,4,7,8-PeCDF	18.4 ± 6.3	11.5-29.5	9/9		16.8 ± 7.6	4.2-45.0	46/46
1,2,3,7,8-PeCDD	13.2 ± 4.0	10.5-21.4	10/10		10.4 ± 5.8	1.5-34.5	46/46
1,2,3,4,7,8-/1,2,3,6,7,8-HxCDF	17.3 ± 6.9	13.6-28.8	8/9		17.3 ± 10.9	ND, 6.3-71.2	32/46
1,2,3,6,7,8-HxCDD	90.5 ± 38.9	50-177	10/10		79.6 ± 47.0	19.2-291	46/46
1,2,3,4,6,7,8-HpCDF	39.4 ± 19.6	12.8-67	7/9		32.7 ± 15.9	ND, 9.7-110	43/46
1,2,3,4,6,7,8-HpCDD	116 ± 41.8	53-208	10/10		137 ± 79	34.4-520	46/46
1,2,3,4,6,7,8,9-OCDD	611 ± 226	317-985	10/10		796 ± 458	202-2961	46/46

[a]From Ryan et al. (1985)

[b]PeCDF, pentachlorodibenzofuran; PeCDD, pentachlorodibenzodioxin; HxCDF, hexachlorodibenzofuran; HxCDD, hexachlorodibenzodioxin; HpCDF, heptachlorodibenzofuran; HpCDD, heptachlorodibenzodioxin; OCDD, octachlorodibenzodioxin

[c]ND, not detected

Table 13. Levels of PCDDs and PCDFs (pg/g) found in human adipose tissue (wet-weight basis)[a]

Analyte[b]	Sweden			Mean value non-exposed (n=18)	Range	Mean value cancer patients (n=17)	Range	Mean value non-cancer patients (n=14)	Range	German worker	Chemist
	Mean value (n=31)	Range	Mean value exposed (n=13)								
TCDD	3	0-9	2	3	2-6	3	2-9	3	2-6	100	NA
1,2,3,7,8-PeCDD	10	3-24	6	9	4-18	9	4-24	9	3-18	18	5
1,2,3,4,7,8-HxCDD	ND									ND	ND
1,2,3,6,7,8-HxCDD	15	3-55	19	12	3-18	18	3-55	12	8-18	48	12
1,2,3,7,8,9-HxCDD	4	3-5	5	4	3-5	4	3-13	4	3-5	12	5
1,2,3,4,6,7,9-HpCDD	ND									ND	ND
1,2,3,4,6,7,8-HpCDD	97	12-380	104	85	12-176	100	12-380	85	20-168	20	100
OCDD	414	90-763	398	421	98-679	408	90-620	421	182-763	80	374
2,3,7,8-TCDF	3.9	0.3-11	3.7	4.2	0.3-11	3.4	0.3-7.2	4.6	0-11	<3	7
1,2,3,7,8-PeCDF	ND									ND	ND
2,3,4,7,8-PeCDF	54	9-87	50	32	9-54	45	9-87	33	11-65	32	26
1,2,3,4,7,8-/1,2,3,4,7,9-HxCDF	6	1-15	7	5	1-6	6	1-15	5	2-7	11	12
1,2,3,6,7,8-HxCDF	5	1-13	5	4	1-5	5	1-13	4	2-7	5	7
2,3,4,6,7,8-HxCDF	2	1-7	2	2	1-4	2	1-7	2	1-4	2	38
1,2,3,4,6,7,8-HpCDF	11	1-49	14	10	1-18	13	1-49	10	5-16	37	17
1,2,3,4,6,7,9-HpCDF	ND									ND	ND
1,2,3,4,6,8,9-HpCDF	ND									ND	ND
1,2,3,4,7,8,9-HpCDF	ND									ND	ND
OCDF	4									NA	240

[a]From Nygren et al. (1986); ND, not detected (<1); NA, not analysed
[b]For abbreviations used, see footnote to Table 12; TCDF, tetrachlorodibenzofuran; OCDF, octachlorodibenzofuran

(a) *Production plants*

A mean urinary concentration of 1.37 mg/l 2,4-D was measured in workers involved in the production and formulation of 2,4-D herbicides (Vural & Burgaz, 1984). No data on levels of chlorophenoxy herbicides to which workers were exposed in industrial accidents were available to the Working Group.

Nygren *et al.* (1986) also analysed adipose tissue from two occupationally exposed persons (a German chemical factory worker and a laboratory chemist) and found a dramatically different pattern of PCDDs and PCDFs from that seen in the general population (see Table 13). The fat sample from the German worker was obtained more than 30 years after he was highly exposed to TCDD in a German factory, in November 1953. The level of TCDD in this sample was 25-30 times higher than in the normal Swedish population. Moreover the ratio of TCDD:1,2,3,7,8-penta-CDD was >5, whereas it is usually approximately 0.5 or lower, and the ratio of TCDD:2,3,7,8-tetra-CDF was >30, whereas it is usually close to 1.0. The chemist had synthesized more than 80 different PCDF isomers in the few years before the biopsy was taken.

(b) *Forestry and agriculture*

The highest urinary levels of 2,4-D, 2,4,5-T and MCPA are reported from ground spraying operations in forestry work. According to measurements made in Australia, Canada, Finland, New Zealand, Sweden and the USA, mean concentrations ranging from 0.3 to 8 mg/l are common during this type of herbicide application (see Table 11). During aerial spraying, the exposure levels were lower — 0.01-0.33 mg/l on average.

Some studies have reported exposure data as estimated dose per body weight. Leng *et al.* (1982) summarized the exposure of forestry workers in the USA to 2,4,5-T as follows: mixers, 12-138 μg/kg bw; backpack sprayers, 19-104 μg/kg bw; spray tractor drivers, 33-49 μg/kg bw; helicopter pilots, <1-44 μg/kg bw; supervisors, 2-30 μg/kg bw; and flagmen <1-3 μg/kg bw. Lavy *et al.* (1982) reported mean doses from <1-56 μg/kg bw among aerial applicatiors of 2,4-D in the USA. These results are similar to those reported for 2,4-D in Canadian studies: 4-39 μg/kg bw (Franklin *et al.*, 1982) and <1-22 μg/kg bw (Frank *et al.*, 1985). Nash *et al.* (1982) estimated 2,4-D exposures during agricultural use of the herbicide as 20 μg/kg bw for mixers and loaders and <10 μg/kg bw for pilots and ground applicators.

In agriculture, herbicides are often used as mixtures or in combination, and workers may therefore be exposed both to chlorophenoxy herbicides and other pesticides as well as to emulsifiers, solvents and other additives. In a study of 24 farmers, median urinary concentrations of 0.31 mg/l MCPA, 0.23 mg/l dichlorprop and 0.28 mg/l mecoprop were measured. In professional spraymen, the median levels of the three compounds were 1.7, 0.74 and 2.0 mg/l, respectively (Kolmodin-Hedman *et al.*, 1983a).

During ground spraying using hand-held sprayers, mist blowers or tractor-driven equipment, the airborne concentrations of 2,4-D, MCPA and dichlorprop ranged from 10 to 300 μg/m³ (Kolmodin-Hedman & Erne, 1980; Kangas *et al.*, 1984; Libich *et al.*, 1984). During mixing and spraying operations along power line rights-of-way, airborne concentration of 2,4,5-T esters were <10-60 μg/m³ (Hervin & Smith, 1978). During aerial application of herbicides, exposure levels to 2,4-D were <20 µg/m³ (Franklin et al., 1982; Lavy et al., 1982). In one case, a breathing zone concentration of 143 µg/m³ was measured (Franklin et al., 1982).

The study by Nygren et al. (1986) included a total of 31 persons (see Table 13). Of these, 13 had been exposed to chlorophenoxy herbicides and 18 were nonexposed controls. No difference was seen between these two groups in the levels and patterns of PCDDs and PCDFs in adipose tissue, although a long time may have elapsed between exposure and sampling. In addition, no difference in PCDD and PCDF levels was found between 17 persons with cancer (soft-tissue sarcomas, lymphomas) and 14 noncancer patients.

(c) Miscellaneous

In one extensive occupational monitoring programme undertaken in New South Wales, Australia, in 1979-1980, urine samples were analysed for herbicide residues. The subjects included pesticide factory staff, pest control operators, farmers, park workers and others potentially exposed to 2,4-D or 2,4,5-T. No 2,4-D or 2,4,5-T was detected (<0.001 mg/l) in 735 and 377 of 973 samples, respectively. Most of the other samples contained <0.1 mg/l, and only 27 contained >1 mg/l 2,4-D and 40, >1 mg/l 2,4,5-T (Simpson, 1982).

Exposure of soldiers and the general population to dioxin in connection with the military use of chlorophenoxy herbicides in Viet Nam has been the subject of much concern among veterans in the USA and Australia and the population in Viet Nam. Shepard and Young (1983) reported a study in which very low levels of tetra-CDD, believed to be the 2,3,7,8-isomer, were detected in adipose tissue from some Viet Nam veterans; however, the levels were not believed to correlate well with known exposure data or with health status. Experimental conditions were not described in this report. A more detailed description of these data is provided by Gross et al. (1984), who give the analytical methods and the quality control programmes used. The results of the total study are given in Table 14. The authors concluded that Viet Nam veterans designated by the Veterans' Administration as 'heavily exposed' to Agent Orange had detectable amounts of TCDD in adipose tissue: the levels found for two of the three 'heavily exposed' veterans were higher than those for other Viet Nam veterans or for the controls.

2.6 Analysis

Selected methods for the analysis of chlorophenoxy herbicides in the air and in the urine of exposed workers are summarized in Table 15.

Analysis of dermal exposure pads has been used in several studies to evaluate skin contact with herbicides (see, e.g., Franklin et al., 1982; Lavy et al., 1982; Sell & Maitlen, 1983). Chlorophenoxy herbicides have also been measured in plasma (see, e.g., Kolmodin-Hedman et al., 1979; Åkerblom et al., 1983).

Methods for the analysis of chlorinated dibenzodioxins and dibenzofurans have been reviewed (Rappe & Buser, 1981); see also the monograph on occupational exposures to chlorophenols, section 2.6, p. 337).

Table 14. TCDD levels (ng/kg) in adipose tissue of US veterans[a]

Group (code number)	Concentration	Limit of detection
Heavily exposed veterans		
10	23	4
10	35	9
19	ND	3
26	99	10
26	63	6
Lightly exposed veterans		
1	ND	5
13	ND	2
15	7	4
28	7	5
28	8	6
34	5	3
Possibly exposed veterans		
6	5	3
8	5	3
9	ND	3
11	3	2
12	9	3
14	4	3
16	ND	4
24	5	3
24	5	4
25	12	4
25	10	3
27	ND	6
29	13	5
30	ND	3
Controls		
5	4	4
7	3	2
17	4.3[b]	3
18	ND	4
20	5	4
21	6	3
23	8	2
23	6	3
31	7	4
32	4	4
33	14	7
US Air Force scientists		
2	5	2
3	4	1
4	6	2

[a]From Gross et al. (1984); sample sizes ranged from 2.2 to 11.6 g for each extraction; amounts of internal standard used varied from 2.0-2.6 ng/extraction; ND, not detected

[b]Duplicate analyses of same extract

Table 15. Methods for the analysis of chlorophenoxy herbicides

Sample matrix	Substance measured	Sample preparation	Assay procedure[a]	Limit of detection[b]	Reference
Air	2,4-D	Collect in Amberlite XAD-2 tube; desorb (hexane or sodium hydroxide); acidify; methylate (boron trifluoride in methanol); extract (benzene)	GC	0.01 $\mu g/m^3$ (small samples) 0.05 $\mu g/m^3$ (larger tubes)	Johnson et al. (1977)
	2,4-D/ 2,4,5-T	Collect in bubbler (ethanol); hydrolyse	TLC/spectrophotometry	50 $\mu g/m^3$	Kolmodin-Hedman et al. (1979)
	2,4-D/ 2,4,5-T	Collect on filter; extract (methanol)	HPLC/UV	150 $\mu g/m^3$	Eller (1984)
	2,4-D	Collect on filter/ XAD-4 resin; Soxhlet extract (acetone); methylate (diazomethane)	GC/EC (confirmed by TLC/GC-MS)	0.01 $\mu g/m^3$	Draper & Street (1982)
	2,4-D/ MCPA, dichlorprop, mecoprop	Collect on bubbler (distilled water); acidify	HPLC/UV	1 $\mu g/m^3$	Åkerblom et al. (1983)
	2,4-D, dichlorprop	Collect in Fluorisil tubes; desorb (methanol); hydrolyse; methylate (boron trifluoride in methanol); extract (benzene)	GC/EC	0.2 $\mu g/m^3$	Libich et al. (1984)
	2,4-D/ MCPA	Collect on filter/bubbler (ethanol); hydrolyse; extract (chloroform); dissolve (methanol)	HPLC/UV	–	Kangas et al. (1984)
Urine	2,4,5-T	Hydrolyse (alkali); extract (benzene); methylate (diazomethane); clean-up (silica gel)	GC/EC	–	Nony et al. (1976)
	2,4-D	Basify; acidify; clean-up (XAD-2); extract (sodium bicarbonate in acetonitrile); methylate (diazomethane); dissolve (hexane)	GC/EC	–	Smith & Hayden (1979)
	2,4-D/ 2,4,5-T, silvex	Hydrolyse (acid); extract (diethylether); acid/base partition; methylate (diazomethane)	GC/EC	50 $\mu g/l$	Draper (1982)
	2,4,5-T	Hydrolyse; extract	HPLC/UV	–	Ferry et al. (1982)

Table 15 (contd)

Sample matrix	Substance measured	Sample preparation	Assay procedure[a]	Limit of detection[b]	Reference
Urine (contd)	2,4-D/ MCPA, mecoprop, dichlorprop	Hydrolyse (acid); clean-up (Sep-Pak); extract (phosphate buffer); derivatize (pentafluorobenzyl bromide); dissolve (hexane)	GC/EC	50 μg/l	Åkerblom et al. (1983)
	2,4-D	Hydrolyse (alkali); acidify; clean-up (Sep-Pak); methylate (boron trifluoride in methanol); extract (hexane)	GC/EC	70 μg/l	Sell & Maitlen (1983)
	2,4-D	Hydrolyse (acid); extract (benzene); methylate (dimethyl sulphate); clean-up (silica gel)	GC/EC	30 μg/l	Vural & Burgaz (1984)
	2,4-D	Hydrolyse (alkali); wash (dichloromethane); acidify; extract (diethylether); methylate (boron trifluoride in methanol); extract (benzene)	GC/EC	0.5 μg/l	Frank et al. (1985)

[a]GC, gas chromatography; TLC, thin-layer chromatography; HPLC, high-performance liquid chromatography; UV, ultraviolet detection; EC, electron capture detection; MS, mass spectrometry

[b]The limits of detection are not always comparable, because they are listed as given by the authors.

Because dermal absorption is an important route of exposure to chlorophenoxy herbicides, biological monitoring is useful in estimating the absorbed dose. Urine concentrations provide the most accurate estimate of body burden; because of slow absorption, it is best to collect samples after a few days of exposure. [See also pp. 383-384, on which kinetic studies are discussed.]

3. Biological Data Relevant to the Evaluation of Carcinogenic Risk to Humans

3.1 Experimental data

Evaluations of the degrees of evidence for the carcinogenicity in animals and for activity in short-term tests of the chlorophenoxy herbicides considered in previous *IARC Monographs* are listed in Table 16. No attempt has been made to update these data.

Table 16. Chlorophenoxy herbicides and their major impurity considered in this monograph that have previously been evaluated in the *IARC Monographs*[a]

Chemical	Evidence for carcinogenicity in animals	Evidence for genetic activity in short-term tests
2,4-D	inadequate	inadequate
MCPA[b]	no data	inadequate
2,4,5-T	inadequate	inadequate
TCDD	sufficient	inadequate

[a]From IARC (1982a,b), except where noted
[b]From IARC (1983)

3.2 Biological effects in humans other than cancer

(a) Toxic effects

The literature on acute poisonings and on the health effects of occupational exposures to chlorophenoxy herbicides have been reviewed recently (IARC, 1977a, 1983; International Programme on Chemical Safety, 1984; Suskind & Hertzberg, 1984). Most of the toxicological information is derived from cases of acute poisoning.

Suicide patients who ingested chlorophenoxy acids died from circulatory collapse without distinct post-mortem findings. Non-fatal intoxications with 2,4-D have resulted in acute parasympathetic nervous system symptoms and, particularly, persistent neurological dysfunction, according to case reports. A surviving case of MCPA intoxication recovered without reported sequelae. The subacute effects reported frequently from health surveys of 2,4,5-T manufacturing workers, including those exposed during accidents, are acneform eruptions (chloracne), fatigue, nervousness and irritability; chloracne is a persistent and consistent clinical marker. Exposure to TCDD has also been associated with impairment of liver function, peripheral neuropathy, personality changes, porphyria cutanea tarda, and hypertrichosis and hyperpigmentation (IARC, 1977b).

(b) Effects on reproduction and prenatal toxicity

Field and Kerr (1979) found a positive correlation between the annual usage of 2,4,5-T in Australia in 1965-1976 and the prevalence rate of neural tube defects at birth in subsequent years in New South Wales. During 1969-1975, the use of commercial 2,4,5-T in Hungary increased from 46 to 1200 tonnes; however, over the period 1970-1976, the incidences of stillbirths, spina bifida and anencephalus declined, and the incidences of cleft palate, cleft lip and cystic kidney disease remained relatively stable (Thomas, 1980).

A study in Arkansas, USA, involved dividing the state into high, medium and low use of 2,4,5-T between 1948 and 1974 on the basis of rice acreage. No significant difference in rates of facial cleft was found among the different areas between 1943 and 1974 (Nelson *et al.*, 1979).

The US Environmental Protection Agency (1979b) investigated spontaneous abortion rates in three areas of Oregon, USA, in relation to 2,4,5-T spray practices in 1972-1977. Significantly higher rates were found in the area in which 2,4,5-T was used. [The Working Group noted that the methods used for case ascertainment were inadequate.]

The possible effects of aerial spraying were studied in the Northland region of New Zealand by dividing it into seven areas according to the extent of 2,4,5-T spraying, as assessed by a detailed review of the records of the companies involved. Maternal exposure was determined by area of residence. An association was found for all birth malformations combined and for club foot, hypospadias and epispadias and heart defects separately. No association was found with central nervous system defects, nor with cleft lip or palate (Hanify et al., 1981).

[The Working Group noted that the above studies were ecological surveys, and suffered from the usual limitations.]

A survey of the occupations of the fathers of children recorded in the Office of Population Censuses and Surveys register of congenital malformations in England and Wales (1974-1979) showed an increased risk of facial clefts in children of gardeners, groundsmen and agricultural workers, who were regarded as potentially exposed to herbicides (Balarajan & McDowall, 1983). No increased risk of these malformations was seen in children of fathers in agriculture or forestry in Oxfordshire and West Berkshire in the years 1965-1974 (Golding & Sladden, 1983).

Two case-control studies were carried out on possible reproductive effects in soldiers who had served in Viet Nam and had had potential exposure to Agent Orange, a mixture of butyl esters of 2,4,5-T and 2,4-D contaminated with TCDD. An Australian study, involving 8517 case-control pairs, found a relative risk of 1.02 (95% confidence limits, 0.78-1.32) for veterans fathering children with birth anomalies compared to non-veterans (Donovan et al., 1984). A US study, involving 7133 babies born to Viet Nam veterans, found an overall relative risk of 0.97 (Erickson et al., 1984).

A study of pregnancy outcome of wives of professional pesticide sprayers was conducted in New Zealand (Smith et al., 1981). The herbicide sprayed predominantly was 2,4,5-T. There were 1172 births among applicator families in the study period (1969-1979 for spraying of 2,4,5-T; 1960-1979 for spraying of any pesticide) and 1122 births among a comparison group of agricultural contractors. Information was gained by postal questionnaire, with an overall response rate of 89% among applicators and 83% of agricultural contractors. Major congenital defects were reported in 2% (24) of births to applicators and 1.6% (18) of births to agricultural contractors; the difference was not significant, and the rates were similar to those for the general population. Similar rates were seen for the two groups for stillbirth (0.9% versus 1.0%) and miscarriage (8.6% versus 9.3%).

In a further analysis of a subset of these data, those pregnancy outcomes associated with spraying of 2,4,5-T by the father in the same year as the birth or the year before (427) were selected and compared with pregnancy outcomes not associated with spraying of any herbicide in that period of time (352) (Smith et al., 1982a). The relative risk for congenital defects among children of exposed fathers was 1.19 (90% confidence limits, 0.58-2.45) and that for miscarriage 0.89 (90% confidence limits, 0.61-1.30).

(c) *Absorption, distribution, excretion and metabolism*

(i) *2,4,5-T*

In five male volunteers who received an oral dose of 5 mg/kg bw 2,4,5-T, there was almost complete gastrointestinal absorption. Disappearance from blood and appearance in urine followed first-order kinetics and showed a half-time of 23 h. An average of 88.5% of the dose was excreted in the urine within 96 h of administration, and the renal clearance was 180-260 ml/min. No acid-labile conjugate or free trichlorophenol was detected in the urine. 2,4,5-T bound reversibly to plasma proteins (98.7%), and the relative volume of distribution was 0.079 l/kg. Faecal excretion was <1% of the dose (Gehring *et al.*, 1973).

In a similar study, in which 2, 3 or 5 mg/kg bw 2,4,5-T were administered orally, maximum plasma concentrations were detected 7-24 h after administration. Following a 5-mg/kg dose, the disappearance half-time averaged 18.8 h, and the average relative volume of distribution was 0.157 l/kg. For all doses, an average of 63-79% of the dose was recovered in the urine within 96 h of administration (Kohli *et al.*, 1974a).

Potential inhalation exposure could account for only approximately 1% of the total amount of 2,4,5-T recovered within four days in the urine of spray applicators, whereas the estimated dermal exposure was potentially 1000 times greater, indicating the relative importance of exposure *via* the dermal route (Lavy *et al.*, 1980).

(ii) *2,4,-D*

In a study on the kinetics of 2,4-D, five male volunteers received an oral dose of 5 mg/kg bw. Absorption was almost complete, as indicated by the recovery of 88-106% of the dose in the urine within 144 h. When elimination of 2,4-D from plasma was followed in three subjects, the average disappearance half-time was 11.6 h. For two subjects, the relative volumes of distribution were 238 and 294 ml/kg, respectively; an apparent biphasic clearance was exhibited by a third subject. Approximately 80% of the 2,4-D was excreted unchanged in the urine and the remainder as an acid-labile conjugate (Sauerhoff *et al.*, 1977a).

Rapid (half-time, 2.5 h) and extensive gastrointestinal absorption of 2,4-D was also found by Kohli *et al.* (1974b), who observed an elimination half-time of 33 h and a volume of distribution of 0.1 l/kg.

From a comparison of urinary excretion of 2,4-D after intravenous administration and application of 4 μg/cm^2 on forearm skin, the dermal absorption of 2,4-D was calculated to be 5.8%. Dermal absorption was protracted, with peak concentrations detected in the urine three days after exposure (Feldman & Maibach, 1974). After exposure of ground sprayers to 2,4-D, peak concentrations in the blood and urine were detected after 0 to at least three (blood) or four (urine) days, and the apparent half-time for urinary excretion of 2,4-D was 14-79 h (calculated from data presented by Nash *et al.*, 1982; Nash *et al.*, 1982; Taskar *et al.*, 1982). Frank *et al.* (1985) calculated that a maximum of 4.5% of the amount of 2,4-D deposited on the bare skin of a bystander directly sprayed with 2,4-D was absorbed. In occupational exposures, skin appears to be the most important route of absorption (Draper & Street, 1982; Lavy *et al.*, 1982; Kolmodin-Hedman *et al.*, 1983a; Kangas *et al.*, 1984; Libich *et al.*, 1984; Frank *et al.*, 1985).

(iii) *MCPA*

In five volunteers (three men, two women) given 15 µg/kg bw MCPA orally, the highest plasma concentrations were seen after 1 h. Urinary excretion was almost complete by 24 h, at which time approximately 40% of the dose had been recovered (Kolmodin-Hedman *et al.*, 1983b). In a similar study (four men), an average of 55% of a 5-mg oral dose was recovered in the urine within 96 h (Fjeldstad & Wannag, 1977).

In a study on dermal absorption in five volunteers (three men, two women), soft-paper pads saturated with 10 ml of a 10% aqueous solution of MCPA were applied on the skin of the thigh for 2 h and the pad covered with surgical tape. Peak plasma levels occurred at about 24 h (two subjects). By 144 h, approximately 2 mg MCPA (0.2% of the applied dose) had been recovered in the urine; the peak of urinary excretion was seen 24-48 h after application (Kolmodin-Hedman *et al.*, 1983b).

Dermal absorption is an important factor in occupational exposure to MCPA, since time-weighted average concentrations of the compound in breathing-zone air samples were only <0.1 mg/m^3, whereas concentrations in the urine of exposed workers reached 12 µg/ml (Kolmodin-Hedman *et al.*, 1983a).

(iv) *Silvex*

Silvex, in powder form, was given orally to eight volunteers (seven men, one woman) at a dose of 1 mg/kg, and its concentration in plasma and urine was studied. Silvex was almost completely absorbed; peak concentrations in plasma were reached in 2-4 h. The disappearance of silvex from the plasma was best described by a two-compartment model, with relative volumes of distribution for each compartment of 81-158 ml/kg and 45-163 ml/kg, respectively. Each compartment followed first-order kinetics, and the two successive half-times for plasma disappearance were 0.9-6.9 h, and 9.0-33.0 h, respectively. Within 144 h, urinary excretion of silvex, which decreased bi-exponentially, amounted to 66-95% of the dose (as silvex and silvex conjugates). In the urine, 29-80% of silvex was excreted unchanged and the rest as acid- and base-labile conjugates; glycine conjugates were not detected. Up to 3% of the dose was detected in the faeces (Sauerhoff *et al.*, 1977b).

(v) *Other*

Dichlorprop and mecoprop have been detected in the urine of exposed farmers and spraymen (Kolmodin-Hedman *et al.*, 1983a).

(*d*) *Mutagenicity and chromosomal effects*

The genetic effects of chlorophenoxy herbicides and their contaminants have been reviewed (Wassom *et al.*, 1977/1978; Seiler, 1978; Grant, 1979).

Mulcahy (1980) examined the incidences of chromosomal aberrations and sister chromatid exchanges (SCEs) in 15 soldiers ten years after serving in the Australian armed forces in Viet Nam for periods ranging from six to 15 months. Eight control subjects were matched for age and sex who had no history of industrial or agricultural exposure to herbicides. The mean frequency of chromosomal aberrations in peripheral lymphocytes was 5.06 per 100 cells in the exposed group *versus* 4.38 in the controls; SCE frequencies were 5.25 and 5.45 per cell, respectively. The differences were not significant.

A study which reported that exposure to chlorophenoxy herbicides during spraying in South Viet Nam induced chromosomal aberrations in humans was judged to be based on inadequate data (reviewed by National Academy of Sciences, 1974).

An analysis of lymphocyte chromosomes of agricultural workers in Idaho, USA, with extensive occupational exposure to pesticides was reported by Yoder *et al.* (1973). A group of 26 herbicide-exposed workers was compared with 16 controls. The list of the most commonly used herbicides comprised 14 formulations, but the predominant exposures were to amitrole, 2,4-D and atrazine. Blood samples were drawn both off-season and mid-season and cultured for 48 h. Only 25 metaphases were examined from each. From off-season to mid-season the mean number of chromatid gaps in the herbicide-exposed group increased four fold (from 0.38 ± 0.10 to 1.38 ± 0.22 per person per 25 cells). In the same group, chromatid breaks increased 25 fold (from 0.07 ± 0.05 to 1.81 ± 0.35). The off-season aberration frequencies were, however, very low as compared with the control group (off-season, 0.63 ± 0.22 for gaps and 0.31 ± 0.12 for breaks), although it is possible to use subjects as their own controls. The authors concluded that the increased incidence of chromosomal aberrations was probably due to exposure to herbicides, although it was not possible to distinguish which herbicide formulations were responsible. [The Working Group noted the small number of cells examined and that possible confounding factors were not taken into consideration.]

SCE frequency was studied in 57 herbicide and pesticide sprayers in New Zealand (Crossen *et al.*, 1978). 2,4,5-T and 2,4-D were mentioned as two of the 30 formulations most commonly encountered in the study. Overall, there was no difference in SCE frequency between the control group and the sprayers (mean rates, 7.65 *versus* 8.48). The sprayers were divided into three groups: those with no protection, those with some protection (either clothing, gloves or respirator) and those with full protection. Those with no protection had a significantly higher mean SCE rate than the control group (mean rate, 9.03 *versus* 7.65), but the authors noted that the length and level of exposure could also have influenced the findings. There was no difference in SCE rate between those who used herbicides exclusively and those using both herbicides and pesticides. [The Working Group noted that confounding factors such as smoking were not taken into consideration.]

Högstedt *et al.* (1980) studied peripheral lymphocyte chromosomes from ten Swedish workers who had worked with several pesticides for two to 29 years (mean, 13 years). Among the pesticides used were MCPA, mecoprop and 2,4,D; none had used 2,4,5-T, dinoseb (2-*sec*-butyl-4,6-dinitrophenol) or amitrole during the last two years. The control group consisted of seven farm workers who had never worked with pesticides. Cells were cultured for 72 h, and 200 cells from each subject were analysed. No significant difference in the frequency of chromosomal aberrations was found between the two groups (gaps, 1.7 per 100 cells *versus* 1.6 in the controls; breaks and exchanges, 2.4 *versus* 2.8).

Linnainmaa (1983) studied SCEs in lymphocytes of workers in Finland spraying forest foliage with chlorophenoxy herbicides containing amine salts and esters of 2,4-D and MCPA, or mixtures of the two. Three successive blood samples were taken from 50 male sprayers (who had used protective clothing); the first before the spraying season, the second in the middle of the spraying season and the third within two days after the subject had finished spraying, in order to follow possible exposure-related changes. Urine samples were taken at the same time as the second blood sample. Levels of 2,4-D and MCPA in the urine, which were used as an indicator of exposure, varied from 0.00 to 10.99 mg/l (mean, 1.80 mg/l). Suitable chromosome preparations were obtained from 35 herbicide workers and 15 controls not working with herbicides. No significant difference in SCE frequency was observed in samples taken before, during or after the exposure; the nonexposed control group fell in the same range. Smokers in both groups had significantly higher mean values than nonsmokers. The average frequencies of SCEs/cell in nonsmoking sprayers were 8.6, 8.0, and 8.8 before, during, and after spraying, respectively, *versus* 8.1 in nonsmoking controls. The corresponding values for the smokers were 9.7, 9.5, 9.9 *versus* 10.0.

Some individuals with similar exposure were also studied for frequency of chromosomal aberrations (Mustonen *et al.*, 1986). Where possible, 100 first-division metaphases were examined from each of 19 workers and 15 controls. No difference was found between the two groups. The percentage of aberrant cells (gaps included) in controls was 1.5 ± 0.3 in nonsmokers and 1.9 ± 0.4 in smokers. In exposed subjects, the respective percentages were 1.2 ± 0.5 and 1.8 ± 0.4.

3.3 Case reports and epidemiological studies of carcinogenicity to humans

(*a*) *Case reports*

A number of reports describe the occurrence of cancer in workers exposed to TCDD (see IARC, 1977b). A case report describes three cases of soft-tissue sarcoma in US veterans 10-13 years after exposure to Agent Orange (Sarma & Jacobs, 1982). Palva *et al.* (1975) described a case of aplastic anaemia in a farmer three months after exposure to MCPA in Finland (he had also used herbicides during the previous five years); a year later he developed acute myelomonocytic leukaemia (Timonen & Palva, 1980).

A clinical study of 123 male patients with non-Hodgkin's lymphoma in Sweden found that four of five patients with cutaneous lesions reported spraying large areas with chlorophenoxy herbicides (Olsson & Brandt, 1981).

(*b*) *Cohort studies*

A cohort of 348 railroad workers in Sweden exposed for 45 days or more during 1957-1978 to 2,4-D, 2,4,5-T or amitrole were investigated in a follow-up study (Axelson *et al.*, 1980) [see also monograph on amitrole in this volume, p. 309]. There was a deficit of deaths from all causes (45 observed, 49 expected) but an excess from malignant neoplasms

(17 observed, 11.9 expected). In a subcohort exposed to 2,4-D or 2,4,5-T but not amitrole, there were six deaths from cancer with 5.6 expected, all of which occurred in those first exposed ten years or more before death (3.1 expected). In a subcohort exposed to both amitrole and chlorophenoxy herbicides (2,4-D or 2,4,5-T), there were six deaths from cancer with 2.9 expected, of which all six (with 1.8 expected; $p < 0.005$) occurred in those first exposed ten years or more before death; there were altogether three deaths from stomach cancer with 0.5 expected in men first exposed to chlorophenoxy herbicides ten years before death. The men were also exposed to other organic (e.g., monuron and diuron) and inorganic chemicals (e.g., potassium chlorate).

Hogstedt and Westerlund (1980) studied the mortality of 142 male forestry workers in Sweden exposed to 2,4-D and 2,4,5-T in 1954-1967, and 244 male forestry workers without such exposures (follow-up, 1954-1978). Five deaths from cancer were observed among those exposed *versus* 6.4 expected from national rates. Among unexposed workers there were 10 observed deaths from cancer *versus* 14.7 expected. Among 16 exposed foremen, five incident cases of cancer were found in the Swedish Cancer Registry against 1.4 expected ($p < 0.02$), while three cases of cancer were seen in 126 exposed workers with 8.4 expected ($p < 0.05$). These eight cases were localized in the stomach (1), pancreas (2), lung (1), skin (1), prostate (2) and bladder (1). The expected value for soft-tissue sarcoma among the exposed was about 0.1. Foremen were exposed to chlorophenoxy herbicides for an average of 176 days and workers for 30 days.

Barthel (1981) performed a study encompassing the 14 districts of the German Democratic Republic (excluding Berlin); 1658 male subjects who had been active for at least five years as agricultural workers or agronomists between 1948 and 1972 were potentially exposed to 2,4-D and MCPA. Cancer incidence in the group was assessed through county tumour reference centres and death certificates; 124/169 neoplasms were histologically verified. Fifty cases of bronchial carcinoma occurred between 1970 and 1978 *versus* 27.5 expected from national morbidity rates. One case of soft-tissue sarcoma and five of lymphatic neoplasms were observed. [The Working Group noted that smoking was not taken into account in this study, but that differences in smoking habits are unlikely to explain a relative risk of the magnitude observed.]

Riihimäki *et al.* (1982, 1983) examined a cohort of 1926 male Finnish workers involved in brush control for at least two weeks between 1951-1971. These workers were exposed to 2,4-D and 2,4,5-T, among other agents. The follow-up period was from 1972-1980, and 16 694 person-years were represented. The observed rates were compared with expected numbers from national death rates. Only 26 cancer deaths were noted with 36.5 expected. In the subgroup with over 10 years' latency, 20 cancers were observed whereas 24.3 were expected [standardized mortality rate (SMR), 82; 95% confidence interval (CI), 50-127], including 12 lung cancers (expected, 11.1) [SMR, 108; 95% CI, 56-189] and four cancers of the stomach and oesophagus (expected, 3.7) [SMR, 108; 95% CI, 30-277]. No lymphoma or soft-tissue sarcoma was observed. Cancer incidence was also studied, with similar results. The authors point out, however, that the small size of the cohort, the brief follow-up period and the low exposure limited the utility of this study.

Lynge (1984, 1985) studied 4563 persons employed by two chemical plants in Denmark which produced 2,4-D, dichlorprop, MCPA, mecoprop and 2,4,5-T in the period 1947-1981. 2,4,5-T was produced mainly from 1951-1959 from 2,4,5-trichlorophenol made externally, and 2,4,5-T esters were made from 2,4,5-T produced externally up to 1981. 2,4-D and MCPA were manufactured by chlorination of the phenol or cresol; during 1960-1970, up to 50% of the MCPA was produced as spray-dried MCPA sodium salt. Incident cases of cancer were identified by record linkage with data in the Danish Cancer Registry. From the year production of chlorophenoxy herbicides started in the two plants (1947 and 1951, respectively) until 1982, there were, for the total cohort, 159 cancers of all sites *versus* 160.6 expected in men and 49 observed *versus* 55.9 expected in women. The observed numbers of cases for individual sites of cancer did not differ statistically significantly from the expected numbers. Among persons first exposed ten years or more before cancer diagnosis, there were 11 cases of stomach cancer *versus* 6.3 expected (nonsignificant). No case occurred among women. Two cases occurred in men employed in the manufacture and packaging of chlorophenoxy herbicides *versus* 1.5 expected. Seven cases of malignant lymphoma were observed among men with 5.4 cases expected; among women, there was one case with 1.2 expected. No case was observed among men or women manufacturing and packaging chlorophenoxy herbicides. Among men, there were five cases of soft-tissue sarcoma *versus* 1.8 expected (relative risk, 2.72; 95% CI, 0.88-6.34); no case was diagnosed among women (0.75 expected). For the four men with more than ten years' latency since first exposure, the relative risk estimate was increased to 3.67 (95% CI, 1.00-9.39). Only one had been assigned to the manufacture and packaging of chlorophenoxy herbicides, two were working in shipping and one in pigment milling. The durations of employment of these men were 90, 30, three and 0.5 months. At these plants, 59% of men and 50% of women had been employed for less than one year. Among the subgroup of persons employed in the manufacture and packaging of chlorophenoxy herbicides, there were 11 cases of lung cancer among men *versus* 5.3 expected ($p < 0.05$); there was no excess in women, but five cases of cervical cancer were seen *versus* 1.8 expected (nonsignificant).

Mortality odds ratios for service in Viet Nam were estimated for 1496 veterans discharged in 1970-1973 and who died in New York State, USA, between 1970 and 1980. Of these, 555 had served in Viet Nam. A ratio of 1.09 (95% CI, 0.18-6.70) was found for soft-tissue sarcoma in Viet Nam veterans *versus* non-Viet Nam veterans (Lawrence *et al.*, 1985).

As pointed out in the monograph on occupational exposures to chlorophenols (p. 319), workers involved in manufacturing 2,4,5-T itself (rather than the precursor 2,4,5-trichlorophenol) may be exposed to both 2,4,5-trichlorophenol and 2,4,5-T. Some of the manufacturing cohort studies referred to in that monograph are also, therefore, relevant to 2,4,5-T itself.

Potential exposure to 2,4,5-T was reported by Ott *et al.* (1980) in their mortality study of a small cohort of 204 workers which identified one cancer death, with 1.3 expected, in workers exposed for more than one year (see monograph on occupational exposures to chlorophenols, p. 342).

Fingerhut et al. (1984) carried out a review of cases reported in cohort studies and of several additional case reports of soft-tissue sarcoma (see monograph on occupational exposures to chlorophenols, p. 343). [The Working Group considered that some over- or underascertainment of soft-tissue sarcoma is possible.]

The mortality study reported by Cook et al. (1986) involved potential exposure to both trichlorophenol and 2,4,5-T, although the report focused on TCDD exposure in trichlorophenol manufacture. There were five cases of non-Hodgkin's lymphoma (SMR, 238; 95% CI, 77-556) but no dose-response relationship to TCDD exposure. No dose-response analysis was presented of potential 2,4,5-T exposure (see monograph on occupational exposures to chlorophenols, p. 343). [The Working Group noted that it was difficult to determine the exposure of the workers to 2,4,5-T.]

(c) *Case-control studies*

(i) *Soft-tissue sarcoma*

The first case-control study of soft-tissue sarcoma followed the observation of a number of patients at a cancer clinic in Sweden who had reported previous exposure to chlorophenoxy herbicides (Hardell & Sandström, 1979). A total of 52 male patients, 21 living and 31 deceased, were identified from records of the Department of Oncology of the University Hospital of Umeå as having been admitted with a diagnosis of soft-tissue sarcoma between 1970 and 1977. Four matched controls were selected for each case, from the National Population Registry for living patients, and from the National Registry for Causes of Death for deceased patients. Exposure was ascertained through a postal questionnaire with a variety of questions about exposure; the answers were supplemented by telephone without knowledge of case or referent status. For deceased patients and controls, the procedure was the same, but contact was made with the next of kin. In an attempt to verify occupational exposures, questionnaires were also sent to the employers of persons stating work in forestry, saw mills and pulp industries. According to the authors, employers' statements from the latter two industries agreed closely with the statements given by the interviewed persons and the same was therefore assumed to be true for exposure to chlorophenoxy herbicides — but there were many non-respondents from the forestry companies. A requirement for being classified as exposed was at least one full day of exposure more than five years before the tumour was diagnosed. When patients and controls with exposure to chlorophenols were excluded, the relative risk estimate was 5.3 (95% CI, 2.4-11.5), with 13 cases exposed. Of the 13 cases, 12 had been exposed to 2,4,5-T or 2,4-D, and one to MCPA alone; combined exposure to 2,4,5-T and 2,4-D was reported by nine cases. The exposure of two cases consisted solely of working on ground that was wet from earlier spraying. Latency from first exposure was predominantly in the range of 10-20 years. The median duration of exposure was three to four months (range, two days to 49 months). As three of these patients were those that had been identified earlier and initiated the study, a calculation was made after exclusion of three cases and their controls; this did not affect the results (relative risk, 4.7; 95% CI, 2.0-10.7). [The Working Group noted that, according to the authors, a further four exposed patients had been reported as cases before the case-control study was published; but no calculation was presented based on their exclusion.]

A second study (Eriksson *et al.*, 1981) of soft-tissue sarcoma was undertaken in southern Sweden, where MCPA and 2,4-D have been used widely in agriculture. The study involved 110 living and deceased cases reported in 1974-1978 and 220 referents selected by methods similar to those used in the first study (Hardell & Sandström, 1979), and with the same assessments and requirements for exposure. A relative risk estimate of 8.5 was obtained for exposure to chlorophenoxy herbicides alone for more than 30 days (seven cases), and one of 5.7 for exposures of less than or equal to 30 days (seven cases). The odds ratio for exposure to chlorophenoxy herbicides other than 2,4,5-T was 4.2 [95% CI, 1.3-15.8].

An initial analysis of occupations recorded on the National New Zealand Cancer Registry between 1976 and 1980 (102 cases compared to 306 controls) did not find an excess of soft-tissue sarcoma cases in agriculture and forestry workers compared with patients with cancers of other sites on the cancer registry (Smith *et al.*, 1982b). Subsequently, 82 cases (or their next-of-kin) of soft-tissue sarcoma were interviewed by telephone regarding past occupations and specific use of chlorophenoxy herbicides by an interviewer who was unaware of the case/referent status of the person, and the results were compared with those of 92 randomly selected referents with cancers at other sites (Smith *et al.*, 1984). In 43% of the cases and 34% of the referents, the patient gave the data himself or herself. Since the results were not affected by such stratification, only results of unstratified analyses were reported. An odds ratio of 1.6 (90% CI, 0.7-3.3) was calculated for those who had probably or definitely been exposed for more than one day more than five years prior to diagnosis of the tumour (17 cases). None of the cases was in a professional applicator.

Another study was undertaken with interviews of 51 further cases appearing on the Cancer Registry up to 1982 (Smith & Pearce, 1986). Cases were identified as in the first study, and histology reports were reviewed for each case. Eligible patients or their next-of-kin were interviewed by telephone by the same experienced interviewer, who was unaware of the case/referent status of the person. Referents were selected from a large series used for another study on lymphoma and multiple myeloma (Pearce *et al.*, 1986), who had been interviewed in the same manner and comprised 315 cancer patients (excluding lymphoma, multiple myeloma and soft-tissue sarcoma). The odds ratio for exposure for more than one day more than five years prior to registration was 0.7 (90% CI, 0.3-1.5). The combined estimate, using data from both the previous study (Smith *et al.*, 1983) and the present one, for 133 cases compared with 407 referents, was 1.1 (90% CI, 0.7-1.8). The proportions of exposed referents were 0.14 for the first study and 0.15 for the second study. None of the cases occurred in a professional applicator.

A population-based case-referent study was conducted in northern Italy (Vineis *et al.*, 1986). Thirty-seven male and 31 female patients with soft-tissue sarcoma formed the case series and 85 males and 73 females made up the referent series. Of the cases, 24 were deceased, and they were matched by municipality of residence with 36 deceased referents. The living referents were drawn randomly from the population of each province. Cases and referents, or their next-of-kin, were interviewed by a trained interviewer, who was unaware of their case/referent status, by means of a personal visit; 16 cases and 37 referents gave information by post. Information on various uses of herbicides, mainly 2,4-D, MCPA and 2,4,5-T (until 1970), was collected in detail and assessed by two experts in agricultural chemistry, who were unaware of the case/referent status of the person. Exposure to herbicides in these provinces was mainly associated with rice weeding, an activity traditionally performed by women. The highest exposure to herbicides occurred in the early 1950s, when rice weeding was still performed manually but chlorophenoxy herbicides were being tested and introduced. An age-adjusted odds ratio of 0.91 was found for living men with suspected exposure, and one of 2.7 (90% one-tailed CI, 0.59-12.37) for living women. An age-adjusted odds ratio of 15.5 (1.3-180.3) was found for living women under the age of 75 years, exposed between 1950-1955. For deceased females, the odds ratio, based on four exposed cases of likely and certain exposure, was 1.05 (0.21-5.1). Only one dead male case had been exposed. The authors suggested that geographical overmatching had occurred for deceased cases and referents. [The Working Group noted that two of the exposed cases were Kaposi's sarcoma.]

A case-control study ascertained Viet Nam service for 281 men with soft-tissue sarcoma and for a matched control group (Greenwald *et al.*, 1984). Cases were diagnosed between 1962 and 1980 and identified on the New York State Cancer Registry. The odds ratio for service in Viet Nam was 0.53 (95% CI, 0.21-1.31), with ten cases reporting service in Viet Nam. An odds ratio of 0.70 was obtained for those reporting contact with Agent Orange, TCDD or 2,4,5-T. [The Working Group noted that bias may have arisen as a result of the choice of control group. Furthermore, exposure may have occurred to many other chemicals, and the time between exposure and disease was short.]

(ii) *Malignant lymphoma*

Concerns about chlorophenoxy herbicides and lymphoma arose in Sweden when a number of male patients with histiocytic lymphoma reported past exposure to chlorophenoxy herbicides. A case-control study of 169 cases of malignant lymphoma (60 Hodgkin's disease, 105 non-Hodgkin's lymphoma, four unclassifiable) was then undertaken, including 338 controls (Hardell *et al.*, 1981). The study design, including ascertainment of exposure, was similar to that of the Swedish soft-tissue sarcoma studies (see Hardell & Sandström, 1979). A relative risk estimate of 4.8 (95% CI, 2.9-8.1) was obtained for exposure to chlorophenoxy herbicides, excluding cases and controls exposed to chlorophenols. Stratifying by duration of exposure, the relative risk estimate was 4.3 for less than 90 days, and 7.0 for 90 days or more exposure to chlorophenoxy herbicides. Most chlorophenoxy herbicide-exposed cases reported exposure to both 2,4,5-T and 2,4-D (25 cases); two reported exposure to 2,4,5-T, 2,4-D and MCPA, seven to 2,4-D alone, and five to MCPA alone (Hardell, 1981a). No 'noticeable difference' in excess risk could be demonstrated between Hodgkin's disease and non-Hodgkin's lymphoma.

In New Zealand, an analysis of reported occupation appearing on the New Zealand Cancer Registry indicated an excess of malignant lymphoma and multiple myeloma among men in agricultural occupations during 1977-1981. Although 734 cases and four controls per case were selected from the Registry, the main findings concerned a subgroup of 88 cases of malignant lymphoma, classified as ICD 202 which covers non-Hodgkin's lymphoma other than lymphosarcoma and reticulosarcoma. An odds ratio of 1.76 (95% CI, 1.03-3.02) was obtained for those under the age of 65 working in agriculture (Pearce *et al.*, 1985). However, a subsequent interview study of 83 cases classified as ICD 202 (see monograph on occupational exposures to chlorophenols, p. 344) did not suggest that exposure to chlorophenoxy herbicides was the explanation, since an odds ratio of 1.3 (90% CI, 0.7-2.5) was obtained when the controls were people with other cancers, and an odds ratio of 1.0 (0.5-2.1) when general population controls were used for people probably or definitely exposed for more than one day not in the five years before cancer registration (Pearce *et al.*, 1986).

(iii) *Nasal and nasopharyngeal cancer*

In the study of Hardell *et al.* (1982), described in the monograph on occupational exposures to chlorophenols (p. 344), an odds ratio of 2.1 (95% CI, 0.9-4.7) was found for exposure to chlorophenoxy herbicides.

(iv) *Colon and liver cancer*

The same Swedish authors tested the assumption that a greater recall of past herbicide exposure by patients with soft-tissue sarcoma and lymphoma than by the controls may have produced the earlier findings. A study was conducted involving 157 male colon cancer patients (Hardell, 1981b), who were interviewed in the same manner as in the earlier studies, and whose exposure was compared with that of the combined controls from the two earlier studies from the same region (Hardell & Sandström, 1979; Eriksson *et al.*, 1981). A relative risk estimate of 1.3 (95% CI, 0.6-2.8) was obtained based on 11 exposed cases (out of 154) and 43 exposed referents (out of 541).

Hardell *et al.* (1984) performed a case-control study in the northern region of Sweden on 103 primary liver cancer cases diagnosed in 1974-1981 and 206 controls, with a study design similar to that of the earlier studies (Hardell & Sandström, 1979; Eriksson *et al.*, 1981). Of the cases, 8.2% reported exposure to chlorophenoxy herbicides *versus* 6.5% among the referents (odds ratio, 1.7; 95% CI, 0.7-4.4). The study indicated an association between exposure to organic solvents and primary liver cancer.

Vân (1984) investigated previous exposure to herbicides during wartime for 21 male cases of primary hepatic carcinoma admitted to the Viet Duc Huu Ngai Hospital of Hanoi, Viet Nam, in January to September 1982, and for 42 control subjects admitted in the same months for gastrointestinal diseases. Cases and controls were aged 18-50 years. Six of 21 cases and 3/42 controls had been living, working or fighting in sprayed regions of South Viet Nam, at the time of spraying or subsequently, for a length of time ranging from eight to 77 months. No information was available on possible confounding factors. [The Working Group noted that the possibility of bias, exposures to many unknown chemicals and the short reported latency make the study uninformative.]

[The possibility of recall bias in the Swedish case-referent studies has been discussed by the authors (Axelson, 1980; Hardell, 1981a,b; Hardell *et al.*, 1981; Hardell & Axelson, 1982; Hardell *et al.*, 1984). Had a significant recall bias existed, the studies on colon cancer (Hardell, 1981b) and liver cancer (Hardell *et al.*, 1984) would have been expected to give significantly elevated odds ratios for herbicide exposure; such findings were not reported.

[The New Zealand studies used other cancer patients as controls. However, a comparison of such controls with general population controls also interviewed by telephone did not reveal differences in past exposure frequency (Pearce *et al.*, 1986). The use of other cancers as controls does not therefore explain the difference in findings between the Swedish and New Zealand studies.

[The Working Group noted that none of the cases of soft-tissue sarcoma seen in New Zealand occurred in a professional herbicide applicator. The Working Group noted differences in the use patterns of chlorophenoxy herbicides in Sweden and New Zealand, e.g., hand-notching and application of amine salts to trees was used in Sweden. Also, no information was available on the solvents, emulsifiers and other additives in the chlorophenoxy herbicide formulations used in Sweden and New Zealand. Limited information is available on the levels of TCDD in formulations used in Sweden and New Zealand, but the data indicate contamination at about the same level.]

4. Summary of Data Reported and Evaluation

4.1 Exposure data

Chlorophenoxy herbicides have been produced extensively since the 1950s for use in agriculture and as defoliants, although production and use are now decreasing in many countries. Widespread occupational exposure to chlorophenoxy herbicides and their chlorinated dibenzodioxin impurities is known to have occurred during their production, formulation, application and disposal. Increased urinary levels of chlorophenoxy compounds and increased concentrations of some chlorinated dibenzodioxins in adipose tissue have been measured in highly exposed persons. The presence of dibenzodioxins and dibenzofurans has been demonstrated in the adipose tissue of nonoccupationally exposed people in many countries.

During occupational exposure, such as ground spraying and other manual application of these herbicides, dermal absorption is a major route of entry into the body.

In manufacturing plants, exposures occur during the handling of raw materials, intermediates, finished products and process wastes. High-level short-term occupational exposures have also been caused by industrial accidents.

4.2 Experimental data

Previous IARC evaluations of the carcinogenicity to experimental animals of several individual chlorophenoxy herbicides and of 2,3,7,8-tetrachlorodibenzo-*para*-dioxin (TCDD), an impurity found in some of these herbicides, are summarized in section 3.1.

4.3 Human data

Studies comparing the occurrence of congenital malformations in areas and periods characterized by different usage of chlorophenoxy herbicides were uninformative with regard to the teratogenicity of these agents. Two case-control studies on birth anomalies among the children of Australian and US veterans and of New Zealand pesticide sprayers showed no excess risk associated with paternal exposure to herbicides.

No study was available of pregnancy outcomes of women exposed occupationally to chlorophenoxy herbicides.

In one study of persons exposed to chlorophenoxy herbicides during military operations in Viet Nam, conducted ten years after exposure, no increase in the incidence of chromosomal aberrations or sister chromatid exchanges was observed.

Cytogenetic studies have been carried out on workers occupationally exposed to chlorophenoxy herbicides during spraying. In three of the studies, there was also exposure to other herbicides, and the effect of chlorophenoxy herbicides could not be assessed. Studies in which occupational exposure was only to chlorophenoxy herbicides showed no increased incidence of chromosomal aberrations or sister chromatid exchanges.

In a large Danish cohort study of chemical workers exposed to chlorophenoxy herbicides [particularly (4-chloro-2-methylphenoxy)acetic acid (MCPA), 2-(4-chloro-2-methylphenoxy)propanoic acid (mecoprop), 2,4-dichlorophenoxyacetic acid (2,4-D) and 2-(2,4-dichlorophenoxy)propanoic acid (dichlorprop)], as well as other chemicals, no overall increase in cancer incidence rate was observed, but there were significantly increased risks of soft-tissue sarcoma and lung cancer in different subcohorts, which were not necessarily those with the highest exposures to chlorophenoxy herbicide preparations. A Finnish cohort study of brush control workers with short follow-up time showed no increased risk. A small Swedish cohort study of railroad workers who sprayed herbicides showed an increased risk of cancers at all sites combined for those exposed to both chlorophenoxy herbicide preparations and other herbicides. An excess incidence of all cancers was also reported from a very small cohort of Swedish forestry foremen exposed to chlorophenoxy herbicide preparations and other herbicides. A study of long-term pesticide applicators in the German Democratic Republic, heavily exposed to a number of chemicals, including 2,4-D and MCPA, demonstrated an increased risk of bronchial carcinoma.

A population-based case-control study conducted in northern Sweden showed a statistically significant association between exposure to chlorophenoxy herbicides, especially in forestry, and the occurrence of soft-tissue sarcomas. A second study on this type of tumour was conducted in southern Sweden, where a significant increase in the risk of developing soft-tissue sarcomas was associated with previous exposures to chlorophenoxy herbicides, mainly in agriculture. An increased risk of soft-tissue sarcoma was described among highly exposed Italian rice weeders in a population-based case-control study. A case-control study from New Zealand did not demonstrate an increased risk of soft-tissue sarcoma in people exposed to chlorophenoxy herbicides.

A statistically significant association between malignant lymphoma and exposure to chlorophenoxy herbicides was found in a Swedish case-control study; however, no such association was seen in a case-control study of these tumours from New Zealand. In a Danish cohort of chemical workers exposed to chlorophenoxy herbicides, there was also no increased risk of malignant lymphoma.

Three Swedish case-control studies of colon, liver and nasal cancer, respectively, which used the same study design and methods as in the studies on soft-tissue sarcoma and malignant lymphoma, did not demonstrate significantly increased risks. Exposure recall bias of cancer patients thus does not seem to explain the differences between the results of the Swedish and the New Zealand case-control studies of soft-tissue tumours and lymphomas.

In summary, well-conducted case-control studies have provided the most information on the association between cancer and occupational exposure to chlorophenoxy herbicides. Statistically significant elevated odds ratios have been observed for cancers at some sites, but not consistently, in independent studies. The results of one cohort study on the incidence of soft-tissue sarcoma support the finding in case-control studies of an increased relative risk for these tumours. Other cohort studies have added little information. No consistent exposure-response relationship emerged from the different studies, and, in the studies that found an association, exposures were shorter than those usually associated with occupation-related cancers.

4.4 Evaluation[1]

There is *limited evidence* that occupational exposures to chlorophenoxy herbicides are carcinogenic to humans.

5. References

Åkerblom, M., Kolmodin-Hedman, B. & Höglund, S. (1983) *Studies on occupational exposure to phenoxy acid herbicides*. In: Miyamoto, J., ed., *IUPAC Pesticide Chemistry*, Oxford, Pergamon Press, pp. 228-232

American Conference of Governmental Industrial Hygienists (1985) *TLVs Threshold Limit Values and Biological Exposure Indices for 1985-86*, 2nd ed., Cincinnati, OH, pp. 15, 30

Anon. (1974) EPA exonerates 2,4,5-T. *Chem. Week*, 3 July, p. 10

Anon. (1983) Boehringer Ingelheim to stop 2,4,5-T production. *Eur. Chem. News*, *40*, 20

Anon. (1985) *Farm Chemicals Handbook '85*, Willoughby, OH, Meister Publishing Co., pp. C70-C71, C80, C146-C148, C223

Axelson, O. (1980) A note on observational bias in case-referent studies in occupational health epidemiology. *Scand. J. Work Environ. Health*, *6*, 80-82

Axelson, O., Sundell, L., Andersson, K., Edling, C., Hogstedt, C. & Kling, H. (1980) Herbicide exposure and tumor mortality. An updated epidemiologic investigation on Swedish railroad workers. *Scand. J. Work Environ. Health*, *6*, 73-79

---

[1]For definition of the italicized term, see Preamble, p. 22.

Balarajan, R. & McDowall, M. (1983) Congenital malformations and agricultural workers. *Lancet, i*, 1112-1113

Barthel, E. (1981) Increased risk of lung cancer in pesticide-exposed male agricultural workers. *J. Toxicol. environ. Health, 8*, 1027-1040

Buser, H.R. & Rappe, C. (1978) Identification of substitution patterns in polychlorinated dibenzo-*p*-dioxins (PCDDs) by mass spectrometry. *Chemosphere, 7*, 199-211

California Department of Agriculture (1984) *Pesticide Use Report, Annual 1983*, Sacramento, CA

Cochrane, W.P., Singh, J., Miles, W., Wakeford, B. & Scott, J. (1982) *Analysis of technical and formulated products of 2,4-dichlorophenoxy acetic acid for the presence of chlorinated dibenzo-p-dioxins*. In: Hutzinger, O., Frei, R.W., Merian, E. & Pocchiari, F., eds, *Chlorinated Dioxins and Related Products. Impact on the Environment*, Oxford, Pergamon Press, pp. 209-213

Cook, R.R., Bond, G.G., Olsen, R.A., Ott, M.G. & Gondek, M.R. (1986) Evaluation of the mortality experience of workers exposed to the chlorinated dioxins. *Chemosphere* (in press)

Crossen, P.E., Morgan, W.F., Horan, J.J. & Stewart, J. (1978) Cytogenetic studies of pesticide and herbicide sprayers. *N. Z. med. J., 88*, 192-195

Deutsche Forschungsgemeinschaft (German Research Community) (1985) *Maximal Concentrations in the Workplace and Biological Tolerance Values for Substances in the Work Environment 1985* (Ger.), Vol. 21, Weinheim, Verlagsgesellschaft mbH, pp. 29, 54

Direktoratet for Arbeidstilsynet (Directorate of Labour Inspection) (1981) *Administrative Norms for Pollution in Work Atmosphere, No. 361* (Norw.), Oslo, pp. 10, 20

Donovan, J.W., MacLennan, R. & Adena, N. (1984) Vietnam service and the risk of congenital anomalies: a case-control study. *Med. J. Aust., 140*, 394-397

Draper, W.M. (1982) A multiresidue procedure for the determination and confirmation of acidic herbicide residues in human urine. *J. agric. Food Chem., 30*, 227-231

Draper, W.F. & Street, J.C. (1982) Applicator exposure to 2,4-D, dicamba, and a dicamba isomer. *J. environ. Sci. Health, B17*, 321-339

Edmunds, J.W., Lee, D.F. & Nickels, C.M.L. (1973) Determination of 2,3,7,8-tetrachlorodibenzo-1,4-dioxin and 2,4,5-trichlorophenoxyacetic acid and 2,4,5-T alkyl ester herbicides. *Pestic. Sci., 4*, 101-105

Eller, P.M., ed. (1984) *NIOSH Manual of Analytical Methods*, 3rd ed., Cincinnati, OH, National Institute for Occupational Safety and Health, pp. 5001-1 — 5001-4

Erickson, J.D., Mulinare, J., McClain, P.W., Fitch, T.G., James, L.M., McClearn, A.B. & Adams, M.J., Jr (1984) Vietnam veterans' risks for fathering babies with birth defects. *J. Am. med. Assoc., 252*, 903-912

Eriksson, M., Hardell, L., Berg, N.O., Möller, T. & Axelson, O. (1981) Soft-tissue sarcomas and exposure to chemical substances: a case-referent study. *Br. J. ind. Med., 38*, 27-33

Esposito, M.P., Tiernan, T.O. & Dryden, F.E. (1980) *Dioxins*, Cincinnati, OH, US Environmental Protection Agency, Industrial Environmental Research Laboratory

Feldman, R.J. & Maibach, H.I. (1974) Percutaneous penetration of some pesticides and herbicides in man. *Toxicol. appl. Pharmacol.*, *28*, 126-132

Ferry, D.G., Gazeley, L.R. & Edwards, I.R. (1982) 2,4,5-T absorption in chemical applicators. *Proc. Univ. Otago med. Sch.*, *60*, 31-34

Field, B. & Kerr, C. (1979) Herbicide use and incidence of neural-tube defects. *Lancet*, *i*, 1341-1342

Fingerhut, M.A., Halperin, W.E., Honchar, P.A., Smith, A.B., Groth, D.H. & Russell, W.O. (1984) An evaluation of reports of dioxin exposure and soft tissue sarcoma pathology among chemical workers in the United States. *Scand. J. Work Environ. Health*, *10*, 299-303

Fjeldstad, P. & Wannag, A. (1977) Human urinary excretion of the herbicide 2-methyl-4-chlorophenoxyacetic acid. *Scand. J. Work Environ. Health*, *3*, 100-103

Food and Agriculture Organization (1984) *FAO Production Yearbook 1983*, Vol. 37, Rome, p. 298

Frank, R., Campbell, R.A. & Sirons, G.J. (1985) Forestry workers involved in aerial application of 2,4-dichlorophenoxyacetic acid (2,4-D): exposure and urinary excretion. *Arch. environ. Contam. Toxicol.*, *14*, 427-435

Franklin, C.A., Grover, R., Markham, J.W., Smith, A.E. & Yoshida, K. (1982) Effect of various factors on exposure of workers involved in the aerial application of herbicides. *Am. Conf. gov. ind. Hyg. Trans.*, *43*, 97-117

Gehring, P.J., Kramer, C.G., Schwetz, B.A., Rose, J.Q. & Rowe, V.K. (1973) The fate of 2,4,5-trichlorophenoxyacetic acid (2,4,5-T) following oral administration to man. *Toxicol. appl. Pharmacol.*, *26*, 352-361

Golding, J. & Sladden, T. (1983) Congenital malformations and agricultural workers. *Lancet*, *i*, 1393

Grant, W.F. (1979) The genotoxic effects of 2,4,5-T. *Mutat. Res.*, *65*, 83-119

Greenwald, P., Kovasznay, B., Collins, D.N. & Therriault, G. (1984) Sarcomas of soft tissues after Vietnam service. *J. natl Cancer Inst.*, *73*, 1107-1109

Gross, M.L., Lay, J.O., Jr, Lyon, P.A., Lippstreu, D., Kangas, N., Harless, R.L., Taylor, S.E. & Dupuy, A.E., Jr (1984) 2,3,7,8-Tetrachlorodibenzo-*p*-dioxin levels in adipose tissue of Vietnam veterans. *Environ. Res.*, *33*, 261-268

Hamner, C.L. & Tukey, H.B. (1944) The herbicidal action of 2,4-dichlorophenoxyacetic and 2,4,5-trichlorophenoxyacetic acid on bindweed. *Science*, *100*, 154-155

Hanify, J.A., Metcalf, P., Nobbs, C.L. & Worsley, K.J. (1981) Aerial spraying of 2,4,5-T and human birth malformations: an epidemiological investigation. *Science*, *212*, 349-351

Hardell, L. (1981a) *Epidemiological Studies on Soft-Tissue Sarcoma and Malignant Lymphoma and Their Relation to Phenoxy Acid or Chlorophenol Exposure* (*University Medical Dissertations, New Series No. 65*), Umeå, University of Umeå, Department of Oncology

Hardell, L. (1981b) Relation of soft-tissue sarcoma, malignant lymphoma and colon cancer to phenoxy acids, chlorophenols and other agents. *Scand. J. Work Environ. Health*, 7, 119-130

Hardell, L. & Axelson, O. (1982) Soft-tissue sarcoma, malignant lymphoma and exposure to phenoxyacids or chlorophenols. *Lancet, i*, 1408-1409

Hardell, L. & Sandström, A. (1979) Case-control study: soft-tissue sarcomas and exposure to phenoxyacetic acids or chlorophenols. *Br. J. Cancer*, 39, 711-717

Hardell, L., Eriksson, M., Lenner, P. & Lundgren, E. (1981) Malignant lymphoma and exposure to chemicals, especially organic solvents, chlorophenols and phenoxy acids: a case-control study. *Br. J. Cancer*, 43, 169-176

Hardell, L., Johansson, B. & Axelson, O. (1982) Epidemiological study of nasal and nasopharyngeal cancer and their relation to phenoxy acid or chlorophenol exposure. *Am. J. ind. Med.*, 3, 247-257

Hardell, L., Bengtsson, N.O., Johnsson, U., Eriksson, S. & Larsson, L.G. (1984) Aetiological aspects on primary liver cancer with special regard to alcohol, organic solvents and acute intermittent porphyria — an epidemiological investigation. *Br. J. Cancer*, 50, 389-397

Hattula, M.L., Reunanen, H. & Arstila, A.U. (1979) Toxicity of 4-chloro-*o*-cresol to rat: 1. Light microscopy and chemical observations. *Bull. environ. Contam. Toxicol.*, 21, 492-497

Hayes, W.J. (1982) *Pesticides Studied in Man*, Baltimore, MD, Williams & Wilkins, pp. 520-577

Hazardous Materials Advisory Committee (1974) *Herbicide Report. Chemistry and Analysis, Environmental Effects, Agricultural and Other Applied Uses*, Washington DC, US Environmental Protection Agency, pp. 18-20, 66

Health and Safety Executive (1985) *Occupational Exposure Limits 1985* (*Guidance Note EH 40/85*), London, Her Majesty's Stationery Office, pp. 11, 19

Hervin, R.L. & Smith, A.B. (1978) *Hazard Evaluation Determination Report No. 77-7A & 7B-486, Union Electric Company, St Louis, Missouri*, Cincinnati, OH, National Institute for Occupational Safety and Health

Högstedt, B., Kolnig, A.-M., Mitelman, F. & Skerfving, S. (1980) Cytogenetic study of pesticides in agricultural work. *Hereditas*, 92, 177-178

Hogstedt, C. & Westerlund, B. (1980) Cohort studies of cause of death of forest workers with and without exposure to phenoxy acid preparations (Swed.). *Läkartidningen*, 77, 1828-1831

Holtorf, R.C. (1982) *Preliminary Quantitative Usage Analysis of MCPA*, Washington DC, US Environmental Protection Agency, Benefits and Field Studies Division, Office of Pesticide Programs

IARC (1977a) *IARC Monographs on the Evaluation of the Carcinogenic Risk of Chemicals to Man*, Vol. 15, *Some Fumigants, the Herbicides 2,4-D and 2,4,5-T, Chlorinated Dibenzodioxins and Miscellaneous Industrial Chemicals*, Lyon, pp. 111-138, 273-299

IARC (1977b) *IARC Monographs on the Evaluation of the Carcinogenic Risk of Chemicals to Man*, Vol. 15, *Some Fumigants, the Herbicides 2,4-D and 2,4,5-T, Chlorinated Dibenzodioxins and Miscellaneous Industrial Chemicals*, Lyon, pp. 41-102

IARC (1977c) *IARC Monographs on the Evaluation of the Carcinogenic Risk of Chemicals to Man*, Vol. 15, *Some Fumigants, the Herbicides 2,4-D and 2,4,5-T, Chlorinated Dibenzodioxins and Miscellaneous Industrial Chemicals*, Lyon, pp. 307-340

IARC (1978) *IARC Monographs on the Evaluation of the Carcinogenic Risk of Chemicals to Humans*, Vol. 17, *Some N-Nitroso Compounds*, Lyon, pp. 125-175

IARC (1982a) *IARC Monographs on the Evaluation of the Carcinogenic Risk of Chemicals to Humans*, Suppl. 4, *Chemicals, Industrial Processes and Industries Associated with Cancer in Humans, IARC Monographs, Volumes 1 to 29*, Lyon, pp. 211-212

IARC (1982b) *IARC Monographs on the Evaluation of the Carcinogenic Risk of Chemicals to Humans*, Suppl. 4, *Chemicals, Industrial Processes and Industries Associated with Cancer in Humans, IARC Monographs, Volumes 1 to 29*, Lyon, pp. 101-103, 235-238

IARC (1982c) *IARC Monographs on the Evaluation of the Carcinogenic Risk of Chemicals to Humans*, Suppl. 4, *Chemicals, Industrial Processes and Industries Associated with Cancer in Humans, IARC Monographs, Volumes 1 to 29*, Lyon, pp. 238-243

IARC (1983) *IARC Monographs on the Evaluation of the Carcinogenic Risk of Chemicals to Humans*, Vol. 30, *Miscellaneous Pesticides*, Lyon, pp. 255-269

International Labour Office (1980) *Occupational Exposure Limits for Airborne Toxic Substances, A Tabular Compilation of Values from Selected Countries*, 2nd (rev.) ed. (*Occupational Safety and Health Series No. 37*), Geneva, pp. 84-85, 206-207

International Programme on Chemical Safety (1984) *2,4-Dichlorophenoxyacetic Acid (2,4-D) (Environmental Health Criteria 29)*, Geneva, World Health Organization

Johnson, E.R., Yu, T.C. & Montgomery, M.L. (1977) Trapping and analysis of atmospheric residues of 2,4-D. *Bull. environ. Contam. Toxicol.*, *17*, 369-372

Kangas, J., Soininen, H., Klen, T. & Riihimäki, V. (1984) Applicators' exposure to phenoxy acid in forestry (Finn.). *Työterveyslaitoksen tutkimuksia* (Publication Series of the Institute of Occupational Health), *2*, 118-126

Kohli, J.D., Khanna, R.N., Gupta, B.N., Dhar, M.M., Tandon, J.S. & Sircar, K.P. (1974a) Absorption and excretion of 2,4,5-trichlorophenoxy acetic acid in man. *Arch. int. Pharmacodyn. Ther.*, *210*, 250-255

Kohli, J.D., Khanna, R.N., Gupta, B.N., Dhar, M.M., Tandon, J.S. & Sircar, K.P. (1974b) Absorption and excretion of 2,4-dichlorophenoxyacetic acid in man. *Xenobiotica, 4,* 97-100

Kolmodin-Hedman, B. & Erne, K. (1980) Estimation of occupational exposure to phenoxy acids (2,4-D and 2,4,5-T). *Arch. Toxicol., Suppl. 4,* 318-321

Kolmodin-Hedman, B., Erne, K., Håkansson, M. & Engqvist, A. (1979) *Testing of Occupational Exposure to Phenoxy Acids (2,4-D and 2,4,5-T) (Arbete och Hälsa* [Work and Health] *No. 17)* (Swed.), Stockholm, Swedish Board of Labour Protection

Kolmodin-Hedman, B., Höglund, S. & Åkerblom, M. (1983a) Studies on phenoxy acid herbicides. I. Field study. Occupational exposure to phenoxy acid herbicides (MCPA, dichlorprop, mecoprop and 2,4-D) in agriculture. *Arch. Toxicol., 54,* 257-265

Kolmodin-Hedman, B., Höglund, S., Swensson, Å. & Åkerblom, M. (1983b) Studies on phenoxy acid herbicides. II. Oral and dermal uptake and elimination in urine of MCPA in humans. *Arch. Toxicol., 54,* 267-273

Lavy, T.L., Shepard, J.S. & Mattice, J.D. (1980) Exposure measurements of applicators spraying (2,4,5-trichlorophenoxy)acetic acid in the forest. *J. agric. Food Chem., 28,* 626-630

Lavy, T.L., Waldstad, J.D., Flynn, R.R. & Mattice, J.D. (1982) (2,4-Dichlorophenoxy)-acetic acid exposure received by aerial application crews during forest spray operations. *J. agric. Food Chem., 30,* 375-381

Lawrence, C.E., Reilly, A.A., Quickenton, P., Greenwald, P., Page, W.F. & Kuntz, A.J. (1985) Mortality patterns of New York State Vietnam veterans. *Am. J. publ. Health, 75,* 277-279

Leng, M.L. (1986) *Chlorophenoxyalkanoic acids.* In: Gerhartz, W. & Yamamoto, S.Y., eds, *Ullmann's Encyclopaedia of Industrial Chemistry,* Weinheim, VCH Publisher Inc. (in press)

Leng, M.L., Ramsey, J.C., Braun, W.H. & Lavy, T.L. (1982) *Review of studies with 2,4,5-trichlorophenoxyacetic acid in humans including applicators under field conditions.* In: Plimmer, J.R., ed., *Pesticide Residues and Exposure (ACS Symposium Series Vol. 182),* Washington DC, American Chemical Society, pp. 133-156

Libich, S., To, J.C., Frank, R. & Sirons, G.J. (1984) Occupational exposure of herbicide applicators to herbicides used along electric power transmission line right-of-way. *Am. ind. Hyg. Assoc. J., 45,* 56-62

Linnainmaa, K. (1983) Sister chromatid exchanges among workers occupationally exposed to phenoxy acid herbicides, 2,4-D and MCPA. *Teratog. Carcinog. Mutagenesis, 3,* 269-279

Lynge, E. (1984) *Weed Killers and Cancer. Cohort Study of Cancer Outcome Among Workers in Production of Phenoxyacids in a Chemical Factory* (Dan.), Copenhagen, Arbejdsmiljøfondet

Lynge, E. (1985) A follow-up study of cancer incidence among workers in manufacture of phenoxy herbicides in Denmark. *Br. J. Cancer, 52,* 259-270

Meister, R.T., ed. (1983) *1983 Weed Control Manual*, Willoughby, OH, Meister Publishing Co.

Mulcahy, M.T. (1980) Chromosome aberrations and 'Agent Orange'. *Med. J. Aust.*, November 15, 573-574

Mustonen, R., Kangas, J., Vuojolahti, P. & Linnainmaa, K. (1986) Effect of phenoxyacetic acids on the induction of chromosome aberrations *in vitro* and *in vivo*. *Mutagenesis*, *1*, 241-255

Nash, R.G., Kearney, P.C., Maitlen, J.C., Sell, C.R. & Fertig, S.N. (1982) *Agricultural applicators exposure to 2,4-dichlorophenoxyacetic acid*. In: Plimmer, J.R., ed., *Pesticide Residues and Exposure (ACS Symposium Series Vol. 182)*, Washington DC, American Chemical Society, pp. 119-132

National Academy of Sciences (1974) *The Effects of Herbicides in South Vietnam. Part A. Summary and Conclusions*, Washington DC, Committee on the Effects of Herbicides in Vietnam, Division of Biological Sciences, Assembly of Life Sciences, National Research Council

National Research Council (1977) *Drinking Water and Health*, Washington DC, National Academy of Sciences, pp. 509-519

Nelson, C.J., Holson, J.F., Green, H.G. & Gaylor, D.W. (1979) Retrospective study of the relationship between agricultural use of 2,4,5-T and cleft palate occurrence in Arkansas. *Teratology*, *19*, 377-383

Nigg, H.N. & Stamper, J.H. (1983) Exposure of Florida airboat aquatic weed applicators to 2,4-dichlorophenoxyacetic acid (2,4-D). *Chemosphere*, *12*, 209-215

Nony, C.R., Bowman, M.C., Holder, C.L., Young, J.F. & Oller, W.L. (1976) Trace analysis of 2,4,5-trichlorophenoxyacetic acid, its glycineamide and their alkaline hydrolyzable conjugates in mouse blood, urine and feces. *J. pharm. Sci.*, *65*, 1810-1816

Norström, Å., Rappe, C., Lindahl, P. & Buser, H.-R. (1979) Analysis of some older Scandinavian formulations of 2,4-dichlorophenoxy acetic acid and 2,4,5-trichlorophenoxy acetic acid for contents of chlorinated dibenzo-*p*-dioxins and dibenzofurans. *Scand. J. Work Environ. Health*, *5*, 375-378

Nygren, M., Rappe, C., Lindström, G., Hansson, M., Bergqvist, P.-A., Marklund, S., Domellöf, L., Hardell, L. & Olsson, M. (1986) *Identification of 2,3,7,8-substituted polychlorinated dioxins (PCDDs) and dibenzofurans (PCDFs) in environmental and human samples*. In: Rappe, C., Chouhary, G. & Keith, L., eds, *Chlorinated Dioxins and Dibenzofurans in the Total Environment*, Vol. III, New York, A.F. Lewis & Co. (in press)

Olsson, H. & Brandt, L. (1981) Non-Hodgkin's lymphoma of the skin and occupational exposure to herbicides. *Lancet*, *ii*, 579

Ott, M.G., Holder, B.B. & Olsen, R.D. (1980) A mortality analysis of employees engaged in the manufacture of 2,4,5-trichlorophenoxyacetic acid. *J. occup. Med.*, *22*, 47-50

Palva, H.L.A., Koivisto, O. & Palva, I.P. (1975) Aplastic anaemia after exposure to a weed killer, 2-methyl-4-chlorophenoxyacetic acid. *Acta haematol.*, *53*, 105-108

Pearce, N.E., Smith, A.H. & Fisher, D.O. (1985) Malignant lymphoma and multiple myeloma linked with agricultural occupations in a New Zealand cancer registry-based study. *Am. J. Epidemiol.*, *121*, 225-237

Pearce, N.E., Smith, A.H., Howard, J.K., Sheppard, R.A., Giles, H.J. & Teague, C.A. (1986) Non-Hodgkin's lymphoma and exposure to phenoxy herbicides, chlorophenols, fencing work and meat works employment. A case-control study. *Br. J. ind. Med.*, *43*, 75-83

Plimmer, J.R. (1980) *Herbicides*. In: Mark, H.F., Othmer, D.F., Overberger, C.G. & Seaborg, G.T., eds, *Kirk-Othmer Encyclopedia of Chemical Technology*, 3rd ed., Vol. 12, New York, Wiley Interscience, pp. 297-351

Pokorny, R. (1941) Some chlorophenoxyacetic acids. *Science*, *63*, 1768

Que Hee, S.S. & Sutherland, R.G. (1981) *The Phenoxyalkanoic Herbicides*, Vol. 1, *Chemistry, Analysis, and Environmental Pollution*, Boca Raton, FL, CRC Press

Rappe, C. & Buser, H.R. (1981) *Chemical properties and analytical methods*. In: Kimbrough, R.D., ed., *Halogenated Biphenyls, Terphenyls, Naphthalenes, Dibenzodioxins and Related Products*, Amsterdam, Elsevier, pp. 41-75

Rappe, C., Buser, H.R. & Bosshardt, H.-P. (1978) Identification and quantification of polychlorinated dibenzo-p-dioxins (PCDDs) and dibenzofurans (PCDFs) in 2,4,5-T ester formulations and Herbicide Orange. *Chemosphere*, *7*, 431-438

Riihimäki, V., Asp, S. & Hernberg, S. (1982) Mortality of 2,4-dichlorophenoxyacetic acid and 2,4,5-trichlorophenoxyacetic acid herbicide applicators in Finland. First report of an ongoing prospective cohort study. *Scand. J. Work Environ. Health*, *8*, 37-42

Riihimäki, V., Asp, S., Pukkala, E. & Hernberg, S. (1983) Mortality and cancer morbidity among chlorinated phenoxyacid applicators in Finland. *Chemosphere*, *12*, 779-784

Ross, R.D., Morrison, J., Rounbehler, D.P., Fan, S. & Fine, D.H. (1977) *N*-Nitroso compound impurities in herbicide formulations. *J. agric. Food Chem.*, *25*, 1416-1418

Ryan, J.J., Lizotte, R. & Lau, B.P.-Y. (1985) Chlorinated dibenzo-*p*-dioxins and chlorinated dibenzofurans in Canadian human adipose tissue. *Chemosphere*, *14*, 697-706

Sarma, P.R. & Jacobs, J. (1982) Thoracic soft-tissue sarcoma in Vietnam veterans exposed to Agent Orange. *New Engl. J. Med.*, *306*, 1109

Sauerhoff, M.W., Braun, W.H., Blau, G.E. & Gehring, P.J. (1977a) The fate of 2,4-dichlorophenoxyacetic acid (2,4-D) following oral administration to man. *Toxicology*, *8*, 3-11

Sauerhoff, M.W., Chenoweth, M.B., Karbowski, R.J., Braun, W.H., Ramsey, J.C., Gehring, P.J. & Blau, E.G. (1977b) Fate of silvex following oral administration to humans. *J. Toxicol. environ. Health*, *3*, 941-952

Seiler, J.P. (1978) The genetic toxicology of phenoxy acids other than 2,4,5-T. *Mutat. Res.*, *55*, 197-226

Sell, C.R. & Maitlen, J.C. (1983) Procedure for the determination of residues of (2,4-dichlorophenoxy)acetic acid in dermal exposure pads, hand rinses, urine, and perspiration from agricultural workers exposed to the herbicide. *J. agric. Food Chem.*, *31*, 572-575

Shepard, B.M. & Young, A.L. (1983) *Dioxins as contaminants in herbicides: the US perspective*. In: Tucker, R.E., Young, A.L. & Gray, A.P., eds, *Human and Environmental Risks of Chlorinated Dioxins and Related Compounds*, New York, Plenum Press, pp. 3-11

Simpson, G.R. (1982) *Pesticide exposure — New South Wales Services*: In: Van Heemstra-Legui, A.H. & Tordoir, W.I., eds, *Education and Safe Handling in Pesticide Application, Studies in Environmental Science*, Vol. 18, Amsterdam, Elsevier, pp. 209-214

Simpson, G.R., Higgins, V., Chapman, J. & Bermingham, S. (1978) Exposure of council and forestry workers to 2,4,5-T. *Med. J. Aust.*, *2*, 536-537

Sittig, M., ed. (1980) *Pesticide Manufacturing and Toxic Materials Control Encyclopedia*, Park Ridge, NJ, Noyes Data Corp., pp. 229-234, 292-293, 490-492, 494, 697-700

Smith, A.E. & Hayden, B.J. (1979) Method for the determination of 2,4-dichlorophenoxyacetic acid residues in urine. *J. Chromatogr.*, *171*, 482-485

Smith, A.H. & Pearce, N.E. (1986) Update on soft tissue sarcoma and phenoxy herbicides in New Zealand. *Chemosphere* (in press)

Smith, A.H., Matheson, D.P., Fisher, D.O. & Chapman, C.J. (1981) Preliminary report of reproductive outcomes among pesticide applicators using 2,4,5-T. *N.Z. med. J.*, *93*, 177-179

Smith, A.H., Fisher, D.O., Pearce, N. & Chapman, C.J. (1982a) Congenital defects and miscarriages among New Zealand 2,4,5-T sprayers. *Arch. environ. Health*, *37*, 197-200

Smith, A.H., Fisher, D.O., Pearce, N. & Teague, C.A. (1982b) Do agricultural chemicals cause soft tissue sarcoma? Initial findings of a case-control study in New Zealand. *Commun. Health Stud.*, *6*, 114-119

Smith, A.H., Fisher, D.O., Giles, H.J. & Pearce, N. (1983) The New Zealand soft tissue sarcoma case-control study: interview findings concerning phenoxyacetic acid exposure. *Chemosphere*, *12*, 565-571

Smith, A.H., Pearce, N.E., Fisher, D.O., Giles, H.J., Teague, C.A. & Howard, J.K. (1984) Soft tissue sarcoma and exposure to phenoxyherbicides and chlorophenols in New Zealand. *J. natl Cancer Inst.*, *73*, 1111-1117

Soderquist, C.J. & Crosby, D.G. (1975) Dissipation of 4-chloro-2-methylphenoxyacetic acid (MCPA) in a rice field. *Pestic. Sci.* 6, 17-33

Sundström, G., Jensen, S., Jansson, B. & Erne, K. (1979) Chlorinated phenoxyacetic acid derivatives and tetrachlorodibenzo-*p*-dioxin in foliage after application of 2,4,5-trichlorophenoxyacetic acid esters. *Arch. environ. Contam. Toxicol.*, *8*, 441-448

Suskind, R.R. & Hertzberg, V.S. (1984) Human health effects of 2,4,5-T and its toxic contaminants. *J. Am. med. Assoc.*, *251*, 2372-2380

Taskar, P.K., Das, Y.T., Trout, J.R., Chattopadhyay, S.K. & Brown, H.D. (1982) Measurement of 2,4-dichlorophenoxyacetic acid (2,4-D) after occupational exposure. *Bull. environ. Contam. Toxicol.*, *29*, 586-591

Thomas, H.F. (1980) 2,4,5-T use and congenital malformation rates in Hungary. *Lancet*, *ii*, 214-215

Timonen, T.T.T. & Palva, I.P. (1980) Acute leukaemia after exposure to a weed killer, 2-methyl-4-chlorophenoxyacetic acid. *Acta haematol.*, *63*, 170-171

Työsuojeluhallitus (National Finnish Board of Occupational Safety and Health) (1981) *Airborne Contaminants in the Work Places* (*Safety Bull. 3*) (Finn.), Tampere, p. 10

US Department of Commerce (1975) *US Exports, Schedule B, Commodity by Country* (*FT 410/December 1975*), Washington DC, US Government Printing Office, p. 2-74

US Department of Commerce (1979) *US Exports, Schedule B, Commodity by Country* (*FT 446/Annual 1978*), Washington DC, US Government Printing Office, p. 1-155

US Department of Commerce (1980) *US Exports, Schedule B, Commodity by Country* (*FT 446/Annual 1979*), Washington DC, US Government Printing Office, p. 1-171

US Department of Commerce (1981) *US Exports, Schedule B, Commodity by Country* (*FT 446/Annual 1980*), Washington DC, US Government Printing Office, pp. 1-180 —1-181

US Department of Commerce (1982) *US Exports, Schedule B, Commodity by Country* (*FT 446/Annual 1981*), Washington DC, US Government Printing Office, p. 1-184

US Department of Commerce (1983) *US Exports, Schedule B, Commodity by Country* (*FT 446/Annual 1982*), Washington DC, US Government Printing Office, p. 1-180

US Department of Commerce (1984) *US Exports, Schedule B, Commodity by Country* (*FT 446/Annual 1983*), Washington DC, US Government Printing Office, pp. 1-168 —1-169

US Department of Commerce (1985) *US Exports, Schedule B, Commodity by Country* (*FT 446/Annual 1984*), Washington DC, US Government Printing Office, p. 2-84

US Environmental Protection Agency (1979a) *2,4,5-T: Position Document 1* (*EPA/SPRD-80/76; PB 212665*), Arlington, VA, Special Pesticide Review Division

US Environmental Protection Agency (1979b) *Report of Assessment of a Field Investigation of Six-year Spontaneous Abortion Rates in Three Oregon Areas in Relation to Forest 2,4,5-T Spray Practices* (*Alsea II Report*), Washington DC, Benefits and Field Studies Division

US Environmental Protection Agency (1983) 2,4,5-T and silvex products: intent to cancel registration of pesticide products containing 2,4,5-T and silvex; revocation of notices of intent to hold a hearing to determine whether certain uses of 2,4,5-T or silvex should be cancelled. Enhancement policy on transfer, distribution, sale or importation of unregistered products. *Fed. Regist.*, *48*, 48434-48437

US International Trade Commission (1977) *Imports of Benzenoid Chemicals and Products, 1976* (*USITC Publ. 828*), Washington DC, US Government Printing Office, p. 101

US International Trade Commission (1978a) *Synthetic Organic Chemicals, US Production and Sales, 1977 (USITC Publ. 920)*, Washington DC, US Government Printing Office, p. 321

US International Trade Commission (1978b) *Imports of Benzenoid Chemicals and Products, 1977 (USITC Publ. 900)*, Washington DC, US Government Printing Office, pp. 96-97

US International Trade Commission (1979a) *Synthetic Organic Chemicals, US Production and Sales, 1978 (USITC Publ. 1001)*, Washington DC, US Government Printing Office, p. 279

US International Trade Commission (1979b) *Imports of Benzenoid Chemicals and Products, 1978 (USITC Publ. 990)*, Washington DC, US Government Printing Office, pp. 97-99

US International Trade Commission (1980a) *Synthetic Organic Chemicals, US Production and Sales, 1979 (USITC Publ. 1099)*, Washington DC, US Government Printing Office, p. 235

US International Trade Commission (1980b) *Imports of Benzenoid Chemicals and Products, 1979 (USITC Publ. 1083)*, Washington DC, US Government Printing Office, pp. 110-113

US International Trade Commission (1981a) *Synthetic Organic Chemicals, US Production and Sales, 1980 (USITC Publ. 1183)*, Washington DC, US Government Printing Office, p. 231

US International Trade Commission (1981b) *Imports of Benzenoid Chemicals and Products, 1980 (USITC Publ. 1163)*, Washington DC, US Government Printing Office, pp. 104-108

US International Trade Commission (1982a) *Synthetic Organic Chemicals, US Production and Sales, 1981 (USITC Publ. 1292)*, Washington DC, US Government Printing Office, p. 211

US International Trade Commission (1982b) *Imports of Benzenoid Chemicals and Products, 1981 (USITC Publ. 1272)*, Washington DC, US Government Printing Office, pp. 97-100

US International Trade Commission (1983a) *Synthetic Organic Chemicals, US Production and Sales, 1982 (USITC Publ. 1422)*, Washington DC, US Government Printing Office, p. 225

US International Trade Commission (1983b) *Imports of Benzenoid Chemicals and Products, 1982 (USITC Publ. 1401)*, Washington DC, US Government Printing Office, pp. 94-97

US International Trade Commission (1984a) *Synthetic Organic Chemicals, US Production and Sales, 1983 (USITC Publ. 1588)*, Washington DC, US Government Printing Office, p. 223

US International Trade Commission (1984b) *Imports of Benzenoid Chemicals and Products, 1983* (*USITC Publ. 1548*), Washington DC, US Government Printing Office, pp. 96-99

US International Trade Commission (1985) *Synthetic Organic Chemicals, US Production and Sales, 1984* (*USITC Publ. 1745*), Washington DC, US Government Printing Office, p. 221

US Occupational Safety and Health Administration (1985) Labor. *US Code fed. Regul.*, *Title 29*, Part 1910.1000

US Tariff Commission (1946) *Synthetic Orgnic Chemicals, US Production and Sales, 1944* (*Report No. 155, Second Series*), Washington DC, US Government Printing Office, p. 116

Vân, D.D. (1984) *Herbicides as a possible cause of liver cancer*. In: Westing, A.H., ed., *Herbicides in War, The Long-Term Ecological and Human Consequences*, London, Taylor & Francis, pp. 119-121

Vineis, P., Terracini, B., Ciccone, G., Cignetti, A., Colombo, E., Donna, A., Maffi, L., Pisa, R., Ricci, P., Zanini, E. & Comba, P. (1986) Phenoxy herbicides and soft-tissue sarcomas in female rice-weeders: a population-based case control study. *Scand. J. Work Environ. Health* (in press)

Vural, N. & Burgaz, S. (1984) A gas chromatographic method for determination of 2,4-D residues in urine after occupational exposure. *Bull. environ. Contam. Toxicol.*, *33*, 518-524

Wassom, J.S., Huff, J.E. & Loprieno, N. (1977/1978) A review of the genetic toxicology of chlorinated dibenzo-*p*-dioxins. *Mutat. Res.*, *47*, 141-160

Westing, A.H. (1971) Ecological effects of military defoliation on the forests of South Vietnam. *Bioscience*, *21*, 893-898

Windholz, M., ed. (1983) *The Merck Index*, 10th ed., Rahway, NJ, Merck & Co., pp. 405, 447, 821, 1224, 1297

Worthing, C.R. (1977) *The Pesticide Manual*, 5th ed., London, British Crop Protection Council, pp. 268, 334

Worthing, C.R. (1983) *The Pesticide Manual*, 7th ed., London, British Crop Protection Council, pp. 184, 262, 341, 345

Yoder, J., Watson, M. & Benson, W.W. (1973) Lymphocyte chromosome analysis of agricultural workers during extensive occupational exposure to pesticides. *Mutat. Res.*, *21*, 335-340

Young, A.L., Calcagni, J.A., Thalken, C.E. & Tremblay, J.W. (1978) *The Toxicology, Environmental Fate, and Human Risk of Herbicide Orange and Its Associated Dioxin* (*US AF Technical Report OEHL TR-78-92*), Brooks Air Force Base, TX, US Air Force, Occupational and Environmental Health Laboratory

APPENDIX: SUMMARY OF FINAL EVALUATIONS

Compound	Degree of evidence[a]		
	Humans	Animals	Short-term tests
Dichloromethane	I	S	S
1,1,1,2-Tetrachloroethane	ND	L	I
Pentachloroethane	ND	L	I
1,3-Dichloropropene	I	S	L
1,2-Dichloropropane	ND	L	L
Bis(2-chloro-1-methylethyl)ether	ND	L	I
Methyl chloride	I	I	S
Methyl bromide	I	L	S
Methyl iodide	ND	L	S
Chlorofluoromethane	ND	L	L
Chlorodifluoromethane	I	L	I
2-Chloro-1,1,1-trifluoroethane	ND	L	I
Polybrominated biphenyls	I	S	NE
Amitrole	I	S	L
Chlorophenols (occupational exposures to)	L	—	—
Chlorophenoxy herbicides (occupational exposures to)	L	—	—

[a]I, inadequate; S, sufficient; ND, no data; L, limited; NE, no evidence. For definitions of the degrees of evidence, see pp. 18, 20, 21 and 22 of the Preamble to this volume.

CUMULATIVE INDEX TO IARC MONOGRAPHS ON THE EVALUATION OF THE CARCINOGENIC RISK OF CHEMICALS TO HUMANS

Numbers in italics indicate volume, and other numbers indicate page. References to corrigenda are given in parentheses. Compounds marked with an asterisk(*) were considered by the working groups in the year indicated, but monographs were not prepared because adequate data on carcinogenicity were not available.

A

A-α-C (2-Amino-9H-pyrido[2,3-b]indole)	*40*, 245
Acetaldehyde	*36*, 101 (corr. *40*, 419)
Acetaldehyde formylmethylhydrazone	*31*, 163
Acetamide	*7*, 197
Acetylsalicylic acid (1976)*	
Acridine orange	*16*, 145
Acriflavinium chloride	*13*, 31
Acrolein	*19*, 479
	36, 133
Acrylamide	*39*, 41
Acrylic acid	*19*, 47
Acrylic fibres	*19*, 86
Acrylonitrile	*19*, 73
	Suppl. 4, 25
Acrylonitrile-butadiene-styrene copolymers	*19*, 9
Actinomycins	*10*, 29 (corr. *29*, 399; *34*, 197)
	Suppl. 4, 27
Adipic acid (1978)*	
Adriamycin	*10*, 43
	Suppl. 4, 29
AF-2	*31*, 47
Aflatoxins	*1*, 145 (corr. *7*, 319; *8*, 349)
	10, 51
	Suppl. 4, 31
Agaritine	*31*, 63
Aldrin	*5*, 25
	Suppl. 4, 35
Allyl chloride	*36*, 39

Allyl isothiocyanate	*36*, 55
Allyl isovalerate	*36*, 69
Aluminium production	*34*, 37
Amaranth	*8*, 41
5-Aminoacenaphthene	*16*, 243
2-Aminoanthraquinone	*27*, 191
para-Aminoazobenzene	*8*, 53
ortho-Aminoazotoluene	*8*, 61 (corr. *11*, 295)
para-Aminobenzoic acid	*16*, 249
4-Aminobiphenyl	*1*, 74 (corr. *10*, 343)
	Suppl. 4, 37
1-Amino-2-methylanthraquinone	*27*, 199
2-Amino-5-(5-nitro-2-furyl)-1,3,4-thiadiazole	*7*, 143
4-Amino-2-nitrophenol	*16*, 43
2-Amino-4-nitrophenol (1977)*	
2-Amino-5-nitrophenol (1977)*	
2-Amino-5-nitrothiazole	*31*, 71
6-Aminopenicillanic acid (1975)*	
11-Aminoundecanoic acid	*39*, 239
Amitrole	*7*, 31
	Suppl. 4, 38
	41, 293
Amobarbital sodium (1976)*	
Anaesthetics, volatile	*11*, 285
	Suppl. 4, 41
Angelicin and some synthetic derivatives	*40*, 291
5-Methylangelicin	
4,4'-Dimethylangelicin	
4,5'-Dimethylangelicin	
4,4',6-Trimethylangelicin	
Aniline	*4*, 27 (corr. *7*, 320)
	27, 39
	Suppl. 4, 49
Aniline hydrochloride	*27*, 40
ortho-Anisidine and its hydrochloride	*27*, 63
para-Anisidine and its hydrochloride	*27*, 65
Anthanthrene	*32*, 95
Anthracene	*32*, 105
Anthranilic acid	*16*, 265

Apholate	*9*, 31
Aramite®	*5*, 39
Arsenic and arsenic compounds	*1*, 41
Arsanilic acid	*2*, 48
Arsenic pentoxide	*23*, 39
Arsenic sulphide	Suppl. *4*, 50
Arsenic trioxide	
Arsine	
Calcium arsenate	
Dimethylarsinic acid	
Lead arsenate	
Methanearsonic acid, disodium salt	
Methanearsonic acid, monosodium salt	
Potassium arsenate	
Potassium arsenite	
Sodium arsenate	
Sodium arsenite	
Sodium cacodylate	
Asbestos	*2*, 17 (corr. *7*, 319)
Actinolite	*14* (corr. *15*, 341; *17*, 351)
Amosite	Suppl. *4*, 52
Anthophyllite	
Chrysotile	
Crocidolite	
Tremolite	
Asiaticoside (1975)*	
Auramine	*1*, 69 (corr. *7*, 319)
	Suppl. *4*, 53 (corr. *33*, 223)
Aurothioglucose	*13*, 39
5-Azacytidine	*26*, 37
Azaserine	*10*, 73 (corr. *12*, 271)
Azathioprine	*26*, 47
	Suppl. *4*, 55
Aziridine	*9*, 37
2-(1-Aziridinyl)ethanol	*9*, 47
Aziridyl benzoquinone	*9*, 51
Azobenzene	*8*, 75

B

Benz[*a*]acridine	*32*, 123
Benz[*c*]acridine	*3*, 241
	32, 129
Benzal chloride	*29*, 65
	Suppl. *4*, 84
Benz[*a*]anthracene	*3*, 45
	32, 135

Benzene	*7*, 203 (corr. *11*, 295)
	29, 93, 391
	Suppl. *4*, 56 (corr. *35*, 249)
Benzidine and its salts	*1*, 80
	29, 149, 391
	Suppl. *4*, 57
Benzo[*b*]fluoranthene	*3*, 69
	32, 147
Benzo[*j*]fluoranthene	*3*, 82
	32, 155
Benzo[*k*]fluoranthene	*32*, 163
Benzo[*ghi*]fluoranthene	*32*, 171
Benzo[*a*]fluorene	*32*, 177
Benzo[*b*]fluorene	*32*, 183
Benzo[*c*]fluorene	*32*, 189
Benzo[*ghi*]perylene	*32*, 195
Benzo[*c*]phenanthrene	*32*, 205
Benzo[*a*]pyrene	*3*, 91
	Suppl. *4*, 227
	32, 211
Benzo[*e*]pyrene	*3*, 137
	32, 225
para-Benzoquinone dioxime	*29*, 185
Benzotrichloride	*29*, 73
	Suppl. *4*, 84
Benzoyl chloride	*29*, 83
	Suppl. *4*, 84
Benzoyl peroxide	*36*, 267
Benzyl acetate	*40*, 109
Benzyl chloride	*11*, 217 (corr. *13*, 243)
	29, 49 (corr. *30*, 407)
	Suppl. *4*, 84
Benzyl violet 4B	*16*, 153
Beryllium and beryllium compounds	*1*, 17
Bertrandite	*23*, 143 (corr. *25*, 392)
Beryllium acetate	Suppl. *4*, 60
Beryllium acetate, basic	
Beryllium-aluminium alloy	
Beryllium carbonate	
Beryllium chloride	
Beryllium-copper alloy	
Beryllium-copper-cobalt alloy	
Beryllium fluoride	
Beryllium hydroxide	
Beryllium-nickel alloy	
Beryllium oxide	
Beryllium phosphate	

Beryllium silicate	
Beryllium sulphate and its tetrahydrate	
Beryl ore	
Zinc beryllium silicate	
Betel-quid and areca-nut chewing	*37*, 141
Bis(1-aziridinyl)morpholinophosphine sulphide	*9*, 55
Bis(2-chloroethyl)ether	*9*, 117
N,N-Bis(2-chloroethyl)-2-naphthylamine (Chlornaphazine)	*4*, 119 (corr. *30*, 407)
	Suppl. *4*, 62
Bischloroethyl nitrosourea (BCNU)	*26*, 79
	Suppl. *4*, 63
Bis-(2-chloroisopropyl)ether (1976)*	
1,2-Bis(chloromethoxy)ethane	*15*, 31
1,4-Bis(chloromethoxymethyl)benzene	*15*, 37
Bis(chloromethyl)ether	*4*, 231 (corr. *13*, 243)
	Suppl. *4*, 64
Bis(2-chloro-1-methylethyl)ether	*41*, 149
Bitumens	*35*, 39
Bleomycins	*26*, 97
	Suppl. *4*, 66
Blue VRS	*16*, 163
Boot and shoe manufacture and repair	*25*, 249
	Suppl. *4*, 138
Bracken fern	*40*, 47
Brilliant blue FCF diammonium and disodium salts	*16*, 171 (corr. *30*, 407)
1,3-Butadiene	*39*, 155 (corr. *40*, 418)
1,4-Butanediol dimethanesulphonate (Myleran)	*4*, 247
	Suppl. *4*, 68
n-Butyl acrylate	*39*, 67
Butylated hydroxyanisole (BHA)	*40*, 123
Butylated hydroxytoluene (BHT)	*40*, 161
Butyl benzyl phthalate	*29*, 194 (corr. *32*, 455)
Butyl-cis-9,10-epoxystearate (1976)*	
β-Butyrolactone	*11*, 225
γ-Butyrolactone	*11*, 231

C

Cadmium and cadmium compounds	*2*, 74
Cadmium acetate	*11*, 39 (corr. *27*, 320)
Cadmium chloride	Suppl. *4*, 71
Cadmium oxide	
Cadmium sulphate	
Cadmium sulphide	
Calcium cyclamate	*22*, 58 (corr. *25*, 391)
	Suppl. *4*, 97

Calcium saccharin	*22*, 120 (corr. *25*, 391)
	Suppl. 4, 225
Cantharidin	*10*, 79
Caprolactam	*19*, 115 (corr. *31*, 293)
	39, 247 (corr. *40*, 418)
Captan	*30*, 295
Carbaryl	*12*, 37
Carbazole	*32*, 239
3-Carbethoxypsoralen	*40*, 317
Carbon blacks	*3*, 22
	33, 35
Carbon tetrachloride	*1*, 53
	20, 371
	Suppl. 4, 74
Carmoisine	*8*, 83
Carpentry and joinery	*25*, 139
	Suppl. 4, 139
Carrageenan	*10*, 181 (corr. *11*, 295)
	31, 79
Catechol	*15*, 155
Chloramben (1982)*	
Chlorambucil	*9*, 125
	26, 115
	Suppl. 4, 77
Chloramphenicol	*10*, 85
	Suppl. 4, 79
Chlordane	*20*, 45 (corr. *25*, 391)
	Suppl. 4, 80
Chlordecone (Kepone)	*20*, 67
Chlordimeform	*30*, 61
Chlorinated dibenzodioxins	*15*, 41
	Suppl. 4, 211, 238
Chlorinated toluenes (production of)	*Suppl. 4*, 84
Chlormadinone acetate	*6*, 149
	21, 365
	Suppl. 4, 192
Chlorobenzilate	*5*, 75
	30, 73
Chlorodifluoromethane	*41*, 237
1-(2-Chloroethyl)-3-cyclohexyl-1-nitrosourea (CCNU)	*26*, 173 (corr. *35*, 249)
	Suppl. 4, 83
Chlorofluoromethane	*41*, 229

Chloroform	*1*, 61
	20, 401
	Suppl. *4*, 87
Chloromethyl methyl ether	*4*, 239
	Suppl. *4*, 64
Chlorophenols (occupational exposures to)	Suppl. *4*, 88
	41, 319
Chlorophenoxy herbicides (occupational exposures to)	Suppl. *4*, 211
(see also Phenoxyacetic acid herbicides, occupational exposure to)	*41*, 357
4-Chloro-*ortho*-phenylenediamine	*27*, 81
4-Chloro-*meta*-phenylenediamine	*27*, 82
Chloroprene	*19*, 131
	Suppl. *4*, 89
Chloropropham	*12*, 55
Chloroquine	*13*, 47
Chlorothalonil	*30*, 319
para-Chloro-*ortho*-toluidine and its hydrochloride	*16*, 277
	30, 61
5-Chloro-*ortho*-toluidine (1977)*	
Chlorotrianisene	*21*, 139
2-Chloro-1,1,1-trifluoroethane	*41*, 253
Chlorpromazine (1976)*	
Cholesterol	*10*, 99
	31, 95
Chromium and chromium compounds	*2*, 100
Barium chromate	*23*, 205
Basic chromic sulphate	Suppl. *4*, 91
Calcium chromate	
Chromic acetate	
Chromic chloride	
Chromic oxide	
Chromic phosphate	
Chromite ore	
Chromium carbonyl	
Chromium potassium sulphate	
Chromium sulphate	
Chromium trioxide	
Cobalt-chromium alloy	
Ferrochromium	
Lead chromate	
Lead chromate oxide	
Potassium chromate	
Potassium dichromate	
Sodium chromate	

Sodium dichromate
Strontium chromate
Zinc chromate
Zinc chromate hydroxide
Zinc potassium chromate
Zinc yellow
Chrysene   *3*, 159
                                            32, 247
Chrysoidine   *8*, 91
C.I. Disperse Yellow 3   *8*, 97
Cinnamyl anthranilate   *16*, 287
                                            31, 133
Cisplatin   *26*, 151
                                            Suppl. 4, 93
Citrinin   *40*, 67
Citrus Red No. 2   *8*, 101 (corr. *19*, 495)
Clofibrate   *24*, 39
                                            Suppl. 4, 95
Clomiphene and its citrate   *21*, 551
                                            Suppl. 4, 96
Coal gasification   *34*, 65
Coal-tars   *35*, 83
Coal-tar pitches (*see* Coal-tars)
Coke production   *34*, 101
Conjugated oestrogens   *21*, 147
                                            Suppl. 4, 179
Copper 8-hydroxyquinoline   *15*, 103
Coronene   *32*, 263
Coumarin   *10*, 113
Creosotes (*see* Coal-tars)
meta-Cresidine   *27*, 91
para-Cresidine   *27*, 92
Cycasin   *1*, 157 (corr. *7*, 319)
                                            10, 121
Cyclamic acid   *22*, 55 (corr. *25*, 391)
                                            Suppl. 4, 97
Cyclochlorotine   *10*, 139
Cyclohexylamine   *22*, 59 (corr. *25*, 391)
                                            Suppl. 4, 97
Cyclopenta[*cd*]pyrene   *32*, 269
Cyclophosphamide   *9*, 135
                                            26, 165
                                            Suppl. 4, 99

D
2,4-D and esters   *15*, 111
                                            Suppl. 4, 101, 211
Dacarbazine   *26*, 203
                                            Suppl. 4, 103
D and C Red No. 9   *8*, 107

Dapsone	*24*, 59
	Suppl. 4, 104
Daunomycin	*10*, 145
DDT and associated substances	*5*, 83 (corr. *7*, 320)
DDD (TDE)	*Suppl. 4*, 105
DDE	
Diacetylaminoazotoluene	*8*, 113
N,N'-Diacetylbenzidine	*16*, 293
Diallate	*12*, 69
	30, 235
2,4-Diaminoanisole and its sulphate	*16*, 51
	27, 103
2,5-Diaminoanisole (1977)*	
4,4'-Diaminodiphenyl ether	*16*, 301
	29, 203
1,2-Diamino-4-nitrobenzene	*16*, 63
1,4-Diamino-2-nitrobenzene	*16*, 73
2,4-Diaminotoluene (*see also* Toluene diisocyanate)	*16*, 83
2,5-Diaminotoluene and its sulphate	*16*, 97
Diazepam	*13*, 57
Diazomethane	*7*, 223
Dibenz[*a,h*]acridine	*3*, 247
	32, 277
Dibenz[*a,j*]acridine	*3*, 254
	32, 283
Dibenz[*a,c*]anthracene	*32*, 289 (corr. *34*, 197)
Dibenz[*a,h*]anthracene	*3*, 178
	32, 299
Dibenz[*a,j*]anthracene	*32*, 309
7*H*-Dibenzo[*c,g*]carbazole	*3*, 260
	32, 315
Dibenzo[*a,e*]fluoranthene	*32*, 321
Dibenzo[*h,rst*]pentaphene	*3*, 197
Dibenzo[*a,e*]pyrene	*3*, 201
	32, 327
Dibenzo[*a,h*]pyrene	*3*, 207
	32, 331
Dibenzo[*a,i*]pyrene	*3*, 215
	32, 337
Dibenzo[*a,l*]pyrene	*3*, 224
	32, 343
1,2-Dibromo-3-chloropropane	*15*, 139
	20, 83

Dichloroacetylene	*39*, 369
ortho-Dichlorobenzene	*7*, 231
	29, 213
	Suppl. 4, 108
para-Dichlorobenzene	*7*, 231
	29, 215
	Suppl. 4, 108
3,3'-Dichlorobenzidine and its dihydrochloride	*4*, 49
	29, 239
	Suppl. 4, 110
trans-1,4-Dichlorobutene	*15*, 149
3,3'-Dichloro-4,4'-diaminodiphenyl ether	*16*, 309
1,2-Dichloroethane	*20*, 429
Dichloromethane	*20*, 449
	Suppl. 4, 111
	41, 43
2,6-Dichloro-*para*-phenylenediamine	*39*, 325
1,2-Dichloropropane	*41*, 131
1,3-Dichloropropene	*41*, 113
Dichlorvos	*20*, 97
Dicofol	*30*, 87
Dicyclohexylamine	*22*, 60 (corr. *25*, 391)
Dieldrin	*5*, 125
	Suppl. 4, 112
Dienoestrol	*21*, 161
	Suppl. 4, 183
Diepoxybutane	*11*, 115 (corr. *12*, 271)
Di-(2-ethylhexyl)adipate	*29*, 257
Di-(2-ethylhexyl)phthalate	*29*, 269 (corr. *32*, 455)
1,2-Diethylhydrazine	*4*, 153
Diethylstilboestrol	*6*, 55
	21, 172 (corr. *23*, 417)
	Suppl. 4, 184
Diethylstilboestrol dipropionate	*21*, 175
Diethyl sulphate	*4*, 277
	Suppl. 4, 115
Diglycidyl resorcinol ether	*11*, 125
	36, 181
Dihydrosafrole	*1*, 170
	10, 233
Dihydroxybenzenes	*15*, 155
Dihydroxymethylfuratrizine	*24*, 77

Dimethisterone	6, 167
	21, 377
	Suppl. 4, 193
Dimethoate (1977)*	
Dimethoxane	15, 177
3,3'-Dimethoxybenzidine (ortho-Dianisidine)	4, 41
	Suppl. 4, 116
3,3'-Dimethoxybenzidine-4,4'-diisocyanate	39, 279
para-Dimethylaminoazobenzene	8, 125 (corr. 31, 293)
para-Dimethylaminobenzenediazo sodium sulphonate	8, 147
trans-2[(Dimethylamino)methylimino]-5-[2-(5-nitro-2-furyl)vinyl]-1,3,4-oxadiazole	7, 147 (corr. 30, 407)
3,3'-Dimethylbenzidine (ortho-Tolidine)	1, 87
Dimethylcarbamoyl chloride	12, 77
	Suppl. 4, 118
1,1-Dimethylhydrazine	4, 137
1,2-Dimethylhydrazine	4, 145 (corr. 7, 320)
1,4-Dimethylphenanthrene	32, 349
Dimethyl sulphate	4, 271
	Suppl. 4, 119
Dimethylterephthalate (1978)*	
1,8-Dinitropyrene	33, 171
Dinitrosopentamethylenetetramine	11, 241
1,4-Dioxane	11, 247
	Suppl. 4, 121
2,4'-Diphenyldiamine	16, 313
Diphenylthiohydantoin (1976)*	
Direct Black 38	29, 295 (corr. 32, 455)
	Suppl. 4, 59
Direct Blue 6	29, 311
	Suppl. 4, 59
Direct Brown 95	29, 321
	Suppl. 4, 59
Disulfiram	12, 85
Dithranol	13, 75
Dulcin	12, 97

E

Endrin	5, 157
Enflurane (1976)*	
Eosin and its disodium salt	15, 183

Epichlorohydrin	*11*, 131 (corr. *18*, 125; *26*, 387)
	Suppl. *4*, 122 (corr. *33*, 223)
1-Epoxyethyl-3,4-epoxycyclohexane	*11*, 141
3,4-Epoxy-6-methylcyclohexylmethyl-3,4-epoxy-6-methylcyclohexane carboxylate	*11*, 147
cis-9,10-Epoxystearic acid	*11*, 153
Ethinyloestradiol	*6*, 77
	21, 233
	Suppl. *4*, 186
Ethionamide	*13*, 83
Ethyl acrylate	*19*, 57
	39, 81
Ethylene	*19*, 157
Ethylene dibromide	*15*, 195
	Suppl. *4*, 124
Ethylene oxide	*11*, 157
	Suppl. *4*, 126
	36, 189 (corr. *40*, 419)
Ethylene sulphide	*11*, 257
Ethylene thiourea	*7*, 45
	Suppl. *4*, 128
Ethyl methanesulphonate	*7*, 245
Ethyl selenac	*12*, 107
Ethyl tellurac	*12*, 115
Ethynodiol diacetate	*6*, 173
	21, 387
	Suppl. *4*, 194
Eugenol	*36*, 75
Evans blue	*8*, 151

F

Fast green FCF	*16*, 187
Ferbam	*12*, 121 (corr. *13*, 243)
Fluometuron	*30*, 245
Fluoranthene	*32*, 355
Fluorene	*32*, 365
Fluorescein and its disodium salt (1977)*	
Fluorides (inorganic used in drinking-water and dental preparations)	*27*, 237
Fluorspar	
Fluosilicic acid	
Sodium fluoride	

Sodium monofluorophosphate
Sodium silicofluoride
Stannous fluoride
5-Fluorouracil	*26*, 217
	Suppl. 4, 130
Formaldehyde	*29*, 345
	Suppl. 4, 131
2-(2-Formylhydrazino)-4-(5-nitro-2-furyl)thiazole	*7*, 151 (corr. *11*, 295)
Furazolidone	*31*, 141
The furniture and cabinet-making industry	*25*, 99
	Suppl. 4, 140
2-(2-Furyl)-3-(5-nitro-2-furyl)acrylamide	*31*, 47
Fusarenon-X	*11*, 169
	31, 153

G

Glu-P-1 (2-Amino-6-methyldipyrido[1,2-*a*:3′,2′-*d*]imidazole)	*40*, 223
Glu-P-2 (2-Aminodipyrido[1,2-*a*:3′,2′-*d*]imidazole	*40*, 235
L-Glutamic acid-5-[2-(4-hydroxymethyl)phenylhydrazide]	*31*, 63
Glycidaldehyde	*11*, 175
Glycidyl oleate	*11*, 183
Glycidyl stearate	*11*, 187
Griseofulvin	*10*, 153
Guinea green B	*16*, 199
Gyromitrin	*31*, 163

H

Haematite	*1*, 29
	Suppl. 4, 254
Haematoxylin (1977)*	
Hair dyes, epidemiology of	*16*, 29
	27, 307
Halothane (1976)*	
Heptachlor and its epoxide	*5*, 173
	20, 129
	Suppl. 4, 80
Hexachlorobenzene	*20*, 155
Hexachlorobutadiene	*20*, 179
Hexachlorocyclohexane (α-, β-, δ-, ϵ-, technical HCH and lindane)	*5*, 47
	20, 195 (corr. *32*, 455)
	Suppl. 4, 133

Hexachloroethane	*20*, 467
Hexachlorophene	*20*, 241
Hexamethylenediamine (1978)*	
Hexamethylphosphoramide	*15*, 211
Hycanthone and its mesylate	*13*, 91
Hydralazine and its hydrochloride	*24*, 85
	Suppl. 4, 135
Hydrazine	*4*, 127
	Suppl. 4, 136
Hydrogen peroxide	*36*, 285
Hydroquinone	*15*, 155
4-Hydroxyazobenzene	*8*, 157
17α-Hydroxyprogesterone caproate	*21*, 399 (corr. *31*, 293)
	Suppl. 4, 195
8-Hydroxyquinoline	*13*, 101
Hydroxysenkirkine	*10*, 265

I

Indeno[1,2,3-*cd*]pyrene	*3*, 229
	32, 373
IQ (2-Amino-3-methylimidazo[4,5-*f*]quinoline)	*40*, 261
Iron and steel founding	*34*, 133
Iron-dextran complex	*2*, 161
	Suppl. 4, 145
Iron-dextrin complex	*2*, 161 (corr. *7*, 319)
Iron oxide	*1*, 29
Iron sorbitol-citric acid complex	*2*, 161
Isatidine	*10*, 269
Isoflurane (1976)*	
Isonicotinic acid hydrazide	*4*, 159
	Suppl. 4, 146
Isophosphamide	*26*, 237
Isoprene (1978)*	
Isopropyl alcohol	*15*, 223
	Suppl. 4, 151
Isopropyl oils	*15*, 223
	Suppl. 4, 151
Isosafrole	*1*, 169
	10, 232

J

Jacobine	*10*, 275

K

Kaempferol *31*, 171

L

Lasiocarpine *10*, 281
Lauroyl peroxide *36*, 315
Lead and lead compounds *1*, 40 (corr. *7*, 319)
    Lead acetate and its trihydrate *2*, 52 (corr. *8*, 349)
    Lead carbonate *2*, 150
    Lead chloride *23*, 39, 205, 325
    Lead naphthenate *Suppl. 4*, 149 (corr. *40*, 417)
    Lead nitrate
    Lead oxide
    Lead phosphate
    Lead subacetate
    Lead tetroxide
    Tetraethyllead
    Tetramethyllead
The leather goods manufacturing industry *25*, 279
  (other than boot and shoe manufacture and tanning) *Suppl. 4*, 142
The leather tanning and processing industries *25*, 201
    Suppl. 4, 142
Ledate *12*, 131
Light green SF *16*, 209
Lindane *5*, 47
    20, 196
    Suppl. 4, 133
The lumber and sawmill industries (including logging) *25*, 49
    Suppl. 4, 143
Luteoskyrin *10*, 163
Lynoestrenol *21*, 407
    Suppl. 4, 195
Lysergide (1976)*

M

Magenta *4*, 57 (corr. *7*, 320)
    Suppl. 4, 152
Malathion *30*, 103
Maleic hydrazide *4*, 173 (corr. *18*, 125)
Malonaldehyde *36*, 163
Maneb *12*, 137
Mannomustine and its dihydrochloride *9*, 157

MCPA (*see also* Chlorophenoxy herbicides, occupational exposures to)	*Suppl. 4*, 211
	30, 255
MeA-α-C (2-Amino-3-methyl-9*H*-pyrido[2,3-*b*]indole)	*40*, 253
Medphalan	*9*, 168
Medroxyprogesterone acetate	*6*, 157
	21, 417 (corr. *25*, 391)
	Suppl. 4, 196
Megestrol acetate	*21*, 431
	Suppl. 4, 198
MeIQ (2-Amino-3,4-dimethylimidazo[4,5-*f*]quinoline)	*40*, 275
MeIQx (2-Amino-3,8-dimethylimidazo[4,5-*f*]quinoxaline)	*40*, 283
Melamine	*39*, 333
Melphalan	*9*, 167
	Suppl. 4, 154
6-Mercaptopurine	*26*, 249
	Suppl. 4, 155
Merphalan	*9*, 169
Mestranol	*6*, 87
	21, 257 (corr. *25*, 391)
	Suppl. 4, 188
Methacrylic acid (1978)*	
Methallenoestril (1978)*	
Methotrexate	*26*, 267
	Suppl. 4, 157
Methoxsalen	*24*, 101
	Suppl. 4, 158
Methoxychlor	*5*, 193
	20, 259
Methoxyflurane (1976)*	
5-Methoxypsoralen	*40*, 327
Methyl acrylate	*19*, 52
	39, 99
2-Methylaziridine	*9*, 61
Methylazoxymethanol	*10*, 121
Methylazoxymethanol acetate	*1*, 164
	10, 131
Methyl bromide	*41*, 187
Methyl carbamate	*12*, 151
Methyl chloride	*41*, 161
1-, 2-, 3-, 4-, 5- and 6-Methylchrysenes	*32*, 379
N-Methyl-*N*,4-dinitrosoaniline	*1*, 141
4,4'-Methylene bis(2-chloroaniline)	*4*, 65 (corr. *7*, 320)
4,4'-Methylene bis(*N,N*-dimethyl)benzenamine	*27*, 119

4,4'-Methylene bis(2-methylaniline)	*4*, 73
4,4'-Methylenedianiline and its dihydrochloride	*4*, 79 (corr. *7*, 320)
	39, 347
4,4'-Methylenediphenyl diisocyanate	*19*, 314
2- and 3-Methylfluoranthenes	*32*, 399
Methyl iodide	*15*, 245
	41, 213
Methyl methacrylate	*19*, 187
Methyl methanesulphonate	*7*, 253
2-Methyl-1-nitroanthraquinone	*27*, 205
N-Methyl-*N'*-nitro-*N*-nitrosoguanidine	*4*, 183
3-Methylnitrosaminopropionaldehyde	*37*, 263
3-Methylnitrosaminopropionitrile	*37*, 263
4-(Methylnitrosamino)-4-(3-pyridyl)butanal (NNA)	*37*, 205
4-(Methylnitrosamino)-1-(3-pyridyl)-1-butanone (NNK)	*37*, 209
Methyl parathion	*30*, 131
1-Methylphenanthrene	*32*, 405
Methyl protoanemonin (1975)*	
7-Methylpyrido[3,4-*c*]psoralen	*40*, 349
Methyl red	*8*, 161
Methyl selenac	*12*, 161
Methylthiouracil	*7*, 53
Metronidazole	*13*, 113
	Suppl. 4, 160
Mineral oils	*3*, 30
	Suppl. 4, 227
	33, 87 (corr. *37*, 269)
Mirex	*5*, 203
	20, 283 (corr. *30*, 407)
Miristicin (1982)*	
Mitomycin C	*10*, 171
Modacrylic fibres	*19*, 86
Monocrotaline	*10*, 291
Monuron	*12*, 167
5-(Morpholinomethyl)-3-[(5-nitrofurfurylidene)-amino]-2-oxazolidinone	*7*, 161
Mustard gas	*9*, 181 (corr. *13*, 243)
	Suppl. 4, 163

N

Nafenopin	*24*, 125
1,5-Naphthalenediamine	*27*, 127

1,5-Naphthalene diisocyanate	*19*, 311
1-Naphthylamine	*4*, 87 (corr. *8*, 349; *22*, 187)
	Suppl. *4*, 164
2-Naphthylamine	*4*, 97
	Suppl. *4*, 166
1-Naphthylthiourea (ANTU)	*30*, 347
Nickel and nickel compounds	*2*, 126 (corr. *7*, 319)
Nickel acetate and its tetrahydrate	*11*, 75
Nickel ammonium sulphate	Suppl. *4*, 167
Nickel carbonate	
Nickel carbonyl	
Nickel chloride	
Nickel-gallium alloy	
Nickel hydroxide	
Nickelocene	
Nickel oxide	
Nickel subsulphide	
Nickel sulphate	
Nihydrazone (1982)*	
Niridazole	*13*, 123
Nithiazide	*31*, 179
5-Nitroacenaphthene	*16*, 319
5-Nitro-*ortho*-anisidine	*27*, 133
9-Nitroanthracene	*33*, 179
6-Nitrobenzo[*a*]pyrene	*33*, 187
4-Nitrobiphenyl	*4*, 113
6-Nitrochrysene	*33*, 195
Nitrofen	*30*, 271
3-Nitrofluoranthene	*33*, 201
5-Nitro-2-furaldehyde semicarbazone	*7*, 171
1-[(5-Nitrofurfurylidene)amino]-2-imidazolidinone	*7*, 181
N-[4-(5-Nitro-2-furyl)-2-thiazolyl]acetamide	*1*, 181
	7, 185
Nitrogen mustard and its hydrochloride	*9*, 193
	Suppl. *4*, 170
Nitrogen mustard *N*-oxide and its hydrochloride	*9*, 209
2-Nitropropane	*29*, 331
1-Nitropyrene	*33*, 209
N-Nitrosatable drugs	*24*, 297 (corr. *30*, 307)
N-Nitrosatable pesticides	*30*, 359
N'-Nitrosoanabasine	*37*, 225
N'-Nitrosoanatabine	*37*, 233

N-Nitrosodi-n-butylamine	*4*, 197
	17, 51
N-Nitrosodiethanolamine	*17*, 77
N-Nitrosodiethylamine	*1*, 107 (corr. *11*, 295)
	17, 83 (corr. *23*, 417)
N-Nitrosodimethylamine	*1*, 95
	17, 125 (corr. *25*, 391)
N-Nitrosodiphenylamine	*27*, 213
para-Nitrosodiphenylamine	*27*, 227 (corr. *31*, 293)
N-Nitrosodi-n-propylamine	*17*, 177
N-Nitroso-N-ethylurea	*1*, 135
	17, 191
N-Nitrosofolic acid	*17*, 217
N-Nitrosoguvacine	*37*, 263
N-Nitrosoguvacoline	*37*, 263
N-Nitrosohydroxyproline	*17*, 304
N-Nitrosomethylethylamine	*17*, 221
N-Nitroso-N-methylurea	*1*, 125
	17, 227
N-Nitroso-N-methylurethane	*4*, 211
N-Nitrosomethylvinylamine	*17*, 257
N-Nitrosomorpholine	*17*, 263
N'-Nitrosonornicotine	*17*, 281
	37, 241
N-Nitrosopiperidine	*17*, 287
N-Nitrosoproline	*17*, 303
N-Nitrosopyrrolidine	*17*, 313
N-Nitrososarcosine	*17*, 327
N-Nitrososarcosine ethyl ester (1977)*	
Nitrovin	*31*, 185
Nitroxoline (1976)*	
Nivalenol (1976)*	
Noresthisterone and its acetate	*6*, 179
	21, 441
	Suppl. 4, 199
Norethynodrel	*6*, 191
	21, 461 (corr. *25*, 391)
	Suppl. 4, 201
Norgestrel	*6*, 201
	21, 479
	Suppl. 4, 202
Nylon 6	*19*, 120
Nylon 6/6 (1978)*	

O

Ochratoxin A	*10*, 191
	31, 191 (corr. *34*, 197)

Oestradiol-17β	*6*, 99
	21, 279
	Suppl. 4, 190
Oestradiol 3-benzoate	*21*, 281
Oestradiol dipropionate	*21*, 283
Oestradiol mustard	*9*, 217
Oestradiol-17β-valerate	*21*, 284
Oestriol	*6*, 117
	21, 327
Oestrone	*6*, 123
	21, 343 (corr. *25*, 391)
	Suppl. 4, 191
Oestrone benzoate	*21*, 345
	Suppl. 4, 191
Oil Orange SS	*8*, 165
Oral contraceptives	
Combined	*21*, 103,133
	Suppl. 4, 173
Sequential	*21*, 111
	Suppl. 4, 177
Orange I	*8*, 173
Orange G	*8*, 181
Oxazepam	*13*, 58
Oxymetholone	*13*, 131
	Suppl. 4, 203
Oxyphenbutazone	*13*, 185

P

Panfuran S (Dihydroxymethylfuratrizine)	*24*, 77
Parasorbic acid	*10*, 199 (corr. *12*, 271)
Parathion	*30*, 153
Patulin	*10*, 205
	40, 83
Penicillic acid	*10*, 211
Pentachloroethane	*41*, 99
Pentachlorophenol (*see also* Chlorophenols,	*20*, 203
occupational exposures to)	*Suppl. 4*, 88, 205
Pentobarbital sodium (1976)*	
Perylene	*32*, 411
Petasitenine	*31*, 207
Phenacetin	*3*, 141
	24,135
	Suppl. 4, 47

Phenanthrene	*32*, 419
Phenazopyridine [2,6-Diamino-3-(phenylazo)-pyridine] and its hydrochloride	*8*, 117 *24*, 163 (corr. *29*, 399) Suppl. *4*, 207
Phenelzine and its sulphate	*24*, 175 Suppl. *4*, 207
Phenicarbazide	*12*, 177
Phenobarbital and its sodium salt	*13*, 157 Suppl. *4*, 208
Phenoxyacetic acid herbicides (occupational exposure to) (*see also* Chlorophenoxy herbicides, occupational exposures to)	Suppl. *4*, 211
Phenoxybenzamine and its hydrochloride	*9*, 223 *24*, 185
Phenylbutazone	*13*, 183 Suppl. *4*, 212
ortho-Phenylenediamine (1977)*	
meta-Phenylenediamine and its hydrochloride	*16*, 111
para-Phenylenediamine and its hydrochloride	*16*, 125
N-Phenyl-2-naphthylamine	*16*, 325 (corr. *25*, 391) Suppl. *4*, 213
ortho-Phenylphenol and its sodium salt	*30*, 329
N-Phenyl-*para*-phenylenediamine (1977)*	
Phenytoin and its sodium salt	*13*, 201 Suppl. *4*, 215
Piperazine oestrone sulphate	*21*, 148
Piperonyl butoxide	*30*, 183
Polyacrylic acid	*19*, 62
Polybrominated biphenyls	*18*, 107 *41*, 261
Polychlorinated biphenyls	*7*, 261 *18*, 43 (corr. *40*, 419) Suppl. *4*, 217
Polychloroprene	*19*, 141
Polyethylene (low-density and high-density)	*19*, 164
Polyethylene terephthalate (1978)*	
Polyisoprene (1978)*	
Polymethylene polyphenyl isocyanate	*19*, 314
Polymethyl methacrylate	*19*, 195
Polyoestradiol phosphate	*21*, 286
Polypropylene	*19*, 218
Polystyrene	*19*, 245
Polytetrafluoroethylene	*19*, 288
Polyurethane foams (flexible and rigid)	*19*, 320
Polyvinyl acetate	*19*, 346
Polyvinyl alcohol	*19*, 351

Polyvinyl chloride	7, 306
	19, 402
Polyvinylidene fluoride (1978)*	
Polyvinyl pyrrolidone	19, 463
Ponceau MX	8, 189
Ponceau 3R	8, 199
Ponceau SX	8, 207
Potassium bis(2-hydroxyethyl)dithiocarbamate	12, 183
Potassium bromate	40, 207
Prednisone	26, 293
	Suppl. 4, 219
Procarbazine hydrochloride	26, 311
	Suppl. 4, 220
Proflavine and its salts	24, 195
Progesterone	6, 135
	21, 491 (corr. 25, 391)
	Suppl. 4, 202
Pronetalol hydrochloride	13, 227 (corr. 16, 387)
1,3-Propane sultone	4, 253 (corr. 13, 243; 20, 591)
Propham	12, 189
β-Propiolactone	4, 259 (corr. 15, 341)
n-Propyl carbamate	12, 201
Propylene	19, 213
Propylene oxide	11, 191
	36, 227 (corr. 38, 397)
Propylthiouracil	7, 67
	Suppl. 4, 222
The pulp and paper industry	25, 157
	Suppl. 4, 144
Pyrazinamide (1976)*	
Pyrene	32, 431
Pyrido[3,4-c]psoralen	40, 349
Pyrimethamine	13, 233
Pyrrolizidine alkaloids	10, 333

Q

Quercetin	31, 213
Quinoestradol (1978)*	
Quinoestrol (1978)*	
para-Quinone	15, 255
Quintozene (Pentachloronitrobenzene)	5, 211

R

Reserpine	10, 217
	24, 211 (corr. 26, 387; 30, 407)
	Suppl. 4, 222

Resorcinol	*15*, 155
Retrorsine	*10*, 303
Rhodamine B	*16*, 221
Rhodamine 6G	*16*, 233
Riddelliine	*10*, 313
Rifampicin	*24*, 243
Rotenone (1982)*	
The rubber industry	*28* (corr. *30*, 407)
	Suppl. 4, 144
Rugulosin	*40*, 99

S

Saccharated iron oxide	*2*, 161
Saccharin	*22*, 111 (corr. *25*, 391)
	Suppl. 4, 224
Safrole	*1*, 169
	10, 231
Scarlet red	*8*, 217
Selenium and selenium compounds	*9*, 245 (corr. *12*, 271; *30*, 407)
Semicarbazide hydrochloride	*12*, 209 (corr. *16*, 387)
Seneciphylline	*10*, 319
Senkirkine	*10*, 327
	31, 231
Shale-oils	*35*, 161
Simazine (1982)*	
Sodium cyclamate	*22*, 56 (corr. *25*, 391)
	Suppl. 4, 97
Sodium diethyldithiocarbamate	*12*, 217
Sodium equilin sulphate	*21*, 148
Sodium oestrone sulphate	*21*, 147
Sodium saccharin	*22*, 113 (corr. *25*, 391)
	Suppl. 4, 224
Soots	*35*, 219
Soots and tars	*3*, 22
	Suppl. 4, 227
Spironolactone	*24*, 259
	Suppl. 4, 229
Sterigmatocystin	*1*, 175
	10, 245
Streptozotocin	*4*, 221
	17, 337
Styrene	*19*, 231
	Suppl. 4, 229
Styrene-acrylonitrile copolymers	*19*, 97
Styrene-butadiene copolymers	*19*, 252

Styrene oxide	*11*, 201
	19, 275
	Suppl. 4, 229
	36, 245
Succinic anhydride	*15*, 265
Sudan I	*8*, 225
Sudan II	*8*, 233
Sudan III	*8*, 241
Sudan brown RR	*8*, 249
Sudan red 7B	*8*, 253
Sulfafurazole (Sulphisoxazole)	*24*, 275
	Suppl. 4, 233
Sulfallate	*30*, 283
Sulfamethoxazole	*24*, 285
	Suppl. 4, 234
Sulphamethazine (1982)*	
Sunset yellow FCF	*8*, 257
Symphytine	*31*, 239

T

2,4,5-T and esters	*15*, 273
	Suppl. 4, 211, 235
Tannic acid	*10*, 253 (corr. *16*, 387)
Tannins	*10*, 254
Terephthalic acid (1978)*	
Terpene polychlorinates (Strobane®)	*5*, 219
Testosterone	*6*, 209
	21, 519
Testosterone oenanthate	*21*, 521
Testosterone propionate	*21*, 522
2,2′,5,5′-Tetrachlorobenzidine	*27*, 141
Tetrachlorodibenzo-*para*-dioxin (TCDD)	*15*, 41
	Suppl. 4, 211, 238
1,1,2,2-Tetrachloroethane	*20*, 477
Tetrachloroethylene	*20*, 491
	Suppl. 4, 243
Tetrachlorvinphos	*30*, 197
Tetrafluoroethylene	*19*, 285
Thioacetamide	*7*, 77
4,4′-Thiodianiline	*16*, 343
	27, 147
Thiouracil	*7*, 85
Thiourea	*7*, 95
Thiram	*12*, 225
Tobacco habits other than smoking	*37* (corr. *40*, 417)

Tobacco smoking	*38* (corr. *40*, 418)
Toluene diisocyanate	*39*, 287 (corr. *40*, 418)
2,4-Toluene diisocyanate	*19*, 303
	39, 287
2,6-Toluene diisocyanate	*19*, 303
	39, 287
ortho-Toluenesulphonamide	*22*, 121
	Suppl. 4, 224
ortho-Toluidine and its hydrochloride	*16*, 349
	27, 155
	Suppl. 4, 245
Toxaphene (Polychlorinated camphenes)	*20*, 327
Treosulphan	*26*, 341
	Suppl. 4, 246
1,1,1-Trichloroethane	*20*, 515
1,1,2-Trichloroethane	*20*, 533
Trichloroethylene	*11*, 263
	20, 545
	Suppl. 4, 247
2,4,5- and 2,4,6-Trichlorophenols	*20*, 349
	Suppl. 4, 88, 249
Trichlorotriethylamine hydrochloride	*9*, 229
Trichlorphon	*30*, 207
T$_2$-Trichothecene	*31*, 265
Triethylene glycol diglycidyl ether	*11*, 209
Trifluralin (1982)*	
2,4,5-Trimethylaniline and its hydrochloride	*27*, 177
2,4,6-Trimethylaniline and its hydrochloride	*27*, 178
4,5′,8-Trimethylpsoralen	*40*, 357
Triphenylene	*32*, 447
Tris(aziridinyl)-*para*-benzoquinone (Triaziquone)	*9*, 67
	Suppl. 4, 251
Tris(1-aziridinyl)phosphine oxide	*9*, 75
Tris(1-aziridinyl)phosphine sulphide (Thiotepa)	*9*, 85
	Suppl. 4, 252
2,4,6-Tris(1-aziridinyl)-*s*-triazine	*9*, 95
1,2,3-Tris(chloromethoxy)propane	*15*, 301
Tris(2,3-dibromopropyl)phosphate	*20*, 575
Tris(2-methyl-1-aziridinyl)phosphine oxide	*9*, 107
Trp-P-1 (3-Amino-1,4-dimethyl-5*H*-pyrido[4,3-*b*]indole and its acetate	*31*, 247
Trp-P-2 (3-Amino-1-methyl-5*H*-pyrido[4,3-*b*]indole and its acetate	*31*, 255
Trypan blue	*8*, 267

U

Ultraviolet radiation	*40*, 379
Uracil mustard	*9*, 235
	Suppl. 4, 256

Urethane   7, 111

V

Vinblastine sulphate   26, 349 (corr. 34, 197)
                                                  Suppl. 4, 257
Vincristine sulphate   26, 365
                                                  Suppl. 4, 259
Vinyl acetate   19, 341
                                                  39, 113
Vinyl bromide   19, 367
                                                  39, 133
Vinyl chloride   7, 291
                                                  19, 377
                                                  Suppl. 4, 260
Vinyl chloride-vinyl acetate copolymers   7, 311
                                                  19, 412
4-Vinylcyclohexene   11, 277
                                                  39, 181
Vinyl fluoride   39, 147
Vinylidene chloride   19, 439
                                                  Suppl. 4, 262 (corr. 31, 293)
                                                  39, 195
Vinylidene chloride-vinyl chloride copolymers   19, 448

Vinylidene fluoride   39, 227
N-Vinyl-2-pyrrolidine   19, 461

X

2,4-Xylidine and its hydrochloride   16, 367
2,5-Xylidine and its hydrochloride   16, 377
2,6-Xylidine (1977)*

Y

Yellow AB   8, 279
Yellow OB   8, 287

Z

Zearalenone   31, 279
Zectran   12, 237
Zineb   12, 245
Ziram   12, 259

PUBLICATIONS OF THE INTERNATIONAL AGENCY FOR RESEARCH ON CANCER

SCIENTIFIC PUBLICATIONS SERIES

(Available from Oxford University Press)

No. 1 LIVER CANCER (1971)
176 pages; out of print

No. 2 ONCOGENESIS AND HERPES VIRUSES (1972)
Edited by P.M. Biggs, G. de-Thé & L.N. Payne
515 pages; out of print

No. 3 N-NITROSO COMPOUNDS - ANALYSIS AND FORMATION (1972)
Edited by P. Bogovski, R. Preussmann & E.A. Walker
140 pages

No. 4 TRANSPLACENTAL CARCINOGENESIS (1973)
Edited by L. Tomatis & U. Mohr,
181 pages; out of print

No. 5 PATHOLOGY OF TUMOURS IN LABORATORY ANIMALS. VOLUME I. TUMOURS OF THE RAT. PART 1 (1973)
Editor-in-Chief V.S. Turusov
214 pages

No. 6 PATHOLOGY OF TUMOURS IN LABORATORY ANIMALS. VOLUME I. TUMOURS OF THE RAT. PART 2 (1976)
Editor-in-Chief V.S. Turusov
319 pages

No. 7 HOST ENVIRONMENT INTERACTIONS IN THE ETIOLOGY OF CANCER IN MAN (1973)
Edited by R. Doll & I. Vodopija,
464 pages

No. 8 BIOLOGICAL EFFECTS OF ASBESTOS (1973)
Edited by P. Bogovski, J.C. Gilson, V. Timbrell & J.C. Wagner,
346 pages; out of print

No. 9 N-NITROSO COMPOUNDS IN THE ENVIRONMENT (1974)
Edited by P. Bogovski & E.A. Walker
243 pages

No. 10 CHEMICAL CARCINOGENESIS ESSAYS (1974)
Edited by R. Montesano & L. Tomatis,
230 pages

No. 11 ONCOGENESIS AND HERPES-VIRUSES II (1975)
Edited by G. de-Thé, M.A. Epstein & H. zur Hausen
Part 1, 511 pages
Part 2, 403 pages

No. 12 SCREENING TESTS IN CHEMICAL CARCINOGENESIS (1976)
Edited by R. Montesano, H. Bartsch & L. Tomatis
666 pages

No. 13 ENVIRONMENTAL POLLUTION AND CARCINOGENIC RISKS (1976)
Edited by C. Rosenfeld & W. Davis
454 pages; out of print

No. 14 ENVIRONMENTAL N-NITROSO COMPOUNDS — ANALYSIS AND FORMATION (1976)
Edited by E.A. Walker, P. Bogovski & L. Griciute
512 pages

No. 15 CANCER INCIDENCE IN FIVE CONTINENTS. VOL. III (1976)
Edited by J. Waterhouse, C.S. Muir, P. Correa & J. Powell
584 pages

No. 16 AIR POLLUTION AND CANCER IN MAN (1977)
Edited by U. Mohr, D. Schmahl & L. Tomatis
331 pages; out of print

No. 17 DIRECTORY OF ON-GOING RESEARCH IN CANCER EPIDEMIOLOGY 1977 (1977)
Edited by C.S. Muir & G. Wagner,
599 pages; out of print

No. 18 ENVIRONMENTAL CARCINOGENS. SELECTED METHODS OF ANALYSIS
Editor-in-Chief H. Egan
Vol. 1. ANALYSIS OF VOLATILE NITROSAMINES IN FOOD (1978)
Edited by R. Preussmann, M. Castegnaro, E.A. Walker & A.E. Wassermann
212 pages; out of print

SCIENTIFIC PUBLICATIONS SERIES

No. 19 ENVIRONMENTAL ASPECTS
OF N-NITROSO COMPOUNDS (1978)
Edited by E.A. Walker, M. Castegnaro,
L. Griciute & R.E. Lyle
566 pages

No. 20 NASOPHARYNGEAL
CARCINOMA: ETIOLOGY AND
CONTROL (1978)
Edited by G. de-Thé & Y. Ito,
610 pages; out of print

No. 21 CANCER REGISTRATION
AND ITS TECHNIQUES (1978)
Edited by R. MacLennan, C.S. Muir,
R. Steinitz & A. Winkler
235 pages

No. 22 ENVIRONMENTAL CARCINOGENS.
SELECTED METHODS OF ANALYSIS
Editor-in-Chief H. Egan
Vol. 2. METHODS FOR THE MEASURE-
MENT OF VINYL CHLORIDE IN
POLY(VINYL CHLORIDE), AIR, WATER
AND FOODSTUFFS (1978)
Edited by D.C.M. Squirrell & W. Thain,
142 pages; out of print

No. 23 PATHOLOGY OF TUMOURS IN
LABORATORY ANIMALS. VOLUME II.
TUMOURS OF THE MOUSE (1979)
Editor-in-Chief V.S. Turusov
669 pages

No. 24 ONCOGENESIS AND HERPES-
VIRUSES III (1978)
Edited by G. de-Thé, W. Henle & F. Rapp
Part 1, 580 pages
Part 2, 522 pages; out of print

No. 25 CARCINOGENIC RISKS -
STRATEGIES FOR INTERVENTION
(1979)
Edited by W. Davis & C. Rosenfeld,
283 pages; out of print

No. 26 DIRECTORY OF ON-GOING
RESEARCH IN CANCER EPI-
DEMIOLOGY 1978 (1978)
Edited by C.S. Muir & G. Wagner,
550 pages; out of print

No. 27 MOLECULAR AND CELLULAR
ASPECTS OF CARCINOGEN
SCREENING TESTS (1980)
Edited by R. Montesano, H. Bartsch & L. Tomatis
371 pages

No. 28 DIRECTORY OF ON-GOING
RESEARCH IN CANCER EPI-
DEMIOLOGY 1979 (1979)
Edited by C.S. Muir & G. Wagner,
672 pages; out of print

No. 29 ENVIRONMENTAL CARCINOGENS.
SELECTED METHODS OF ANALYSIS
Editor-in-Chief H. Egan
Vol. 3. ANALYSIS OF POLYCYCLIC
AROMATIC HYDROCARBONS IN
ENVIRONMENTAL SAMPLES (1979)
Edited by M. Castegnaro, P. Bogovski,
H. Kunte & E.A. Walker
240 pages; out of print

No. 30 BIOLOGICAL EFFECTS OF
MINERAL FIBRES (1980)
Editor-in-Chief J.C. Wagner
Volume 1, 494 pages
Volume 2, 513 pages

No. 31 N-NITROSO COMPOUNDS:
ANALYSIS, FORMATION AND
OCCURRENCE (1980)
Edited by E.A. Walker, M. Castegnaro,
L. Griciute & M. Börzsönyi
841 pages; out of print

No. 32 STATISTICAL METHODS IN
CANCER RESEARCH
Vol. 1. THE ANALYSIS OF CASE-
CONTROL STUDIES (1980)
By N.E. Breslow & N.E. Day
338 pages

No. 33 HANDLING CHEMICAL
CARCINOGENS IN THE LABORATORY -
PROBLEMS OF SAFETY (1979)
Edited by R. Montesano, H. Bartsch,
E. Boyland, G. Della Porta, L. Fishbein,
R.A. Griesemer, A.B. Swan & L. Tomatis,
32 pages

No. 34 PATHOLOGY OF TUMOURS
IN LABORATORY ANIMALS. VOLUME
III. TUMOURS OF THE HAMSTER
(1982)
Editor-in-Chief V.S. Turusov,
461 pages

No. 35 DIRECTORY OF ON-GOING
RESEARCH IN CANCER EPIDEMIOLOGY
1980 (1980)
Edited by C.S. Muir & G. Wagner,
660 pages; out of print

SCIENTIFIC PUBLICATIONS SERIES

No. 36 CANCER MORTALITY BY
OCCUPATION AND SOCIAL CLASS
1851-1971 (1982)
By W.P.D. Logan
253 pages

No. 37 LABORATORY DECONTAMI-
NATION AND DESTRUCTION OF
AFLATOXINS B_1, B_2, G_1, G_2 IN
LABORATORY WASTES (1980)
Edited by M. Castegnaro, D.C. Hunt,
E.B. Sansone, P.L. Schuller,
M.G. Siriwardana, G.M. Telling,
H.P. Van Egmond & E.A. Walker,
59 pages

No. 38 DIRECTORY OF ON-GOING
RESEARCH IN CANCER EPI-
DEMIOLOGY 1981 (1981)
Edited by C.S. Muir & G. Wagner,
696 pages; out of print

No. 39 HOST FACTORS IN HUMAN
CARCINOGENESIS (1982)
Edited by H. Bartsch & B. Armstrong
583 pages

No. 40 ENVIRONMENTAL CARCINOGENS
SELECTED METHODS OF ANALYSIS
Editor-in-Chief H. Egan
Vol. 4. SOME AROMATIC AMINES AND
AZO DYES IN THE GENERAL AND
INDUSTRIAL ENVIRONMENT (1981)
Edited by L. Fishbein, M. Castegnaro,
I.K. O'Neill & H. Bartsch
347 pages

No. 41 N-NITROSO COMPOUNDS:
OCCURRENCE AND BIOLOGICAL
EFFECTS (1982)
Edited by H. Bartsch, I.K. O'Neill,
M. Castegnaro & M. Okada,
755 pages

No. 42 CANCER INCIDENCE IN FIVE
CONTINENTS. VOLUME IV (1982)
Edited by J. Waterhouse, C. Muir,
K. Shanmugaratnam & J. Powell,
811 pages

No. 43 LABORATORY DECONTAMI-
NATION AND DESTRUCTION OF
CARCINOGENS IN LABORATORY
WASTES: SOME N-NITROSAMINES
(1982) Edited by M. Castegnaro,
G. Eisenbrand, G. Ellen, L. Keefer,
D. Klein, E.B. Sansone, D. Spincer,
G. Telling & K. Webb
73 pages

No. 44 ENVIRONMENTAL CARCINOGENS.
SELECTED METHODS OF ANALYSIS
Editor-in-Chief H. Egan
Vol. 5. SOME MYCOTOXINS (1983)
Edited by L. Stoloff, M. Castegnaro,
P. Scott, I.K. O'Neill & H. Bartsch,
455 pages

No. 45 ENVIRONMENTAL CARCINOGENS.
SELECTED METHODS OF ANALYSIS
Editor-in-Chief H. Egan
Vol. 6: N-NITROSO COMPOUNDS
(1983)
Edited by R. Preussmann, I.K. O'Neill,
G. Eisenbrand, B. Spiegelhalder &
H. Bartsch
508 pages

No. 46 DIRECTORY OF ON-GOING
RESEARCH IN CANCER EPI-
DEMIOLOGY 1982 (1982)
Edited by C.S. Muir & G. Wagner,
722 pages; out of print

No. 47 CANCER INCIDENCE IN
SINGAPORE (1982)
Edited by K. Shanmugaratnam, H.P. Lee
& N.E. Day
174 pages; out of print

No. 48 CANCER INCIDENCE IN
THE USSR Second Revised
Edition (1983)
Edited by N.P. Napalkov,
G.F. Tserkovny, V.M. Merabishvili,
D.M. Parkin, M. Smans & C.S. Muir,
75 pages

No. 49 LABORATORY DECONTAMI-
NATION AND DESTRUCTION OF
CARCINOGENS IN LABORATORY
WASTES: SOME POLYCYCLIC
AROMATIC HYDROCARBONS (1983)
Edited by M. Castegnaro, G. Grimmer,
O. Hutzinger, W. Karcher, H. Kunte,
M. Lafontaine, E.B. Sansone, G. Telling
& S.P. Tucker
81 pages

No. 50 DIRECTORY OF ON-GOING
RESEARCH IN CANCER EPI-
DEMIOLOGY 1983 (1983)
Edited by C.S. Muir & G. Wagner,
740 pages; out of print

SCIENTIFIC PUBLICATIONS SERIES

No. 51 MODULATORS OF EXPERI-
MENTAL CARCINOGENESIS (1983)
Edited by V. Turusov & R. Montesano
307 pages

No. 52 SECOND CANCER IN
RELATION TO RADIATION
TREATMENT FOR CERVICAL
CANCER: RESULTS OF A CANCER
REGISTRY COLLABORATION (1984)
Edited by N.E. Day & J.C. Boice, Jr,
207 pages

No. 53 NICKEL IN THE HUMAN
ENVIRONMENT (1984)
Editor-in-Chief, F.W. Sunderman, Jr,
529 pages

No. 54 LABORATORY DECONTAMI-
NATION AND DESTRUCTION OF
CARCINOGENS IN LABORATORY WASTES:
SOME HYDRAZINES (1983)
Edited by M. Castegnaro, G. Ellen,
M. Lafontaine, H.C. van der Plas,
E.B. Sansone & S.P. Tucker,
87 pages

No. 55 LABORATORY DECONTAMI-
NATION AND DESTRUCTION OF
CARCINOGENS IN LABORATORY WASTES:
SOME N-NITROSAMIDES (1984)
Edited by M. Castegnaro,
M. Benard, L.W. van Broekhoven,
D. Fine, R. Massey, E.B. Sansone,
P.L.R. Smith, B. Spiegelhalder,
A. Stacchini, G. Telling & J.J. Vallon,
65 pages

No. 56 MODELS, MECHANISMS AND
ETIOLOGY OF TUMOUR PROMOTION
(1984)
Edited by M. Börszönyi, N.E. Day,
K. Lapis & H. Yamasaki
532 pages

No. 57 N-NITROSO COMPOUNDS:
OCCURRENCE, BIOLOGICAL EFFECTS
AND RELEVANCE TO HUMAN
CANCER (1984)
Edited by I.K. O'Neill, R.C. von Borstel,
C.T. Miller, J. Long & H. Bartsch,
1013 pages

No. 58 AGE-RELATED FACTORS
IN CARCINOGENESIS (1985)
Edited by A. Likhachev, V. Anisimov
& R. Montesano
288 pages

No. 59 MONITORING HUMAN
EXPOSURE TO CARCINOGENIC AND
MUTAGENIC AGENTS (1984)
Edited by A. Berlin, M. Draper,
K. Hemminki & H. Vainio
457 pages

No. 60 BURKITT'S LYMPHOMA: A
HUMAN CANCER MODEL (1985)
Edited by G. Lenoir, G. O'Conor
& C.L.M. Olweny
484 pages

No. 61 LABORATORY DECONTAMI-
NATION AND DESTRUCTION OF
CARCINOGENS IN LABORATORY
WASTES: SOME HALOETHERS (1984)
Edited by M. Castegnaro,
M. Alvarez, M. Iovu, E.B. Sansone,
G.M. Telling & D.T. Williams
55 pages

No. 62 DIRECTORY OF ON-GOING
RESEARCH IN CANCER EPI-
DEMIOLOGY 1984 (1984)
Edited by C.S. Muir & G. Wagner 728 pages

No. 63 VIRUS-ASSOCIATED CANCERS
IN AFRICA (1984)
Edited by A.O. Williams, G.T. O'Conor,
G.B. de-Thé & C.A. Johnson,
773 pages

No. 64 LABORATORY DECONTAMI-
NATION AND DESTRUCTION OF
CARCINOGENS IN LABORATORY
WASTES: SOME AROMATIC AMINES
AND 4-NITROBIPHENYL (1985)
Edited by M. Castegnaro, J. Barek,
J. Dennis, G. Ellen, M. Klibanov,
M. Lafontaine, R. Mitchum,
P. Van Roosmalen, E.B. Sansone,
L.A. Sternson & M. Vahl
85 pages

No. 65 INTERPRETATION OF NEGATIVE
EPIDEMIOLOGICAL EVIDENCE FOR
CARCINOGENICITY (1985)
Edited by N.J. Wald & R. Doll
232 pages

No. 66 THE ROLE OF THE REGISTRY
IN CANCER CONTROL (1985)
Edited by D.M. Parkin, G. Wagner
& C.S. Muir
155 pages

SCIENTIFIC PUBLICATIONS SERIES

No. 67 TRANSFORMATION ASSAY OF ESTABLISHED CELL LINES: MECHANISMS AND APPLICATIONS (1985)
Edited by T. Kakunaga & H. Yamasaki
225 pages

No. 68 ENVIRONMENTAL CARCINOGENS. SELECTED METHODS OF ANALYSIS VOL. 7 — SOME VOLATILE HALOGENATED HYDROCARBONS (1985)
Edited by L. Fishbein & I.K. O'Neill
479 pages

No. 69 DIRECTORY OF ON-GOING RESEARCH IN CANCER EPIDEMIOLOGY 1985 (1985)
Edited by C.S. Muir & G. Wagner
756 pages

No. 70 THE ROLE OF CYCLIC NUCLEIC ACID ADDUCTS IN CARCINOGENESIS AND MUTAGENESIS (1986)
Edited by B. Singer & H. Bartsch
467 pages

No. 71 ENVIRONMENTAL CARCINOGENS. SELECTED METHODS OF ANALYSIS VOL. 8. SOME METALS: As, Be, Cd, Cr, Ni, Pb, Se, Zn (1986)
Edited by I.K. O'Neill, P. Schuller & L. Fishbein
485 pages

No. 72 ATLAS OF CANCER IN SCOTLAND 1975-1980: INCIDENCE AND EPIDEMIOLOGICAL PERSPECTIVE (1985)
Edited by I. Kemp, P. Boyle, M. Smans & C. Muir
282 pages

No. 73 LABORATORY DECONTAMINATION AND DESTRUCTION OF CARCINOGENS IN LABORATORY WASTES: SOME ANTINEOPLASTIC AGENTS (1985)
Edited by M. Castegnaro, J. Adams, M. Armour, J. Barek, J. Benvenuto, C. Confalonieri, U. Goff, S. Ludeman, D. Reed, E.B. Sansone & G. Telling
163 pages

No. 74 TOBACCO: A MAJOR INTERNATIONAL HEALTH HAZARD (1986)
Edited by D. Zaridze and R. Peto
325 pages

No. 75 CANCER OCCURRENCE IN DEVELOPING COUNTRIES (1986)
Edited by D.M. Parkin
339 pages

No. 76 SCREENING FOR CANCER OF THE UTERINE CERVIX (1986)
Edited by M. Hakama, A.B. Miller & N.E. Day
311 pages

No. 77 HEXACHLOROBENZENE: PROCEEDINGS OF AN INTERNATIONAL SYMPOSIUM (1986)
Edited by C.R. Morris & J.R.P. Cabral
668 pages

No. 78 CARCINOGENICITY OF CYTOSTATIC DRUGS (1986)
Edited by D. Schmähl & J. Kaldor
330 pages

No. 79 STATISTICAL METHODS IN CANCER RESEARCH, VOL. 3, THE DESIGN AND ANALYSIS OF LONG-TERM ANIMAL EXPERIMENTS (1986)
By J.J. Gart, D. Krewski, P.N. Lee, R.E. Tarone & J. Wahrendorf
220 pages

No. 80 DIRECTORY OF ON-GOING RESEARCH IN CANCER EPIDEMIOLOGY 1986 (1986)
Edited by G. Wagner & C. Muir
805 pages

No. 81 ENVIRONMENTAL CARCINOGENS. METHODS OF ANALYSIS AND EXPOSURE MEASUREMENT VOL. 9, PASSIVE SMOKING (1986)
Edited by I.K. O'Neill, K.D. Brunnemann, B. Dodet & D. Hoffmann
(in press)

No. 82 STATISTICAL METHODS IN CANCER RESEARCH, VOL. 2 THE DESIGN AND ANALYSIS OF COHORT STUDIES (1987)
By N.E. Breslow & N.E. Day
(in press)

No. 83 LONG-TERM AND SHORT-TERM ASSAYS FOR CARCINOGENS: A CRITICAL APPRAISAL (1986)
Edited by R. Montesano, H. Bartsch, H. Vainio, J. Wilbourn & H. Yamasaki
561 pages

No. 84 THE RELEVANCE OF N-NITROSO COMPOUNDS TO HUMAN CANCER: EXPOSURES AND MECHANISMS (1987)
Edited by H. Bartsch, I.K. O'Neill & R. Schulte-Hermann
(in press)

NON-SERIAL PUBLICATIONS

(Available from IARC)

ALCOOL ET CANCER (1978)
By A.J. Tuyns (in French only)
42 pages

CANCER MORBIDITY AND CAUSES OF
DEATH AMONG DANISH BREWERY
WORKERS (1980)
By O.M. Jensen
145 pages

DIRECTORY OF COMPUTER SYSTEMS
USED IN CANCER REGISTRIES (1986)
By H.R. Menck & D.M. Parkin
236 pages

IARC MONOGRAPHS ON THE EVALUATION OF THE CARCINOGENIC RISK OF CHEMICALS TO HUMANS
(English editions only)

(Available from WHO Sales Agents)

Volume 1
Some inorganic substances, chlorinated hydrocarbons, aromatic amines, N-nitroso compounds, and natural products (1972)
184 pp.; out of print

Volume 2
Some inorganic and organometallic compounds (1973)
181 pp.; out of print

Volume 3
Certain polycyclic aromatic hydrocarbons and heterocyclic compounds (1973)
271 pp.; out of print

Volume 4
Some aromatic amines, hydrazine and related substances, N-nitroso compounds and miscellaneous alkylating agents (1974)
286 pp.

Volume 5
Some organochlorine pesticides (1974)
241 pp.; out of print

Volume 6
Sex hormones (1974)
243 pp.

Volume 7
Some anti-thyroid and related substances, nitrofurans and industrial chemicals (1974)
326 pp.; out of print

Volume 8
Some aromatic azo compounds (1975)
357 pp.

Volume 9
Some aziridines, N-, S- and O-mustards and selenium (1975)
268 pp.

Volume 10
Some naturally occurring substances (1976)
353 pp.; out of print

Volume 11
Cadmium, nickel, some epoxides, miscellaneous industrial chemicals and general considerations on volatile anaesthetics (1976)
306 pp.

Volume 12
Some carbamates, thiocarbamates and carbazides (1976)
282 pp.

Volume 13
Some miscellaneous pharmaceutical substances (1977)
255 pp.

Volume 14
Asbestos (1977)
106 pp.

Volume 15
Some fumigants, the herbicides 2,4-D and 2,4,5-T, chlorinated dibenzodioxins and miscellaneous industrial chemicals (1977)
354 pp.

Volume 16
Some aromatic amines and related nitro compounds - hair dyes, colouring agents and miscellaneous industrial chemicals (1978)
400 pp.

Volume 17
Some N-nitroso compounds (1978)
365 pp.

Volume 18
Polychlorinated biphenyls and polybrominated biphenyls (1978)
140 pp.

Volume 19
Some monomers, plastics and synthetic elastomers, and acrolein (1979)
513 pp.

Volume 20
Some halogenated hydrocarbons (1979)
609 pp.

Volume 21
Sex hormones (II) (1979)
583 pp.

Volume 22
Some non-nutritive sweetening agents (1980)
208 pp.

IARC MONOGRAPHS SERIES

Volume 23
Some metals and metallic compounds (1980)
438 pp.

Volume 24
Some pharmaceutical drugs (1980)
337 pp.

Volume 25
Wood, leather and some associated industries (1981)
412 pp.

Volume 26
Some antineoplastic and immuno-suppressive agents (1981)
411 pp.

Volume 27
Some aromatic amines, anthraquinones and nitroso compounds, and inorganic fluorides used in drinking-water and dental preparations (1982)
341 pp.

Volume 28
The rubber industry (1982)
486 pp.

Volume 29
Some industrial chemicals and dyestuffs (1982)
416 pp.

Volume 30
Miscellaneous pesticides (1983)
424 pp.

Volume 31
Some food additives, feed additives and naturally occurring substances (1983)
314 pp.

Volume 32
Polynuclear aromatic compounds, Part 1, Environmental and experimental data (1984)
477 pp.

Volume 33
Polynuclear aromatic compounds, Part 2, Carbon blacks, mineral oils and some nitroarene compounds (1984)
245 pp.

Volume 34
Polynuclear aromatic compounds, Part 3, Industrial exposures in aluminium production, coal gasification, coke production, and iron and steel founding (1984)
219 pp.

Volume 35
Polynuclear aromatic compounds, Part 4, Bitumens, coal-tar and derived products, shale-oils and soots (1985)
271 pp.

Volume 36
Allyl Compounds, aldehydes, epoxides and peroxides (1985)
369 pp.

Volume 37
Tobacco habits other than smoking; betel-quid and areca-nut chewing; and some related nitrosamines (1985)
291 pp.

Volume 38
Tobacco smoking (1986)
421 pp.

Volume 39
Some chemicals used in plastics and elastomers (1986)
403 pp.

Volume 40
Some naturally occurring and synthetic food components, furocoumarins and ultra-violet radiation (1986)
444 pp.

Volume 41
Some halogenated hydrocarbons and pesticide exposures (1986)
434 pp.

Supplement No. 1
Chemicals and industrial processes associated with cancer in humans (IARC Monographs, Volumes 1 to 20) (1979)
71 pp.; out of print

Supplement No. 2
Long-term and short-term screening assays for carcinogens: a critical appraisal (1980)
426 pp.

Supplement No. 3
Cross index of synonyms and trade names in Volumes 1 to 26 (1982)
199 pp.

Supplement No. 4
Chemicals, industrial processes and industries associated with cancer in humans (IARC Monographs, Volumes 1 to 29) (1982)
292 pp.

IARC MONOGRAPHS SERIES (contd)

Supplement No. 5
Cross index of synonyms and trade names in Volumes 1 to 36 (1985)
259 pp.

INFORMATION BULLETINS ON THE SURVEY OF CHEMICALS BEING TESTED FOR CARCINOGENICITY

(Available from IARC)

No. 8 (1979)
Edited by M.-J. Ghess, H. Bartsch
& L. Tomatis
604 pp.

No. 9 (1981)
Edited by M.-J. Ghess, J.D. Wilbourn,
H. Bartsch & L. Tomatis
294 pp.

No. 10 (1982)
Edited by M.-J. Ghess, J.D. Wilbourn
H. Bartsch
326 pp.

No. 11 (1984)
Edited by M.-J. Ghess, J.D. Wilbourn,
H. Vainio & H. Bartsch
336 pp.

No. 12 (1986)
Edited by M.-J. Ghess, J.D. Wilbourn,
A. Tossavainen & H. Vainio
389 pp.

THE LIBRARY
UNIVERSITY OF CALIFORNIA
San Francisco
(415) 476-2335

THIS BOOK IS DUE ON THE LAST DATE STAMPED BELOW
Books not returned on time are subject to fines according to the Library Lending Code. A renewal may be made on certain materials. For details consult Lending Code.

14 DAY	14 DAY	
AUG 27 1987	MAY 20 1989	
RETURNED	June 3	
AUG 29 1987	RETURNED	
14 DAY	JUN -6 1989	
APR 26 1988	14 DAY	
RETURNED	OCT 12 1990	
APR 15 1988	RETURNED	
14 DAY	OCT 12 1990	
JAN 19 1989		
RETURNED		
JAN 10 1989		